D1199117

HUMAN FACTORS IN ENGINEERING AND DESIGN

HUMAN FACTORS IN ENGINEERING AND DESIGN

FIFTH EDITION

Ernest J. McCormick, Ph.D.
Professor Emeritus of Psychological Sciences
Purdue University

Mark S. Sanders, Ph.D.
California State University, Northridge

McGRAW-HILL BOOK COMPANY
New York St. Louis San Francisco Auckland Bogotá
Hamburg Johannesburg London Madrid Mexico Montreal New Delhi
Panama Paris São Paulo Singapore Sydney Tokyo Toronto

TO

EMILY, WYNNE, AND JAN

AND

SHEILA, GENELLE, AND MELANIE

This book was set in Times Roman by Allen Wayne Communications, Inc.
The editors were Patricia S. Nave and Susan Gamer;
the production supervisor was Leroy A. Young.
The cover was designed by Anne Canevari Green.
New drawings were done by J & R Services, Inc.
R. R. Donnelley & Sons Company was printer and binder.

HUMAN FACTORS IN ENGINEERING AND DESIGN

4 5 6 7 8 9 0 DODO 8 9 8 7 6 5 4 3

ISBN 0-07-044902-3

Library of Congress Cataloging in Publication Data

McCormick, Ernest James.
Human factors in engineering and design.

Bibliography: p.
Includes indexes.
1. Human engineering. I. Sanders, Mark S.
II. Title.
TA166.M39 1982 620.8′2 81-12355
ISBN 0-07-044902-3 AACR2

CONTENTS

PREFACE

The field of human factors–referred to as *ergonomics* in Europe and elsewhere–deals with the consideration of human characteristics, expectations, and behaviors in the design of the things people use in their work and everyday lives and of the environments in which they work and live. In simple terms, human factors has been referred to as *designing for human use.*

The history of humankind has of course always included efforts to design tools and other devices to better serve human needs and to provide protection from adverse environments. However, it has been only during the industrial revolution and more particularly the present century that such efforts have taken on any systematic patterns.

During the first three or four decades of this century the early industrial engineers (sometimes called *efficiency experts*) carried out some studies and developed certain principles aimed at improving work efficiency. During World War I a few investigators, especially some psychologists in Great Britain in what is now the Medical Research Council, carried out some exploratory studies dealing with human work, such research being continued to the present.

It was during World War II, however, that what we now call *human factors* or *ergonomics* started to become crystallized as a somewhat distinct discipline. The instigation for such efforts was the fact that new and complicated types of military equipment could not be operated safely or effectively, or maintained adequately, by many well-trained military personnel. This problem rather specifically focused attention on the design of equipment and resulted in some efforts aimed at designing equipment that would be more suitable for human use. This direct focus gave rise to the coining of the first "name" for the discipline, *human engineering.* The term *human factors* emerged somewhat later.

The initial focus on designing military equipment for human use later oozed into certain nonmilitary areas such as the design of certain transportation equipment, communication equipment, and computers, and a few other fields, but the inroads into these areas was relatively moderate for many years. In recent years, primarily in the 1970s, there has developed a fairly widespread recognition (at least among human factors people) of the importance of considering human factors in the design of virtually all man-made things and environments that people use, including the built environment of buildings and communities, consumer products, health services, recreation equipment and facilities, production processes, transportation and communication systems, etc.

This text is intended as a survey of the human factors field. Its major thesis is the one just espoused: that the man-made features and facilities of our civilization should be designed with due consideration to the human use thereof. In terms of coverage the book's major divisions include information input, human output and control, work space and arrangement, environment, and selected human factors topics. In terms of content material the book includes discussions of certain basic human abilities and characteristics (such as vision, hearing, psychomotor skills, and anthropometric characteristics) and numerous examples of investigations that demonstrate the effects of design features on human performance and human welfare (safety, health, job satisfaction, etc.).

The specific research material included in the text represents only a minute fraction of the vast amount of research that has been carried out in the various specific areas. It has been our intent to use as illustrative material examples of research that are relatively important or that adequately illustrate the central points in question. Although much of the specific material may not be forever remembered by the reader, we hope that the reader will at least develop a deep appreciation of the importance of considering human factors in the design of the man-made features of the world in which we work and live.

Appreciation is expressed to the many investigators whose research is cited. References to their work are included at the end of each chapter.

Ernest J. McCormick

Mark S. Sanders

PART ONE

INTRODUCTION

CHAPTER 1

PROLOGUE

In the bygone millenia our ancestors lived in an essentially "natural" environment in which their existence virtually depended on what they could do directly with their hands (as in obtaining food) and with their feet (as in chasing prey, getting to food sources, and escaping from predators). Over the centuries they developed simple tools and utensils, and constructed shelter for themselves toward the end of aiding and abetting the process of keeping alive and making life a bit more tolerable. Such developments were the beginning of what we now call human factors, that is, the design of things and facilities so that they can reasonably well serve human needs. (In European and most other countries this field is called *ergonomics*.)

The human race has come a long way since the days of primitive life to the present day with our tremendous array of man-made products and facilities that have been made possible with current technology, including physical accoutrements and facilities that simply could not have been imagined by our ancestors in their wildest dreams. In at least many civilizations of our present world, the majority of the "things" people use are man-made. In other words, many people live in very much of a man-made world. Even those engaged in activities close to nature–fishing, farming, camping, bird watching–use many man-made devices. An abbreviated listing of man-made facilities could include hand tools, kitchen utensils, vehicles, highways, machinery, computers, houses and other buildings, TV sets, telephones, and space capsules (plus the binoculars for the bird watchers).

The current interest in human factors arises from the fact that technological developments have focused attention (in some instances dramatically) on the need to consider human beings in such developments. Munipov (1979) expresses this point as follows:

> Under the conditions of the scientific and technological revolution, an increasingly important social and economic value is attached to ergonomics (i.e., human factors), because of its implications for man and society. Ergonomics is to contribute not only to the creation of optimal conditions for work and leisure, but also to the development of new cultural values and social conditions for an overall development of the human being.

Technology, of course, is a mixed bag in that it can contribute either to the improvement or to the detriment of our lot in life. In a sense, the goal of human factors is to guide the applications of technology in the direction of benefiting humanity—or, to put it another way, to facilitate the development of a symbiotic relationship between technology and human beings. This text is intended as an overview of the human factors field; its various sections and chapters deal with at least some of the most important aspects of the field as they apply to such objectives.

HUMAN FACTORS DEFINED

Definitions sometimes represent treacherous exercises in semantics, but, for better or worse, they probably are necessary exercises. We shall approach the definition of human factors in three stages, as follows:

- The central *focus* of human factors relates to the consideration of human beings in carrying out such functions as (1) the design and creation of man-made objects, products, equipment, facilities, and environments that people use; (2) the development of procedures for performing work and other human activities; (3) the provision of services to people; and (4) the evaluation of the things people use in terms of their suitability for people.
- The *objectives* of human factors in these functions are twofold, as follows: (1) to enhance the effectiveness and efficiency with which work and other human activities are carried out; and (2) to maintain or enhance certain desirable human values (e.g., health, safety, satisfaction). The second objective is essentially one of human welfare and well-being.
- The central *approach* of human factors is the systematic application of relevant information about human abilities, characteristics, behavior, and motivation in the execution of such functions.

Although no short catch phrase can adequately characterize the scope of the burgeoning field of human factors, such expressions as *designing for human use* and *optimizing working and living conditions* may at least lend a partial impression of what human factors is about.

Aside from involvement in the functions referred to above, the human factors field also embraces certain related functions, such as testing and evaluating equipment, facilities, procedures, etc., in terms of human factors aspects: job design; development of job aids and training materials; and the selection and training of personnel who would be involved in the use of equipment, products, facilities, procedures, etc. Further, the human factors discipline usually is viewed as embracing relevant supportive research toward the end of developing appropriate data and guidelines to be applied in the carrying out of the functions referred to.

HUMAN FACTORS: PAST, PRESENT, AND FUTURE

The development of the human factors field has been inextricably intertwined with developments in technology.

Human Factors in the Past

As indicated before, human factors had its origins in the development by early humans of simple tools, utensils, shelters, etc. Over the intervening millions of years, there were of course improvements in the design of the things people used, these improvements being the result of an evolutionary process; if a particular tool or device did not adequately serve its purpose, succeeding generations would tend to improve the design. Human work was of course performed primarily by hand, with the use of hand tools, until the industrial revolution which was brought about by the development of machines. Christensen (1976) characterized the machine age in terms of three phases.

Phase I: The Age of Machines (1750–1890) This period witnessed the transition from what Christensen called the *eons of tools* to the *age of machines.* It was characterized particularly by brilliant inventions in the textile industry and the broad application of steam power. Jacquard, in France, drawing upon previous inventions, developed the use of punched cards to program weaving machines. In Great Britain, Watt designed a self-regulating governor for the steam engine, such control serving as the beginning of automation.

Phase II: The Power Revolution (1870–1945) This period was characterized primarily by major expansion in the use of power, such as in manufacturing, transportation, and agriculture, including the development and use of electric power for such purposes, as well as for communications.

It was during the latter part of this phase that the behavioral sciences became established. During World War I, for example, there was considerable attention to the matter of selection and training of military personnel, in both the United States and Great Britain, this effort being aimed at "fitting the man to the job." Following the war, further attention was given in these and other countries to personnel selection and training. During the first decades of the century there also were a few developments that would now be viewed as being related to human factors, such as methods analysis and time studies by industrial engineers, and some investigations relating to human work that were carried out by the (then) Industrial Health Research Board of Great Britain.

It was during World War II that the human factors field started to become more delineated. In particular, it was found that some of the newly developed items of military equipment could not be operated effectively or safely by many people. This realization triggered efforts to design such equipment in terms of human considerations; this objective then represented a shift from "fitting the man to the job" to "fitting the job to the man."

Phase III: Machines for Minds (1945–?) Christensen points out that the first two phases of the industrial revolution had the effect of aiding, relieving, and extending human muscles, while the third phase deals more with efforts to aid, relieve, and extend human mental capabilities. There is of course no clear dividing line of time between these two objectives, but in general terms the first two phases were periods in which major emphasis was placed on human control of power, as in manufacturing processes and transportation. In the third phase there have of course been continued efforts to enhance human control of power, but in addition a distinct focus developed on the use of machines for performing functions that can be considered as *mental*, primarily by the use of computers.

It was during this phase that the human factors field became a well-recognized discipline. Actually, since World War II the field has undergone rather significant changes. The period during and shortly after the war has been called the "knobs and dials" era because of the attention at that time to the design of control devices and visual instruments that could be used most rapidly and accurately; many of the military and nonmilitary human factors problems at that time were associated with such devices. Many knob-and-dial problems are still with us, but the field has broadened considerably since then. After the war the human factors field started to ooze out into certain nonmilitary areas (such as in certain areas of manufacturing, communications, and transportation) with the primary thrust in the direction of simplification of the operation and maintenance of the "hardware" people used in their work activities. In more recent years the discussions of the human factors clan have been characterized by a couple of themes. One of these deals with the notion of optimizing the combination of human beings and physical equipment. In this regard the central theme was that in the design of systems one should take full advantage of the capabilities that are unique to human beings (such as reasoning and decision making, and certain sensory and perceptual skills) and of those that are unique to machines (such as performing highly repetitive functions). In connection with the objective of optimizing the combination of human beings and physical equipment, it should be pointed out that there has been (and still is) an issue regarding the trade-off between designing systems that virtually anyone can use with minimum training versus the design of systems that would require considerable training on the part of the operators. In part this becomes an economic issue, since highly automatic systems that require little training may add substantially to the cost of the systems themselves, whereas systems that depend more on human involvement in their operation might require substantially more training on the part of the operators and therefore add to training and personnel costs in general.

The other (related) theme deals with the interest in improving the quality of working life through appropriate job design. Although the concern for the quality of working life developed outside the human factors disciplines, some of the people in the human factors area who shared this concern began focusing the attention of the human factors disciplines on the matter of job satisfaction with a view toward the design of jobs that, in general, would provide greater opportunity for job satisfaction on the part of workers.

Although there is current concern for both of these objectives, it probably needs to be stated that the guidelines for achieving such objectives still are items of unfinished business for the human factors disciplines.

The Current Applications of Human Factors

The human factors field received its initial impetus in the military services, and it is probable that the primary systematic application of human factors is still within the military services. Over the years, however, consideration of human factors has been extended, in varying degrees, to other areas of application, such as to the design of transportation equipment, production machinery and processes, communications equipment, agricultural equipment, highway systems, living facilities, recreation equipment, facilities and equipment for the handicapped, and consumer products and services. However, it should be emphasized that the applications in most of these areas have so far been sporadic and limited, in certain instances being extremely limited.

The Future of Human Factors

The clouded crystal ball makes it risky to predict what will happen in the future. However, it may be reasonable to speculate about the directions that human factors *might* take, even if we cannot predict with reasonable certainty the directions that it *will* take.

To begin with, let it be said that many of the things people create for their own use have human factors implications, and that, to date, there has been only limited attention (or no attention) to the human factors aspects of their design.

Considering the entire range of the things that our present man-made world comprises, there is a huge backlog of unfinished business—areas in which human factors to date have made only modest inroads, and that offer at least the possibility of broad-scale applications. Such areas include the design of transportation equipment and systems, buildings of all kinds, many consumer products, and health services that would better serve the needs of users. In addition, human factors could contribute to the improvement of safety in many facets of life, and further there is much that remains to be done in the application of human factors principles and data to the design of jobs that would contribute to increased job satisfaction; such improvement could be through improved design of machines and equipment used in work, the modification of work processes and procedures, and job enlargement. In addition, the increasing use of computers in industry (as in the control of production processes as well as in information processing activities) probably will require new types of interfaces between humans and machines that will have important implications in connection with the design of computer systems.

Aside from these possible areas of future (expanded) applications, however, there seem to be looming on the horizon more vaguely conceived areas of potential applications relating generally to society as a whole and to the changes taking place in society. In this regard Chapanis (1976a) suggests that these changes can be grouped into two categories. First are those changes that are related to the world itself, that are external to us, but that are forcing us to adjust to rapidly altering conditions. In this category are such factors as the energy crisis, the population explosion, and environmental pollution. The second category of changes includes those that can be traced to changes in our mores, our social values and expectations, and our ways of living. This group would include such things as attitudes about the relative responsibilities of producers and consumers in designing for safety, equal employment opportunities for all persons, and attitudes about the quality of life.

To speculate about the possible impact of at least one of these factors, let us consider the energy crisis and its human factors implications. The energy crisis very likely will have a profound impact on transportation, with increased dependence on public transportation, the development of smaller, more efficient automobiles, and more emphasis on improvement of flow of all types of traffic. There may be more dependence on the use of telecommunications media and devices as travel substitutes for conferences, and for shopping, banking, and education. In addition one might speculate about the reexamination of the trend toward the increased use of power-assisted tools, machines, and devices; there might even be a tendency toward reverting to greater dependence upon human energy sources. Further, there might be changes in the nature of our living facilities.

In discussing the process of change in human life, Chapanis (1979) makes the point that we still tend to view "man-machine systems" (such as specific machines or combinations thereof) as isolated and independent units within society. He envisions the next step in sophistication (and one that he sees developing within the human factors field) as one in which we begin to consider the effects of system design upon society as a whole. As he states:

> When we do that we often come to realize that what we do may have far-reaching effects on society, on its productivity, and on its economy. Moreover, at this stage we come face to face with value judgments that are difficult to quantify and to reach agreement on.

Continuing, he expresses the opinion that for every change, every innovation, and every invention, there are possible benefits to society as a whole, and possible costs. Aside from the expected economic costs, there are other possible costs to society, such as pollution, waste of energy, adverse effects on health, and risks of accidents. The risks of accidents include those in the air (from pollution), in transportation systems, in nuclear power plants, at sea (e.g., collisions of oil tankers), and in power systems (e.g., blackouts).

The fact of there being both benefits and costs associated with virtually every innovation imposes the requirement on society for making judgments regarding the cost-benefit ratio that, collectively, would be reasonably optimum for society. Considering only the matter of costs associated with accidents, for example, since there is no way to make any given system or procedure absolutely safe, society somehow has to make value judgments about the level of risk that can be accepted as the "cost" of the benefits that can accrue from the system or procedure. In this frame of reference a major role of human factors specialists in the future might well be that of helping in the design of things so that risks are minimized to the point where they become acceptable—in other words to ensure that the benefits of whatever is designed outweigh the risks that still remain in their use.

In looking ahead, one could envision the possible changes in human life that might occur as the consequence of increased automation, increased population density, increased demand for food, increased pollution, higher energy costs, recycling, the development of the developing countries with accompanying pressures on the world's resources, and the possibility of increased leisure time. It is very possible that the changes in life style that might accompany such developments could pose challenges to the human factors discipline that we simply cannot now even envision.

As one looks ahead, the potential range of applications of human factors principles and data is tremendous, embracing virtually every facet of life. In the best of all worlds, human factors could be viewed as contributing to the quality of life and work, consistent with the constraints that are imposed by the many realities that are part and parcel of human life. The extent to which the promise of human factors becomes fulfilled (in improving the future quality of life and work) will have to be documented by future historians.

OUR INVOLVEMENT WITH WHAT WE USE

Human factors experts generally deal with design problems relating to specific items of the facilities and environments that people use in their work and everyday living. However, as discussed above, the application of human factors principles and data to specific problems should be viewed from the point of view of the implications for society as a whole. Keeping this point of view in mind, the specific areas of application of human factors generally fall into the three classes discussed below.

Man-Machine Systems

We can consider a *man-machine system* as a combination of one or more human beings and one or more physical components interacting to bring about, from given inputs, some desired output. In this frame of reference, the common concept of *machine* is too restricted, and we should rather consider a "machine" to consist of virtually any type of physical object, device, equipment, facility, thing, or what have you that people use in carrying out some activity that is directed toward achieving some desired purpose or in performing some function. In a relatively simple form a man-machine system (or what we will sometimes refer to as a *system*) can be a person with a hoe, a hammer, a hod, or a hair curler. Going up the scale of complexity, one can regard as systems the family automobile, an office machine, a lawn mower, and a roulette wheel, each equipped with its operator. More complex systems include aircraft, bottling machines, conveyor systems, telephone systems, and automated oil refineries, along with their personnel. Some systems are less delineated and more "amorphous" than these, such as the servicing systems of gasoline stations and hospitals and other health services, the operation of an amusement park or a highway and traffic system, and the rescue operations for locating an aircraft downed at sea.

The essential nature of people's involvement in a system is an active one, interacting with the system to fulfill the function for which the system is designed.

Physical Environment

The physical environments which people "use" include two general categories. The first consists of the physical space and related facilities which people use, ranging from the immediate environment (such as a work station, a lounge chair, or a typing desk) through the intermediate (such as a home, an office, a factory, a school, or a football stadium) to the general (such as a neighborhood, a community, a city, or a highway system). The second category consists of the various aspects of the ambient environment, such as illumination, atmospheric conditions (including pollution), and noise. It should be noted that some aspects of the physical environment in which we live and work are part of the natural environment and may not be amenable to modification (although one can provide protection from certain undesirable environmental conditions such as heat or cold). Although the nature of people's involvement with their physical environment is essentially passive, the environment tends to impose certain constraints on their behavior (such as limiting the range of their movements or restricting their field of view) or to predetermine certain aspects of behavior (such as stooping down to look into a file cabinet, wandering through a labyrinth in a supermarket to find where the bread is, or trying to see the edge of the road when driving on a rainy night).

Personal and Protective Items

The third class of man-made things people use consists of many types of personal items (such as apparel, handbags, and skis) and protective equipment and gear (such as safety shoes and hats, safety goggles, astronaut suits, gloves, and earplugs). The human involvement with such items is typically passive, although their design can also impose certain constraints on behavior or predetermine the nature of certain aspects of behavior.

Discussion

Aside from these three categories of the man-made accoutrements and trappings of civilization, there are some other odds and ends that tend to defy nice, neat categorization, such as morning newspapers, playing cards, and postage stamps. Aside from the complications of classification, however, we would like to reinforce the central point we have been alluding to, namely, that whatever the nature of the human involvement with the man-made features of civilization, the specific design features thereof can influence, for better or worse, their functional utility or some relevant human value.

THE ROLES OF PEOPLE IN A MAN-MADE WORLD

It has been traditional in the human factors field to view the activities of people in man-machine systems with human beings and mechanical devices operating in an interactive fashion to bring about some desired objective. The discussion of such interactions logically brings up the question as to the roles or functions that people perform in such systems.

Types of Man-Machine Systems

Before discussing such roles or functions, however, let us refer to various types of man-machine systems. In this regard let us first clarify the distinction between closed- and open-loop systems. A closed-loop system is continuous, performing some process which requires continuous control (such as in vehicular operation and the operation of certain continuously controlled chemical processes), and requires continuous feedback for its successful operation. The feedback provides information about any error that should be taken into account in the continuing control process. An open-loop system is one which, when activated, needs no further control or at least cannot be further controlled. In this type of system the "die is cast" once the system has been put into operation, and no further control can be exercised, such as in firing a rocket that has no guidance system. Although feedback with such systems obviously cannot serve continuous control, feedback can, if provided, serve to improve subsequent operations of the system. It should also be noted that in most open-loop systems there is almost inevitably some internal feedback within the operator, even if not provided for outside the operator.

Aside from the distinction between closed- and open-loop systems, man-machine systems can be characterized by the degree of manual versus machine control. Although the distinctions between and among systems in terms of such control are far from clear-cut, we can generally consider systems in three classes: manual, mechanical, and automatic.

Manual Systems A manual system consists of hand tools and other aids which are coupled together by the human operator (usually a craft worker) who controls the operation. Operators of such systems use their own physical energy as a power source, transmit to their tools—and receive from them—a great deal of information, typically operate at their own speed, and can readily exploit their ability to act as a "high variety" system.

Mechanical Systems These systems (also referred to as *semiautomatic* systems) consist of well-integrated physical parts, such as various types of powered machine tools. They are generally designed so as to perform their functions with little variation. The power typically is provided by the machine, and the operator's function is then essentially one of control, usually by the use of control devices.

Automated Systems When a system is fully automated, it performs all operational functions, including sensing, information processing and decision making, and action. Such a system needs to be fully programmed in order to take appropriate action for all possible contingencies that are sensed. Most automatic systems are of a closed-loop nature. If such a system were perfectly reliable, it conceivably could offer the possibility of taking over all functions and leaving humans to twiddle their thumbs or take off for the golf course. No one would have to stay to tend the store. Since perfectly reliable automated systems are not a likely possibility, however, at least in our lifetime, it is probable that certain primary human functions in such systems will be those of monitoring, programming, and maintenance.

Discussion of Systems The distinctions made between manual, mechanical, and automatic systems are not really clear-cut. In fact, within any given system, the different components (which can be considered subsystems) can vary in the degree of their manual versus automatic features.

Operational Functions and Components of Systems

In order for a system (or a subsystem) to fulfill its purposes, certain operational functions need to be performed. For example, in a postal operation it is necessary to perform such functions as mail collection, stamp cancellation, mail sorting, and delivery. Each such operational function, in turn, needs to be performed by an individual or by a physical component. In system-design processes it is sometimes the practice to specify these operational functions and to set them forth in blocks in a block diagram, with the tentative expectation that each function can be *allocated* to a corresponding physical component or to a human being. More will be said about this allocation process later. But it should be noted here, as Jones (1967) points out, that this assumption of a one-to-one correspondence between functional blocks and separate physical (and presumably human) components is most applicable to *flow* systems, in which there is a sequence of clearly discriminable operations. But although all systems do involve the execution of functions by human or physical components, or by both, in some systems the functions are, as Jones puts it, less "determinate" and may not be clearly discriminable from each other. In such instances the assumption of a one-to-one correspondence between functions and system components simply may not apply; Jones cites, as an example, the operation of an automobile by its driver, indicating that there is no quick and unambiguous way of knowing that any particular block diagram (of functions and components) represents the constantly changing interaction between the driver, the vehicle, and the environment. The moral of these observations seems to be that the function-component frame of reference may serve a useful purpose in the development of some systems but may be of less value for other systems.

Basic Functions of Systems

The execution of any operational function, in turn, typically involves a combination of four more basic functions, as follows: sensing (information receiving), information storage, information processing and decision, and action functions; they are depicted graphically in Figure 1-1. Since information storage interacts with all the other functions, it is shown above the others. The other three functions occur in sequence.

1 *Sensing (information receiving):* One of these functions is sensing, or information receiving. Some of the information entering a system is from outside the system, for example, airplanes entering the area of control of a control-tower operator, an order

FIGURE 1-1
Types of basic functions performed by human or machine components of man-machine systems.

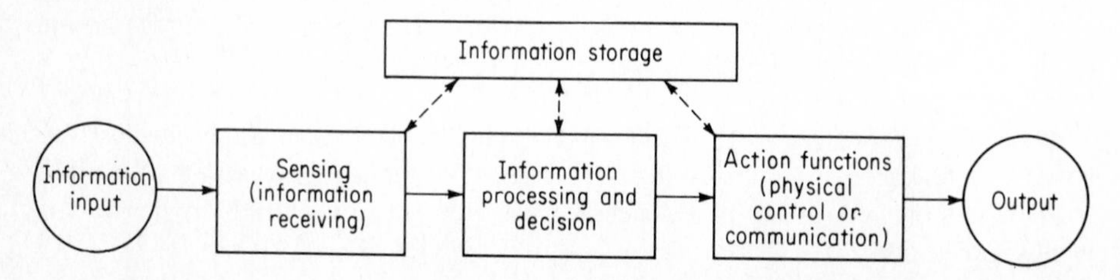

for the production of a product, the heat that sets off an automatic fire alarm, various cues regarding the presence of schools of fish, and telegraph communications. Some information, however, may originate from inside the system itself. Such information can be of a feedback nature (such as the reading on the speedometer of the action of the accelerator or the feel of a control lever), or it can be information that is stored in the system.

The sensing, if by a human being, would be through the use of the various sense modalities, such as vision, audition, and touch. There are various types of machine sensing devices, such as electronic, photographic, and mechanical. Sensing by a machine in some cases is simply a substitute for the same sensing function by a human. The electronic device in an automated post office which identifies the location of a stamp on an envelope is simply doing the same thing that a person would otherwise do to place the envelope in proper position for canceling the stamp. The sonar used for detecting schools of fish, however, involves "sensing" the fish in a manner that humans are not capable of.

2 *Information storage:* For human beings, information storage is synonymous with memory of learned material. Information can be stored in physical components in many ways, as on punch cards, magnetic tapes, templates, records, and tables of data. Most of the information that is stored for later use is in coded or symbolic form.

3 *Information processing and decision:* Information processing embraces various types of operations performed with information that is received (sensed) and information that is stored. When human beings are involved in information processing, this process, simple or complex, typically results in a decision to act (or in some instances, a decision *not* to act). When mechanized or automated machine components are used, their information processing must be programmed in some way in order to cause the component to respond in some predetermined manner to each possible input. Such programming is, of course, readily understood if a computer is used. Other methods of programming involve the use of various types of schemes, such as gears, cams, electrical and electronic circuits, and levers.

4 *Action functions:* What we shall call the *action* functions of a system generally are the operations which occur as a consequence of the decisions that are made. These functions fall roughly into two classes. The first of these is some type of physical control action or process, such as the activation of certain control mechanisms or the handling, movement, modification, or alteration of materials or objects. The other is essentially a communication action, be it by voice (in human beings), signals, records, or other methods. Such functions also involve some physical actions, but these are in a sense incidental to the communication function.

When dealing with human activities, three of these functions (sensing, information processing, and action) represent what is conventionally referred to by psychologists as the S → O → R (stimulus-organism-response) paradigm. These three functions are part and parcel of most human activities in the sense that a *stimulus* acts upon an *organism* to bring about a *response*. Frequently the stimulus is external to the individual (such as a traffic light seen by a driver, a part coming down an assembly-line conveyor belt, or the spotting of a defect by an inspector). In some instances the stimulus is generated within, or by, the individual (such as a recognition that it is time to perform a spec-

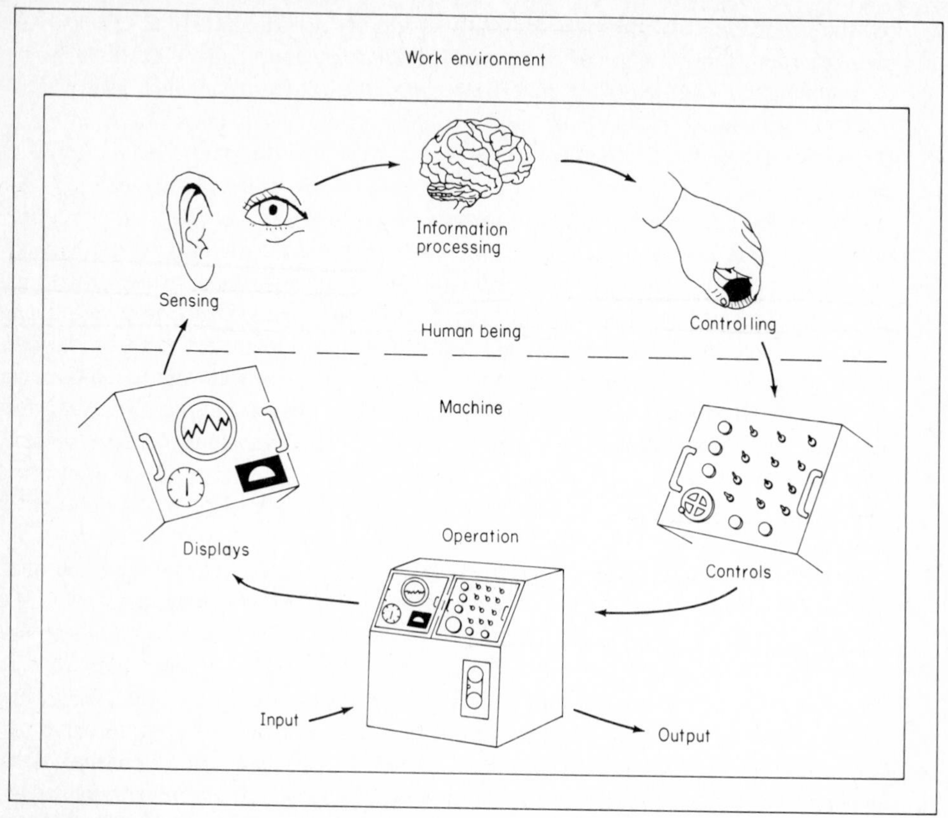

FIGURE 1-2
Schematic representation of a man-machine system. [*Source: Chapanis, 1976(b). From Chapanis, Alphonse. Engineering psychology. In Marvin D. Dunnette (Ed.),* Handbook of industrial and organizational psychology, p. 701. *Copyright © 1976 by Rand McNally College Publishing Company. Used by permission of Houghton Mifflin Company.*]

ified activity, or the completion of one step of a process that is to be followed by another).

The typical type of interaction between a person and a machine is illustrated in Figure 1-2. This shows how the displays of a machine serve as the stimuli for an operator, which trigger some type of information processing on the part of the operator (including decision making), which in turn results in some action (as in the operation of a control mechanism) that controls the operation of the machine. This is essentially an example of a closed-loop mechanical system as discussed above, since the effects of the machine operation (resulting from the control action) continually affect the information presented on the displays (such as in the case of an air-speed indicator in an aircraft).

Discussion

A critical premise of the human factors discipline is that the nature of the physical equipment and facilities that people use in their work and everyday living directly affects the nature of the roles and functions with which people are involved in their use. This premise, in turn, leads directly to the objectives of human factors, to design such equipment and facilities in order to enhance work efficiency and human welfare.

THE CASE FOR HUMAN FACTORS

Since humanity has somehow survived for these many thousands of years without people specializing in human factors, one might wonder why—at the present stage of history—it has become desirable to have human factors experts who specialize in worrying about these matters. As indicated before, the objectives of human factors are not new; history is filled with evidence of efforts, both successful and unsuccessful, to create tools and equipment which satisfactorily serve human purposes and to control more adequately the environment within which people live and work. But during most of the centuries of history, the development of tools and equipment depended in large part on the process of evolution, of trial and error. Through the use of a particular device—an ax, an oar, a bow and arrow—it was possible to identify its deficiencies and to modify it accordingly, so that the next "generation" of the device would better serve its purpose.

It has been the increased rate of technological development of recent decades that has generated the need to consider human factors early in the design game, and in a systematic manner. Because of the complexity of many new and modified systems it frequently is impractical (or at least excessively costly) to make changes in them after they are actually produced. The cost of retrofitting frequently is exorbitant. Thus, the initial designs of many items must be as satisfactory as possible in terms of human factors considerations. A few questions that illustrate some of the types of considerations that might be taken into account during the design stage could include the following: Should a mail-sorting system have an optical scanner that automatically activates mechanisms for sorting mail by zip code or should human operators scan the mail and activate keying devices for sorting by zip code? Should a particular warning signal be visual or auditory? How much "feel" should be built into the power-assisted steering mechanism of a car? Would the information load of a given (tentative) assortment of visual displays be within reasonable human limits? How much illumination should be provided for a given operation?

In effect, then, the increased complexities of the things people use (as the consequence of technology) place a premium on having assurance that the item in question will fulfill the two objectives of functional effectiveness and human welfare. The need for such assurance requires that human factors be taken into account early during the (usually long) design and development process.

In reflecting about the things people use it is relevant to add a special comment about a matter that is of increasing concern, that of product safety. In the United States and certain other countries legislation and court decisions are focusing attention on the liability of organizations that offer goods and services to the public regarding the potential hazards to users. One important aspect of such liability is that relating to the human factors aspects of such goods and services.

Looking toward the future (when the human factors disciplines are likely to become more involved with the impact of technology on society as a whole) human factors could well have a very substantial influence on the quality of human life.

A WORD ABOUT INDIVIDUAL DIFFERENCES

In the design of physical systems and accoutrements in terms of human considerations there is an easy reference to the "typical" or "average" human being. However, some words of caution are in order in conceptualizing the "model" of the human beings for whom the designer is designing whatever is being designed: (1) Human beings come in assorted sizes, shapes, and varieties; although there are circumstances in which it is appropriate to design for the somewhat mythical "typical" or "average" individual, the designer should be ever mindful of the fact of individual differences. (2) Some things are to be designed for special groups, such as infants, children, teenagers, the elderly, or the infirm; in such instances the designer obviously should consider such groups as a "model." (3) When certain things or facilities are to be designed for "the public," the designer should provide for almost the entire gamut of human beings; for example, in public places the various features such as entry ways, doors, ramps, escalators, and signs should be suitable not only for the hale and hearty but also for infirm elderly people and for kids tagging along behind their parents.

HUMAN FACTORS DISCIPLINES

Human factors is not a discipline separate and distinct from others; rather, it represents an area of overlapping interest to people in various disciplines. In a very general sense there are two types of disciplines that have common human factors interests. In the first place there are certain *sciences* that generally provide the knowledge and insight relating to human beings that are relevant to the design of the man-made features of our civilization. The primary relevant sciences are psychology; the biological sciences including physiology; sociology; mathematics; statistics; and anthropology. On the other hand there are certain *professions* that are concerned with design processes, in particular industrial engineering as well as the other branches of engineering, industrial design, and architecture.

The line between these two groups is admittedly thin, but in general terms the sciences serve as the source of human factors data through research, and the professions serve as "practitioners" in applying human factors data to design problems.

The organizations concerned with human factors tend to reflect the interdisciplinary nature. In the United States the Human Factors Society has members from all the disciplines mentioned (and others). This is also true of certain dominantly European and international organizations, such as the Ergonomics Society and the International Ergonomics Association. Various countries around the world have their own societies, such as Australia, New Zealand, Canada, Holland, Germany, Italy, India, and Japan. In addition, certain organizations have human factors divisions, such as the Society of Engineering Psychologists of the American Psychological Association and the Systems, Man and Cybernetics group of the Institute of Electrical and Electronics Engineers.

COVERAGE OF THIS TEXT

Since a comprehensive treatment of the entire scope of human factors would fill a small library, this text must be restricted to a rather modest segment of the total human factors domain. The central theme intended is that of illustrating the way in which the fulfillment of the two primary objectives of human factors (i.e., functional effectiveness and human welfare) can be influenced by the extent to which relevant human considerations have been taken into account during the design of the object, facility, or environment in question. Further, this theme will be followed as it relates to at least some of the more commonly recognized human factors "content" areas (such as the design of displays for presenting information to people, human control processes, and physical environment). It is suggested that pursuing this theme across the several subareas would have the further advantage of an overview of the content areas of human factors.

The implications of various perceptual, mental, and physical characteristics as they might affect or influence the objectives of human factors probably can best be reflected by the result of relevant research investigations and of documented operational experience. Therefore the theme of the text will generally be carried out by presenting and discussing the results of illustrative research and by bringing in generalizations or guidelines that are supported by research or experiences that have relevance to the design process in terms of human factors considerations. Thus, in the various subject or content areas much of the material in this text will consist of summaries of research that reflect the relationships between design variables on the one hand and criteria of functional effectiveness or human welfare on the other hand.

It is recognized that the illustrative material that will be brought in to carry out this theme will be in no way comprehensive, but it is hoped that it will represent in most content areas some of the more important facets of those areas.

Although the central theme will, then, deal with the human factors aspects of the design of the many things people use, there will be some modest treatment of certain related topics, such as how human factors fits in with the other phases of design and development processes, and of certain of the personnel-related functions, such as selection and training.

REFERENCES

Chapanis, A. Ergonomics in a world of new values. *Proceedings: 6th Congress of the International Ergonomics Association*, July 11-16, 1976, I-XI. Santa Monica, Calif.: Human Factors Society, 1976(a).

Chapanis, A. Engineering psychology. In M. D. Dunnette (Ed.), *Handbook of industrial and organizational psychology.* Chicago: Rand McNally, 1976(b).

Chapanis, A. Quo vadis, ergonomia? *Ergonomics*, 1979, *22*(6), 595-605.

Christensen, J. M. Ergonomics: Where have we been and where are we going? *Proceedings: 6th Congress of the International Ergonomics Association,* July 11-16, 1976, XXV-XXXIII. Santa Monica, Calif.: Human Factors Society, 1976.

Jones, J. C. The designing of man-machine systems. *Ergonomics,* 1967, *10*(2), 101-111.

Munipov, V. M. Ergonomics as a factor in social and economic development. *Ergonomics,* 1979, *22*(6), 607-611.

CHAPTER 2

THE DATA BASE OF HUMAN FACTORS

The central approach of human factors is the application of relevant information about human characteristics and behavior to the design of objects, facilities, and environments that people use. Most relevant human factors information is based on experimentation and observation. Research plays a central role in this regard, and the research basis of human factors is emphasized in this text. For these reasons, this chapter will deal with a few of the basic concepts of human research as related to human factors, and in addition will cover the somewhat related matters of human performance and systems reliability.

AN OVERVIEW OF RESEARCH METHODS

This chapter is not intended as a text in research methods; rather, it is intended to provide an overview of certain aspects of research methodology which are commonly encountered in human factors related research. Most human factors research involves the use of human beings as subjects and so we will focus our attention on such research. It should be noted that certain investigations deal more with "events" (such as automobile accidents or results of a production process) or with objects (such as different types of control devices) than with people as such; however, people usually are involved with such events or objects.

Human factors research can usually be classified into one of three types: descriptive studies, experimental research, or evaluation research. *Descriptive studies* generally seek to describe a population (usually of people) in terms of certain attributes. Examples of descriptive studies would include an anthropometric survey of truck drivers wherein the height, weight, arm length, etc., of the drivers were measured and tabulated; a survey of the opinions of subway train riders; or a survey of hearing loss among rock band members.

The purpose of *experimental research* is to assess the effects of one or more variables on various aspects of behavior. Some examples would include assessing the effects of various amounts of control resistance on tracking performance, assessing the effects of various levels of environmental heat on mental performance, or assessing the effects of various types of seats on perceived comfort.

Evaluation research is similar to experimental research in that its purpose is to assess the effects of "something"; however, in evaluation research the "something" is usually a complex system. Evaluation research generally is more global and comprehensive than is experimental research. A system is evaluated by comparison with its goals; both intended consequences and unintended outcomes must be assessed. Often evaluation research includes a cost-benefit analysis. Examples of evaluation research include evaluating a new training program, evaluating a new design for rapid transit vehicles, and evaluating a new computer information management system.

Not all human factors research fits neatly into these three categories; often a particular study can be classified into more than one category. No matter which category of research is undertaken, however, there are several fundamental decisions which must be addressed in order to plan and execute the work properly. These include picking a research setting, selecting variables, choosing a sample of subjects, and deciding how the data will be collected and how it will be analyzed. We will discuss each of these briefly.

The Research Setting

Frequently investigators with particular research purposes in mind may have some freedom to choose the situational context of their research—that is, whether to carry it out in a laboratory setting, or in the real world (i.e., field research), or to use simulations of the real world. There are other circumstances, however, in which the nature of the research problem virtually dictates the locale. For example, if investigators wished to study the differential thresholds for discriminating amplitudes of vibration, they would have to do so in a laboratory since there probably are no situations in the field where precise control of vibration amplitudes could be achieved.

The choice of research setting involves complex trade-offs. Research carried out in the field usually has the advantage of realism in terms of relevant task variables, environmental constraints, and subject characteristics including motivation. Thus, there is a better chance that the results obtained from the study can be generalized to the real-world operational environment. The disadvantages, however, include cost (which can be prohibitive), safety hazards for subjects, and lack of experimental control. In field studies often there is no opportunity to replicate the experiment a sufficient number of times; many variables cannot be held constant; and often certain data cannot be collected because the process would be too disruptive.

The laboratory setting has the principal advantage of experimental control; extraneous variables can be controlled and the experiment can be replicated almost at will; data collection can be made more precise. For this, however, the researcher may sacrifice some realism and generalizability. Simulation is an attempt to combine the generalizability of field research with the control of laboratory research.

A distinction should be made between *physical* simulations and *computer* simulations. Physical simulations are usually constructed of hardware and represent (i.e, look like, feel like, or act like) some system, procedure, or environment. Physical simulations can range from very simple items (such as a picture of a control panel) to extremely complex configurations (such as a moving-base jumbo jet flight simulator with elaborate out-of-cockpit visual display capabilities). Some simulators are small enough to fit on a desk top; others can be quite large, such as a 400-square-foot (400-ft^2) underground coal mine simulator built by one of the authors.

Computer simulation involves modeling a process or series of events in a computer. By changing the parameters the model can be run and predicted results can be obtained. For example, manpower needs, periods of overload, and equipment downtime can be predicted from computer simulations of work processes. To develop an accurate computer model requires a thorough understanding of the system being modeled and usually requires the modeler to make some simplifying assumptions about how the real-world system operates.

Types of Variables

Typically, human factors investigations involve manipulating or obtaining information about two types of variables. One of these is the *independent variable* (also called the *experimental variable* or *predictor*). The other is the *dependent variable* (also called the *criterion*). In human factors research, independent variables usually can be classified into one of three types: (1) task-related variables which include equipment variables (such as different coding techniques, length of control levers, etc.) and procedure variables (such as different work-rest cycles); (2) environmental variables such as variations in illumination, noise, and vibration; and (3) subject-related variables such as height, age, mental ability, etc. In the case where the independent variable is a subject-related variable, it is sometimes called a predictor.

The dependent variable, or criterion, usually consists of some measure of the possible "effects" of the independent variable. Frequently it is a measure of performance, such as reaction time. We will discuss criteria in a subsequent section of this chapter. It might be added that in some investigations the distinction between the independent and dependent variable is not always clear. This is particularly the case when both variables are measures of human characteristics such as, say, physical strength and speed of movement; either one of these can be used as a predictor, and the other as a criterion.

Sampling

Most human factors research involves use of human beings as subjects about whom relevant data are obtained. The critical concern when selecting a sample is that the information obtained from the sample be generalizable to some hypothetical population. The population can be quite broad (such as all adult males), or more restrictive (such as all air traffic controllers). For generalizability, a sample must be representative of the population. By representative we mean that the sample should contain all the relevant aspects of the population in the same proportion as found in the population. A sample that is not representative is said to be biased. To help ensure a representative sample and to permit us to use powerful inferential statistics, it is important that the sample be selected randomly from the population. Random selection occurs when each member of the population has an equal chance of being included in the sample.

Data Collection

With the advent of microprocessors and laboratory computers, more and more human factors data are being automatically collected and stored on magnetic or paper-punched tape. There will, however, always be a place for the human observer and recorder in human factors research. It is important, therefore, that observers be trained in what to observe and how to record it.

In field research the problems of collecting data multiply. Murphy's law seems to rule: "If anything can go wrong, it will." Extra attention, therefore, must be paid to designing data collection procedures and devices to perform in the often unpredictable and unaccommodating field setting. One of the authors recalls evaluating the use of a helicopter patrol for fighting crime against railroads. Murphy's law prevailed; the trained observers never seemed to be available when they were needed, the railroad coordination chairman had to resign, the field radios didn't work, the railroads were reluctant to report incidences of crime, and finally a gang of criminals threatened to shoot down the helicopter if they saw it in the air.

Data Analysis

Once an experiment or study has been carried out and the data have been gathered, the experimenter is in a position to analyze the data to see what relationships there are between or among the independent and dependent variables using appropriate statistical analyses. It is not intended in this text to deal extensively with statistics, or to discuss elaborate statistical methods. Probably most readers are already familiar with most of the statistical methods and concepts that will be touched on later, especially frequency distributions and different measures of central tendency (the mean, median, and mode). For those readers who are not familiar with the concepts of the standard deviation, correlations, and statistical significance, these will be described very briefly.

Standard Deviation The quantitative values of cases within a sample–values such as errors made, height of people, or scores on tests–naturally vary from each other. Certain statistical indices can be used to quantify the degree of variability among the cases. One such measure is the *standard deviation.*[1] This is expressed in terms of the original numerical values of the data and reflects the variability in the distribution of the cases from the mean. In relatively normal distributions approximately two-thirds of all the cases fall within one standard deviation above or below the mean, and over 99 percent of the cases fall within three standard deviations above or below the mean. If the standard deviation of the height of one group of boys is, say, 1 inch (1 in), they would be more homogeneous (i.e., less variable) than another group whose standard deviation is 2 in.

Correlation A coefficient of correlation is a measure of the degree of relationship between two variables. Correlations can range from +1.00, a perfect positive correlation, through zero (which is the absence of any relationship) to −1.00, a perfect nega-

[1] The standard deviation, mathematically, is the square root of the average of the squares of the deviations of the individual cases from the mean. Formulas for computing the standard deviation may be found in most elementary statistical texts.

tive correlation. A positive correlation between two variables indicates that high values on one variable tend to be associated with high values on the other. For example, height and weight are positively correlated. A negative correlation between two variables indicates that high values on one variable tend to be associated with low values on the other variable. Stimulus intensity and reaction time, for example, tend to be negatively correlated.

Statistical Significance To evaluate their results, experimenters should determine to what extent the results are *statistically significant*, such as the difference between two means or the size of a correlation. Statistical significance refers to the *probability* that the results (whatever they may be) could have occurred by *chance* (as opposed to being brought about by the experimental variables under investigation). It is common practice to use either the "5 percent" or the "1 percent" level as the acceptable level of statistical significance. If a difference is significant at the 1 percent level, this means that the obtained difference is of such a magnitude that it could have occurred *by chance* only 1 time out of 100.

RELIABILITY OF MEASUREMENT AND VALIDITY

In virtually any research, especially that relating to human beings, the concepts of reliability of measurement and of validity are of major concern.

Reliability of Measurement

Reliability in the context of measurement refers to the consistency or stability of the measurements of a variable over time or across representative samples. Suppose that a human factors specialist in King Arthur's court were commanded to assess the combat skills of the Knights of the Roundtable. To do this he might have each knight shoot a single arrow at a target and record the distance off target as the measure of combat skill. If all the knights were measured one day and again the next, it is quite likely that the scores would be quite different on the two days. The best archer on the first day could be the worst on the second day. We would say that the measure was unreliable. Much, however, could have been done to improve the reliability of the measure, including having each knight shoot 10 arrows each day and using the average distance off target as the measure, being sure all the arrows were straight and the feathers set properly, and performing the archery inside the castle to reduce the variability in wind conditions, lighting, and other conditions that could change a knight's performance from day to day.

Correlating the sets of scores from the two days would yield an estimate of the reliability of the measure. Generally speaking, test-retest reliability correlations around .80 or above are considered to be satisfactory, although with some measures we have to be satisfied with lower levels.

Validity

Although there are different types of validity, they have in common the determination of the extent to which the measurements of different variables actually measure, or predict, that which they are intended to measure or predict. The American Psychological Association recognizes three types of validity. Although these deal primarily with the validity of tests, they are also relevant for human factors research.

The first type of validity, *criterion-related validity*, refers to the extent to which measures of one variable (an independent variable) predict some other variable (a criterion or dependent variable), this frequently being done by the use of a correlation. In the human factors field examples might include the extent to which age is related to visual acuity or the extent to which intensity of noise exposure is related to hearing loss.

The second type of validity, *content validity*, refers to the extent to which a measure of some variable samples a domain such as a field of knowledge or a set of job behaviors. In the field of testing, for example, content validity is typically used to evaluate achievement tests. In the human factors field this type of validity would apply to such circumstances as measuring the performance of air traffic controllers. To have content validity, such a measure would have to include the various facets of the controllers' performance rather than just a single aspect.

The last type of validity, *construct validity*, refers to the extent to which a measure is really tapping the underlying variable of interest. For example, if one is interested in measuring "mental workload" (the underlying variable) there are numerous measures that could be used ranging from brain waves (EEG) to subjective rating scales. Construct validity is based on a judgmental assessment of an accumulation of empirical evidence regarding the measurement of the variable in question.

CRITERIA IN RESEARCH AND SYSTEM DEVELOPMENT

The criterion (or dependent variable) as used in research is a measure of the possible "effects" of the independent variable (that is, the factor being investigated). Thus, we might compare the speed and accuracy with which people can read different road signs, or the heart rates of people working under two conditions of temperature. Aside from the use of criteria in experimental research undertakings, however, criteria are also used in the testing and evaluation of systems or components such as in testing the performance of a dishwasher, a tractor, or a telephone system, i.e., evaluation research.

Types of Criteria

In general terms the criteria used in human factors research and systems development are of two types: human criteria and system criteria.

Human Criteria There are four relatively different types of human criteria: (1) human performance measures, (2) physiological indices, (3) subjective responses, and (4) accident frequency. In a strict sense human performance must be considered in terms of various sensory, mental, and motor activities. In specific work situations, however, it is usually difficult, if not impossible, to measure human performance strictly in terms of human activity, since such performance usually is inextricably intertwined with the performance characteristics of the physical equipment being used. Thus, the typing performance of a typist is not entirely a function of the typist but also in part the consequence of the typewriter (its make, condition, etc.).

For some purposes indices of various physiological conditions are pertinent criteria. Such possible indices include heart rate, blood pressure, composition of the blood, galvanic skin response, brain waves, respiration rate, skin temperature, blood sugar, and many other measures. Some of these and other physiological variables are used as indices of the physiological effects on people of various methods of work, of work performed with equipment of various designs, of work periods, and of work performed under various environmental situations (such as heat and cold).

For some purposes the subjective responses of people can serve as appropriate criteria; examples are ratings of the performance of individuals, of alternative design features of a system, of the judged importance of different types of information for use in a system, and of the comfort of seats.

For still other purposes accident or injury frequency may serve as appropriate criteria. For example, the number of injuries or deaths per million miles traveled gives a comparison (in terms of this criterion) of various types of transportation systems, such as commercial airlines, railways, buses, and automobiles.

System Criteria Basically system criteria are those that relate to the performance of the system (or subsystem or component thereof), or, in other words, those that reflect something about the degree to which the system (or subsystem or component) achieves what it is intended to achieve. For example, a computer keyboard might be evaluated in terms of such criteria as number and accuracy of data entries made per unit of time, and an earth-moving vehicle might be evaluated in terms of the amount of earth moved per unit of time. Other examples of system criteria are the anticipated life of a system; ease of operation or use; maintainability; reliability; operating cost; and human resources requirements. Some such criteria are rather strictly mechanistic, in the sense that they reflect essentially engineering performance (e.g., the maximum rpm of an engine), whereas others reflect more the performance of the system as it is used by the people involved in it (such as errors in cards punched).

It should be obvious from our discussion that the two classes of criteria are not neat dichotomies, but rather tend to form a continuum, ranging from strictly mechanistic system criteria at one end to strictly behavioral criteria at the other end. In the case of both system and human criteria, there are many specific types that can be used. Examples of various types will be given in later chapters.

Requirements of Satisfactory Criteria

Criteria used in research investigations generally should fulfill certain requirements, namely, validity, reliability, freedom from contamination, and sensitivity of measurement.

Validity The validity of a criterion refers to the extent to which the measure in question is considered to be a relevant or pertinent index of the criterion in mind, such as system performance, quality of work, comfort in seating, or job satisfaction.

Reliability Reliability was discussed above, and need not be discussed further here.

Freedom from Contamination A criterion measure should not be influenced by variables that are extraneous to the variable that is being measured. In our example of the Knights of the Roundtable, the wind conditions, illumination, and quality of the arrows could be sources of contamination since they could affect accuracy yet are unrelated to the variable being measured, namely combat skill.

Sensitivity of Measurement A criterion measure should be measured in units that are commensurate with the anticipated differences among subjects. To continue with our example of the knights, if the distance off target were measured to the nearest yard, it is possible that few, if any, differences between the knights' performance would have been found. The scale (to the nearest yard) would have been too gross to detect the subtle differences in skill between the archers.

HUMAN PERFORMANCE AND SYSTEMS RELIABILITY

Much of the data used in the human factors domain consist of measures of some aspect of human performance. Human performance in the context of systems often boils down to considerations of how fast people can perform their functions and how well they can perform them. How well functions are performed leads directly to considerations of human errors. For example, one might want to obtain answers to such questions as: How many defects might occur in the task of soldering electric wires? How many commands might be missed in a noisy communication system? What is the probability of not detecting a signal light at various angles from the operator?

Types of Human Errors in System Tasks

People are quite inventive in the kinds of bloopers they perpetrate, but it can serve some purposes (such as trying to reduce their frequency or severity) to figure out what kinds of mistakes people do make. As a first step in this direction, classification may be helpful. Rook (1962) developed one scheme, to be used in the classification of errors in an operating system, that provides for a cross classification of errors in terms of two bases, as in Table 2-1. In this formulation *intentional* obviously does not refer to intentional errors, but rather to acts that individuals performed intentionally, thinking they were doing the right thing when in fact they were not (such as pushing the "door

TABLE 2-1
ROOK'S SYSTEM OF CLASSIFICATION OF ERRORS

Conscious level or intent in performing act	Behavior component		
	Input I	**Mediation M**	**Output O**
A Intentional	AI	AM	AO
B Unintentional	BI	BM	BO
C Omission	CI	CM	CO

Source: Rook, 1962.

TABLE 2-2
ALTMAN'S SYSTEM OF CLASSIFICATION OF ERRORS

Discrete acts	Continuous actions	Monitoring
Omissions Insertions Sequence Unacceptable performance	Failure to achieve end state in available time Displacement from target condition over time	False detections Failure to detect

Source: Altman, 1964.

close" button in an elevator instead of the "door open" button, thereby clobbering an entering passenger). The input-mediation-output model of this classification system corresponds to a common sequence of psychological functions that is basic to all behavior, namely, S (stimulus), O (organism), R (response). As Meister (1966) points out, human error occurs when any element in this chain of events is broken, such as failure to perceive a stimulus, inability to discriminate among various stimuli, misinterpretation of meaning of stimuli, not knowing what response to make to a particular stimulus, physical inability to make a required response, and responding out of sequence. Identifying the *source* of errors in terms of input-mediation-output behaviors is a first step in developing inklings about how to reduce the likelihood of errors.

Rook's classification system above applies primarily to discrete (i.e., individual, separate, distinguishable) acts. Some variations in classifying errors in discrete acts, as well as in certain other tasks, have been proposed by Altman (1964). An abbreviated recap of Altman's formulation is given in Table 2-2, with types of errors listed in each category.

Besides defining the nature of the task itself, it is sometimes useful, for analytical purposes, to classify the task in terms of its degree of *revocability* as proposed by Altman (1967), such as immediate correction; correction only after intervening steps; no correction within a given "mission" (e.g., within the present operation); and irrevocable consequences.

Human Performance Data

In discussing performance data it is relevant to refer to performance reliability. In this regard let us hasten to distinguish between the concept of *reliability of measurement* as described above in reference to criteria and the use of the term *performance reliability* as used here to mean the dependability of performance of a system or individual in carrying out an intended function. This use of the term stems from engineering practice, in which it refers to quantitative values that characterize the dependability of system or component performance.

The index of performance reliability that is used in certain types of situations is simply the probability of successful performance, such as the probability of detecting defects in a sample of parts being inspected. In such instances human error is actually the complement of reliability. If, for example, we say that an inspector has a .02 probability of error (i.e., a 2 percent error rate), it is the same as saying that the reliability of the inspector is .98 (i.e., 98 times out of 100 the inspector will detect defective items). For many system development processes the types of human performance data that would be useful include the reliability with which individual tasks can be performed and the time required to perform them. In some circumstances, however, data on other parameters of tasks may be relevant, such as the judged criticality of the tasks, training time required, and skill level required.

Although certain human performance and task data banks have been developed, Blanchard (1975) points out that the ones that are available go back over some years, and are not sufficiently comprehensive to serve many purposes. In turn, he argues for the establishment of certain guidelines in the development of future data banks. Despite the limitations of available data banks, it still may be useful to mention certain ones that have been developed and illustrate the types of task data they contain. In general, there are two sources of such information, namely empirical observations and human estimates.

Empirical Task Data Aside from sets of industrial engineering time-study data, there are two particular sources of essentially empirical task data that should be mentioned. One of these, the Data Store, was developed by the American Institute for Research (Munger, 1962; Munger, Smith, & Payne, 1962; Payne & Altman, 1962) particularly for tasks in the operation of electronic equipment. This Data Store was pulled together from many different sources, supplemented by special laboratory studies. Without going into the tale of how the data were developed, an example is given in Table 2-3 specifically related to the use of a joystick. Note that the data include performance time and reliability in performance under the variations of the illustrated dimensions. In using the Data Store, the time that would be estimated for a given

TABLE 2-3
TIME AND RELIABILITY OF OPERATION OF JOYSTICKS OF VARIOUS LENGTHS WHEN MOVED VARIOUS DISTANCES

Dimension	Time to be added to base time,* s	Reliability
Stick length, in		
6–9	1.50	.9963
12–18	0.00	.9967
21–27	1.50	.9963
Stick movement, degrees		
5–20	0.00	.9981
30–40	0.20	.9975
40–60	0.50	.9960

*Base time, 1.93 s.

Note: Other data (not shown here) are given for variations in other dimensions, specifically, control resistance, presence or absence of arm support, and time lag between movement of control and corresponding movement of display.

Source: Munger, 1962.

TABLE 2-4
ERROR RATES AND RELIABILITIES FOR ILLUSTRATIVE INDUSTRIAL TASKS

Type of error	Probability of error, P	Reliability, $1 - P$
Two wires which can be transposed are transposed	.0006	.9994
A component is omitted	.00003	.99997
A component is wired backward	.001	.999

Source: Rook, 1962.

instance would consist of the base time [in this case 1.93 seconds (1.93 s)] plus the time to be added for the particular dimension characteristics that would apply (such as 1.50 s for a joystick of 24 in, plus 0.50 s if it is to be moved 50°, etc.).

The second source is referred to—in this day of acronyms—as SHERB (Sandia Human Error Rate Bank) (Rigby, 1967), which is a compilation of error-rate data based on the THREP (Technique for Human Error Rate Prediction) (Rook, 1962; Swain, 1964). This body of data consists of human error rates for many industrial tasks based on large numbers of observations. Table 2-4 gives a few examples.

Judgmental Task Data Where empirical data on tasks are not lying around loose, it may be useful to resort to human estimates about certain task parameters for use in a system development process. To illustrate such a process, in the estimation of the error rates of tasks, Irwin, Levitz, and Freed (1964, pp. 143-198) asked experienced missile engineers, technicians, and mechanics to rate 60 tasks by sorting them into 10 piles ranging from those that would be performed with the least error to those with the greatest error. By a subsequent manipulation (using empirical reliability data that were available for some of the tasks and were extrapolated to others) it was possible to derive quantitative estimates of the reliability of all 60 tasks. A few illustrations are listed in Table 2-5.

TABLE 2-5
EXAMPLES OF JUDGMENTS OF TASK RELIABILITY

Task	Mean rating, 10-point scale (10 = greatest error)	Derived reliability
Read time (brush recorder)	8.2	.9904
Install gasket	6.0	.9945
Inspect reducing adapter	4.9	.9958
Open hand valves	3.8	.9968
Remove drain tube	2.6	.9976

Source: Irwin, Levitz, and Freed, 1964.

System Performance Reliability

Actually, there are different measures of system performance reliability, each being relevant to certain types of systems or situations. However, most of these measures of reliability can be applied to human performance as well as to system performance. Certain variations on the reliability theme include: (1) probability of successful performance (this is especially applicable when the performance consists of discrete events, such as detecting a defect or starting a car); (2) mean time to failure (abbreviated MTF; there are several possible variations on this, but they all relate to the amount of time a system or individual performs successfully, either until failure or between failures; this index is most applicable to continuous types of activities). There are other variations that could also be mentioned. For our present discussion, let us consider reliability in terms of the probability of successful performance.

If a system includes two or more components (machine or human, or both), the reliability of the composite system will depend on the reliability of the individual components and how they are combined within the system. Components can be combined within a system either in *series* or in *parallel.*

Components in Series In many systems the components are arranged in series (or sequence) in such a manner that successful performance of the total system depends upon successful performance of each and every component, man or machine. By taking some semantic liberties, we could assume components to be *in series* that may, in fact, be functioning concurrently and interdependently, such as a human operator using some type of equipment. In analyzing reliability data in such cases, two conditions must be fulfilled: (1) failure of any given component results in system failure, and (2) the component failures are independent of each other. When these assumptions are fulfilled, the reliability of the system for error-free operation would be the product of the reliabilities of the several components. To estimate system reliability quantitatively then, it is necessary to express the reliability of the individual components quantitatively, specifically as the probability of error-free performance.

If, for example, there are two components in a system, each of which has a reliability of .90 (meaning that it "works" 90 percent of the time), the reliability of the system would be the product of these two, or .81. The basic reliability formula for sequential situations is:

$$R_{\text{system}} = R_1 \times R_2 \times R_3 \cdots R_n$$

in which $R_1, R_2, R_3 \cdots R_n$ are the reliabilities of the individual components, expressed as percentages of successful functioning.

It can thus be seen that each component *decreases* system reliability by its own factor. If a system consists of, say, 100 components, each with a reliability R of .99, the system reliability would be only .365. If the system consisted of 400 such components, the system reliability would be only .03, which indicates almost certain failure.

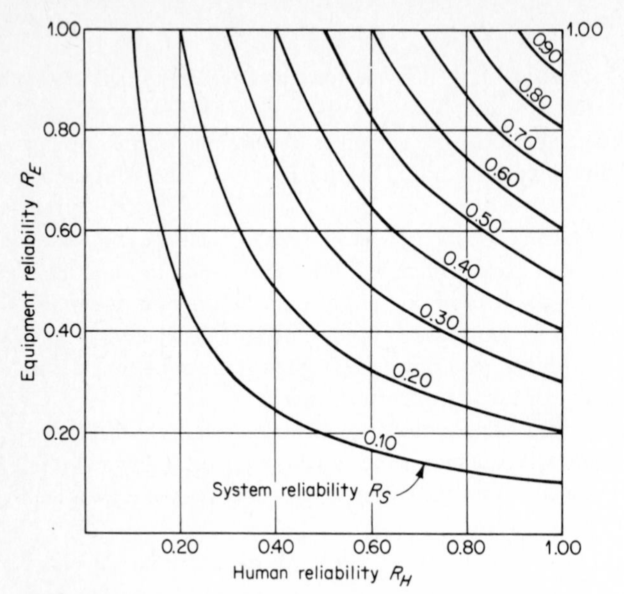

FIGURE 2-1
Effect of human and equipment reliabilities on system reliability. This same relationship applies to two independent physical components or human beings. (*Source: Meister & Rabideau, 1965, p. 16.*)

When successful system performance depends upon successful performance of two components in series, for example, a physical component and a human operator, we can estimate the system reliability from the reliabilities of the two components, as shown in Figure 2-1. That figure is equally applicable for two physical components, or two human beings, operating independently, when system performance depends on performance of both.

Components in Parallel The reliability of a system in which the components are in parallel is entirely different from the situation in which they are in a series. With parallel components there are two or more which in some way are performing the same function. This is sometimes referred to as a *backup* or *redundancy* arrangement—one in which one component backs up another, increasing the probability that the function will be performed. In such a case the system reliability is estimated by combining the probabilities of success (the reliabilities) of the several components, using the following formula:

$$R_{\text{system}} = [1-(1-r)^m]^n$$

where m = number of components in parallel for each function
n = number of functions
r = unit reliability

This formula applies only where the number of components in parallel for each of the

n functions is the same and the unit reliability of each component is the same. Where these conditions do not apply, the derivation becomes somewhat more complex.[2]

In a strictly parallel arrangement, identical components would be used independently, either physical components or people as the case may be. If the reliabilities of individual components are, say, .90, the joint probability for two components (which would be the system reliability) would be .99. Even with relatively low component reliabilities such as .70, system reliability of 99.2 can be achieved with four parallel components. In some instances the parallel components can be of very different types, such as a person backing up a machine component.

In some circumstances if we wish to estimate the effect of human reliability on system reliability, we need to do so in units of human performance that are system-related, these units of behavior being tasks. The steps proposed by Swain (1964, pp. 109–117) to do this are:

1 Define the system or subsystem failure which is to be evaluated.

2 Identify and list all the human operations performed and their relationships to system tasks and functions.

3 Predict error rates for each human operation or group of operations pertinent to the evaluation.

4 Determine the effect of human errors on the system.

5 Recommend changes necessary to reduce the system or subsystem failure rate as a consequence of the estimated effects of the recommended changes.

The particularly sticky wicket of these operations is item 3, the prediction of error rates for human operations (tasks), which would require something other than a Ouija board–it may be hoped, some reasonably quantitative data such as are contained in the empirical task data banks discussed previously.

Although data on human reliability in performing tasks would be useful in estimating total system reliability, it cannot be assumed that errors in performance of all tasks are necessarily equal in their effects on system performance. (To draw from Gilbert and Sullivan, it seems doubtful that to "polish up the handle of the big front door" was a particularly critical task in the British Admiralty system.) Thus, another task parameter might be criticality, and it is this parameter that was used by Pickrel and

[2] Where these conditions do not exist, the reliability can be built up by an iterative technique, as follows: For the components in parallel, select the first. If the system included only that one, the system reliability R_s would equal that of the component R_1. The reliability of the system that included both components 1 and 2 (call it $R_{s(1,2)}$) would be as follows:

$$R_{s(1,2)} = R_1 + (1 - R_1)R_2$$

and, in turn,

$$R_{s(1,2,3)} = R_{s(1,2)} + (1 - R_{s(1,2)})R_3$$

and

$$R_{s(1,2,3,\ldots,n)} = R_{s(1,2,\ldots,n-1)} + (1 - R_{s(1,2,\ldots,n-1)})R_n$$

McDonald (1964) in one phase of their elaboration of human performance reliability. Specifically, they elicited ratings of task criticality, setting ranges of rating values for the following classification:

1 Safe (error will not result in degradation, damage, hazard, or injury).
2 Marginal.
3 Critical.
4 Catastrophic (error will produce severe degradation–loss of system or death or multiple deaths or injuries).

Using such data as a springboard, these investigators urge that a concentration of efforts for failure reduction be made for those errors which are most likely to occur and most likely to have the more serious consequences along the "criticality" scale.

DISCUSSION

Human factors data and principles that are relevant for use in the design of physical equipment and facilities and environments cover a very wide range. Examples of some of the types of such data are scattered throughout this book. In addition, some of the sources referred to include additional sets of data that are relevant to particular topics. The application of such data to practical design problems usually is by people who are concerned with design functions, such as engineers, architects, and industrial designers.

As indicated by Askren and Lintz, the actual design of some items involves some trade-off of advantages and disadvantages. For example, in the location of control devices it may be that it is not possible to place all devices in their most convenient locations, and so some decision needs to be made as to which ones are to be given preference in location. Usually such decisions would be made on the basis of the judged criticality in use of the various control devices, placing the most critical ones in the preferred locations. Many design decisions, however, must be made on the basis of complex combinations of benefit and cost factors. Some of these factors (which may be on either the benefit or cost side of the coin) include system performance and reliability, efficiency, risks of accidents and effects on health, and job satisfaction and dissatisfaction, along with some of the societal values referred to in Chapter 1. In making many design decisions the availability of relevant data can contribute to the making of reasonably optimum decisions in terms of benefit and cost considerations. One major function of human factors efforts, therefore, is to provide pertinent data and to assist in their appropriate application to design problems.

REFERENCES

Alluisi, E. A. *Performance measurement technology: Issues and answers.* Norfolk, Va.: Performance Assessment Laboratory, Old Dominion University, April 1978.

Altman, J. W. Improvements needed in a central store of human performance data. *Human Factors*, 1964, *6*(6), 681–686.

Altman, J. W. Classification of human error. In W. B. Askren (Ed.), *Symposium on reliability of human performance in work,* May 1967 (TR 67-38). Aerospace Medical Research Laboratory (AMRL).

Askren, W. B., & Lintz, L. M. Human performance data in system design trade studies. *Human Factors,* 1975, *17*(1), 4–12.

Blanchard, R. E. Human performance and personnel resource data store design guidelines. *Human Factors,* 1975, *17*(1), 25–34.

Irwin, I. A., Levitz, J. J., & Freed, A. M. Human reliability in the performance of maintenance. *Proceedings of the symposium on quantification of human performance,* Albuquerque, N. Mex.: August 17–19, 1964. M-5.7. Subcommittee on Human Factors, Electronic Industries Association.

Meister, D. Human factors in reliability. In W. G. Ireson, *Reliability handbook* (Sec. 12). New York: McGraw-Hill, 1966.

Meister, D., & Rabideau, G. F. *Human factors evaluation in system development.* New York: Wiley, 1965.

Munger, S. J. *An index of electronic equipment operability: Evaluation booklet.* Pittsburgh: The American Institute for Research, 1962.

Munger, S. J., Smith, R. W., & Payne, D. *An index of electronic equipment operability: Data Store.* Pittsburgh: The American Institute for Research, 1962.

Payne, D., & Altman, J. W. *An index of electronic equipment operability: Report of development.* Pittsburgh: The American Institute for Research, 1962.

Pickrel, E. W., & McDonald, T. A. Quantification of human performance in large, complex systems. *Human Factors,* 1964, *6*(6), 647–663.

Rigby, L. V. *The Sandia Human Error Rate Bank (SHERB).* Paper presented at Symposium on Man-Machine Effectiveness Analysis: Techniques and Requirements. Santa Monica, Calif.: Human Factors Society, Los Angeles Chapter, June 15, 1967.

Rook, L. W., Jr. *Reduction of human error in industrial production* [SCTM 93-62(14)]. Albuquerque, N. Mex.: Sandia Corporation, 1962.

Swain, A. D. THERP (Technique for human error rate prediction). *Proceedings of the symposium on quantification of human performance,* August 17–19, 1964, Albuquerque, N. Mex.: Subcommittee on Human Factors, Electronic Industries Association.

INFORMATION INPUT

CHAPTER **3**

INFORMATION INPUT AND PROCESSING

On our jobs, in our homes, while driving our cars, and in other circumstances we mortals are bombarded with all sorts of stimuli from our environment. These stimuli are sensed by our various sense organs, such as the eyes, ears, nose, taste buds of the tongue, and sensory receptors of the skin. The interpretation of such stimuli (that is, the information they convey) is generally a function of our perceptual processes and of our learned associations (such as learning the alphabet). Our common notion of *information* is reflected by such everyday examples as what we read in the newspapers or hear on TV, the bill for automobile repairs, the gossip over the backyard fence, and the directions on road signs. However, we can view "information" in a much broader frame of reference as embracing the transfer of energy that has meaningful implications in any given situation, such as a driver "communicating" with a car via the control mechanisms; the sensing of air temperature by people or by thermometers; and mechanical, hydraulic, and servo linkages in various types of equipment. Thus, we can envision information flowing along two-way streets involving human beings, the various physical components with which they interact, and the environment, as follows:

Human ↔ Human
Human ↔ Physical components
Human ↔ Environment
Physical components ↔ Physical components
Physical components ↔ Environment

SOURCES AND PATHWAYS OF STIMULI

The stimuli that the sense organs receive actually consist of some form of energy, such as light, sound, heat, mechanical pressure, and chemical energy (such as sensed by the nose or taste buds). As indicated above, the *information* we receive from the stimuli is based on our interpretation of them, such interpretation being rooted in our perceptual processes and learned associations. In considering human information input and processing let us first discuss the types of original sources of information (actually stimuli), the pathways of such information, and the variations in the form of the information that may occur between the original source and the receiver. Perhaps most typically the original source (the *distal* stimuli, if we want to use the long-haired term) is some object, event, or environmental condition. Information from these original sources may come to us *directly* (such as by *direct* observation of an airplane), or it may come to us *indirectly* through some intervening mechanism or device (such as radar or a telescope). In either case, the distal stimuli are sensed by the individual only through the energy that they generate (directly or indirectly) through *proximal* stimuli (light, sound, mechanical energy, etc.). In the case of *indirect* sensing, the *new* distal stimuli may be of two types. In the first place, they may be *coded* stimuli, such as visual or auditory displays. In the second place, they may be *reproduced* stimuli, such as those presented by TV, radio, or photographs, or through such devices as microscopes, microfilm viewers, binoculars, and hearing aids; in such cases the reproduction may be intentionally or unintentionally modified in some way, as by enlargement, miniaturization, amplification, filtering, or enhancement. With either coded or reproduced stimuli, the new, or converted, stimuli become the actual distal stimuli to the human sensory receptors. Figure 3-1 illustrates in a schematic way these various pathways of information reception to the individual, both direct and indirect.

The human factors aspect of design enters into this process in those circumstances in which *indirect* sensing applies, for it is in these circumstances that the designer can design displays for presenting information to people. *Display* is a term that applies to virtually any man-made method of presenting information, such as a highway traffic sign, a family radio, or a page of braille print. Human information input and processing operations depend, of course, upon the sensory reception of relevant stimuli (such as

FIGURE 3-1
Schematic illustration of pathways of information from original sources to sensory receptors. (Although typically the original, basic source is an object, event, condition, the environment, etc., in some situations the effective original source to an individual consists of some man-made coded or reproduced stimuli; office personnel, for example, usually deal with recorded symbols which, for practical purposes, are their "original" distal stimuli.)

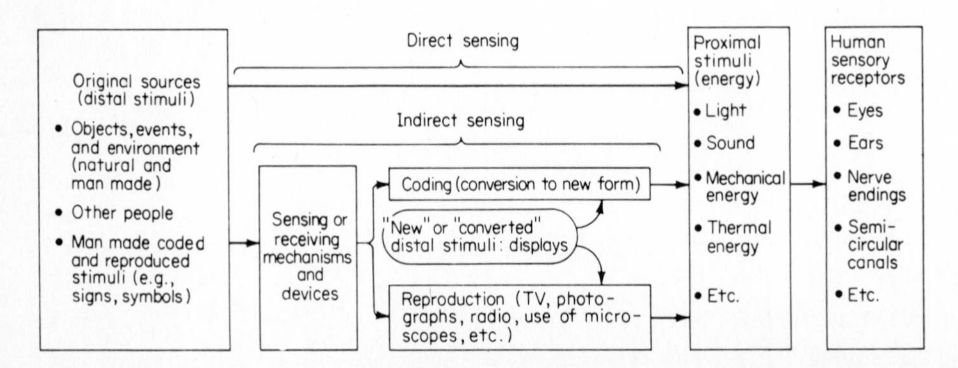

from displays), but they also depend upon perceptual and learning processes. In other words, it is necessary that sensory stimuli be correctly recognized by the receiver and that their "meaning" be understood. If the stimulus is a code symbol of some type (such as a red light), it is necessary that its meaning be understood (in this case, to stop).

In the design of some systems the circumstances may require a determination as to whether a given information input or processing function can best be performed by an individual or by some physical component of a system.

This chapter will deal with certain human factors that may be relevant to the design considerations that relate to human information receiving and processing functions.

DISPLAYS USED IN INFORMATION INPUT

Although displays will be discussed in some detail in Chapters 4 and 5, it is appropriate in our present discussion of information input to take an overview of the different types of displays that are used in various circumstances and of the types of information that can be presented with displays. There are of course many sources of information (i.e., distal stimuli) that people can sense directly without any problem. There are, however, many circumstances in which the information that people need in performing some activity can best be presented indirectly by the use of some type of display. Some such circumstances are given below:

1 When distal stimuli are of the type that humans generally can sense, but cannot sense adequately under the circumstances because of such factors as:

- **a** Stimuli at or below threshold values (e.g., too far, too small, or not sufficiently intense) that need to be amplified by electronic, optical, or other means.
- **b** Stimuli that require reduction for adequate sensing (e.g., large land areas converted to maps).
- **c** Stimuli embedded in excessive noise that generally need to be filtered or amplified.
- **d** Stimuli far beyond human sensing limits that have to be converted to another form of energy for transmission (e.g., by radio and TV) and then reconverted to the original form or converted to another form.
- **e** Stimuli that need to be sensed with greater precision than people can discriminate (e.g., temperature, weights and measures, and sound).
- **f** Stimuli that need to be stored for future reference (e.g., by photography and tape recorder).
- **g** Stimuli of one type that probably can be sensed better or more conveniently if converted to another type in either the same sensory modality (e.g., graphs to represent quantitative data) or a different modality (e.g., auditory warning devices).
- **h** Information about events or circumstances that by their nature virtually require some display presentations (such as emergencies, road signs, and hazardous conditions).

2 When distal stimuli are of the type that humans generally cannot sense or that are beyond the spectrum to which humans are sensitive, and so have to be sensed by sensing devices and converted into coded form for human reception (e.g., certain forms of electromagnetic energy and ultrasonic vibrations).

In these and other types of circumstances, it may be appropriate to transmit relevant information (stimuli) *indirectly* by some type of display. For our purposes, we shall consider a display to be any method of presenting information indirectly, in either *reproduced* or *coded* (symbolic) form. If a decision is made to use a display, there may be some option regarding the sensory modality and the specific type of display to use, since the method of presenting information can influence, for better or worse, the accuracy and speed with which information can be received.

Types of Displays

Displays can be described as either *dynamic* or *static*. Dynamic displays are those that continually change or are subject to change through time, and include the following types: displays that depict the status or condition of some variable, such as temperature and pressure gauges, speedometers, and altimeters; cathode-ray-tube (CRT) displays such as radar, sonar, TV, and radio range signal transmitters; displays that present intentionally transmitted information, such as record players, TV, and movies; and those that are intended to aid the user in the control or setting of some variable, such as the temperature control of an oven. (It might be observed, incidentally, that there are some devices that do double duty as both displays and controls; this is especially the case with devices used for making settings, such as oven controls.) Static displays, in turn, are those that remain fixed over time, such as signs, charts, graphs, labels, and various forms of printed or written material.

Types of Information Presented by Displays

Some of the major types of information presented by displays are described below. Although most displays can be considered as falling into categories that parallel these, there is not a perfect one-to-one relationship, because certain specific displays present two or more "types" of information, or because the user "uses" the displayed information only for certain purposes.

- *Quantitative information:* Display presentations which reflect the quantitative value of some variable, such as temperature or speed. Although in most instances the variable is dynamic, some such information may be static (such as that presented in nomographs and tables).
- *Qualitative information:* Display presentations which reflect the approximate value, trend, rate of change, direction of change, or other aspect of some changeable variable. Such information usually is predicated on some quantitative parameter, but the displayed presentation is "used" more as an indication of the change in the parameter than for obtaining a quantitative value as such.
- *Status information:* Display presentations which reflect the condition or status of a system, such as: "on-off" indications; indications of one of a limited number of conditions, such as "stop-caution-go" lights; and indications of independent conditions

of some class, such as a TV channel. (What is called *status information* here is sometimes referred to in other sources as *qualitative information.* But the terminology used in this text is considered to be more descriptive.)

- *Warning and signal information:* Display presentation used to indicate emergency or unsafe conditions or to indicate the presence or absence of some object or condition (such as aircraft or lighthouse beacons). Displayed information of this type can be static or dynamic.
- *Representational information:* Pictorial or graphic representations of objects, areas, or other configurations. Certain displays may present dynamic images (such as TV or movies) or may present symbolic representations (such as heartbeats shown on an oscilloscope or blips on a cathode-ray tube). Others may present static information (such as photographs, maps, charts, diagrams and blueprints, and graphic representations such as bar graphs and line graphs).
- *Identification information:* Display presentations used to identify some (usually) static condition, situation, or object, such as the identification of hazards, traffic lanes, and color-coded pipes. The identification usually is in coded form.
- *Alphanumeric and symbolic information:* Display presentations of verbal, numerical, and related coded information in many forms, such as signs, labels, placards, instructions, music notes, printed and typed material including braille, and computer printouts. Such information usually is static, but in certain circumstances it may be dynamic, as in the case of news bulletins displayed by moving lights on a building.
- *Time-phased information:* Display presentations of pulsed or time-phased signals, e.g., signals that are controlled in terms of duration of the signals and of intersignal intervals, and of their combinations, such as the Morse code and blinker lights.

The kinds of displays that would be preferable for presenting certain types of information are virtually specified by the nature of the information in question, but for presenting most types of information there are options regarding the kinds of displays and certainly about the specific features of the displays. Examples of some visual and auditory displays will be given in Chapters 4 and 5, along with examples of some of the research that has been carried out regarding the design of displays. Since the coding of stimuli has rather general implications in connection with information input and processing, however, some aspects of coding will be covered later in this chapter.

INFORMATION THEORY

For those interested in human factors, at least a brief exposure to information theory is in order. Its relevance to human factors is based on the fact that it provides for the measurement of information, the unit of measurement being the *bit*. (The term is a boiled-down version of *binary digit*.) Such measurement is used in at least certain human factors research and applications. The *bit* (symbolized by the letter H) is the amount of information necessary to decide between two equally likely alternatives, and is derived with the following formula:

$$H = \log_2 n$$

where n is the number of equally probable alternatives. In case the probabilities of the alternatives are not equal, a more generalized formula can be used, as follows:

$$H = \log_2 \frac{1}{p}$$

in which p is the probability of any given alternative.

When the probabilities of the various alternatives are equal, the amount of information in bits is measured by the logarithm, to the base 2, of the number of such alternatives. With only two alternatives, the information, in bits, is equal to the logarithm of 2 to the base 2, which is then 1. When Paul Revere was to receive a signal from the Old North Church, he was to see 1 signal (1 lantern) if the enemy came by land and 2 if by sea. Assuming that these two alternatives were equally probable, the amount of information available would be 1 bit; a discrimination was to be made between two alternatives.

Let us now take four alternatives, such as four lights on a panel, only one of which may be on at a time. In this case we should have 2 bits of information ($\log_2 4 = 2$). If we had eight such lights, we should have 3 bits of information ($\log_2 8 = 3$), etc. As the number of equally probable alternatives increases geometrically (2, 4, 8, 16, . . .), the amount of information increases arithmetically (1, 2, 3, 4, . . .).

If we were playing "Twenty Questions," it would be possible to determine the correct answer out of 1,048,576 possible alternatives if the questions were properly framed and if all the information were used. The amount of information that would be so obtained with 20 questions would be 20 bits ($\log_2 1{,}048{,}576 = 20$). When the probabilities of the individual events are not equal, the computation of the bit is more complicated, but the basic concept is still applicable.

INFORMATION PROCESSING

Information processing permeates the human factors field because of its influence on human behavior. In this regard Kantowitz and Roediger (1980) emphasize the generally accepted assumption that behavior is determined by the internal flow of information within the organism. Although the flow cannot be observed directly, it is reasonable to infer that, depending on the circumstance, it can involve some combination of such functions as attention, receiving information through the sensory mechanisms, perception, coding and decoding, learning, memory, recall, reasoning, making judgments, making decisions, transmitting information, and executing physical responses.

Information Flow

In some types of human activity the flow of information is rather direct and clear-cut in that there is a direct relationship between some obvious input and a specific output, as when dialing a telephone number (the input being the number in the telephone directory) or when typing from prepared copy. In more complex tasks, however, the relationship between any input and output may not be so distinct, because of the more involved intervening information processing including recall, judgment, and decision-making processes.

Numerous "models" of information processing and of its phases have been proposed to describe how such processes explain behavior; most of these models consist of postulated "black boxes" or processing stages that are assumed to be involved in the theoretical model in question. In connection with alternative models of memory, for example, some researchers suggest that there are different systems that serve as the basis for short-term and for long-term memory, while others maintain there is basically only one kind of memory system. In connection with this issue, and others, the point should be made that even if we are not able to assert definitely what theory is valid it frequently is possible to make some judgment as to which of two or more alternative explanations seems to be most consistent with observed behavior. And in some circumstances one may be able to "explain" behavior operationally, even if there is no substantial theoretical foundation. For example, even though we may not know for sure whether there are different memory "systems," we do know that some things are remembered for short periods of time while others are remembered for longer periods, that there are more definite limits to short-term memory than to long-term memory, that people differ in their short-term and long-term memory abilities, and that there are methods for improving both types.

Examples of Information Processing Theories It is not feasible here to go into detail regarding the various functions involved in information processing or to discuss the various theories of information processing, but we will at least describe a couple of theoretical formulations.[1] Some years ago Broadbent (1958) formulated a model that represented his conception of the flow of information within the nervous system. A diagram of this model is given in Figure 3-2. Although he has modified certain features of his 1958 formulation (Broadbent, 1971), the original formulation was characterized by four features, as follows: (1) The whole nervous system is regarded as a single channel, having a limit to the rate at which it can transmit information. (2) The limited-capacity portion of the nervous system is preceded and protected by a selective filter of some type that operates by selecting those stimulus events which possess some common "feature," and passing on all other features of those events to the limited-capacity system for analysis. (3) The "filter" is preceded by a buffer or temporary (short-term) store which could hold any excess information arriving by channels other than the one selected. (4) A "long-term store" is kept of the information which passes through the limited-capacity system in the form of a record of the conditional probability that events of one kind are followed by events of another kind. Although Broadbent (1971) has modified, and developed further, certain of these features (and has more drastically altered his views regarding certain other features not mentioned above), he maintains that the main outline of the flow of information remains the same.

[1] For more extensive discussions of information processing the reader is referred to the following sources, which are in the list of references at the end of the chapter: Broadbent (1958, 1971); Chase (1978); Kahneman (1973); Kantowitz (1981); Kantowitz and Roediger (1980); Navon and Gopher (1979); and Posner (1978).

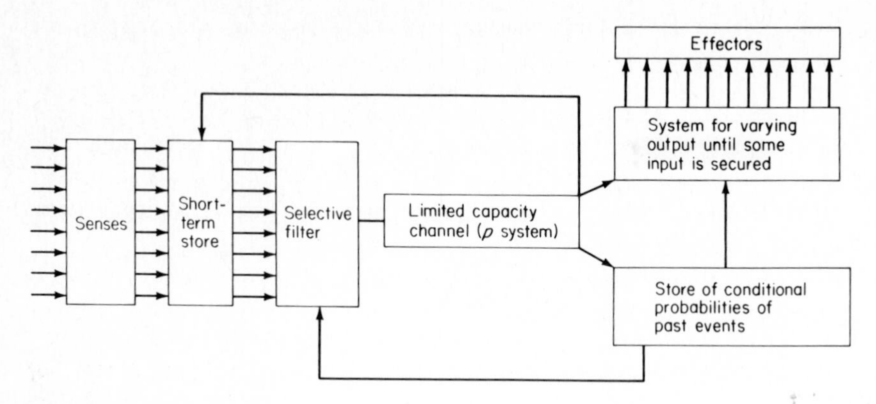

FIGURE 3-2
Broadbent's diagram of the flow of information through the nervous system. (*Source: Broadbent, 1958.*)

However, questions have been raised about Broadbent's model, with its focus on the notion of limited channel capacity. For example, Chase (1978) refers to certain *second-generation* flow diagrams of the human being as an information processor. An example of such a diagram is that presented by Haber and Hershenson (1980) and shown in Figure 3-3. In this model short-term memory replaces Broadbent's limited-capacity communication as the "central structure" of the model. Chase suggests that such a model serves to conceptualize the flow of information within what he refers to as *hypothetical memory structures.*

Discussion of Information Flow The disparities among these and other models of information processing serve to emphasize the point that there is as yet no widely accepted consensus about the identity and nature of the "black boxes" that are involved in the intricate processes leading up to overt human behavior. However, certain gener-

FIGURE 3-3
An information processing model presented by Haber and Hershenson. [*Source: From* The psychology of visual perception *(2nd ed.), by Ralph Norman Haber & Maurice Hershenson, fig. 14.3, p. 298. Copyright © 1973 by Holt, Rinehart and Winston, Inc. Copyright © 1980 by Holt, Rinehart and Winston. Reprinted by permission of Holt, Rinehart and Winston.*]

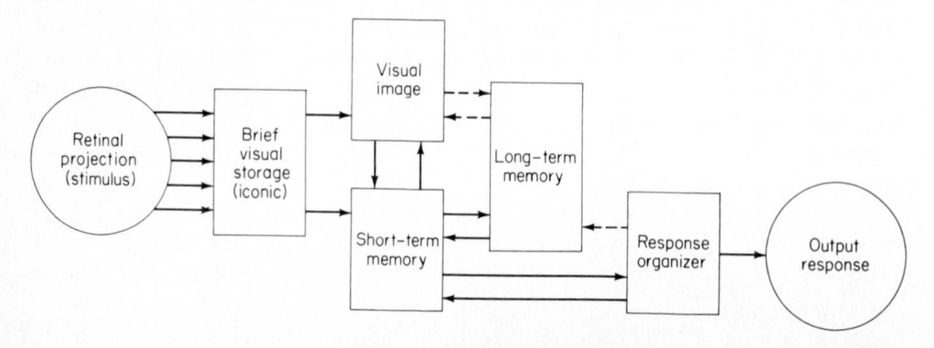

alizations about information processing can be offered on the basis of available evidence or can be inferred from or suggested by such evidence:

- The performance of people in work and elsewhere is the consequence of information processing, including some or all of the functions referred to earlier (attention, sensation, perception, coding and decoding, learning, memory, recall, reasoning, making judgments, making decisions, transmitting information, and executing physical responses).
- The information processing involves a number of stages, each stage consisting of some transformation of the information (such as the interpretation of physical stimuli into meaningful information, as a red light meaning "stop"). The stages can be arranged in different patterns (Kantowitz & Roediger, 1980). The simplest pattern, *serial processing*, occurs when there is a chain of stages with the output of one feeding directly into the next. *Parallel processing* occurs when several stages simultaneously can have access to the same output (of another stage). *Hybrid processing* occurs when serial and parallel processing occur in the same activity.
- The human being generally has a limited capacity for processing information. Although this presumably is not universally applicable to all aspects of information processing, in practical terms the human being must be viewed as having some limits on information processing capacity.
- Time sharing (the performance of two activities simultaneously or in rapid alternation) can result in the degradation of performance of one or both tasks if the collective demands of the activities exceed the capacities of the individual. However, the effects of time sharing depend in part on the nature of the activities. (More will be said about this later.)
- There are ways and means of facilitating certain information processing functions, such as by appropriate design of the stimuli presented (as in displays), various methods of helping people to learn and memorize, and methods of aiding recall.
- Once a decision is made to take a particular action, neural impulses are transmitted to the muscles (the *effectors*) to execute the intended action. Neural feedback from the muscles aids in their control.
- In the flow of information involved in these various functions it is evident that the cortex of the brain is capable of dealing with only a portion of the information that is received by the sensory mechanisms; it is in effect something of a bottleneck. We can "sense" much more information than we can use. Thus, our sensory mechanisms pick up the tremendous variety of stimuli in our environment, such as traffic patterns, landscapes, TV scenes, and football games, even though—at any one moment—we "tune in" only one aspect. In this connection, Steinbuch (1962) has summarized the information reduction that occurs from the initial reception by the sense organs through the intermediate processes to permanent storage (memory) and presents the estimates shown in Table 3-1. He postulates certain as yet unexplained intermediate reduction processes between the neural connections of the sensory organs and the conscious perception of the stimuli. Granting that the estimates in Table 3-1 are rough, we can indeed see that the central-nervous-system processes of consciousness and storage are capable of handling only a fraction of the potentially tremendous information input to the sensory receptors.

TABLE 3-1
ESTIMATED AMOUNTS OF INFORMATION AT VARIOUS STAGES OF INFORMATION PROCESSING

Process	Maximum flow of information, bits/s
Sensory reception	1,000,000,000
Nerve connections	3,000,000
Consciousness	16
Permanent storage	0.7

Source: Steinbuch, 1962.

Information in Sensory Reception

As indicated above, the sensory mechanisms are capable of sensing tremendous quantities of information, although the amount that can be received varies with the sensory modality in question (such as the eyes and ears) and with the nature of the stimuli in question. More will be said about this later in this chapter. (Discussions of certain of the sensory mechanisms are covered in later chapters.)

Information in Human Memory

We saw above that the amount of information that can enter *permanent storage* (i.e., memory) per unit of time is very modest, being about 0.7 bits/s. However, the amount that can be retained in such storage is tremendous, with there being marked individual differences. While the specific processes of learning and storage are not entirely understood, it is generally recognized that they are based on certain changes within the nerve cells (the neurons) of the brain. It has been estimated that there are something like 10 billion of these. Assuming the efficient utilization thereof, it has been estimated that the overall storage capacity of the human memory is somewhere between 100 million and 1 million billion bits (Geyer and Johnson, 1957). This range, of course, is far greater than the storage capacity of any computer now in existence or likely to be developed within any reasonable time. To extend the analogy between human information storage and that of computers, it has been pointed out that storage components of electronic systems are of two types, namely, static (consisting of special patterns of binary data unchanging in time) and dynamic (information in the form of electrical or mechanical impulses). In turn, there is some evidence that the brain also combines both of these schemes (Geyer and Johnson). One of the schemes is provision for storage of *old* information; the other has been described as a *circulatory* conception, accounting for the recording of current or recent information (short-term memory).

Although the bit has been used as a measure of the basic unit of memory, Miller (1956) introduced the concept of a *chunk* as a possible substitute, a chunk consisting of any familiar unit, regardless of size, that can be recalled as an entity, given a single relevant cue. Chase (1978), for example, suggests that Lincoln's Gettysburg Address might be considered as a chunk of information. Because of the possibility of chunking, Chase argues that, in the analysis of human memory capacity, information theory and its associated unit of measure of the bit is not very useful. In view of Chase's reflections one presumably should restrict the use of the bit in connection with human memory to those circumstances in which the relevant units of information as such lend themselves to such measurement. (In other stages of information processing the bit probably would be more relevant than in connection with memory.)

Information in Human Responses

Following the previous discussion of information, we can view human responses as "conveying" information. This notion is most clear-cut in instances in which the output responses are intended to correspond with input stimuli, as in the case of a keypunch operator (who is to "reproduce" the input information by pressing corresponding keys), or in the case of a lathe operator (who is to "reproduce" in the wood or metal being turned the design given on a blueprint). The efficiency with which people can "transmit" information through their responses depends upon the nature of the initial information input and upon the type of response that is required. As an example of the influence of the type of response on the amount of information that can be transmitted, in one study a comparison was made of the verbal versus motor (i.e., key-pressing) responses of subjects who were presented with Arabic numerals (2, 4, or 8 under a number of experimental conditions) at specified rates (1, 2, or 3 per second) and were asked to repeat them *verbally* or to respond by *pressing keys* which corresponded to the numerals to be presented (Alluisi, Muller, and Fitts, 1957). The maximum rate for verbal responses was 7.9 bits/s and for the motor (key-pressing) responses, 2.8 bits/s. In a somewhat more complex task, an analysis was made of the information of placing pegs in holes with varying degrees of tolerance (the difference in diameter of the peg and of the hole) and varying amplitudes (distance of movement) (Fitts, 1954). Information in bits was measured by a special formula, the number of bits for the various experimental conditions ranging from 3 to 10.

Although the amount of information that can be "transmitted" by people through their physical responses thus varies with the situation, it has been estimated that a reasonable ceiling value is of the order of about 10 bits/s (Singleton, 1971). Practice and experience of course increase the amount of information that can be so transmitted.

FACTORS THAT INFLUENCE INFORMATION RECEPTION AND PROCESSING

As would be expected, there are differences between and among individuals in their effectiveness in information reception and processing. Aside from individual differences, however, there are a number of factors that influence the reception and processing of information, for better or worse. We will discuss a few of these.

Noise and the Theory of Signal Detection (TSD)

In many circumstances meaningful stimuli may occur in the presence of "noise" that may interfere with the reception of the meaningful stimuli, whatever they may be, such as warning bells, foghorns, Morse code signals, radar blips, signal lights against their backgrounds, or defects in products that are being inspected. The possible effects of the noise (auditory, visual, or other) on the detection of stimuli have given rise to the formulation of the theory of signal detection (TSD) (Egan, 1975; Swets, 1964).

Basis of TSD To illustrate this theory, let us assume the case of an ambient noise in a factory with an intensity that varies randomly over time, with the occasional occurrence of a warning signal of a crane that increases the total intensity. The varying intensity of the noise by itself at different points in time might be depicted by the normal distribution at the left of Figure 3-4. In turn, the varying intensity of the combined noise and signal might be like the distribution at the right. In the case of very low or very high intensity values, there is no appreciable problem in determining whether only the noise is present or whether the signal is present. It is in the overlapping areas that confusion can occur. The actual probabilities of one or the other are reflected in the figure by the relative proportion of overlap of the two distributions at the overlapping intensity values.

In making a determination about the presence or absence of a signal, there are four response alternatives, as shown in Table 3-2, with a probability (P) associated with each.

We can see in the example in Figure 3-4 that no observer would be able to detect signals 100 percent of the time [$P(Y/sn)$] and have no "false alarms" [$P(Y/n)$]. However, the "criterion" selected by the observer in making a yes-or-no decision has an obvious effect on the frequency of the types of errors made. If the observer is "lenient"

FIGURE 3-4
Illustration of certain of the concepts of the theory of signal detection (TSD), using sound intensity as the parameter. The two distributions reflect the probabilities (at points in time) at which the intensity of the noise (n) or of the signal plus noise (sn) might occur. The figure illustrates a criterion point that might be selected for making a "yes" decision. Other parameters can be viewed in the same manner. See text for discussion, including the meaning of the probability (P) symbols shown.

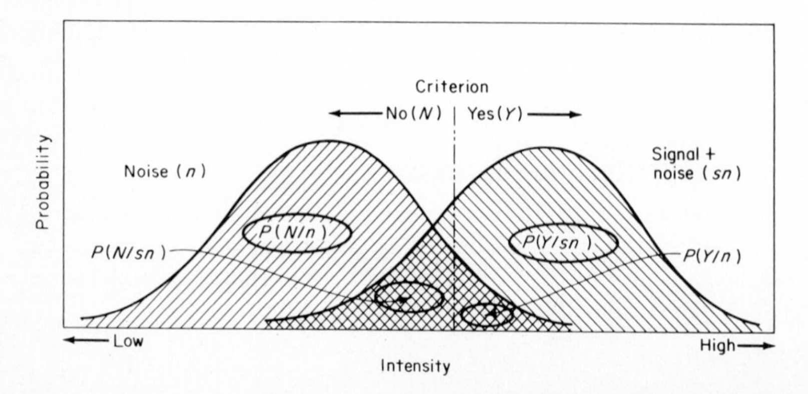

TABLE 3-2
POSSIBLE RESPONSE ALTERNATIVES IN FRAMEWORK OF THEORY OF SIGNAL DETECTION (TSD)

	Stimulus	
Response	**Noise (n)**	**Signal + noise (sn)**
Yes (Y)	$P(Y/n)$	$P(Y/sn)$
No (N)	$P(N/n)$	$P(N/sn)$

in identification of signals, the criterion level would be set toward the left, with an ensuing increase in the number of "false alarms"; whereas if the observer were operating under a "set" to be quite "sure" of a signal, the criterion level would be moved to the right, with a reduction in the "hit" rate [$P(Y/sn)$], the number of correctly identified signals.

Generality of TSD Most of the research with TSD has been carried out in the sanitary confines of laboratories. The hypothetical example illustrated in Figure 3-4 deals with the notion of a sound signal with or without a background noise, with the sensory continuum being intensity. However, the same concepts can be stretched a bit to apply in the cases of other types of "signals" and "noise," such as with voice communications, in visual inspection operations, in detecting traffic signs or lights on a foggy night or in the midst of the clutter of any other visual stimuli, and for that matter in virtually any situation where meaningful stimuli are to be used in circumstances in which there is some form of "noise" that could interfere with accurate sensing of the stimuli.

Although many real-life "signal detection" circumstances do not lend themselves to precise experimental analysis, the important implications of signal detection theory still apply when some meaningful stimulus may be presented in the presence of some type of "noise." For example: (1) Whenever possible, the signal (i.e., the stimulus) should be such that, when combined with whatever noise might occur, the combined stimulus values form a distribution that is *clearly* separated from the noise (this means that, when possible, the meaningful sounds, lights, signs, symbols, objects, etc. should be sufficiently different from the background noise that they can be clearly discriminated). (2) When some overlap cannot be avoided, a decision needs to be made regarding the type of error that can best be tolerated ("false alarms" or failure to detect signals), since this will influence the criterion level that the observer should establish.

Speed and Load of Stimuli

Other features of input stimuli that can affect their reception by human beings are what are called *speed* and *load*. *Load* refers to the variety of stimuli (in type and number) to which the receiver must attend. Thus, if there are several different types of

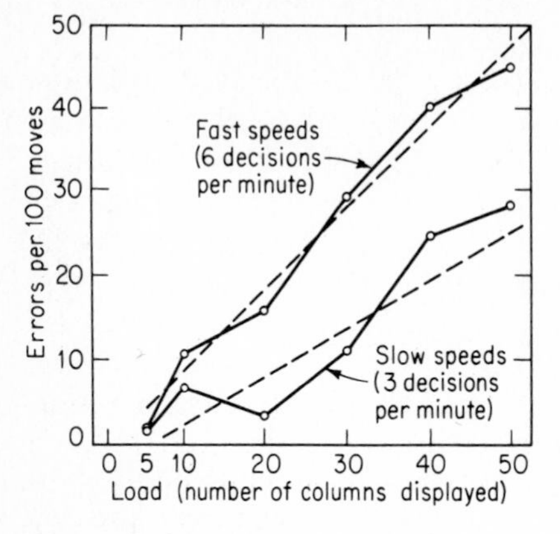

FIGURE 3-5
Effects of increasing load (number of channels used) upon errors. These data come from the study using a panel of various numbers of columns of comparison numbers. The fast and slow speeds were, respectively, 6 and 3 decisions per minute. (*Source: Mackworth & Mackworth, 1958.*)

visual displays, or several displays of the same type, to which a person must give attention, the load on the visual system will be greater than if there were fewer. On the other hand, *speed*, when used in this context, relates to the number of stimuli per unit of time or, conversely, to the time available per stimulus. Incidentally, one could contemplate speed and load for a single sensory modality (such as vision or hearing) or for combinations of the sensory modalities.

As an example of studies dealing with the effects of speed and load, Mackworth and Mackworth (1958) used a visual search task that utilized a display with 50 columns with numbers in certain positions in each column. Load was varied by using 5, 10, 15, 20, 30, 40, or 50 columns. The subjects compared a frequently changing number on three clocks with the numbers in the columns and reported the column in which the most nearly corresponding number appeared. Speed was controlled by the rate at which the number shown on the clocks kept changing. Some of the results are shown in Figure 3-5. This shows for fast speeds (six decisions per minute) and slow speeds (three decisions) the relationship between load and errors. The dotted lines (which reflect the general relationships) indicate that errors increase with load (even though the number of decisions per minute remains constant) and that speed also is related to errors.

Speed Stress and Load Stress The results of various studies, including his own, led Conrad (1951) to postulate the existence of *speed stress* and of *load stress* in certain types of tasks. He suggests that *speed stress* is essentially a reaction on the part of a person working on a task that has the effect of worsening performance beyond what might be expected from the physical characteristics of the display. (What Conrad referred to as *speed stress* would now be considered more a form of *strain* since it is characterized as a reaction of the individual as opposed to a condition that induces a reaction. The distinction between stress and strain is discussed further in Chapter 7.) *Load stress,* on the other hand, changes the character of the task. As the number of signal sources (visual displays) is increased, more time is needed to make judgments simply because of the greater scanning coverage required.

Time Sharing

As described before, time sharing refers to situations in which a person is to perform two activities simultaneously or in rapid alternation. If the limited single-channel capacity theory were universally applicable, one would postulate that, if the combined demands of two tasks exceeded a person's capacity, performance on one or both of the tasks would go to pot. Although, as Kantowitz (1981) points out, numerous studies have demonstrated such effects, in other instances the combination of tasks has not resulted in loss of efficiency in either (see, for example, Allport, Antonis, & Reynolds, 1972). The explanation for such differences presumably lies in the nature of the two tasks used in combination. The fact that some combinations of tasks result in performance degradation and others do not raises some question about the generality of the limited-channel-capacity concept, and has been used as an argument for the existence of several independent channels of information processing. In this regard, however, Kantowitz suggests that a hybrid model—one that is neither strictly serial nor parallel but that contains elements of both—may explain such disparities and still be reasonably compatible with a general limited-capacity concept as applied to information processing.

As indicated above, the effects of time sharing seem to be influenced by the nature of the tasks used in combination. Although it is not possible to indicate the combinations of tasks that are more, or less, influenced by time sharing, certain examples of time sharing of information input will at least give some such hints.

Time Sharing of Visual Inputs In one study Weisz and McElroy (1964) investigated the effects of time sharing on tasks that might be representative of human operator activities in systems, including tasks that required storage and integration of information (which most laboratory tasks have not required). We will skip the details of the study except to say that it involved responses by subjects to five types of geometric forms presented simultaneously with frames shown on a cathode-ray tube (CRT) at three speeds. Two tasks required a response for every frame (the subjects knew for *certain* that they were to respond). Two tasks required responses only when specific forms or combinations appeared (these involved *uncertainty*). And one task required the recognition of pairs of similar forms in similar locations on successive frames (this required *short-term memory*).

The percents of omission errors are shown in Figure 3-6. In general it was found that omission errors under fast speed conditions were greatest for those tasks that involved uncertainty and that required short-term memory. In other words, it appears from this study that in time sharing of tasks such as these speed stress does not affect all tasks equally, but rather affects particularly those with greater uncertainty and those that depend on short-term memory.

Another relevant study is that conducted by Olson (1959). The experimental task was sharing visual attention between two tasks. One of these was a tracking task, consisting of a moving "road" that was curved and a wheel that controlled a pointer on the road; the object was to keep the pointer on the road. The other task was the identification of dial pointers (on anywhere from 6 to 18 dials) which deviated randomly

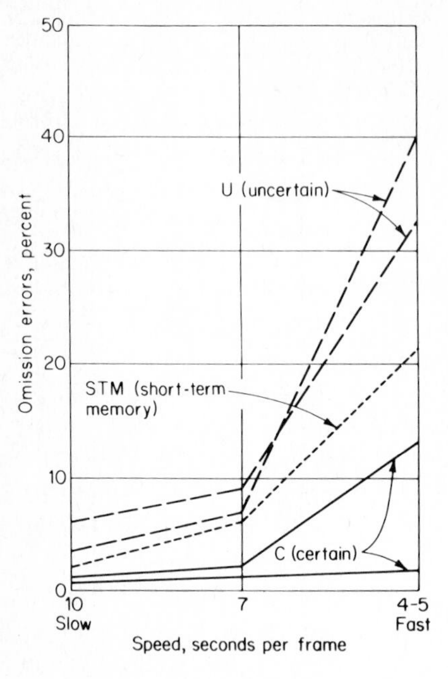

FIGURE 3-6
Percent of omission errors in performing five time-shared tasks of responding to geometric-form stimuli presented with frames on a cathode-ray tube at three speeds. The tasks varied in their demands as follows:

C	Certain: response required for every frame
U	Uncertain: response required only when specified stimuli or combinations appeared
STM	Short-term memory: response required remembering stimuli from one frame to the next

(*Source: Adapted from Weisz & McElroy, 1964.*)

from a neutral position. While the primary concern was more with other variables (dial arrangement, load, speed, etc.), one aspect of the study was related to the present topic of time sharing. Performance on the two tasks was correlated only to the extent of .20, which suggests the hypothesis that individuals tended to adopt their own priority strategies of giving primary attention to one task or the other. Whether the priority strategies are fairly common to most subjects because of the intrinsic nature of the task (as in the case of the study by Weisz and McElroy) or are selected by individuals (as presumably in the study by Olson), it nonetheless appears evident that when the pressures of speed stress tax the capacities of people, something has to give, specifically performance on some of the time-shared tasks.

Time Sharing of Auditory Inputs Essentially the same adverse effect of time sharing (i.e., of something having to give) is apparent in the case of auditory inputs, such as when two or more inputs occur simultaneously, overlap each other to some degree, or occur very close together in time. If an individual is listening, say, for verbal messages, and two messages occur at the same time, only one of them usually will get through. If, however, there is a slight lag in one, the first typically is identified more accurately than the second (Webster & Thompson, 1954). But if there is a distinct intensity difference, the second being the more intense, it will tend to have priority on the receiver's attention, even though it may lag after the first by as much as 2 s.

The adverse effect of simultaneous messages occurs even when only one is relevant and needs to be attended to (Broadbent, 1952). There is also some evidence that when a competing (and irrelevant) message is relatively similar in content and words to a relevant message, the interference effects on the relevant message are greater than if the irrelevant message is rather different (Peters, 1954).

Time Sharing of Auditory and Visual Sensory Channels Since vision and audition are the most important senses for receiving information from displays, it would be particularly in order to compare these two senses in the degree to which they are influenced by interference from each other and from other possible work activities. Present inklings suggest that when both visual and auditory inputs are being time shared, the auditory channel is more resistant to interference effects than the visual channel (Mowbray, 1952).

Discussion of Time Sharing It is fairly evident that there are bounds beyond which the time sharing of sensory inputs typically results in some degradation of performance. When circumstances permit (and they sometimes do not), efforts should, of course, be made to so manipulate the situation that degradation will at least be minimized if not eliminated. As in other contexts, we need to accept generalizations with caution. From this point of view, research evidence–some cited above and some not–seems to suggest a few general guidelines, although these may not be applicable across the board:

1 Where possible, the number of potentially competing sources should be minimized.

2 Where time sharing is likely to impose speed or load stress, the receiver should be provided with some inklings about priorities, so that the receiver's strategy in giving attention to first things first can take these into account.

3 Where possible, the requirements for use of short-term memory and for dealing with low-probability events should be minimized.

4 Where possible, input stimuli which require individual responses should be separated temporally and presented at such a rate that they may be responded to individually. Extremely short intervals (say, less than 0.5 or 0.25 s) should be avoided if at all possible. Where possible, the receiver should be permitted to control the input rate.

5 Where a choice of sensory modalities is feasible in a situation where a sensory input has competition, the auditory sense is generally more durable and is less influenced by other inputs.

6 Some means of directing attention to relevant and more important sources will increase the likelihood of their priority in the receiver's attention; for example, in some situations visual stimuli (such as lights) might be used as advance cues to the location of relevant auditory sources, or vice versa.

7 Where two or more auditory inputs might have to be time-shared, it would be desirable to schedule relevant messages or signals so that they do not occur simultaneously, to separate physically the sources (such as speakers) of relevant versus irrelevant messages, in order to filter out (if possible) any irrelevant messages, and where they cannot be filtered out, to make them as different as possible from those that are relevant such as by making the relevant stimuli more intense or using clearly distinct spectral characteristics.

8 Especially when repetitive manual tasks are time-shared with nonrelated sensory inputs, the greater the learning of the manual task, the less will be its possible effect on the reception of the sensory input.

Use of Redundant Sensory Channels

Two or more sensory channels, such as vision and audition, can be used in a redundant fashion to present the same information. While the evidence about some aspects of human behavior is a bit inconclusive, there is virtually no question but that the use of redundant visual and auditory coding (the simultaneous presentation of identical information to both senses) increases the odds of reception of the information. Different specific studies could be used to illustrate this point. We shall use a vigilance study to illustrate it (Colquhoun, 1975). In this experiment 12 men performed a vigilance task over a period of 10 days, each day performing the task under three display conditions as follows:

1 Auditory (A)
2 Visual (V)
3 Dual-mode (AV) in which both types of display signals were used

Performance was measured in terms of percent of signals detected, the results over the 10 days being shown in Figure 3-7. It is evident that the dual-mode condition resulted in systematically better performance. Incidentally, the auditory signals were also superior to the visual signals, indicating that, for a vigilance type of task, auditory displays may tend to be superior to visual displays.

In another study which reinforces the general relevance of using two sensory modalities in presenting information, Klemmer (1958) had subjects press one of three keys (left, center, or right) in response to visual signals (red, orange, or green, respectively), auditory signals [100, 700, or 5000 hertz (Hz)], or combined signals. The percentages of correct responses were as follows: visual signals, 89 percent; auditory signals, 91 percent; and combined visual and auditory signals, 95 percent.

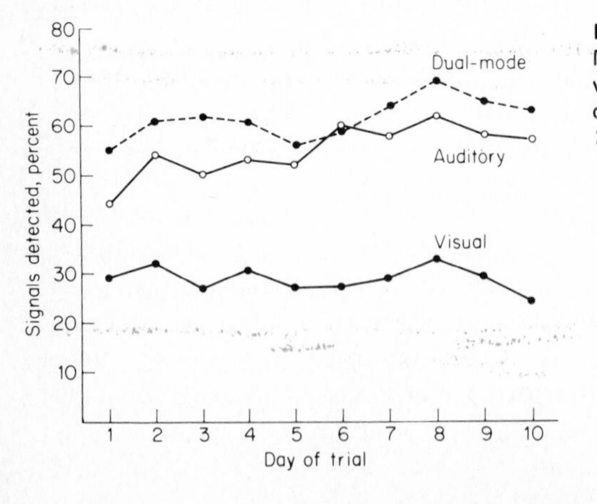

FIGURE 3-7
Mean percent of signals detected in a vigilance task with auditory, visual, and dual-mode signals. (*Source: Colquhoun, 1975, fig. 1, p. 429.*)

Compatibility

One cannot discuss the subject of information input and processing in the human factors context without reference to the concept of *compatibility*. *Compatibility* is a very generalized concept that has substantial applicability to various aspects of human factors. For our purposes it might be defined thus: *Compatibility* refers to the spatial, movement, or conceptual relationships of stimuli and of responses, individually or in combination, which are consistent with human expectations. This concept is most straightforward in the case of stimulus-response (S-R) compatibility, but the concept of compatibility is also applicable where there is simply information transfer (as in the use of certain coding systems) in the absence of any corresponding physical response.

Although there are many different manifestations of compatibility, most instances probably can be considered to fall in one of three groups, namely, (1) *spatial* compatibility, i.e., the compatibility of the physical features, or arrangement in space, of certain items, especially displays and controls; (2) *movement* compatibility, the direction of movement of displays, controls, and system responses; and (3) *conceptual* compatibility, the conceptual associations that people have, such as green representing "go" in certain codes. In the context of perceptual-motor activities there is some presumption of compatibility of stimulus and response in combination. The term *stimulus-response compatibility* (S-R compatibility) was first used by Fitts and Seeger (1953), following the earlier use of the term *compatibility* by A. M. Small. Fitts and Seeger characterized S-R compatibility as follows: "A task involves compatible S-R relations to the extent that the ensemble of stimulus and response combinations comprising the task results in a high rate of information transfer." In this information-theory context, the concept of compatibility implies a hypothetical process of information transformation, or recoding, in the activity, and is predicated on the assumption that the degree of compatibility is at a maximum when the recoding processes are at a minimum.

Origins of Compatibility Relationships Compatibility relationships stem from two possible origins. In the first place, certain compatible relationships are intrinsic in the situation, for example, turning a steering wheel to the right in order to turn to the right. In certain combinations of displays and controls, for example, the degree of compatibility is associated with the extent to which they are isomorphic or have similar spatial relationships. Other compatible relationships are culturally acquired, stemming from habits or associations that are characteristic of the culture in question. For example, in the United States a light switch is usually pushed up to turn it on, but in certain other countries it is pushed down. How such culturally acquired patterns develop is perhaps the consequence of fortuitous circumstances.

The Identification of Compatibility Relationships If one wishes to take advantage of compatible relationships in designing equipment or other items, it is of course necessary to know *what* relationships are *compatible*. There generally are two ways in which these can be ascertained or inferred. In the first place, certain such relationships are obvious or manifest; this is particularly true with many relationships that are intrinsic in the situation, such as the arrangement of corresponding displays and controls in juxtaposition to each other. In addition, certain culturally acquired relationships are so pervasive that they, too, are obvious, such as the red, yellow, and green symbols of traffic lights. But when the most compatible relationships are not obvious, it is necessary to identify them on the basis of empirical experiments. Certain examples will be cited in later chapters, but in general, such experiments produce information on the proportion of subjects who choose each specific relationship of different possible relationships.

Discussion of Compatibility Although different versions of compatibility involve the processes of sensation and perception and also response, the tie-in between these—the bridge between them—is a mediation process. Where compatible relationships can be utilized, the probability of improved performance usually is increased. As with many aspects of human performance, however, there are certain constraints or limitations that need to be considered in connection with compatibility relationships. For example, some such relationships are not self-evident; they need to be ascertained empirically. When this is done, it sometimes turns out that a given relationship is not universally perceived by people; in such instances it may be necessary to "figure the odds," that is, to determine the proportion of people with each possible "association" or response tendency and make a design determination on this basis. In addition, there are some circumstances where trade-off considerations may require that one forgo the use of a given compatible relationship for some other benefit.

SELECTION OF SENSORY MODALITY

In selecting or designing displays for transmission of information in some situations, the selection of the sensory modality (and even the stimulus dimension) is virtually a foregone conclusion, as in the use of vision for road signs and the use of audition for many various purposes. Where there is some option, however, the intrinsic advantages of one over the other may depend upon any of a number of considerations. As indicated above, for example, audition tends to have an advantage over vision in vigilance types of tasks, because of its attention-getting qualities. A more extensive comparison of audition and vision is given in Table 3-3, indicating the kinds of circumstances in which each of these modalities tends to be more useful. These comparisons are based on considerations of substantial amounts of research and experience relating to these two sensory modalities.

The tactual sense is not used very extensively as a means of transmission of information, but does have relevance in certain specific circumstances, such as with blind persons, and in other special circumstances in which the visual and auditory sensory modalities are overloaded. Further reference to such displays will be made in Chapter 5, but it might be added here that special tactual devices attached to the surface of the skin have been found to have particular value as warning devices when the visual and auditory modalities are overloaded (Heard, Johnson, & Mayyasi, 1970).

CODING OF SENSORY INPUTS

Many displays involve the coding of information (actually, of stimuli), as opposed to some form of direct representation or "reproduction" of the total stimulus. Any such

TABLE 3-3
WHEN TO USE THE AUDITORY OR VISUAL FORM OF PRESENTATION

Use auditory presentation if:	Use visual presentation if:
1 The message is simple.	1 The message is complex.
2 The message is short.	2 The message is long.
3 The message will not be referred to later.	3 The message will be referred to later.
4 The message deals with events in time.	4 The message deals with location in space.
5 The message calls for immediate action.	5 The message does not call for immediate action.
6 The visual system of the person is overburdened.	6 The auditory system of the person is overburdened.
7 The receiving location is too bright or dark-adaptation integrity is necessary.	7 The receiving location is too noisy.
8 The person's job requires him to move about continually.	8 The person's job allows him to remain in one position.

Source: Deatherage, 1972, p. 124, table 4-1.

coding implies first the selection of the particular sensory modality (as discussed earlier in this chapter), and second the selection of the particular stimulus (or coding) dimension to use within that sensory modality.

Stimulus Dimensions

The stimulus inputs that people receive from their environment via any sensory modality (vision, audition, etc.) differ in terms of their characteristics. For example, we make visual discriminations in terms of shape, configuration, size, position, color, etc., and auditory discriminations in terms of frequency, intensity, etc. The "natural" environment we sense is of course very complex. But as information is presented to people via displays, the nature of the stimuli used usually is simplified, typically consisting of variations in a given "class" of stimuli which can be considered as a *stimulus dimension*. Thus, the positions of the hands of a clock are used to tell time (especially when we cannot read the numbers); the number and time spacing of Morse code signals represent letters; and the shapes and designs of road signs have different meanings. The utility of any given stimulus dimension to convey information, however, depends upon the ability of people to make the sensory and perceptual distinctions that are required in differentiating one stimulus of a given class from another (such as telling one color from another). Such discriminations, however, usually have to be made on an *absolute* basis rather than on a *relative* basis. A relative judgment is one which is made when there is an opportunity to compare two or more stimuli; thus, one might compare two or more sounds in terms of loudness or two or more lights in terms of brightness. In absolute judgments, there is no opportunity to make comparisons, such as identifying a given note on the piano (say, middle C) without being able to compare it with any others, or identifying a given color out of several possible colors when it is presented by itself.

As one might expect, people are generally able to make fewer discriminations on an absolute basis than on a relative basis. For example, it has been estimated that most people can differentiate as many as 100,000 to 300,000 different colors on a relative basis when comparing two at a time (taking into account variations in hue, lightness, and saturation). On the other hand, the number of colors that can be identified on an absolute basis is limited to no more than a dozen or two.

Absolute judgments can be required in either of two types of circumstances. In the first place, several discrete positions (levels or values) along a stimulus dimension might be used as codes, each position representing a different item of information. If the stimuli consist of tones of different frequencies, the receiver is supposed to identify the *particular* tone. In the second place, the stimulus may be of any value along the stimulus dimension, and the individual needs to make some judgment regarding its value or position along the dimension. Such judgments might be used in various operational ways. A radio operator, for example, might use subjective judgment of loudness of a radio signal to adjust the gain (volume) up or down to some subjective standard. On the other hand an inspector might use subjective judgment to classify an item as either "pass" or "fail," or possibly by grades, such as A, B, and C.

The ability of people to make absolute discriminations among individual stimuli of most stimulus dimensions is really not very large, being generally in the range from 4 to 9 or 10, with corresponding bits from about 2.0 to 3.0 or 3.4, as illustrated by the examples in Table 3-4.

In this connection Miller (1956) refers to the "magical number seven, plus or minus two," meaning that the range of such discriminations is somewhere around 7 ± 2 (5 to 9); for some dimensions the number is greater, and for some less, than this specific range. Seven discriminations would transmit 2.8 bits. With certain types of stimulus dimensions people can differentiate more differences. In the case of geometric forms, for example, most people can identify as many as 15 or more different forms, which is equivalent to 3.9 or more bits. With *combinations* of dimensions the information that can be transmitted sometimes is noticeably greater than with individual dimensions.

A very unfortunate (and costly) example in the United States of failure to recognize the importance of human limitations in making sensory discriminations is the issuance of a new dollar coin (the Susan B. Anthony dollar). After the coin was produced in large quantities and issued, it was found that people had trouble differentiating it by sight and by feel from the quarter (a 25-cent coin) because it was only

TABLE 3-4
AVERAGE NUMBER OF DISCRIMINATIONS AND CORRESPONDING NUMBER OF BITS TRANSMITTED BY CERTAIN STIMULUS DIMENSIONS

Stimulus dimension	Average number of discriminations	Number of bits
Pure tones	5	2.3
Loudness	4-5	2 to 2.3
Size of viewed objects	5-7	2.3 to 2.8
Brightness	3-5	1.7 to 2.3

very slightly larger than the quarter. Because of this problem the coin was widely rejected by people.

Selection of the Stimulus Dimension When the stimulus dimension is not clearly dictated by the nature of the situation at hand, the selection can be based on the relative pros and cons associated with the various dimensions that logically could be considered. Discussion of some of these will be included in later chapters. In considering various alternatives, however, it should be pointed out that sometimes two or more stimulus dimensions can be used in combination. Most codes are undimensional, which means that a given code symbol has a given meaning, such as red, yellow, and green colors of traffic lights, the unique "You're out!" arm signal of the baseball umpire, or the $, ¢, ?, &, =, and ! signs of the printed page. On the other hand, in some contexts two or more coding dimensions are used in combination, such combinations varying in the degree of *redundancy*. A completely redundant code is one in which each thing to be identified has two (or more) unique attributes (such as a unique color *and* a unique shape). In turn, a completely nonredundant code is one in which each and every unique combination of specific stimuli of each of two (or more) stimulus dimensions represents a specific item. For example, each combination of a color *and* a shape might be used to code items in a warehouse. In some circumstances a partially redundant coding scheme might be used.

General Guidelines in the Use of Coding Systems

Generalizations in an area such as the use of codes are pretty treacherous, especially since this ball park is a very complicated one. (We will see later a few instances of apparent inconsistencies in research findings.) Further, in any given situation it may be necessary to effect a trade-off of one advantage for another. There may be, however, a few guidelines that clearly stick out, and others that might be teased out of the available evidence, which may have some generality.

Detectability of Codes To begin with, any stimulus used in coding information needs to be detectable; specifically, it has to be of such a nature that it can be sensed by the sensing mechanism in the situation at hand. This is an *absolute threshold*. But *absolute* thresholds vary from individual to individual, and even for the same individual at various times, and vary with the situation.

Discriminability of Codes Further, every code symbol, even though detectable by the sensory mechanism, needs to be discriminable from other code symbols. This may be a problem where differences along some stimulus dimension are to be used, such as sound frequency and brightness. As discussed above, stimulus dimensions vary in the number of levels that can be identified on an absolute basis. In this connection, the *degree* of difference between adjacent stimuli may have some influence on the effectiveness of a coding system, even if all differences used are reasonably discriminable. The results of one study will illustrate this (Mudd, 1961). In that study three different levels of interstimulus difference were used with each of the four auditory cuing dimensions, the differences between adjacent stimuli being varied. In the case of certain dimensions—especially intensity—the mean response times were related to the magnitude of the differences between stimuli as follows: largest stimulus difference: 4.6 s; average stimulus difference: 5.7 s; and smallest stimulus difference: 6.8 s. The effectiveness of such "separation" of specific stimuli of some class, however, presumably is a function of the nature of the stimulus dimension; in this study, for example, increasing the difference between frequency stimuli did not reduce response time appreciably.

Compatibility of Codes The general concept of compatibility, discussed above, is of particular importance in the selection or design of coding systems for displays. This importance stems from the fact that a high degree of compatibility in a coding system tends to minimize the amount of information transformation, that is, the amount of encoding and decoding required of the receiver, thereby facilitating the information reception process. Thus, we should take advantage of associations that are already built into the repertoire of people; these may be either natural or learned associations. To refer again to traffic lights, if someone were foolish enough to change the system from red, yellow, and green to, say, violet, chartreuse, and azure, we would really be loused up—*because* of the need to recode from one system to another, at least until we learned the new one.

Meaningfulness of Codes The use of any code implies that its meaningfulness to the user should be manifest. This can be predicated on either of two bases: (1) the fact that the code is actually symbolic of that which it represents (such as a road sign showing a curve rather than saying "turn right"); and (2) the learning of the association between the display and what it represents.

Standardization of Codes When coding systems are to be used by different people in different situations (as in the case of traffic signs), their standardization will facilitate their use when people shift from one situation to another.

Use of Multidimensional Codes In very general terms, the use of two or more coding dimensions in combination tends to facilitate the transfer of information to human beings, probably especially so where there is complete redundancy, and perhaps less so with partial redundancy.

DISCUSSION

In any operational situation (as in performing some function in a job), the information input from the environment (whether sensed directly, or sensed indirectly via displays) serves, in combination with the information stored in memory, as the grist for any of numerous types of mediation processes as the basis for decisions about actions to be taken. In this regard, the reception and processing of information by people can be influenced by various features of the input stimuli people receive from their environment. Therefore, when information is to be transmitted indirectly to people by the use of displays, the displays generally should be designed to enhance the reception of relevant information by the receivers. This is especially important in the case of information that is critical.

In the design of a display one should use an appropriate sensory modality, and should select the most appropriate stimulus dimension as the basis for the "coding" system that would be used in the display. Chapters 4 and 5 will include, respectively, discussions of visual and of auditory and tactual displays, along with examples.

REFERENCES

Allport, D. A., Antonis, B., & Reynolds, P. On the division of attention: A disproof of the single channel hypothesis. *Quarterly Journal of Experimental Psychology*, 1972, *34*, 225–235.

Alluisi, E. A., Muller, P. F., Jr., & Fitts, P. M. An information analysis of verbal and motor responses in a forced-paced serial task. *Journal of Experimental Psychology*, 1957, *53*, 153–158.

Broadbent, D. E. Listening to one of two synchronous messages. *Journal of Experimental Psychology*, 1952, *44*, 51–55.

Broadbent, D. E. *Perception and communication.* New York: Pergamon, 1958.

Broadbent, D. E. *Decision and stress.* New York: Academic Press, 1971.

Chase, W. G. Elementary information processing. In W. K. Estes (Ed.), *Handbook of learning and cognitive processes* (Vol. 5). Hillsdale, N. J.: Lawrence Erlbaum Associates, 1978.

Colquhoun, W. P. Evaluation of auditory, visual, and dual-mode displays for prolonged sonar monitoring in repeated sessions. *Human Factors*, 1975, *17*(5), 425–437.

Conrad, R. Speed and load stress in a sensori-motor skill. *British Journal of Industrial Medicine*, 1951, *8*, 1–7.

Deatherage, B. H. Auditory and other sensory forms of information presentation. In H. P. Van Cott and R. G. Kinkade (Eds.), *Human engineering guide to equipment design* (Rev. ed.). Washington, D. C.: U.S. Government Printing Office, 1972.

Egan, J. P. *Signal detection theory and ROC analysis.* New York: Academic Press, 1975.

Fitts, P. M. The information capacity of the human motor system in controlling the amplitude of movement. *Journal of Experimental Psychology*, 1954, *47*, 381–391.

Fitts, P. M., & Seeger, C. M. S-R compatibility: Spatial characteristics of stimulus and response codes. *Journal of Experimental Psychology*, 1953, *46*, 199–210.

Geyer, B. H., & Johnson, C. W. Memory in man and machines. *General Electric Review*, March 1957, *60*(2), 29–33.

Haber, R. N., & Hershenson, M. *The psychology of visual perception* (2d ed.). New York: Holt, 1980.

Heard, M. F., Johnson, W. L., & Mayyasi, A. M. *The effectiveness of tactual, visual, and auditory warning signals under conditions of auditory and visual loading.* Texarcana: Texas A and M University, Red River Army Depot, Graduate Extension Division, August 1970.

Kahneman, D. *Attention and effort.* Englewood Cliffs, N.J.: Prentice-Hall, 1973.

Kantowitz, B. H. Interfacing human information processing and engineering psychology. In W. Howells and E. A. Fleishman (Eds.), *Human performance and productivity.* Hillsdale, N.J.: Lawrence Erlbaum Associates, 1981.

Kantowitz, B. H., & Roediger, H. L., III. Memory and information processing. In G. M. Gazda and R. J. Corsini (Eds.), *Theories of learning.* Itaska, Ill.: F. E. Peacock, 1980.

Klemmer, E. T. Time sharing frequency-coded auditory and visual channels. *Journal of Experimental Psychology*, 1958, *55*, 229–235.

Mackworth, N. H., and Mackworth, J. F. Visual search for successive decisions. *British Journal of Psychology*, 1958, *49*, 210–221.

Miller, G. A. The magical number seven, plus or minus two: Some limits on our capacity for processing information. *Psychological Review*, 1956, *63*, 81–97.

Mowbray, G. H. Simultaneous vision and audition: The detection of elements missing from overlearned sequences. *Journal of Experimental Psychology*, 1952, *44*, 292–300.

Mudd, S. A. *The scaling and experimental investigation of four dimensions of pure tone and their use in an audio-visual monitoring problem.* Unpublished Ph.D. thesis, Lafayette, Ind.: Purdue University, June 1961.

Navon, D., & Gopher, D. On the economy of the human-processing system. *Psychological Review*, 1979, *86*, 214–255.

Olson, P. L. *Display arrangement, number of channels and information speed as related to operator performance.* Unpublished Ph.D. thesis, Lafayette, Ind.: Purdue University, July 1959.

Peters, R. W. *Competing messages: The effect of interfering messages upon the reception of primary messages* (NM 001 064.01.27). U.S. Navy, School of Aviation Medicine, 1954.

Posner, M. I. Chronometric explorations of mind. Hillsdale, N.J.: Lawrence Erlbaum Associates, 1978.

Singleton, W. T. The ergonomics of information presentation. *Applied Ergonomics,* 1971, *2*(4), 213–220.

Steinbuch, K. *Information processing in man.* Paper presented at IRE International Congress on Human Factors in Electronics, Long Beach, Calif., May 1962.

Swets, J. A. (Ed.). *Signal detection and recognition by human observers.* New York: Wiley, 1964.

Webster, J. C., & Thompson, P. O. Responding to both of two overlapping messages. *Journal of the Acoustical Society of America,* 1954, *26*, 396–402.

Weisz, A. Z., & McElroy, L. S. *Information processing in a complex task under speed stress* (Report ESD-TDR 64-391). Decision Sciences Laboratory, Electronics System Division, AFSC, USAF, May 1964.

CHAPTER **4**

VISUAL DISPLAYS

Since the use of visual displays depends upon the visual capabilities of people, we will first discuss the process of seeing and certain types of visual skills.

THE PROCESS OF SEEING

The eye is very much like a camera in that it has an adjustable lens through which light rays are transmitted and focused, and a sensitive area (the retina) upon which the light falls. Figure 4-1 illustrates the principal features of the eye in cross section. Light rays that are reflected from an object enter the transparent *cornea* and pass through the clear fluid (*aqueous humor*) that fills the space between the cornea and the pupil and lens behind the cornea. The *pupil* is a circular variable aperture whose size is changed through action of the muscles of the *iris*, becoming larger in dark surroundings and smaller in bright surroundings. The light rays that are transmitted through the opening of the pupil to the lens are refracted by the adjustable *lens*, and then transverse the clear jellylike *vitreous humor* that fills the eyeball back of the lens. In persons with normal or corrected vision the light rays are brought to appropriate focus on the

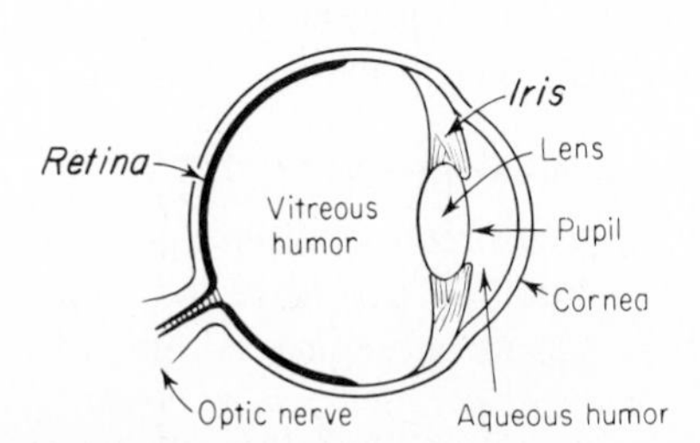

FIGURE 4-1
Principal features of the human eye in cross section. Light passes through the pupil, is refracted by the lens, and is brought to a focus on the retina. The retina receives the light stimulus and transmits an impulse to the brain through the optic nerve.

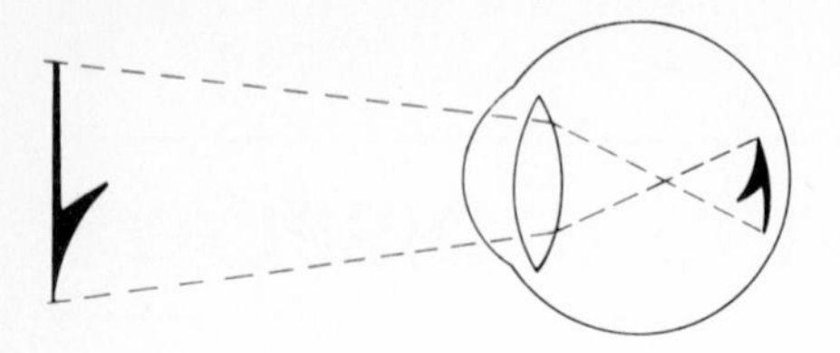

FIGURE 4-2
Illustration of the manner in which the image of an object is reproduced in inverted form on the retina of the eye.

retina. The adjustment of the lens to focus the light rays on the retina is automatically controlled by the *ciliary muscles*. The relaxation of the muscles results in a thickening of the lens, with a consequent shortening of the focal length so that nearby objects are brought into focus. A tightening of the lens adjusts it to a longer focal length. The image of the object on the retina is reversed and inverted, just as in a camera, as illustrated in Figure 4-2. The retina consists of two types of sensitive areas, namely *cones* and *rods*. There are approximately 6 or 7 million cones in the eye; these generally predominate in the center section of the retina. The cones are sensitive to differences in the wavelengths of light, and give rise to the subjective sensation of color. They also are sensitive to variations in brightness, and are the dominant receptors for daytime vision. There are about 130 million rods, and they tend to predominate toward the outer reaches of the retina around the sides of the eyeball. They are primarily sensitive to variations in the amount of light and are not particularly sensitive to differences in wavelength. The rods are particularly important for night vision. The cones and rods, when receiving light through the lens, trigger nerve impulses that are transmitted through the optic nerve to the brain, where the meaningfulness of the stimuli occurs.

The visibility of objects is of course influenced by the amount of light that is reflected from them. The opening of the pupil can aid in increasing the amount of light that passes through the lens, but in very dark surroundings the ability to distinguish objects or specific features of them may be affected, especially if the detailed features are small and if the contrast of those features with their background is slight.

Visual Acuity

Visual acuity is the ability of the eyes to differentiate between the detailed features of what we see, such as reading the fine print in an insurance contract or identifying a person across the street. Acuity depends very largely on the *accommodation* of the eyes, which is the adjustment of the lens of the eye to bring about proper focusing of the light rays on the retina. In normal accommodation, if one is looking at a far object, the lens flattens, and if one looks at a near object, the lens tends to bulge, in order to bring about proper focusing of the image on the retina. This is illustrated in Figure 4-3*a* for far objects, and 4-3*b* for near objects.

In some individuals the accommodation of the eyes is inadequate. This causes the conditions that we sometimes call *nearsightedness* and *farsightedness*. When a person is nearsighted, the lens tends to remain in a bulged condition, so that while the person may achieve a proper focus of near objects, a proper focus of far objects cannot be achieved, as shown in Figure 4-3*c*. Farsightedness, in turn, is a condition in which the

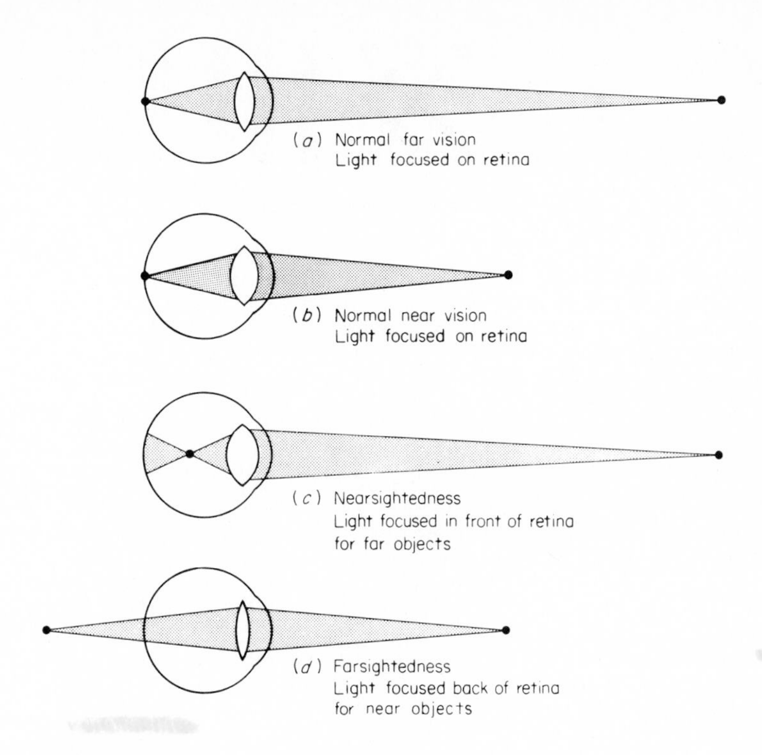

FIGURE 4-3
Illustration of normal accommodation at far and near distances, *a* and *b*, and of nearsightedness *c*, and farsightedness *d*.

lens tends to remain too flat. A farsighted person may see clearly at far distance but encounters difficulty in seeing properly at near distance, as shown in Figure 4-3*d*. Such conditions can sometimes be corrected by appropriate lenses which change the direction of the light rays before they reach the lens of the eye and thereby bring about proper focusing on the retina.

Types of Visual Acuity

The ability to make visual discriminations regarding certain types of visual stimuli does not necessarily mean that an individual can make discriminations regarding other types of stimuli. Because of this there are different types of visual acuity. The most commonly used measure of acuity, *minimum separable acuity*, refers to the smallest feature, or the smallest space between the parts of a target, that the eye can detect. The visual targets used in measuring minimum separable acuity include letters and various geometric forms, such as those illustrated in Figure 4-4. Such targets can be varied in size and distance, the acuity of the subject being determined by the smallest target he can properly identify. Such acuity usually is measured in terms of the reciprocal of the visual angle subtended at the eye by the smallest detail that can be discriminated (i.e., the angular subtense of that detail). The reciprocal of a visual angle of 1 minute of arc

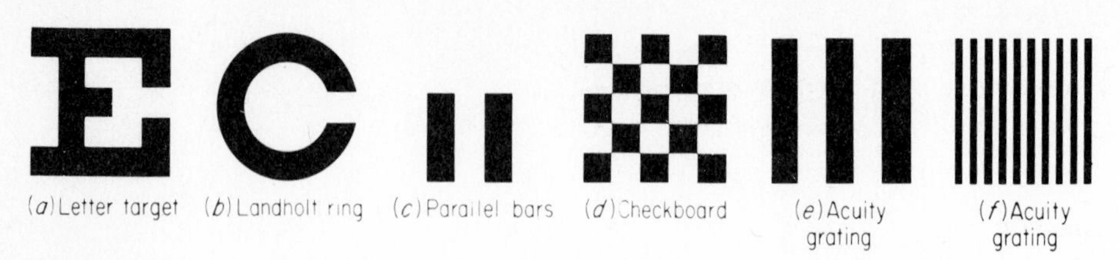

FIGURE 4-4
Illustrations of various types of targets used in visual acuity tests and experiments. The features to be differentiated in targets *a, b, c, d,* and *e* are all the same size and would, therefore, subtend the same visual angle at the eye. With target *a* the subject is to identify each letter; with *c, e,* and *f* the subject is to identify the orientation (such as vertical or horizontal); and with *b* the subject is to identify any of four orientations. With target *d* the subject is to identify one checkerboard target from three others with smaller squares.

usually is used as a standard in scoring. This reciprocal, being unity, provides a base with which poorer or better levels of acuity can be compared. If, for example, one person can identify only a detail that subtends an arc of 1.5 minutes, the acuity score for that person would be the reciprocal of 1.5 minutes, or 0.67. On the other hand, for a person who can identify a detail that subtends an arc of 0.8 minute, the score—the reciprocal of 0.8 minute—would be 1.25.

Vernier acuity refers to the ability to differentiate the lateral displacement, or slight offset, of one line from another that, if not so offset, would form a single continuous line (such as in lining up the "ends" of lines in certain optical devices). *Minimum perceptible acuity* is the ability to detect a spot (such as a round dot) from its background. In turn, *stereoscopic acuity* refers to the ability to differentiate the different images—or pictures—received by the retinas of the two eyes of a single object that has depth. (These two images differ most when the object is near the eyes, and differ least when the object is far away.)

Convergence and Phoria

As we direct our visual attention to a particular object, it is necessary that the two eyes converge on the object so that the images of the object on the two retinas are in corresponding positions; in this way we get an impression of a single object. The two images are said to be *fused* if they do so correspond. Convergence is controlled by muscles that surround the eyeball. Normally, as an individual looks at a particular object, these muscles operate automatically to bring about convergence. But some individuals tend to converge too much, and others tend not to converge enough. These conditions are called *phorias.* Since the double images that occur when convergence does not take place are visually uncomfortable, such people usually have compensated for this by learning to bring about convergence; however, muscular stresses and strains can occur in overcoming these muscular imbalances.

Color Discrimination

As indicated before, the cones of the retina are sensitive to differences in wavelength, and thus are the basis for color discrimination. Various tests are available to measure the color discrimination of individuals, including the discrimination among certain colors. Some people, for example, have difficulty in discriminating between red and green or between blue and yellow. Complete color blindness is very rare.

Dark Adaptation

The adaptation of the eye to different levels of light and darkness is brought about by two functions. In the first place, the pupil of the eye increases in size as we go into a darkened room, in order to admit more light to the eyes; it tends to contract in bright light, in order to limit the amount of light that enters the eye. Another function that affects how well we can see as we go from the light into darkness is a physiological process in the retina in which *visual purple* is built up. Under such circumstances the cones (which are color sensitive) lose much of their sensitivity. Since in the dark our vision depends very largely on the rods, color discrimination is limited in the dark. The time required for complete dark adaptation is usually 30 or more minutes (min). The reverse adaptation, from darkness to light, takes place in some seconds, or at most in a minute or two.

Conditions That Affect Visual Discriminations

The ability of individuals to make visual discriminations is of course dependent upon their visual skills, especially their visual acuity. Aside from individual differences, however, there are certain variables (conditions), external to the individual, that affect visual discriminations. Some of these variables are listed or discussed briefly below.

Luminance Contrast *Luminance contrast* (frequently called *brightness contrast* or simply *contrast*) refers to the difference in luminance of the features of the object being viewed, in particular of the feature to be discriminated by contrast with its background (for example, an arrow on a direction sign against the background area of the sign). The luminance contrast is expressed by the following relationship:

$$\text{Contrast} = \frac{B_1 - B_2}{B_1} \times 100$$

where B_1 = brighter of two contrasting areas
B_2 = darker of two contrasting areas

The contrast between the print on this page and its white background is considerable; if we assume that the paper has a reflectance of 80 percent and that the print has a reflectance of 10 percent, the contrast would be

$$\frac{80 - 10}{80} \times 100 = \frac{70}{80} \times 100 = 88 \text{ percent}$$

If the printing were on medium-gray paper rather than on white, the contrast would, of course, be much less.

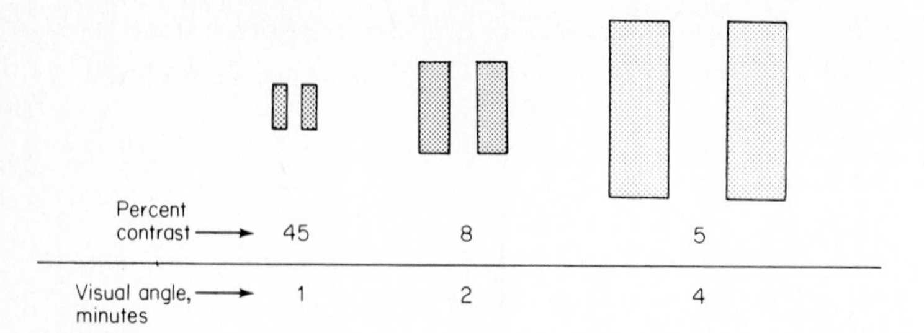

FIGURE 4-5
Relative sizes of visual targets of parallel bars of various levels of contrast that would be required for equal discriminability. (*Source: Adapted from Heglin, 1973, fig. I-10.*)

In case the contrast between a visual target and its background is low, the target must be larger in size for it to be equally discriminable to a target with greater contrast. This is illustrated, for example, by Figure 4-5, which shows, for various levels of contrast of parallel-bar targets, the relative differences in size that would be required for equal discriminability.

Amount of Illumination (This will be discussed in Chapter 13.)

Time Within reasonable limits, the longer the viewing time, the greater the discriminability. (This will be discussed further in Chapter 13, which deals with illumination.)

Luminance Ratio The luminance ratio is the ratio between the luminance of any two areas in the visual field (usually the area of primary visual attention and the surrounding area). (This will also be discussed further in Chapter 13.)

Glare (This will be discussed in Chapter 13.)

Combinations of Variables Available evidence indicates that there are interaction effects on visual performance when various combinations of the above variables exist, such as the combined effects of contrast and motion as reported by Petersen and Dugas (1972). (Some of these effects will be discussed in Chapter 13.)

Movement The movement of a target object or of the observer (or both) decreases the threshold of visual acuity. The ability to make visual discriminations under such circumstances is called *dynamic visual acuity* (DVA). It is usually expressed in degrees of movement per second. Acuity deteriorates rapidly as the rate of motion exceeds 60 degrees per second (60°/s) (Burg, 1966).

Age and Vision

Special mention should be made of the fact that visual skills, especially visual acuity, tend to deteriorate through age. The important implication of this for human factors is that if elderly people are to use visual displays the displays should be so designed that such people can use them.

Discussion

The visual skills people have—especially visual acuity and color discrimination—have a direct bearing upon the design of visual displays, particularly on the ability to *detect* relevant stimuli and to *discriminate* between and among variations thereof (such as positions of pointers on dials, or different letters). But, as discussed in Chapter 3, we can *sense* much more than we can comprehend or remember. The meaningfulness of what we see in visual displays depends in part upon our perceptual processes and the learning of relevant associations (such as learning the alphabet or the shapes of road signs). Thus, the appropriate design of various types of visual displays—which we will now discuss—must be predicated in part upon perceptual and learning factors as well as upon the specific visual skills of people.

QUANTITATIVE VISUAL DISPLAYS

Quantitative displays are used to provide information about the quantitative value of some variable, either a dynamic changeable variable (such as temperature or speed), or what is essentially a static variable (such as a measurement of length, as with a rule). In most uses of such displays there is an explicit or implicit level of precision that is required or desired, such as measurement to the nearest millimeter, centimeter, inch, foot, or mile. A great deal of research has been carried out with quantitative displays, directed toward determining the design features that contribute to speed and accuracy of their use. We will bring in the details of only a few such research undertakings, and will otherwise summarize the implications of such research.

Basic Design of Dynamic Quantitative Displays

There are three basic types of dynamic quantitative displays, as follows: (1) fixed scales with moving pointers; (2) moving scales with fixed pointers (or, in some cases, lubber lines); and (3) digital displays or counters (in which the numbers of mechanical counters click into position, as mileage readings on many speedometers). Examples of these three types are given in Figure 4-6. The first two classes are analog indicators in that the position of the pointer relative to the scale is analogous to the value that is represented. It is evident that there are differences in the effectiveness with which people can use these different designs in different types of circumstances.

Comparison of Different Designs Over the years there have been a number of studies in which certain designs of quantitative scales have been compared. Although the results of some studies are somewhat at odds with each other, certain general implications seem to stand out. For example, there are strong indications that a digital type of display (sometimes called *counters*) is generally superior to an analog display

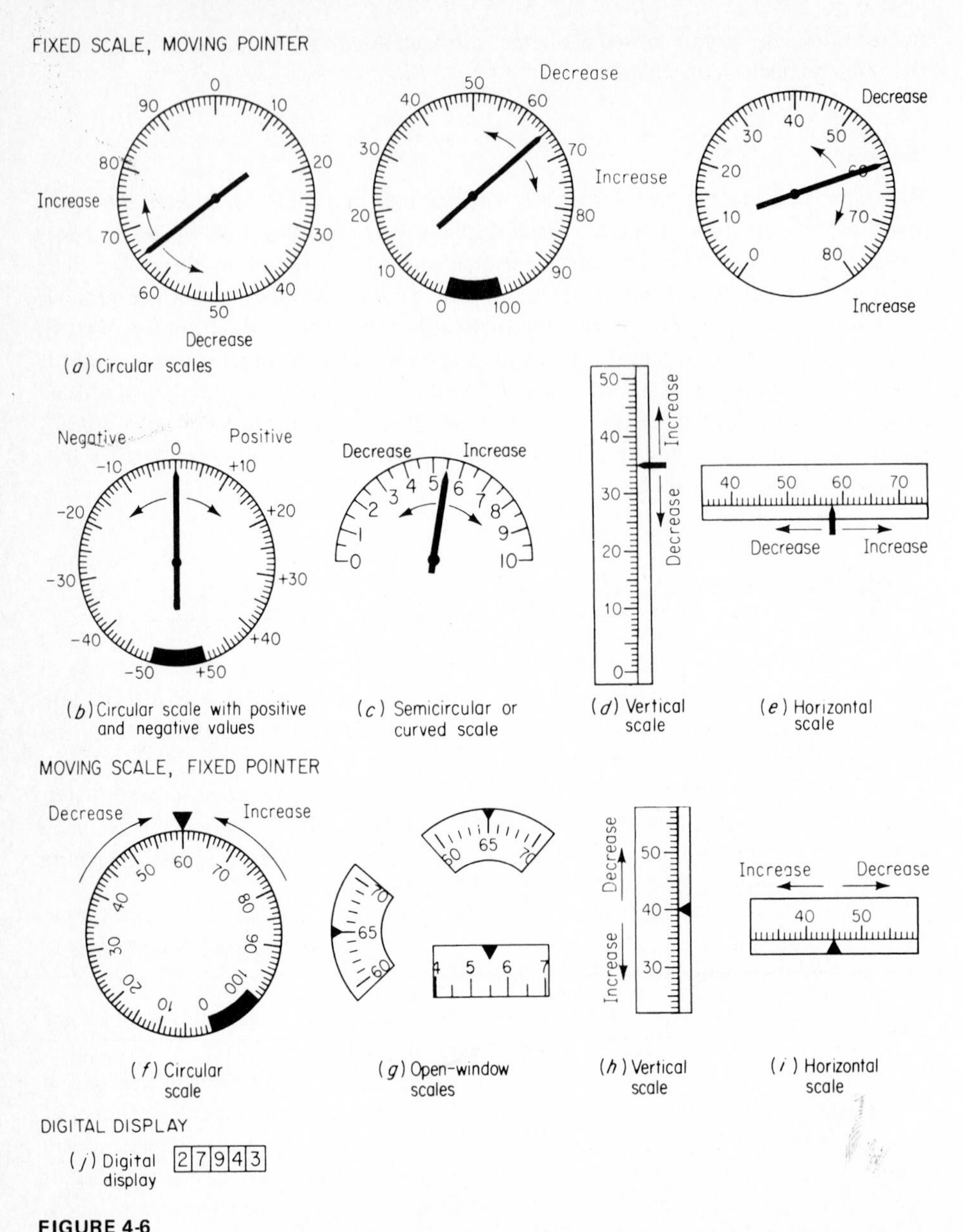

FIGURE 4-6
Examples of certain types of displays used in presenting quantitative information. (Reference will be made later to certain features of these scales.)

if precise numerical values are required, and if the values presented remain visible long enough to be read. This superiority is supported by Sinclair (1971) and is also reflected by the results of a study by Zeff (1965) in which there were only four reading errors out of 800 readings with a digital display, whereas with a circular display there were 50 errors. Further, readings with a digital display were found to be faster, as indicated by the following comparison of mean response times:

Digital display	0.94 s
Circular dial	3.54 s

Although digital displays have a definite advantage for obtaining specific numerical values that tend to remain fixed long enough to be read, analog displays have advantages in other circumstances. Fixed-scale moving-pointer displays, for example, are particularly useful when the values are subject to frequent or continual change that would preclude the use of a digital display (because of limited time for "reading" any given value). In addition, such analog displays have a positive advantage when it is important to observe the direction or rate of change of the values that are presented. By and large, analog displays with fixed scales and moving pointers are superior to those with moving scales and fixed pointers. In this regard Heglin (1973) offers the following list of factors to consider in the selection of analog displays (Heglin, p. II-30):

1 In general, a pointer moving against a fixed scale is preferred.

2 If numerical increase is typically related to some other natural interpretation, such as *more or less*, or *up or down*, it is easier to interpret a straight-line or thermometer scale with a moving pointer (such as *d* and *e* in Figure 4-6) because of the added cue of pointer position relative to the zero or null condition.

3 Normally you should not mix types of pointer-scale (moving-element) indicators when they are used for related functions–to avoid reversal errors in reading.

4 If a manual control over the moving element is expected, there is less ambiguity between the direction of motion of the control and the display if the control moves the pointer rather than the scale.

5 If slight, variable movements or changes in quantity are important to the observer, these will be more apparent if a moving pointer is used.

6 If you wish to have a numerical value readily available, however, a moving scale appearing in an open window can be read more quickly. (Such a scale is essentially a digital or counter display, as discussed above.)

Although fixed scales with moving pointers are generally preferred to moving scales with fixed pointers, they do have their limitations, especially when the range of values is so great that it cannot be shown on the face of a relatively small scale. In such a case certain moving-scale fixed-pointer designs, such as rectangular open-window and horizontal and vertical scales, have the practical advantage of occupying a small panel space, since the scale can be wound around spools behind the panel face, with only the relevant portion of the scale exposed.

Further, research and experience generally tend to favor circular and semicircular scales (*a*, *b*, and *c* in Figure 4-6) over vertical and horizontal scales (*d* and *e* in that figure). However, there are circumstances in which vertical and horizontal scales would have advantages that would argue for their use, as discussed in item 2 above.

Readability of Altimeters The design of altimeters has posed a special problem for aviation for far too long a time. There have been numerous instances in which aircraft accidents have been attributed to misreading of the commonly used model. That model consists of three pointers representing, respectively, 100, 1000, and 10,000 feet (ft), like the second, minute, and hour hands of a watch. In an early experimental study of altimeters, Grether (1949) found that that model was read with more errors and took more time than most other (experimental) designs, presumably because the reader had to *combine* three pieces of information. In turn, a combination counter (for thousands of feet) and pointer (for hundreds) proved to be used most accurately and with fewest errors.

Some further inklings about altimeter designs come from a study by Simon and Roscoe (1956) as reported by Roscoe (1968), in which four types of instruments were compared for use in displaying present altitude, predicted altitude (in 1 min), and command altitude (the altitude at which the plane is supposed to be flying); these are shown in very simplified form in Figure 4-7. The displays were intended to provide comparison of three design variables, namely, (1) vertical versus circular scales, (2) integrated presentations (of the three altitude values mentioned above) versus separate

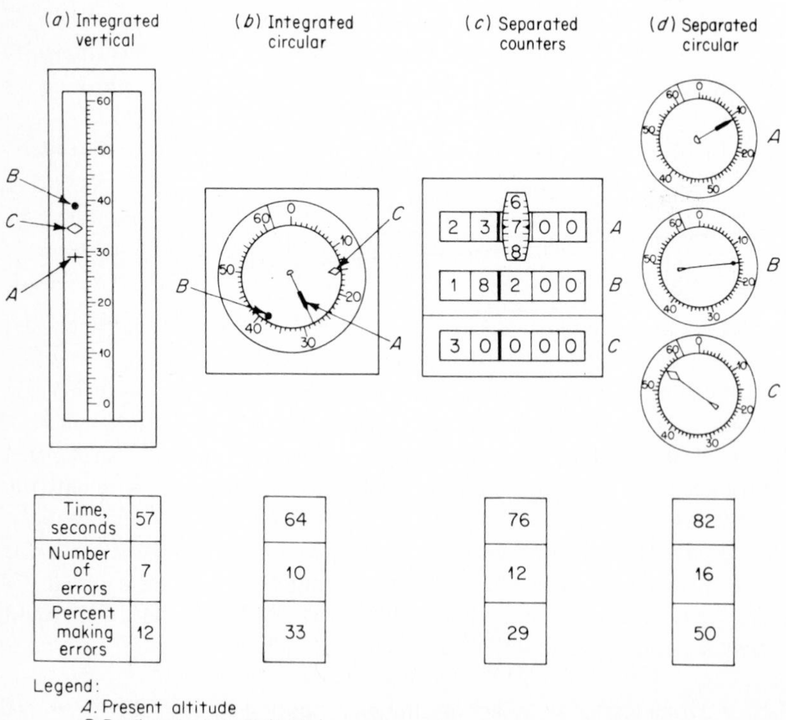

FIGURE 4-7
Four display designs for presenting (*A*) present altitude, (*B*) predicted altitude (in 1 min), and (*C*) command altitude and three criteria (mean time for 10 trials, number of errors, and percent of 24 subjects making errors). The displays are shown in overly simplified form. (*Source: Adapted from Simon & Roscoe, 1956, as presented by Roscoe, 1968.*)

presentations, and (3) spatial-analog presentations versus digital counters. Time and error performance scores of pilots in solving altitude-control decision problems and the percent of pilots making errors in using the four designs are also given in Figure 4-7. The clear and consistent superiority of design *a* (the integrated vertical-scale display) is apparent. The explanation for this given by Roscoe is primarily its pictorial realism in representing relative positions in vertical space by a display in which *up* means *up* and *down* means *down*. (This is again an example of compatibility.) Design *b*, which represents vertical space in a distorted manner (around a circle), did not fare as well as *a*, but it was generally superior to *c* and *d*, both of which consisted of *separate* displays of the three altitude values rather than an *integrated* display. Thus, we can derive a strong hint that integrated displays (where they are indeed appropriate) generally are preferable to displays that have distinctly separate indications for the various values.

As an aside, this example reinforces the point that hard-and-fast generalizations are fairly treacherous. We have seen, for example, that *in this case*, a vertical scale was clearly best and the use of separate counters was manifestly inappropriate, presumably because of the requirement to envision relative positions in vertical space.

Specific Features of Quantitative Scales

The ability of people to make visual discriminations (such as those that are required in the use of quantitative scales) is influenced in part by the specific features that are to be discriminated. Some of the relevant features of quantitative scales are length of scale unit, scale markers (how many and what size), numerical progressions of scales, and the design of pointers.

Length of Scale Unit The length of the scale unit is the length on the scale that represents the numerical value that is the smallest unit to which the scale is to be read. For example, if a pressure gauge is to be read to the nearest 10 pounds (10 lb), then 10 lb would be the smallest unit of measurement; the scale would be so constructed that a given length (in inches, millimeters, etc.) would represent 10 lb of pressure. (Whether there is, or is not, a marker for each such unit is another matter.)

The length of the scale unit should be such that the distinctions between the values to be read can be made with reasonably optimum reliability in terms of human sensory and perceptual skills. Although certain investigators have reported acceptable accuracy in the reading of scales with scale units as low as 0.02 in, most sets of recommendations provide for values ranging from about 0.05 to 0.07, as shown in Figure 4-8. The larger values probably would be warranted when the use of instruments is under less than ideal conditions, such as when used by persons who have below-normal vision, or when used under poor illumination or under pressure of time.

Scale Markers It is generally considered good practice to include a scale marker for every scale unit that is to be read. However, if space or other considerations require the use of scale units shorter than those normally considered desirable (as illustrated in Figure 4-8), the scale markers would be too crowded to be read accurately and rapidly (especially under low illumination). In such a circumstance it is better to design a scale

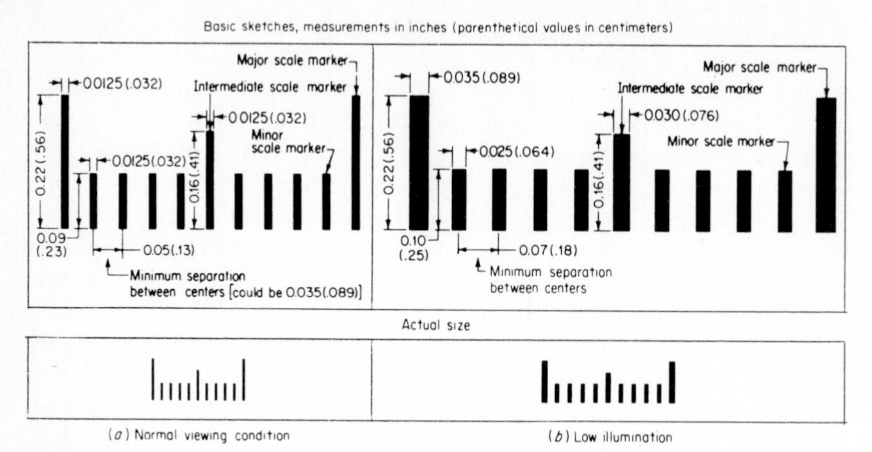

FIGURE 4-8
Recommended format of quantitative scales, considering length of scale unit and graduation markers. Format *a* is proposed for normal illumination conditions under normal viewing conditions, and *b* for low illumination. (*Source: Adapted from Grether & Baker, 1972, p. 88.*)

that does require some interpolation. Actually, people are moderately accurate in interpolation; Cohen and Follert (1970) report that interpolation of fifths, and even of tenths, may yield satisfactory accuracy in many situations. Even so, where high accuracy is required (as with certain test instruments and fine measuring devices), a marker should be placed at every scale unit, even though this requires a larger scale or a closer viewing distance.

Numerical Progressions of Scales Every quantitative scale has some intrinsic numerical-progression system that is characterized by the numerical difference between adjacent graduation markers on the scale and by the numbering of the major scale markers. In general, the garden variety of progression by ones, of 0, 1, 2, 3, etc., is the easiest to use. This lends itself readily to a scale with major markers at 0, 10, 20, etc., with intermediate markers at 5, 15, 25, etc., and with minor markers at individual numbers. Progression by fives is also satisfactory, and by twos is moderately so. Some examples of scales with progressions by ones and fives are shown in Figure 4-9. Where large numerical values are used in the scale, the relative readabilities of the scales are the same if they are all multiplied by 10, 100, 1000, etc. Decimals, however, make scales more difficult to use, although for scales with decimals the same relative advantages and disadvantages hold for the various numerical progressions. The zero in front of the decimal point should be omitted when such scales are used.

The Design of Pointers The few studies that have dealt with pointer design leave some unanswered questions, but, in general, some of the common recommendations are the following: the use of pointed pointers (with a tip angle of about 20°); having the tip of the pointer meet, but not overlap, the smallest scale markers; having the color of the pointer extend from the tip to the center of the scale (in the case of cir-

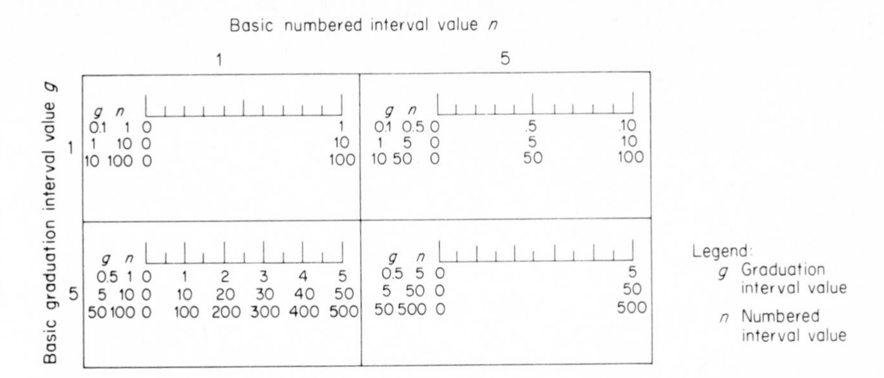

FIGURE 4-9
Examples of certain generally acceptable quantitative scales with different numerical-progression systems (ones/fives). The values to the left in each case are, respectively, the graduation scale interval *g* (the difference between the minor markers), and the numbered scale interval *n* (the difference between numbered markers). For each scale there are variations of the basic values of the system, these being decimal multiples or multiples of one or five.

cular scales); and having the pointer close to the surface of the scale (to avoid parallax).

Combining Scale Features Several of the features of quantitative scales discussed above have been integrated into relatively standard formats for designing scales and their markers, as shown in Figure 4-8. These are based on scale unit lengths of (*a*) 0.05 in for use under normal viewing conditions (with adequate illumination) and (*b*) 0.07 in for low illumination and for viewing at about 28 in. Although the formats in Figure 4-8 are shown in a horizontal scale, the features can of course be incorporated in circular or semicircular scales.

It should be added that the design features shown in this figure should be considered as general guidelines rather than as rigid requirements, and that the advantages of certain features may, in practical situations, have to be traded off for other advantages. The further point should be made that experience and logic in display design suggest the desirability of simple, uncluttered, rather bold designs, as illustrated in Figure 4-10.

Scale Size and Viewing Distance The above discussion of the detailed features of scales is predicated on a normal viewing distance of 28 in. If a display is to be viewed at a greater distance, the features would have to be enlarged in order to maintain, at the eye, the same visual angle of the detailed features. To maintain that same visual angle, the following formula can be applied for any other viewing distance in inches (x):

$$\text{Dimension at } x \text{ in} = \text{dimension at 28 in} \times \frac{x \text{ in}}{28}$$

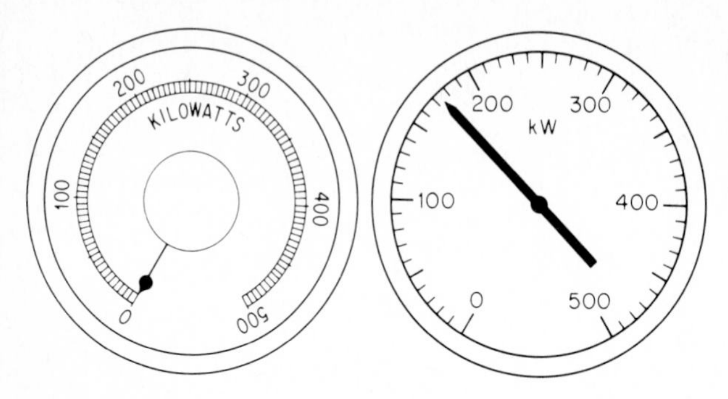

FIGURE 4-10
Illustration of two designs of a meter. The one at the right would be easier to read because it is bolder and less cluttered than the one at the left. It has fewer graduation markers, and the double arc-line has been eliminated. The scale length is increased by placing the markers closer to the perimeter; although this requires that the numerals be placed inside the scale, the clear design and the fact that the numerals are upright probably would partially offset this disadvantage. (*Source: Adapted from* Applied ergonomics handbook, *1974, fig. 3-1.*)

QUALITATIVE VISUAL DISPLAYS

In using displays for obtaining qualitative information, the user is primarily interested in the approximate value of some continuously changeable variable (such as temperature, pressure, or speed) or in its trend or rate of change. The basic underlying data used for such purposes usually are quantitative.

The Quantitative Basis for Qualitative Reading

Quantitative data may be used as the basis for qualitative reading in at least three ways, as follows: (1) for determining the status or condition of the variable in terms of each of a limited number of predetermined ranges (such as determining if the temperature gauge of an automobile is "cold," "normal," or "hot"); (2) for maintaining approximately some desirable range of values [such as maintaining a driving speed between 50 and 55 miles per hour (55 mi/h)] ; and (3) for observing trends, rates of change, etc. (such as noting the rate of change in altitude of an airplane). In the qualitative use of quantitative data, however, there is evidence that suggests that a display that is best for a quantitative reading is not necessarily best for a qualitative reading task.

Some evidence for support of this contention comes from a study in which open-window, circular, and vertical designs are compared (Elkin, 1959). In one phase of this study, subjects made qualitative readings, as shown in Table 4-1. The accuracy of the readings was very high (only 3 errors were made in 1440 readings). The average times taken for the readings, however, are interesting, especially when compared with the *lowest* average reading times for the quantitative reading task (Table 4-2). Thus, while the open-window design took the least time (of the three types) for quantitative reading, it took the longest time for qualitative reading.

TABLE 4-1
RESPONSES TO BE MADE TO POINTER SETTINGS IN QUALITATIVE READING TASK

Pointer setting	Response to be made by subject
Above 60	High
40–60	OK
Below 40	Low

Source: Elkin, 1959.

TABLE 4-2
TIMES FOR QUALITATIVE AND QUANTITATIVE READINGS WITH THREE TYPES OF SCALES

	Average reading time, s	
Type of scale	Qualitative	Quantitative
Open-window	115	102
Circular	107	113
Vertical	101	118

Source: Elkin, 1959.

The optimum designs of displays for qualitative reading, however, depend on *how* they are to be used, that is, the particular type of qualitative reading. If the entire continuum of values can be sliced up into a limited number of ranges—each of which represents some general "level"—the optimum design would be one in which each range of values is separately coded, such as by color, as illustrated in Figure 4-11. When color coding is not feasible (as under certain illumination conditions or with color-deficient individuals), zones on an instrument can be shape-coded. In this connection, it is desirable (if feasible) to take advantage of any natural associations people may have with designs or shapes. One study was directed toward determining what, if any, such associations people had with each of seven different coding designs (Sabeh, Jorve, & Vanderplas, 1958). After having solicited a large number of designs initially, the investigators selected the seven shown in Figure 4-12. These were presented to 140 subjects, along with a list of seven "meanings," as follows: caution; undesirable; mixture–lean; mixture–rich; danger–upper limit; danger–lower limit; and dangerous vibration. Figure 4-12 shows the number of subjects out of 140 who selected the indicated meaning to a statistically significant level. This is another illustration of the concept of compatibility (in this case the compatibility of association with symbol "meanings") as applied to a design problem.

Sometimes what is essentially a quantitative scale is used in what is referred to as *check-reading* manner, namely, simply to determine if the value represented reflects

FIGURE 4-11
Illustration of color coding of sections of instruments that are to be read qualitatively.

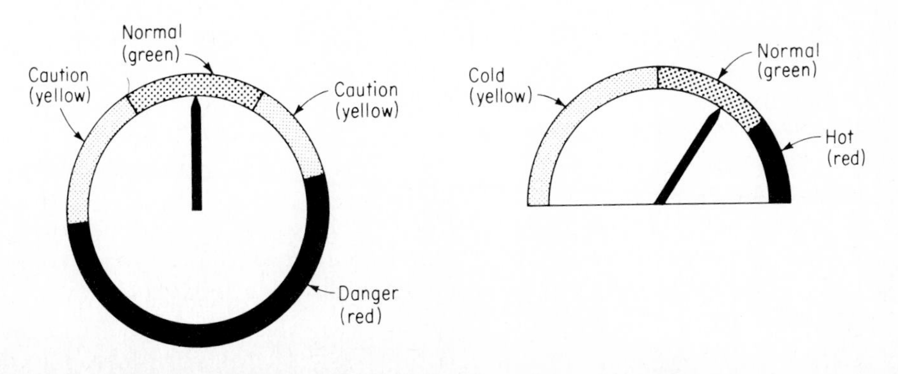

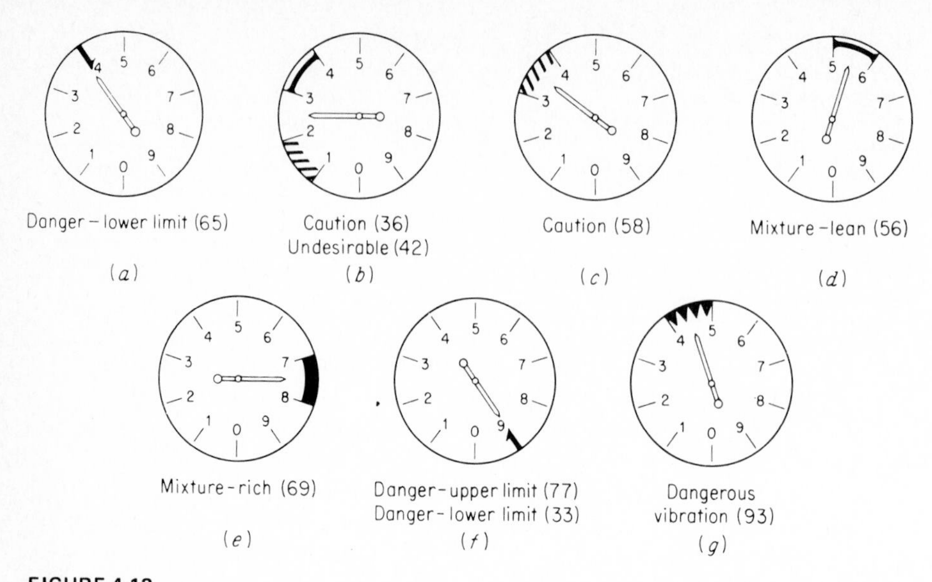

FIGURE 4-12
Association of coded zone markings with subjective "meaning," showing the number of individuals (out of 140) who reported significant associations. (*Source: Adapted from Sabeh, Jorve, & Vanderplas, 1958.*)

what is a normal (satisfactory, neutral, null) condition represented by a single value or a very narrow range of values, or an abnormal condition. In display design this frequently is done by having a mark along the scale that represents that value or range of values.

An interesting variation of design for check-reading a qualitative instrument has been investigated by Kurke (1956). He used simulations of three variations of a quantitative instrument in which a given range of readings indicated a "danger" condition which required attention. These three variations, no indication, a red line, and a red wedge, are shown in Figure 4-13, along with mean-time scores of a group of subjects.

The argument for the use of precoded displays for qualitative reading (when this is feasible) is rooted in the nature of the human perceptual and cognitive processes. To use a strictly quantitative display to determine if a given value is within one range or another range involves an additional cognitive process of allocating the value that is read to one of the possible ranges of values that represent the categories that have operational meaning. The initial perception of a precoded display immediately conveys the meaning of the display indicator.

It should be noted that quantitative displays with coded zones also can be used to reflect trends, directions, and rates of change. Further, they can also be used for quantitative reading if the scale values are included, as in Figure 4-13. If in the use of quantitative data for qualitative reading it is not appropriate to precode certain zones, some conventional form of quantitative display would have to be used. The particular choice, however, would depend upon the relative importance of the qualitative versus quantitative readings, as reflected by the previously cited study by Elkin.

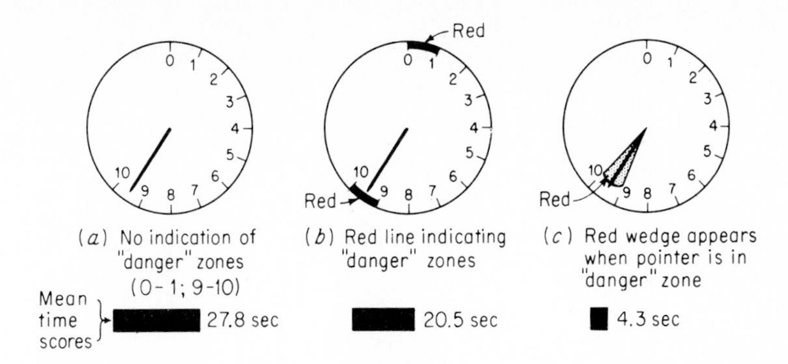

FIGURE 4-13
Three designs of quantitative instruments used in a check-reading situation where the "danger" condition was to be identified. Design *a* has no indication of the danger zone, *b* has a red line at the circumference, and *c* shows a red wedge when the pointer is in the danger zone. (*Source: Kurke, 1956, fig. 2. Copyright © 1956 by the American Psychological Association. Reprinted by permission.*)

Panels of Displays for Check Reading When several or many basically quantitative round dial displays are to be used for check reading (e.g., simply to determine if a condition is normal or not normal), the design and arrangement of the dials can facilitate the identification of any "abnormal" dial. For example, it has been found that dials with the "normal" or "null" condition represented by the pointers at the 9 o'clock or 12 o'clock positions are used more accurately and rapidly than if the location of the normal condition varies from display to display. Also, dials whose normal pointer positions form certain symmetrical patterns can be used with reasonable accuracy and speed. The basis for the advantage of some systematic configuration of the normal conditions of such dials is essentially a function of our perceptual processes, in particular what is referred to as the *gestalt*, that is, the perception of the total configuration; any deviant dial "breaks up" that gestalt, focusing our attention on it. In connection with the use of symmetrical patterns, some patterns are more effective than others.

This was shown by Dashevsky (1964), for example, in a comparison of the various patterns shown in Figure 4-14. Some of those patterns (*d, e,* and *f*) incorporated lines extending from one dial to another to form continuous lines when the pointers were in their null positions. The errors resulting from this comparison are given in Table 4-3. Such results tend to indicate that dials for check reading probably are more effective if the normal positions of the pointers are oriented toward the 12 o'clock position (and probably toward the 9 o'clock position) than if arranged in patterns of various subgroups. The results also suggest that extending the lines between the dials enhances the detection of deviant dials.

Another related study by Oatman (1964) involved a comparison of the detection rates of deviant dials with extended pointers (*c* and *d* of Figure 4-15) and of short pointers (*a* and *b*) in combination with the extended line between the dials (*b* and *c*) and of open patterns (*a* and *d*). He concluded that factors that make the *pointers of the dial* more conspicuous (e.g., length of pointer) are apparently more significant in

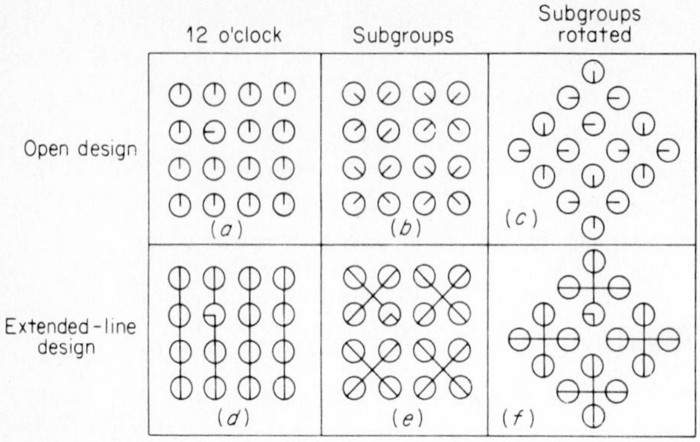

FIGURE 4-14
Patterns of panels of check-reading dials used in study by Dashevsky (1964). In this study the 12 o'clock extended-line pattern *d* resulted in the lowest number of errors. (*Copyright © 1964 by the American Psychological Association. Reprinted by permission.*)

TABLE 4-3
ERRORS IN CHECK-READING DIALS SHOWN IN FIGURE 4-14

	Arrangement		
	12 o'clock	**Subgroups**	**Subgroups rotated**
Open	*a.* 53	*b.* 193	*c.* 201
Extended line	*d.* 8	*e.* 15	*f.* 41

Source: Dashevsky, 1964.

FIGURE 4-15
Patterns of panels of check-reading dials used in a study by Oatman (1964). In this study extended-pointer designs (such as *c* and *d*) resulted in fewer errors than those without extended pointers.

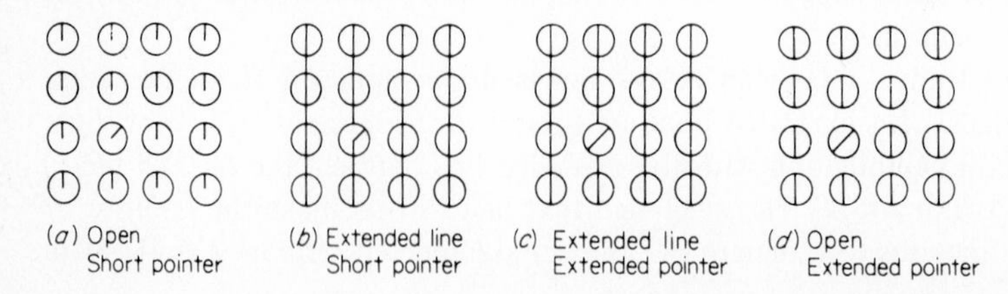

reducing check-reading errors than factors that make the display *pattern* simpler (e.g., extended line between the dials).

Granting that there are differences in the relative importance of certain design features of displays for check reading, collectively the evidence seems to suggest having (1) dials arranged with the normal positions of the pointers aligned in the 9 o'clock or 12 o'clock position, (2) the pointers "extended" across the dial face (rather than from the center to the circumference), and (3) extended lines between the dials.

Although this discussion of check reading has dealt with a handful of rather specific design features, the primary intent has been to illustrate how various research investigations can contribute to the building up of a body of data that can serve as the basis for establishing certain generalizations regarding the design of some aspect or feature of a system. In the case of check reading, the guidelines mentioned above are very much in line with what is known about perceptual processes (which deal with the organization and interpretation of input data), especially regarding the *gestalt* concept. This concept emphasizes the point that people tend to perceive complex configurations as complete entities, with the result that any aspect that does not conform to the configuration is readily perceived (in the frame of reference of check reading, this would apply to any dial that deviates from the basic configuration).

STATUS INDICATORS

In a sense, some "qualitative" information approximates an indication of the "status" of a system or a component, such as the use of some displays for check reading to determine if a condition is normal or abnormal, or the qualitative reading of an automobile thermometer to determine if the condition is hot, normal, or cold. However, what are more strictly status indicators reflect separate, discrete conditions, such as "on and off," or (in the case of traffic lights) "stop, caution, and go." In the case of traffic lights and many other circumstances, the most straightforward type of display is a signal light. However, in other circumstances other types of status indicators can be used. It might be added that if a quantitative instrument is to be used *only* for check-reading purposes, one could use a status indicator instead of the quantitative scale.

SIGNAL AND WARNING LIGHTS

Flashing or steady-state lights are used for various purposes, including the following: as indications of warning (as on highways); as identification of aircraft at night; as navigation aids and beacons; and to attract attention, such as to certain locations on an instrument panel. There apparently has been little research relating to such signals, but we can infer some general principles from our knowledge of human sensory and perceptual processes that might be helpful.

Detectability of Signal and Warning Lights

There are, of course, various factors that influence the detectability of lights. Certain such factors are discussed below.

Size, Luminance, and Exposure Time The absolute threshold for the detection of

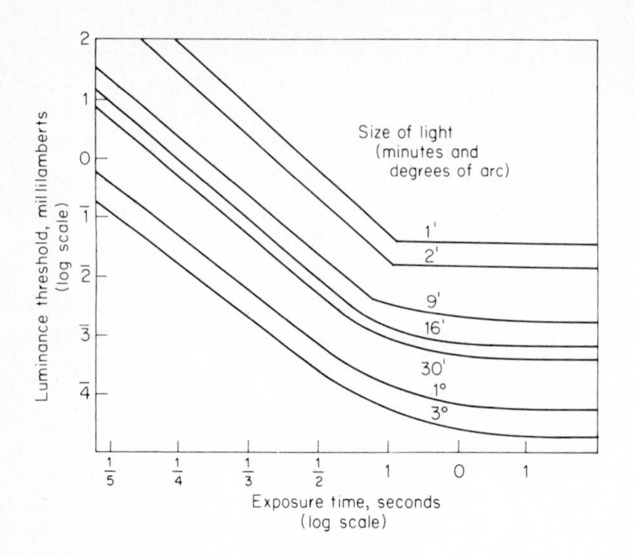

FIGURE 4-16
Minimum sizes of lights (in minutes and degrees of arc) that can be detected 50 percent of the time under varying combinations of exposure time and luminance. (*Source: Adapted from Teichner & Krebs, 1972.*)

a flash of light depends in part on a combination of size, luminance, and exposure time. Drawing in part upon some previous research, Teichner and Krebs (1972) have depicted the minimum sizes of lights (in terms of visual angle of the diameter in minutes of arc) that can be detected 50 percent of the time under various combinations of exposure times (in seconds) and luminance (in millilamberts), these relationships being shown in somewhat simplified form in Figure 4-16. From this it is clear that the luminance threshold decreases linearly as a function of exposure time up to a particular value, but that the exposure time at which the luminance threshold levels off decreases systematically with target size (actually the "area" of the light as implied by its diameter). Thus, the very minimal, lower-bound detectability of lights (50 percent accuracy) is seen to be a function of size, luminance, and exposure time. (Operational values should well exceed those in Figure 4-16.)

Color of Lights Another factor that is related to the effectiveness of signal lights is color. Using response time as an indication of the effectiveness of four different colors, Reynolds, White, and Hilgendorf (1972) report the following order (from fastest to slowest): red, green, yellow, and white. However, the background color and ambient illumination can interact to influence the ability of people to detect and respond to lights of different colors. In general the researchers found that if a signal has good brightness contrast against a dark background, and if the absolute level of brightness of the signal is high, the color of the signal is of minimal importance in attracting attention. But with low signal-to-background brightness contrast, a red signal has a marked advantage, followed by green, yellow, and white in that order.

Flash Rate of Lights In the case of flashing lights, the flash rate should be *well* below that at which a flashing light appears as a steady light (the flicker-fusion frequency), which is approximately 30 times per second (30/s). In this regard, rates of about 3 to 10/s (with duration of at least 0.05 s) have been recommended for attracting attention (Woodson and Conover, 1964, pp. 2-26), and Markowitz (1971) makes the point that the range of 60 to 120 flashes per minute (1 to 2 per second), as now used on highways and in flyways, appears to be compatible with human discrimination capabilities and available hardware constraints.

Background of Lights As might be expected, signal lights cannot be discriminated well when other background lights are somewhat similar. (Traffic lights in areas with neon signs and Christmas tree lights represent very serious deviations from this principle.) And still another background characteristic relates to the steady versus flashing state of any background lights. In an interesting investigation of these, Crawford (1963) used both steady and flashing signal lights against backgrounds of *irrelevant* lights (what we might call *noise*), these being all steady, all flashing, or some admixture of steady and flashing lights. Very briefly, his results indicated that the average time to identify the signal lights was minimal when the background-noise lights were all steady (this was especially so when the signal light was itself flashing); that the advantage of a flashing signal light (contrasted with a steady light) was completely lost if even one background-noise light was flashing; and that steady signals were more effective (could be identified more quickly) than flashing signals if the proportion of the noise lights that were flashing was any greater than 1 out of 10. In other words, flashing lights against other flashing lights really make life difficult for the viewer.

Recommendations Regarding Signal and Warning Lights

On the basis of research and experience, Heglin (1973) offers a number of recommendations regarding the use of signal and warning lights, some of which are given below:

When should they be used? To warn of actual or potential dangerous condition.

How many warning lights? Ordinarily only one. (If several warning lights are required, use a master warning or caution light, and a word panel to indicate specific danger condition.)

Steady-state or flashing? Flashing lights should be reserved for extreme emergencies, since they are distracting.

Flash duration: If flashing lights are used, flash rates should be from 3 to 10 per second (4 is best), with equal intervals of light and dark.

Warning-light intensity: The light should be at least twice as bright as the immediate background.

Location: The warning light should be within 30° of the operator's normal line of sight.

Color: Warning lights are normally red because red means *danger* to most people. (Other signal lights in the area should be of other colors.)

REPRESENTATIONAL DISPLAYS

Representational displays—both static and dynamic—tend to fall into two classes: (1) those that are essentially pictorial (intended to *reproduce* an object or scene, as on a TV scope or in an aerial photograph); and (2) those that are illustrative or symbolic (such as maps or aircraft-position displays). In either case the intent is to convey a visual impression that requires little or no interpretation. Since there are many varieties of such displays, we will here discuss only a few particular types.

Aircraft-Position Displays

The problem of representing the position and movement of aircraft has haunted designers—and pilots—for years. One major facet of this problem relates to the basic movement relationships to be depicted by the display, there being two such relationships, as shown in Figure 4-17 and described as follows:

- *Moving-aircraft:* The earth (specifically, the horizon) is fixed, with the aircraft moving in relation to it (moving-aircraft or outside-in display).
- *Moving-horizon:* The aircraft is fixed, with the horizon moving in relation to it (moving-horizon or inside-out display). (Most aircraft displays are of this type.)

This is basically a problem of visual perception, specifically with respect to what is referred to as the *figure and ground phenomenon*. Is it more compatible with people's perceptions to envision the aircraft relative to a fixed earth or vice versa? As Johnson and Roscoe (1972) lament, the experimental findings relating to this are flimsy and range from "suspect to inconclusive." In some of the studies by Roscoe and Williges and their students at the University of Illinois (such as Beringer et al., 1975), it was found that experienced pilots tended to perform better with moving-horizon displays than with moving-aircraft displays. However, the pilots' superior performance on the moving-horizon displays was undoubtedly due to their previous experience with such displays; aside from the effects of previous experience, there is reasonable evidence to indicate that the moving part of a display normally should be displayed against a fixed scale or coordinate system (i.e., the moving-aircraft principle). (See the discussion of the principle of the moving part, below.)

FIGURE 4-17
Illustration of the two basic movement relationships for depicting aircraft attitude, namely, the moving-aircraft (outside-in) and the moving-horizon (inside-out). (*Source: Adapted from Johnson & Roscoe, 1972.*)

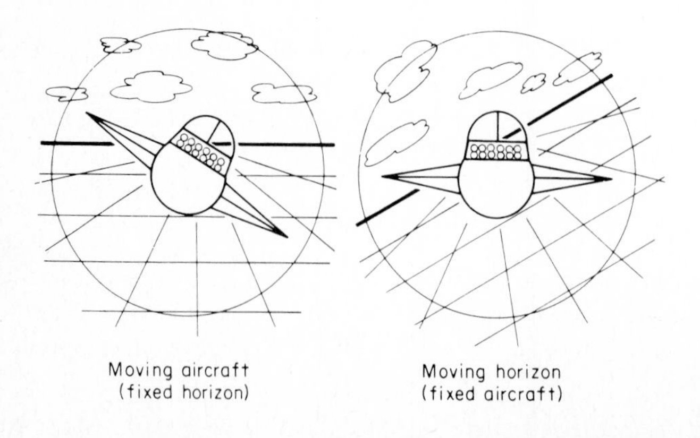

In addition to the question of whether the aircraft or the horizon (or both) should move, aircraft-position displays can be designed in order to require either pursuit or compensatory tracking. (A *pursuit* display shows the movement of both the target and the pursuing aircraft against common reference coordinates, whereas a *compensatory* display shows only the difference, or error, in their relative positions.) In the Illinois experiments, pursuit and compensatory versions of each type of attitude presentation were compared, and, for all tasks in which performance differences occurred, the pursuit-tracking displays produced the superior performance.

Principles of Aircraft-Position Displays Although the evidence regarding certain aspects of aircraft-position displays is still not definitive, Roscoe (1968) has teased out of the relevant research a few principles of display that he believes have substantial validity. Certain of these have been touched on before, in one context or another, but they will nevertheless be reiterated here.

1 *The principle of display integration:* The notion of *display integration* requires that *related* information be presented in a common display system which allows the relationships to be perceived directly (as illustrated in Figure 4-7, design *a*). This principle does *not* apply to the haphazard combining of unrelated information in a common display.

2 *The principle of pictorial realism:* This principle relates to the presentation of graphic relationships in such a manner that the encoded symbols can be readily identified with what they represent; in effect, the symbols are an analog of that which they represent.

3 *The principle of the moving part:* In the use of aircraft displays this principle is in conformity with the outside-in (i.e., moving-aircraft) display rather than with the inside-out (i.e., moving-horizon) display. In more general terms, it seems preferable for the image of the *moving part* (i.e., an aircraft or symbol representing any other moving object) to be displayed against a fixed scale or coordinate system.

4 *The principle of pursuit tracking:* In pursuit tracking, the index of *desired* performance (sometimes referred to as the *target*) and the index of *actual* performance (called the *follower, cursor,* or *controlled element*) move over the display against a common scale or coordinate system. Generally this scheme results in better performance than does compensatory tracking, in which the index of either the desired or actual performance is fixed, with the moving index showing only the *error,* or difference.

5 *The principle of frequency separation:* This principle relates to the relative speed of movement of display indications; when "high-frequency" (suddenly changing or rapidly alternating) information is displayed, the moving element must respond in the expected direction (i.e., compatibility of movement is especially critical); but when "low-frequency" information is displayed, the direction of movement is not as crucial.

6 *The principle of optimum scaling:* This principle deals with the physical relationship (really the ratio) of the physical dimensions of that to be represented (i.e., features of the surface of the earth) to the dimensions on the display that represent such features, such as the number of millimeters or inches on the display that represent 1 mile (1 mi) on the earth. This problem gets all intertwined with precision requirements, but taking these into account, some optimum relationship is possible.

Although the above principles were crystallized because of their relevance to the problems of aircraft flight and navigation displays, they probably are equally valid for numerous other reasonably corresponding display problems.

Cathode-Ray-Tube (CRT) Displays

The natures of the images presented on cathode-ray tubes (CRTs) are of course a function of the purpose of the display, but they include direct representation of scenes (as in TV), *blips* that represent objects (as in radar and aircraft-control-tower displays), graphic representations (as in various types of test and medical equipment), and generated alphanumeric and symbolic characters. Even a listing of the human factors problems involved in the design of CRT displays would be sobering. A few such problems include the size of the display (and this would depend on the purpose and use), the brightness of the displayed configurations and of the background, the contrast, the size of the symbols or configurations to be viewed, viewing distance and angle, and the effects of background noise.

For our illustrative purposes here we will discuss primarily one facet that is associated with the visual resolution of CRT symbols, in particular the number of *raster scan lines*—or simply, *scan lines*. These are the continuous narrow strips of the picture area of varying brightness formed by one horizontal sweep of the scanning spot of a CRT. (These are usually measured in terms of number per inch or number per millimeter, or the number for the "form" presented. Most American TV broadcasts use 525 scan lines; the number per inch would of course depend upon the size of a TV screen.)

As would be expected, the ability of people to recognize images depends in part on the number of scan lines, but some evidence to demonstrate this comes from a study by Wong and Yacoumelos (1973), as depicted in Figure 4-18*a*. This shows the percent-correct identifications of four types of symbols in scanning maps presented on a TV screen with 5, 7, or 9 scan lines per millimeter. The greatest increase in accuracy was from 7 to 9. In this regard Heglin (1973) recommends about 10 raster lines for accurate and rapid discrimination of most symbols, with the individual symbols subtending a visual angle of from 12 to 15 minutes of arc. However, some types of symbols are more easily discriminated than others, and such differences could influence the number of scan lines required for practical use. Figure 4-18*b*, for example, shows the differential identification of four types of images, namely: (1) point symbols (schools, churches, bridges, etc.); (2) alphanumeric; (3) line symbols (highways, railroads, boundaries, etc.); and (4) area symbols (vegetation, urban areas, and water). (Further reference will be made later to alphanumeric characters as presented on CRTs.)

Complex Configurations

Some representational displays present complex configurations of such content as land areas, traffic routes, and wiring or piping diagrams. In the development of such displays the dominant guideline is that of simplicity. Obviously, the application of this principle needs to be within the constraints imposed by the operational requirements for "fidelity" of the configuration. The argument for simplicity arises from the fact that the perceptual processes of "searching" for relevant features take longer (and are subject to higher error rates) if an image is cluttered up with what may be irrelevant material. Within the constraints mentioned above there are two possible directions of

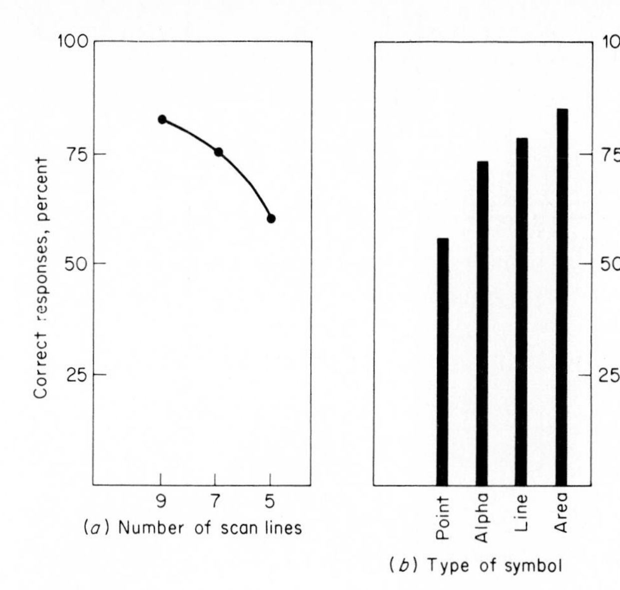

FIGURE 4-18
(*a*) Percent correct identification of various types of images on maps presented on TV with different numbers of scan lines per millimeter; (*b*) percent correct identification of four types of images (point sources, alphanumeric, lines, and areas). (*Source: Wong & Yacoumelos, 1973, figs. 7 and 9.*)

simplification—one consists simply of removal of extraneous detail; the other, of presenting a schematic representation. The latter approach is illustrated by the use of strip maps and by the representation of the London subway system as shown in Figure 4-19. With most display design problems, one should first ask (and answer) the questions: What information does the user need? How can that information best be presented? Obviously the simplification of complex configurations should be guided by the answers to these questions.

Graphic Representations

The format of some of the graphic representations (bar charts, pie charts, line charts, etc.) that find their way into newspapers and other publications leads one to hope that there *must* be better ways of presenting the information that the graphs presumably are intended to convey. Although there has been relatively little research to date regarding the design of graphic representations, one investigation will be summarized to demonstrate that research in this area may actually have some practical application. This study dealt with a comparison of three formats for depicting trend data, as illustrated in Figure 4-20*a*, *b*, and *c*, a line format, a vertical-bar format, and a horizontal-bar format (Schutz, 1961). For each format, there were variations in the numbers of points depicted (6, 12, or 18) and in the number of missing values. The subjects were required to estimate the trend of the data and were scored on the time required to make such estimates and on the accuracy of their estimates. On both criteria the line graph proved to be preferable, as indicated by the mean scores in Table 4-4.

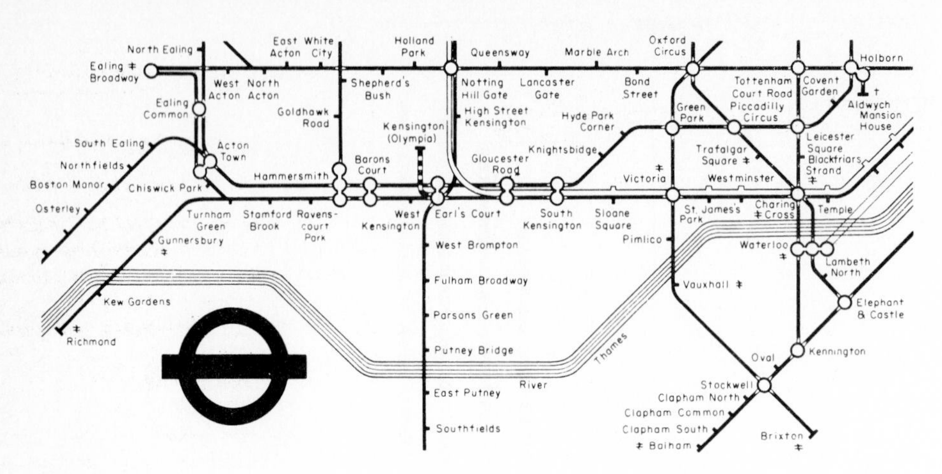

FIGURE 4-19
Part of the London subway system given with a simplified schematic representation. The schematic form, used in the control center, is much easier for people to use. (*Photograph from London Transport.*)

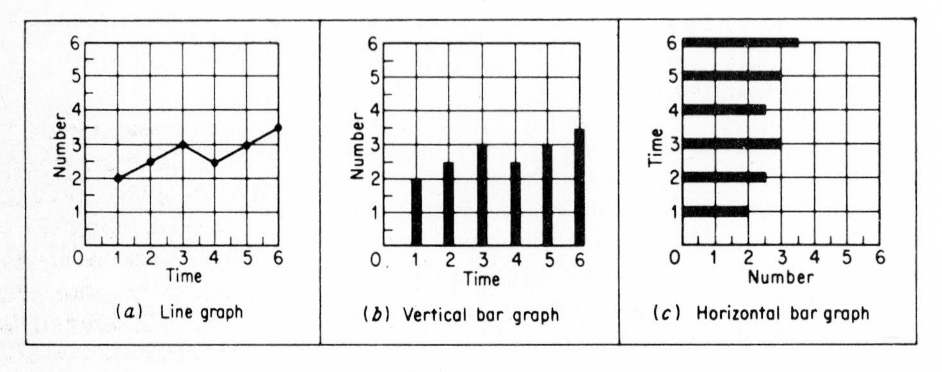

FIGURE 4-20
Illustrations of formats of multiple-trend charts that were compared on time and accuracy of reading; some examples had 12 or 18 points instead of the 6 shown (line graphs generally were superior). See text for discussion. (*Source: Adapted from Schutz, 1961.*)

TABLE 4-4
TIME AND ACCURACY IN INTERPRETING THREE FORMS OF GRAPHS

Format	Mean relative time	Mean accuracy score
Line	6.81	1.72
Vertical bar	7.36	1.64
Horizontal bar	8.91	1.40

Source: Schutz, 1961.

ALPHANUMERIC AND RELATED DISPLAYS

The effectiveness of communications that involve alphanumeric and symbolic characters depends upon various factors, including typography, content, symbolic selection of words, and writing style. Certain aspects of such communications will be discussed to illustrate their effects on the reception of the information presented.[1] Ironically, even in discussions of alphanumeric information there is a fair quota of confusion in the use of words.

For our purposes we will adopt the following definitions:

- *Visibility:* The quality of a character or symbol that makes it separately visible from its surroundings. (This is essentially the same as the term *detectability* as used in Chapter 3.)
- *Legibility:* The attribute of alphanumeric characters that makes it possible for each one to be identifiable from others. This depends on such features as stroke width, form of characters, contrast, and illumination. (This is essentially the same as the term *discriminability* as used in Chapter 3.)
- *Readability:* A quality that makes possible the recognition of the information content of material when represented by alphanumeric characters in meaningful groupings, such as words, sentences, or continuous text. (This depends more on the spacing of characters and groups of characters, on their combination into sentences or other forms, on the spacing between lines, and on margins, than on the specific features of the individual characters.)

Typography

The *typography* of alphanumeric material refers to the various features of the characters and their arrangement.

Stroke Width The *stroke width* of alphanumeric characters is usually expressed as the ratio of the thickness of the stroke to the height of the letters or numerals. Some examples of stroke width-to-height ratios are shown in Figure 4-21. Investigations of the legibility of stroke width still leave some loose ends, although there are certain reasonably stable implications that have emerged from a few studies. Under normal illumination conditions with no particular time constraints most people can discriminate alphanumeric characters with wide differences in stroke width. It is under somewhat adverse reading conditions that stroke width (and other characteristics of characters) become important.

In general terms the optimum ratios for black characters on a white background are somewhat lower than in the case of white characters on a black background. This difference is attributable to a phenomenon called *irradiation*, in which white features appear to "spread" into adjacent black areas, but not the reverse. The phenomenon is especially accentuated with highly illuminated displays. (In one study years ago, for

[1] For an excellent survey of legibility and related aspects of alphanumeric characters and related symbols, the reader is referred to Cornog and Rose (1967).

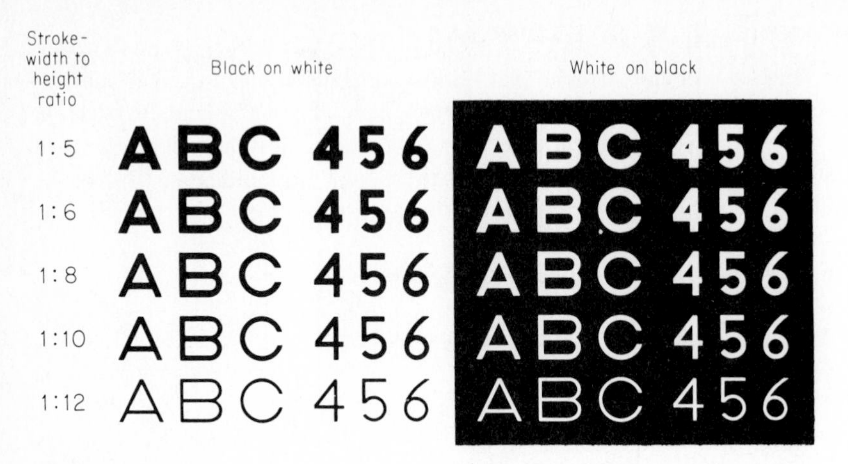

FIGURE 4-21
Illustrations of stroke width-to-height ratios of letters and numerals.

example, the optimum stroke width of white numerals on black was 1:40!) Dark adaptation of individuals also tends to accentuate the effect. Because of this effect, white characters on black should have thinner stroke widths than black on white. Incidentally, when dark adaptation is required, the characters preferably should be white on black, whereas when dark adaptation is not required, black on white is preferable.

On the basis of various studies, it is possible to set forth some generalizations regarding the stroke width-to-height ratios of alphanumeric characters as follows:

Black on white	1:6 to 1:8
White on black	1:8 to 1:10

Width-Height Ratios The relationship between the width and height of alphanumeric characters usually is described as the *width-height ratio* (expressed as a ratio such as 4:5, or as a percent such as 80 percent). In the case of capital letters, the experimental evidence suggests that the ratio be about 1:1, although this can be reduced to about 3:5 without serious loss in legibility. In the case of numerals the rather standard recommendation is that of a ratio of about 3:5. These ratios are illustrated in Figure 4-22.

Font of Alphanumeric Characters Actually, most conventional fonts of alphanumeric characters (and many of the offbeat styles) can be read with reasonable adequacy under normal conditions where size, contrast, illumination, and time permit. There are, however, significant differences in the legibility and readability of different type fonts when viewing conditions are adverse, where time is important, or where accuracy is important. In this connection, the font of capital letters and numerals shown in Figure 4-22 (United States Military Specification No. MIL-M-18012B) has been rather widely tested and found to be generally satisfactory. Although specifically designated for aircrew station displays, the characters have a wider range of applicability. Another set of characters that are widely used by the military services is that

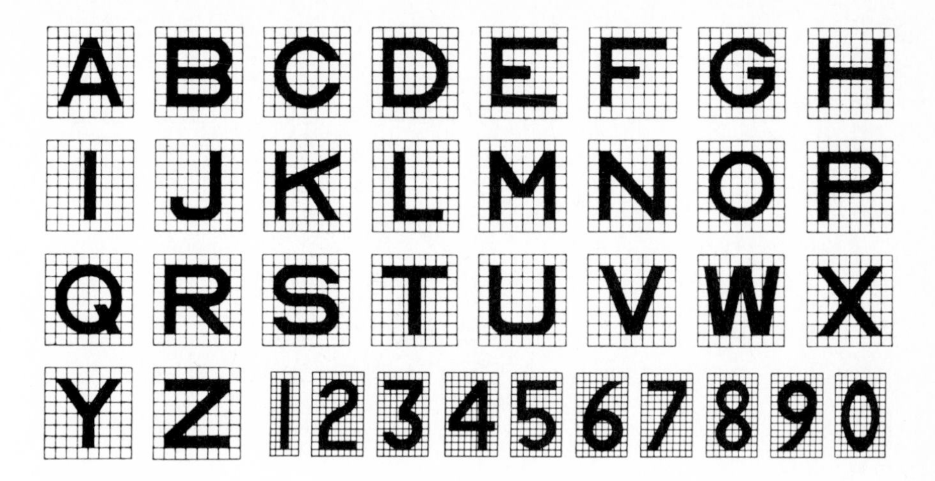

FIGURE 4-22
Letter and numeral font of United States Military Specification No. MIL-M-18012B (July 20, 1964); also referred to as NAMEL (Navy Aeronautical Medical Equipment Laboratory) or as AMEL. The letters as shown have a width-height ratio of 1:1 (except for *I, J, L,* and *W*). These ratios can be reduced to about 3:5 without any appreciable reduction in legibility. The numerals have a width-height ratio of 3:5 (except 1 and 4).

shown in Figure 4-23 [MIL Standard MS 33558 (ASG)]. These are sometimes referred to as AND (Air Force-Navy Drawing 10400).

These fonts (based on military studies) are not specifically represented in conventional fonts used in nonmilitary situations. However, there are a number of modern Gothic styles of print that have been found to have acceptable levels of legibility. As Heglin (1973) points out, a few Gothic styles are archaic and should never be used, these being styles with nonuniform stroke widths, with strokes that do not terminate in a rectangular form, and with serifs all over the letters (*serifs* being flourishes and embellishments of the strokes). The more "modern" Gothic styles have uniform stroke widths and strokes that do not have any serifs. A few examples of such Gothic styles as presented by Heglin are illustrated in Table 4-5. Numerous other Gothic styles are, however, equally acceptable.

FIGURE 4-23
United States Military Standard letters and numerals MIL Standard MS 33558 (ASG) (Dec. 17, 1957). The basic stroke width-to-height ratio is 1:8 and the width is about 70 percent of the height. These are sometimes referred to as AND (Air Force-Navy Drawing 10400).

A B C D E F G H I J K L M
N O P Q R S T U V W X Y Z
0 1 2 3 4 5 6 7 8 9

TABLE 4-5
EXAMPLES OF GOTHIC STYLES

Very light styles	
Futura light	ABCDEFGHIJKLMNOP 1234567890
Sans serif light	ABCDEFGHIJKLM 1234567890
Vogue light	ABCDEFGHIJKLMNOPQR 1234567890
Light styles	
Futura book	ABCDEFGHIJKLMNOPQRST 1234567890
Sans serif medium	**ABCDEFGHIJKLMNOPQ 1234567890**
Tempo medium	ABCDEFGHIJKLMNOPQ 1234567890
Medium styles	
Futura medium	ABCDEFGHIHKLMNOPQRST 1234567890
Sans serif bold	**ABCDEFGHI 1234567890**

Source: Heglin, 1973.

Although many commercially available styles of type have been found through experience and research to be quite legible under normal conditions, Heglin makes the point that there may be other designs that could be developed that would be more legible, especially under adverse reading conditions. He reports one experiment, for example, in which some radically different numerals were used, with virtually no errors in legibility. Examples of this design are given below:

12358

Since this design is, as he puts it, "shocking," he goes on to make the point that, although legibility and beauty are not synonymous, they should be close companions. He proposes that a desirable goal should be to develop styles which are both legible and pleasing in appearance, and suggests that type designs such as those shown below represent a start in this direction:

1234567890
1234567890
1234567890

TABLE 4-6

TABLE OF HEIGHTS H OF LETTERS AND NUMERALS RECOMMENDED FOR LABELS AND MARKINGS ON PANELS, FOR VARYING DISTANCE AND CONDITIONS, DERIVED FROM FORMULA H (IN) = $0.0022D + K_1{}^* + K_2$

Viewing distance, in	0.0022D value	Nonimportant markings, $K_2 = .0$			Important markings, $K_2 = .075$		
		$K_1 = .06$	$K_1 = .16$	$K_1 = .26$	$K_1 = .06$	$K_1 = .16$	$K_1 = .26$
14	0.0308	0.09	0.19	0.29	0.17	0.27	0.37
28	0.0616	0.12	0.22	0.32	0.20	0.30	0.40
42	0.0926	0.15	0.25	0.35	0.23	0.33	0.43
56	0.1232	0.18	0.28	0.38	0.25	0.35	0.45

*Applicability of K_1 values.

$K_1 = .06$ (above 1.0 fc, favorable reading conditions)
$K_1 = .16$ (above 1.0 fc, unfavorable reading conditions)
$K_1 = .16$ (below 1.0 fc, favorable reading conditions)
$K_1 = .26$ (below 1.0 fc, unfavorable reading conditions)

Source: Based on formula of Peters and Adams, 1959; see text.

Size of Alphanumeric Characters The ability of people to make visual discriminations (such as of alphanumeric characters) depends on such factors as size, contrast, illumination, and exposure time. A systematic procedure has been proposed by Peters and Adams (1959) for determining the size of alphanumeric characters that takes into account certain such factors (illumination, viewing conditions, viewing distance, and importance of reading accuracy). This procedure is based on the following formula:

$$H \text{ (height of letter, in)} = 0.0022D + K_1 + K_2$$

where D = viewing distance, in
K_1 = correction factor for illumination and viewing conditions
K_2 = correction for importance (for important items such as emergency labels, $K_2 = .075$; for all other conditions, $K_2 = .0$)

This formula has been applied to various viewing distances, in combination with the other variables, to derive the heights of letters and numerals for those conditions as given in Table 4-6. The lower bounds of these values for a reading distance of 28 in (for K_1 values of 0.06) are 0.12 in for nonimportant and 0.20 in for important markings. These correspond well with the minimum values proposed by Grether and Baker (1972, p. 107), for a 28-in viewing distance of 0.10 in for noncritical and 0.20 in for critical information or adverse reading conditions. For greater viewing distances, of course, the sizes of characters need to be increased.

In a field study that lacked thorough control of the stimulus material and illumination, Smith (1979) found, with over 300 different displays, that when the ratio of letter height to viewing distance[2] was 0.0046 the material was legible 98 percent of the

[2] For small angles, the ratio of letter height to viewing distance is the same in radians (rad), and one minute of visual angle is about 0.00029 rad.

time.[2] Using this ratio yields the following letter heights at various reading distances:

Reading distance, in	14	28	42	56
Letter height, in	0.06	0.13	0.19	0.26

These values are within the approximate range of the values for nonimportant markings in Table 4-6 as proposed by Peters and Adams for favorable reading conditions with reasonably adequate illumination, but generally fall short of values they recommended for important markings, especially those used under unfavorable viewing conditions and with inadequate illumination. Although Smith points out that Peters and Adams provide no empirical evidence to support their recommendations, it would seem reasonable to use values greater than those resulting from Smith's study in the case of alphanumeric characters to be read under adverse viewing conditions and with poor illumination, and especially if the potential viewers might have any visual deficiencies (such as in the case of elderly persons).

One additional point should be made about the size of characters, as demonstrated by Poulton (1972). Some styles of letters have longer "ascenders" and "descenders" than others (i.e., the tips of letters such as *b*, and the tails of letters such as *y*). He found that legibility is largely influenced by what he calls "*x* height," the height of the main body of the letters, and is not influenced much by the length of the ascenders or descenders, nor by the body size (the total height of the block of metal upon which the letter is cast, referred to as *6-point, 10-point,* etc.). Further, an "*x* height" of lowercase letters of about 1.2 mm is very close to the lower bounds of legibility. In a separate study involving the scanning of lists of foods for specific items, Poulton (1969) found that consumers did significantly better with a 6-point "Invers" type (which has an *x* height of about 1.2 mm) than with a smaller typeface, and indicates that this typeface is as small as food manufacturers should use on food containers if the print is to be reasonably legible to consumers of all ages.

Readability The readability of printed or typed text, and its comprehension, is a function of a wide assortment of factors such as type style (font), type form (capital, lowercase, boldface, italics, etc.), size, contrast, leading (spacing) between lines, length of lines, and margins. This is obviously a very broad spectrum of variables, and it is not in our province here to pull together and synthesize the research in this area. However, to illustrate such research, let us summarize the results of one particular study (Poulton, 1959). In the experiment, a comparison was made of the comprehension of material printed in the four types and formats shown and described below:

1 7 Modern Extended No. 1: point size 11; 0.03-in leading between lines, 5.2-in line length; 1 column per page.

2 7 Modern Extended No. 1: point size 9; 0.01-in leading between lines; 2.8-in line length; 2 columns per page.

3 101 Imprint: point size 11; 0.02-in leading between lines; 5.0-in line length; 1 column per page.

4 327 Times New Roman: point size 9; 0.01-in leading between lines; 2.8-in line length; 2 columns per page.

The subjects (275 scientists) were tested on their comprehension of the subject matter. Condition 1 was considerably easier to comprehend than the others, probably owing largely to style of type (compared with 3, which was comparable otherwise in point size, number of columns, etc.) and perhaps in part to point size and number of columns (compared with 2, the same style but in smaller point size and with two columns instead of one). Such a study is, of course, far from conclusive, but at least it suggests some of the variables that may influence readability and comprehension.

As another example of factors that influence readability, Tinker (1955) had some subjects read regular type such as this (actually roman type), *had other subjects read italicized type such as this,* AND HAD OTHER SUBJECTS READ CAPITALIZED TYPE SUCH AS THIS. The regular-type group read significantly faster than those who read material in all capital letters. These results undoubtedly are due in large part to familiarity of people with the conventional upper- and lowercase type in continuous text.

In the use of words as labels (such as on instrument panels, for identification), however, the shoe is on the other foot; words in all capital letters in this type of situation generally are more readable than those in lowercase or mixed type.

In the case of much reading material certain passages are of course more important than others. What is called *typograph cuing* refers to the use of variations in the appearance of text to provide a visual distinction for the more important passages. Foster and Coles (1977) investigated the effects of three typographic formats as they might influence the retention of relevant printed material, these being (1) CAPITAL LETTERS LIKE THIS; (2) **boldface letters like this**; and (3) a control condition with regular type like this. With the use of tests given to people after they had read text material in which the "important" information was treated in these three ways, the investigators found that the boldface format was superior to the all-capital format or the regular-type format. Although excessive use of boldface type for cuing the reader to the most relevant passages should of course be avoided, such a treatment in moderation apparently contributes to the learning of the passages so treated.

CRT and Other Illuminated Symbols

Under some circumstances alphanumeric and symbolic characters are generated by some form of illumination, as on CRTs, by the use of electroluminescent (EL) light, by the miniaturization of incandescent lights, etc. Depending on the particular technique used, the configurations reproduced can be conventional characters or those produced by generating dots, lines, or segmented lines. Various investigators have studied the adequacy of different forms of such characters. For example, Vartabedian (1971) compared uppercase and lowercase letters formed with dots and those formed with lines, as shown in Figure 4-24. Displays consisting of 27 words were searched by the subjects for specific words given by the experimenter. The following conclusions were reported: (1) displays with uppercase letters were searched faster than those with lowercase letters; (2) search time was about the same for the dot and stroke designs, and for each of the three sizes used (0.12, 0.14, and 0.16 in); and (3) the subjects preferred uppercase letters and those formed from dots rather than lines.

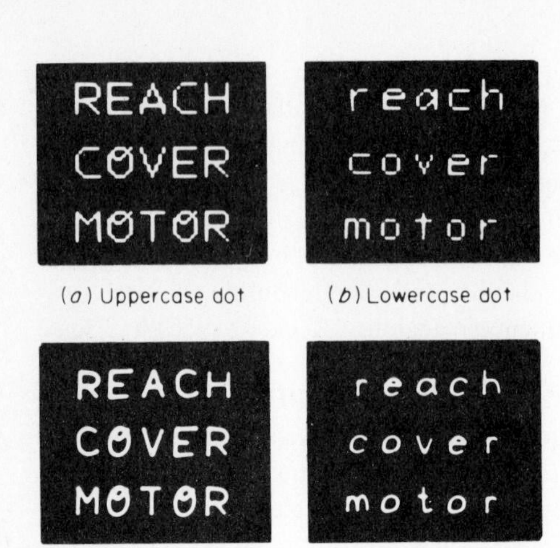

FIGURE 4-24
Examples of words used in cathode-ray-tube (CRT) displays for comparison of four types of generated letters. (*Source: Vartabedian, 1971, fig. 1.*)

In a subsequent study (Vartabedian, 1973) it was found that readability of a 7 X 9 dot pattern of characters was about the same as for conventional stroke symbols (formed with a Leroy lettering set) and that there was even a slight edge in favor of the dot patterns (in 12 out of 21 subjects). Thus, the types of characters that can be generated most economically with CRTs are at least as readable as conventional characters and as acceptable for the users.

The advantage of uppercase letters reflected by the results of the study by Vartabedian (1971) is consistent with the use of uppercase letters in other contexts where the task is more one of searching or identification than one of reading. In another investigation, Plath (1970) compared the legibility of conventional numerals (the NAMEL numerals shown in Figure 4-22) with two forms of segmented numerals, slanted and vertical, as shown in Figure 4-25. The errors in legibility under time-constrained conditions were least for the NAMEL numerals, as shown in that figure. Such results argue against the use of segmented figures where accuracy is critical and when time is

Type of numeral	Examples	Number of errors
NAMEL	345	187
Slanted segmented	345	391
Vertical segmented	345	388

FIGURE 4-25
Forms of numerals investigated by Plath, along with errors made in their identification under time-controlled conditions. (*Source: Adapted from Plath, 1970, fig. 1 and table 1.*)

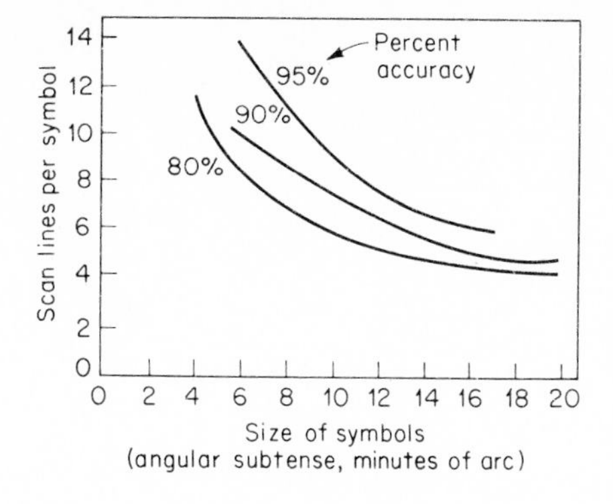

FIGURE 4-26
Number of scan lines and size of alphanumeric and geometric symbols required for each of three levels of identification accuracy. (*Source: Hemingway & Erickson, 1969, fig. 10.*)

severely limited. However, as indicated earlier, if there are no time limitations in viewing alphanumeric symbols and if they are of reasonable size and presented under adequate illumination, virtually any form can be read. Alphanumeric design becomes critical under adverse conditions.

Although the discriminability of symbols used on CRT displays is influenced by their design and size, it is also influenced by the number of scan lines of the CRT (which were referred to earlier in the chapter). Using 16 symbols (squares, stars, triangles, half-circles, pentagons, etc.), Hemingway and Erickson (1969) first explored their legibility when presented in various sizes and with various numbers of scan lines, and then synthesized their results with those of other investigations (albeit those in which alphanumeric characters were used) and ended up with the pattern shown in Figure 4-26. This shows the combinations of scan lines and angular subtense (i.e., the size in visual angle) for each of three levels of accuracy. It is clear from this that there is a trade-off between number of scan lines and size of symbols, but the results still indicate that for a high accuracy it is desirable to provide for 8 or 10 or more scan lines unless the symbols are quite large.

Saying What Is Meant

Your experiences in everyday reading and writing will confirm the fact that recorded verbal material does not always convey reliably the meaning that is intended. The unambiguous, understandable use of language is very pertinent to various tangents of human factors engineering, including the preparation of training materials, job aids, instructions, directions, and labels. Chapanis (1967) ran across one horrible example of a notice beside an elevator that read as shown in the first box of Figure 4-27. That figure also indicates what the sign really meant, and presents an improved version that is both clearer and shorter. Broadbent (1977) reports another interesting example of a confusing instruction, on an electric razor: "This razor should only be used on the 220-V setting in the United Kingdom." This could be interpreted as meaning that "Outside the United Kingdom, this razor should be used on the 110-V setting." [Such an interpretation would of course ruin the razor in most European countries, where 220-V (220-

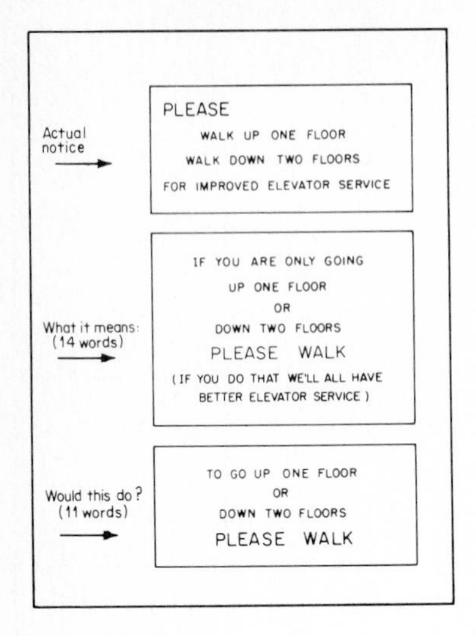

FIGURE 4-27
Examples of an elevator sign and of suggested revisions. (*Source: Adapted from Chapanis, 1967.*)

volt) electricity is common.] The correct phrasing of the instruction should have been "In the United Kingdom, this razor should be used only on the 220-V setting."

In reflecting about the relevance of language in instructions, in presenting information, and in other situations related to human factors, Broadbent makes the point that many statements can be expressed in an affirmative, a passive, or a negative manner, as illustrated below:

Affirmative: "The large lever controls the depth of the cut."
Passive: "The depth of the cut is controlled by the large lever."
Negative: "The small lever does not control the depth of the cut."

Broadbent points out that most linguistic research indicates that active, affirmative statements generally are easier to understand than passive or negative statements. However, he goes on to indicate that there are circumstances in which negative and passive statements may be easier to understand than active statements. Although we will not elaborate about such exceptions, as Broadbent puts it, it is usually best to word instructions and related informational material in an active and affirmative way, and to avoid negatives and passives. Yet it may be quite suitable to use a negative when one has to challenge some presupposition of the reader, and to use negatives or passives if they correspond more closely to the order which would be relevant to the reader. Above all Broadbent states that the "main topic" should come early in the sentence. (The "exceptions" in the case of the preparation of written material emphasize the point that, in this aspect of human factors as well as others, one should make recommendations which depend on the circumstances rather than sticking rigidly to simple rules or guidelines.)

Readability of Groupings of Letters and Numerals

We are frequently required to store in long-term memory sequences of several numerals or letters such as our own identification numbers, license numbers, and account

TABLE 4-7
GROUPINGS OF DIGITS USED IN STUDY OF TOUCH-TONE TELEPHONE

7 digits		5 digits	
Grouping	**Example**	**Grouping**	**Example**
3-4	736-2385	3-2	435-28
1-5-1	7-36238-5	None	43528
None	7362385		

Source: Klemmer and Stocker, 1974.

numbers, or to store in short-term memory such sequences as telephone numbers and numbers to be keypunched. Such long-term or short-term storage usually is facilitated when we slice up the entire sequence into groups. In the case of numerals, there is pretty persuasive evidence that the grouping of numerals into groups of two, three, or four tends to aid short-term memory. This was illustrated, for example, by the results of a study by Klemmer and Stocker (1974) in which subjects entered into a Touch-Tone telephone keyboard the numerals that were presented to them. In various aspects of the experiment the numerals were presented in different groupings, as shown in Table 4-7. The digits presented at any one time were randomly selected. In the experiment, each subject was first provided the opportunity to proceed at his or her own pace (a *self-paced* condition), and then performed under a *forced-paced* condition (in which the pace used was the average of the self-paced rate).

The results for the 7-digit presentations are given in Figure 4-28, showing self-paced times for the three groupings of digits and the percent of errors for these groupings. The superiority of the 3-4 grouping is evident in terms of both time and accuracy, over

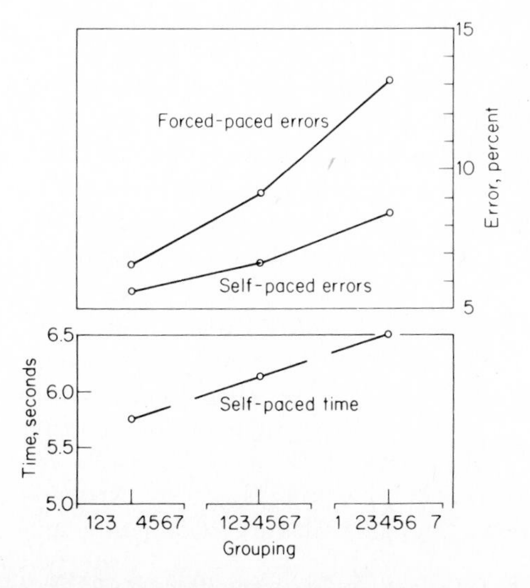

FIGURE 4-28
Time and errors in using a Touch-Tone telephone to enter seven digits presented in various groupings. (*Source: Klemmer & Stocker, 1974, fig. 2, p. 677. Copyright © 1974 by the American Psychological Association. Reprinted by permission.*)

the no-grouping condition and the 1-5-1 condition. In the case of the 5-digit presentations the 3-2 grouping was also found to be superior to the no-grouping condition.

The same general principle of grouping also applies in the case of letters used in combination with digits, as demonstrated by Hull (1976). In his study, the subjects were given brief presentations of slides of six characters, and then recorded what they recalled of the characters. In the phase of the study in which letters and digits were used in combination, the errors of recall were significantly fewer when the letters and the digits were in "patterns" of groups of two or three than when they were randomized. Some examples of the patterned and randomized materials are given below:

Patterned examples: WFC328; 769HRD; WF93RD; 76CH28
Random examples: Y1WJ15; CH9R2R; N3C6FD; HD7WC4

As an aside, Hull reported that letters with "high acoustic similarity" (that tend to sound somewhat alike, such as BCDEG) are less easily recalled than letters with low acoustic similarity (such as DFHNR).

The results of these and other investigations argue for the grouping of numerals, letters, or both into small groups (as two, or three, or four) when they are to be presented to people with the expectation of recall on the basis of short-term memory, or even long-term memory. In this present world of ours there are many such circumstances, as in typing and the use of other keyboards (such as computers), in using the telephone, in remembering license numbers, and in remembering our numerous numerical identifications (such as, in the United States, Social Security numbers).

VISUAL CODES AND SYMBOLS

Our present civilization abounds with a wide assortment of visual codes, symbols, and signs that are intended to convey meaning to us, their use being part of almost all phases of human activity, as: travel, business, medicine, the sciences, religion, engineering, and recreation. Actually the use of graphic symbols as a means of communication goes back into early human history; it became part of the folklore of most cultures. Dreyfus (1972), who developed a data bank of 20,000 graphic symbols used all over the world, considered such symbols as being of one of three types: (1) representational (fairly accurate, simplified pictures of objects, such as a skull and crossbones to represent danger, or of actions, such as a man on a bicycle representing a bicycle path); (2) abstract (symbols that reduce essential elements of a message to graphic terms, retaining only faint resemblance to the original concept, as the signs of the zodiac); and (3) arbitrary (symbols that have been invented and that then need to be learned, such as the triangle "yield" traffic sign).

Visual codes and symbols (which we will simply call *codes*) are used for various purposes, such as: to present road and highway information; to represent hazards; to identify and give directions regarding features and facilities of parks and recreation facilities; to present information on cathode-ray tubes (CRTs); to present information in printed material, such as computer printouts, maps, charts, manuals, newspapers, etc.; and to present information in various types of other displays, such as representational displays. The types of codes that would be appropriate for certain specific situations are probably strongly indicated by the circumstances, but in many circumstances there are options from which to choose.

Codes can be considered as being of one of three classes (single codes and two types of multidimensional codes), as follows:

Single codes (such as red, yellow, and green traffic lights that indicate "stop," "caution," and "go").

Redundant codes (codes that consist of two distinct methods of conveying the same information, such as a red, a yellow, and a green traffic light located, respectively, at the top, center, and bottom locations).

Compound codes (two or more coding systems used in combination, each system representing a different concept, such as a rectangular road sign with a telephone symbol within the rectangle; the rectangular form represents "information" of some type and the telephone represents the particular type of information).

Single Visual Codes

There are many specific types of visual codes, including colors; alphanumeric characters; geometric shapes; area size (of some shapes such as squares); angular orientation (as the hands of a watch); visual number; brightness of lights; flash rate of lights; and line type (as dotted versus solid). Some of the studies of visual codes have involved a comparison of the effectiveness of different types of codes for use in relation to various types of "tasks." One such study is that of Hitt (1961), who used the five different codes shown in Figure 4-29: numerals, letters, geometric shapes, configurations, and colors. The code symbols were used in maplike displays with eight columns and five rows. Eight symbols of each class were used to represent eight types of buildings, facilities, industries, etc. Each display included only one type of code, but under different conditions the number of different symbols of the type was varied (two, four,

FIGURE 4-29
Illustration of code symbols used in comparing coding of targets. (*Source: Hitt, 1961.*)

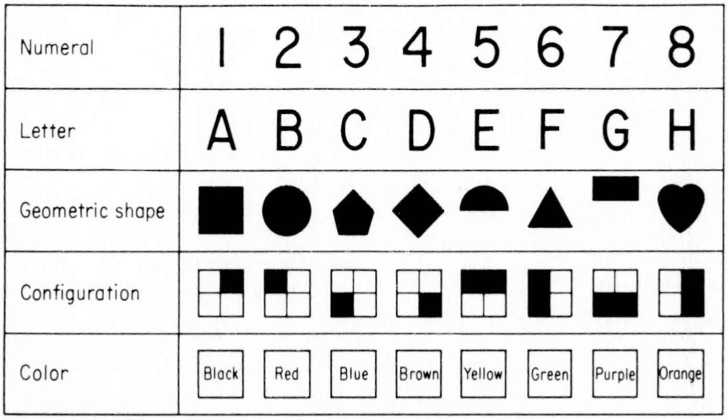

Numeral	1	2	3	4	5	6	7	8
Letter	A	B	C	D	E	F	G	H
Geometric shape								
Configuration								
Color	Black	Red	Blue	Brown	Yellow	Green	Purple	Orange

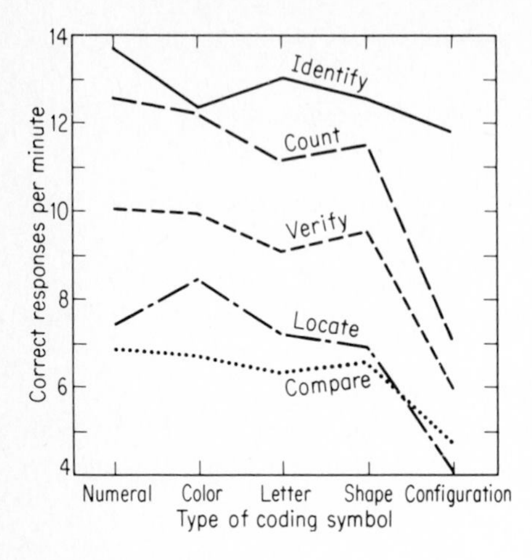

FIGURE 4-30
Relationship between coding method and performance in five tasks comparing targets on a display. (*Source: Adapted from Hitt, 1961.*)

and eight) and the density (the number of different symbols in a cell) was varied (one, two, and three).

The subjects performed five different tasks with these displays: *identification* (e.g., identify type of industry in one cell); *location* (e.g., locate cell that includes only one steel mill); *counting* (e.g., count number of aluminum plants in row C); *comparison* (e.g., compare number of petroleum refineries in one cell with that in another); and *verification* (e.g., the plant in a given cell is a steel mill—true or false). Their chore was to go down a list of questions of these types and record their answers. A summary of the major results is given in Figure 4-30, which shows, for the different codes and tasks, the number of correct responses per minute.

In general terms it can be seen that the numerical and color codes were best for most of the tasks and that the configuration code was consistently the last, with the letter and shape codes falling in between, in some instances comparing reasonably well with the numerical and color codes.

Let us look at one more study in which single visual codes were used, this being a study by Smith and Thomas (1964). They used four codes, namely, colors, military symbols, geometric forms, and aircraft shapes, as shown in Figure 4-31; each code had five symbols. For each of the three shape codes, displays were prepared with 20, 60, or 100 symbols of the type in question, these being randomly allocated to any of 400 positions in a 20 by 20 imaginary matrix. In various parts of the study three sets of displays were used (each set having separate displays for each of the three shape codes), as follows:

1 Sets with shape symbols colored randomly

2 Sets with shape symbols all the same color (but with different displays for each color)

3 Sets in which each of the five symbols of a shape class was coded a unique color, with different displays of each symbol-color combination

The task of the subjects was to count the number of items of a predesignated *target* class, such as *red, gun, circle,* or *B-52,* depending upon the set of displays used in the

Aircraft shapes	C-54	C-47	F-100	F-102	B-52
Geometric forms	Triangle ▲	Diamond ♦	Semicircle	Circle ●	Star ★
Military symbols	Radar	Gun	Aircraft	Missile	Ship
Colors (Munsell notation)	Green (2.5 G 5/8)	Blue (5 BG 4/5)	White (5 Y 8/4)	Red (5 R 4/9)	Yellow (10 YR 6/10)

FIGURE 4-31
Four sets of codes used in a study by Smith and Thomas (1964) (copyright © 1964 by the American Psychological Association and reproduced by permission). The notations under the color labels are the Munsell color matches of the colors used. (*Source:* Munsell book of color, *1959.*)

particular phase of the study. Both time and errors were recorded, the results being shown in Figure 4-32. As would be expected, time and errors increased with density (the number of items in a display); but more important, it was obvious that time and errors differed for the various types of codes, with color generally being the best.

Apart from the specific results of these two studies, it is obvious from such studies that the use of various types of visual coding systems can result in differential effectiveness in their use by people. But we should hasten to add the point that the utility

FIGURE 4-32
Mean time *a* and errors *b* in counting items of four classes of codes as a function of display density (copyright © 1964 by the American Psychological Association and reprinted by permission). The X's indicate comparison data for displays of 100 items with color (or shape) held constant. (*Source: Smith & Thomas, 1964.*)

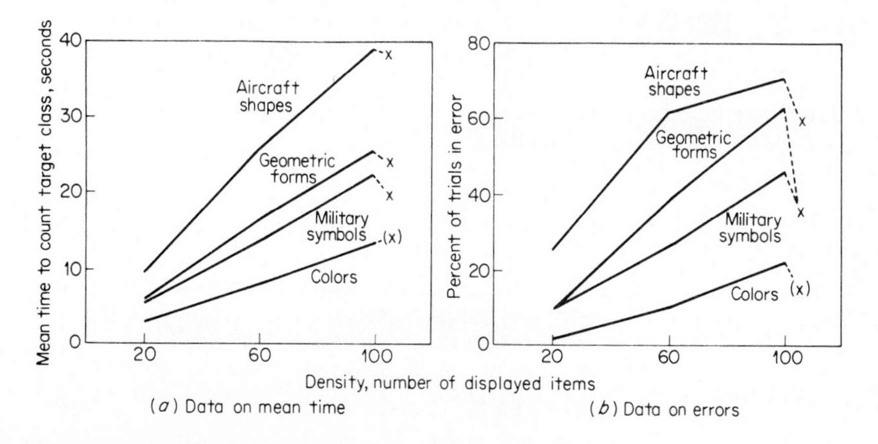

of a given code system can depend upon the context in which it is used–in particular, the nature of the task being performed.

In connection with the comparison of the effectiveness of various types of codes, Christ (1975) analyzed the results of 42 studies in which color codes were compared with other types of codes. Some of the results of his analyses are shown in Figure 4-33. In particular this shows, for two tasks (identification and search tasks), the relative effectiveness of color codes as compared with each of six other codes. Each bar shows the range of percent gain (+) or loss (−) in effectiveness of color codes as compared with another code for the various studies represented. (Note that the number of studies represented by the individual bars varies, in a couple of instances there being only one comparison study.) We can see that color codes were systematically superior to the other codes for the search task. In the case of an identification task, color codes are clearly superior to size, brightness, and "other" shapes, and are generally inferior to alphanumeric codes (letters and digits). The comparison with familiar geometric shapes is relatively ambiguous, although Christ (on the basis of other data) indicates that color codes are generally better for identification tasks than are such shapes.

FIGURE 4-33
Percent range of gain (+) or loss (—) in effectiveness of using color codes for identification and searching tasks as compared to certain other codes. The ranges of percents shown are the ranges reported for the various studies for which comparisons were made. (*Source: Adapted from Christ, 1975, table 1, p. 563.*)

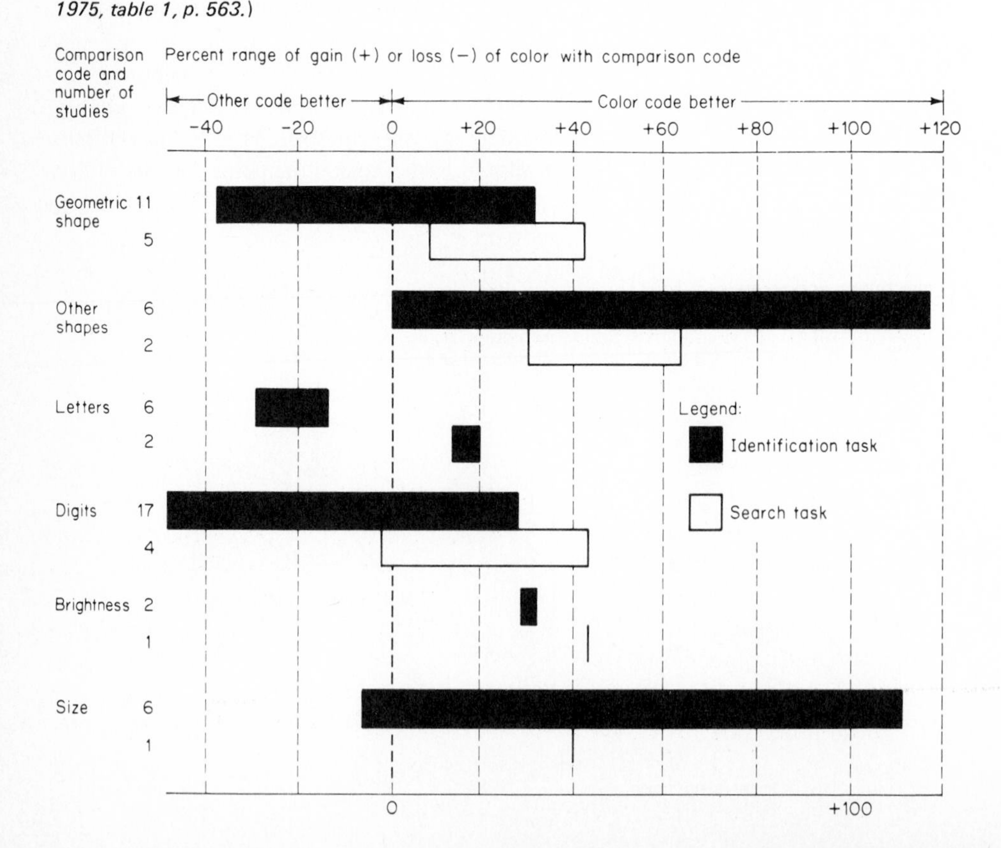

TABLE 4-8
SUMMARY OF CERTAIN VISUAL CODING METHODS
(Numbers refer to number of levels which can be discriminated on an absolute basis under optimum conditions)

Alphanumeric	Single numerals, 10; single letters, 26; combinations, unlimited. Good; especially useful for identification; uses little space if there is good contrast. Certain items easily confused with each other.
Color (of surfaces)	Hues, 9; hue, saturation, and brightness combinations, 24 or more. Preferable limit, 9. Particularly good for searching and counting tasks. Affected by some lights; problem with color-defective individuals.*†
Color (of lights)	10. Preferable limit, 3. Limited space required. Good for qualitative reading.‡
Geometric shapes	15 or more. Preferable limit, 5. Generally useful coding system, particularly in symbolic representation; good for CRTs. Shapes used together need to be discriminable; some sets of shapes more difficult to discriminate than others.‡
Angle of inclination	24. Preferable limit, 12. Generally satisfactory for special purposes such as indicating direction, angle, or position on round instruments like clocks, CRTs, etc.§
Size of forms (such as squares)	5 or 6. Preferable limit, 3. Takes considerable space. Use only when specifically appropriate.
Visual number	6. Preferable limit, 4. Use only when specifically appropriate, such as to represent numbers of items. Takes considerable space; may be confused with other symbols.
Brightness of lights	3–4. Preferable limit, 2. Use only when specifically appropriate. Weaker signals may be masked.‡
Flash rate of lights	Preferable limit, 2. Limited applicability if receiver needs to differentiate different flash rates. Flashing lights, however, have possible use in combination with controlled time intervals (as with lighthouse signals and naval communications) or to attract attention to specific areas.

* Feallock, Southard, Kobayashi, & Howell, 1966.
†M. R. Jones, 1962.
‡Grether and Baker, 1972.
§Muller, Sidorsky, Slivinske, Alluisi, & Fitts, 1955.

Although there are indeed indications that visual codes differ in their relevance for various tasks and in various situations, it must be recognized that definite guidelines regarding the use of such codes still cannot be laid down. Recognizing this, and realizing that good judgment must enter into the selection of visual codes for specific purposes, a comparison such as that given in Table 4-8 can serve as at least partial guidance. In particular, Table 4-8 indicates the approximate number of levels of each of various visual codes that can be discriminated, along with some sideline comments about certain of the methods.

Redundant Visual Codes

Redundant visual codes provide for using two or more different codes to represent each separate item that is to be coded. There is a general tendency to view such redun-

TABLE 4-9
REDUCTION IN TIMES FOR SEARCHING AND COUNTING COLOR-CODED ALPHANUMERIC CHARACTERS AS COMPARED WITH BLACK AND WHITE

	Reduction in mean time, %
Search task	45 to 70
Counting task	63 to 70

Source: Smith, 1963.

dancy as contributing to the accuracy and speed in receiving the intended information. However, there are inklings that the utility of such codes depends in part on the combinations of codes used and possibly on the task involved in using them. In the use of colors in combination with alphanumeric characters in searching and counting tasks, for example, Smith (1963) found the combination to be effective in reducing the times for both tasks with alphanumeric characters in fields of 20, 60, or 100 such items. The percents of *reduction* in mean times when the alphanumeric items were color-coded (as opposed to being black and white) are summarized in Table 4-9, these results showing reductions for both tasks, but particularly for the search task.

On the other hand, in the experimental use of combinations of color and shape as coding dimensions of road signs, S. Jones (1978) found that the use of color did not enhance the correct interpretation of such signs beyond that resulting from the use of shape codes by themselves. And Saenz and Riche (1974) also report that there was no systematic difference in the effectiveness of a redundant color and shape code as contrasted to a color code by itself (although they did find certain color-by-shape interactions).

It is not feasible at this point in time to specify the particular combinations of codes that can be used redundantly for enhancing performance on specific tasks. But aside from saying that redundant codes can be useful in certain (as yet not clearly delineated) tasks, it can also be said that there is no evidence that such coding adversely affects performance, thus suggesting the notion that—barring cost or other factors—the use of redundant coding is on the conservative, "safe" side of the decision fence.

Compound Visual Codes

In compound coding, two or more separate, completely independent, code systems are used in combination. Thus, in an inventory system one might use shape codes to identify different classes of items, and color to identify the month of procurement. Heglin (1973) makes the point that in the use of compound codes no more than two should be used together where rapid interpretation of the display is required. However, certain combinations of code systems either do not "go well" together, or simply cannot be used together. The combinations that Heglin considers to be potentially useful are given in Figure 4-34.

	Color	Numeral and letter	Shape	Size	Brightness	Location	Flash rate	Line length	Angular orientation
Color		X	X	X	X	X	X	X	X
Numeral and letter	X			X		X	X		
Shape	X			X	X		X		
Size	X	X	X		X		X		
Brightness	X		X	X					
Location	X	X						X	X
Flash rate	X	X	X	X					X
Line length	X					X			X
Angular orientation	X					X	X	X	

FIGURE 4-34
Potential combinations of coding systems for use in compound coding. (*Source: Adapted from Heglin, 1973, table VI-6, p. VI-22.*)

Color Coding

Since color is a fairly common visual code, we will discuss it somewhat further. An important question relating to color deals with the number of distinct colors persons with normal color vision can differentiate on an absolute basis. It has generally been presumed that the number was relatively moderate; for example, M. R. Jones (1962) indicated that the normal observer could identify about nine surface colors, varying primarily in hue. However, it appears that, with training, people can learn to make upwards of a couple of dozen such discriminations when combinations of hue, saturation, and lightness are prepared in a nonredundant manner (Feallock, Southard, Kobayashi, & Howell, 1966). On the basis of this study, 24 colors were so chosen that only two of those within the set were confused by subjects with others in the set under four experimental lighting conditions.[3]

When relatively untrained people are to use color codes, however, the better part of wisdom would argue for the use of smaller numbers. In this regard Conover and Kraft (1958) present four sets of colors for coding purposes when absolute recognition is required (barring the use of color-deficient people). These sets include, respectively, eight, seven, six, and five colors.[4]

[3] The identifications of these colors in the new Federal Standard on Colors follow, with an (X) after those that were identifiable under all lighting conditions: 32648 (X), 31433, 30206 (X), 30219, 30257, 30111, 31136 (X), 31158, 32169, 32246, 32356, 33538 (X), 33434, 33695, 34552 (X), 34558, 34325 (X), 34258 (X), 34127, 34108 (X), 35189, 35231 (X), 35109, and 37144 (X).

[4] The Munsell (1959) notations for the colors in these four sets are given below:
8-color: 1R; 9R; 1Y; 7GY; 9G; 5B; 1P; 3RP
7-color: 5R; 3YR; 5Y; 1G; 7BG; 7PB; 3RP
6-color: 1R; 3YR; 9Y; 5G; 5B; 9P
5-color: 1R; 7YR; 7GY; 1B; 5P

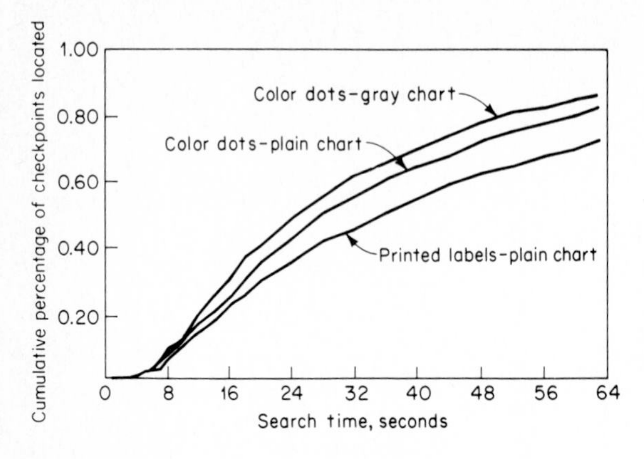

FIGURE 4-35
Search time for various checkpoints (e.g., airfields, bridges, etc.) on aeronautical charts with checkpoints identified by color codes and by printed labels. (*Source: Adapted from Shontz, Trumm, & Williams, 1971, fig. 2.*)

There are not very many work situations in which people have to visually "search" for certain specified items in extensive arrays of many other items. Some examples might include the use of aeronautical and navigational charts and maps, and conceivably certain inspection types of tasks. In such situations color seems to be a particularly useful coding system. This was illustrated, for example, by the results of a study by Shontz, Trumm, and Williams (1971) in which subjects were to locate (i.e., to search for) specific types of 28 checkpoints (e.g., airfields, bridges, etc.) on aeronautical charts. The three experimental conditions used were:

1 Printed labels of checkpoints, plain chart
2 Color dots (3/16 in) to represent checkpoints, plain chart
3 Color dots (3/16 in) to represent checkpoints, gray (achromatic) chart

The results, given in Figure 4-35, show the cumulative percentage of checkpoints located over the experimental period. The figure indicates a systematic difference in favor of the color-coded conditions (2 and 3).

In those circumstances in which such visual search activities are involved, however, certain display features can influence the time required for searching out the relevant stimuli–such features as the density of stimuli (the total number of stimuli from which the relevant ones must be identified), the code size (the number of colors used), and the number of items of each category or type in the display. In this regard Carter and Cahill (1979) present evidence that search times generally increase as density increases, and decrease when color coding is initiated with limited numbers of colors (the *code size*). Search times tend to decrease somewhat further with increases in number of color codes used, up to a point, beyond which increases in the number of colors used tend to result in increased visual search times. These results are illustrated in Figure 4-36, although it should be added that there presumably are some interactions that tend to complicate the combined effects of the different features of the display. In general terms it seems evident that the use of moderate numbers of colors in such visual search circumstances is useful, but that very limited numbers and large numbers of such codes tend to increase search times.

In reflecting about the use of color codes, it appears that they have their greatest utility in circumstances in which there is some type of searching, scanning, or locating

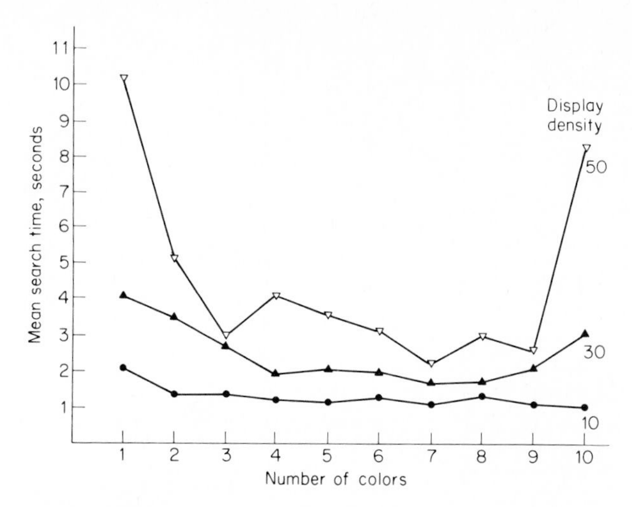

FIGURE 4-36
Mean search times for identifying relevant stimuli in two-dimensional displays of various combinations of display density (10, 30, and 50 items) and color code sizes (1 to 10). (*Source: Carter & Cahill, 1979, fig. 1, p. 295.*)

task, presumably because they "catch the eye" more rapidly than most other visual codes. It can also be added that Christ (1975) provides evidence that the use of color in redundant codes has generally been found to be useful (although, as discussed above, there are at least certain exceptions to this).

The Design of Symbolic Codes

The discussion above has indicated that symbolic codes can be very useful as a means of transmitting certain kinds of information. In this regard there is no particular problem in designing symbols that represent certain specific physical objects (as—in the case of road signs—pedestrians, bicycles, or deer). However, some physical objects are more difficult to depict, and certain "functions" and concepts are even more difficult to depict. When any question arises as to whether a particular symbol would, or would not, be suitable for representing what it is supposed to represent, some procedure for evaluating its appropriateness is in order. Several schemes have been used for this purpose, and we will mention a few of these.

Direct Interpretation With this procedure various experimental symbols are shown, one at a time, to a group of subjects who are then asked to write down or state what they think each symbol represents. This procedure was used by Cahill (1975), for example, in evaluating the interpretability of 10 graphic symbols that were intended to represent various parts or functions involved in the operation of farm and industrial machinery. The symbols are shown in Figure 4-37. (Reference to the numbers under them will be made later.) One group of subjects was shown the symbols after having been given these instructions:

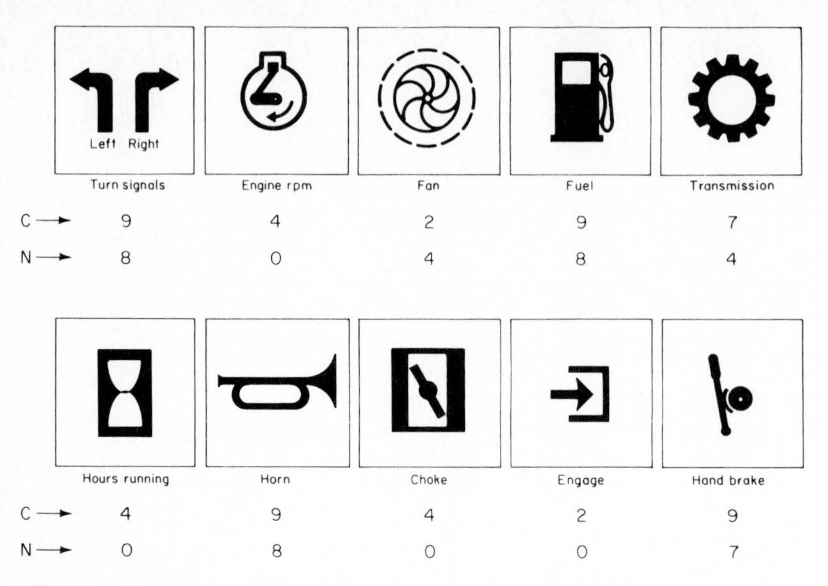

FIGURE 4-37
Symbols representing parts or functions of farm and industrial machinery used in study by Cahill. Numbers below represent numbers of subjects who made correct interpretations. C = "in context" (9 subjects); N = "not in context" (8 subjects). (*Source: Adapted from Cahill, 1975, fig. 1, p. 377. Copyright © 1975 by the American Psychological Association. Reprinted by permission.*)

> You will be shown 10 graphic symbols intended for use in labeling parts or functions of heavy automotive machinery, such as a bulldozer. Your task is to interpret the meaning of each symbol. As each symbol is shown write down the automotive part or function you think it represents.

This was called the "no context" group.

A second group of subjects was given these same instructions *plus* a drawing of a cab interior with numbers on the instruments and controls that they were to match up with the symbols presented to them. This was called the "context" group because they had the diagram of the instrument panel with which to relate the symbols.

The subjects were further differentiated in terms of whether they had or had not had experience of some sort with heavy equipment. Those with such experience did better in both the "context" and "no context" phases than did those without experience, but even so several of the symbols were not systematically recognized. The numbers under the symbols in Figure 4-37 show the number of "experienced" individuals who gave correct identifications in the four conditions. It is evident that only 4 or 5 of the 10 symbols fared very well (even with experienced individuals). This emphasizes the importance of empirical performance testing of symbols if there is any question about their appropriateness.

Confusion Matrix Another procedure that is sometimes used is that of showing several symbols to subjects, with their identifications, and then presenting the symbols one at a time, asking the subjects to write down or state the symbol they think the one

Symbol presented	Response							
	1	2	3	4	5	6	7	Other (8 – 19)
1. Seat belt	50							0
2. Charging circuit		44	2					4
3. Choke			28			12	2	8
4. Horn				49				1
5. Fuel					36		3	11
6. Coolant temperature			4			34	2	10
7. Engine oil			2			5	39	4

FIGURE 4-38
Portion of a confusion matrix showing number of individuals giving specific responses when presented with each of seven symbols. (*Source: Adapted from Green & Pew, 1978, table 2, p. 110.*)

presented is. The results of such a presentation are summarized in a confusion matrix that shows, for each symbol, the frequency of correct responses along with the frequency with which each was "confused" with each other. In one study in which this technique was used, Green and Pew (1978) used 19 symbols representing vehicular controls and displays (some of them being similar to those used by Cahill and shown in Figure 4-37). The 50 subjects first learned the meaning of the 19 symbols. Later, as one phase of the study, the experimenter read the label of one of the symbols and then showed a slide of one of the symbols. The subject was asked to press one of two thumb buttons to indicate whether the label and symbol were the same or different. The results were presented in the form of a confusion matrix as illustrated in Figure 4-38. Figure 4-38 shows the responses for only 7 of the 19 symbols, with the responses for the remaining 12 symbols being grouped under an "other" category. The responses for only one of the symbols (the seat belt) were perfect, there being varying degrees of "confusion" for the others. It is obvious that this technique can aid in identifying the symbols that are most consistently associated with their intended meaning.

Reaction Time Various investigators have used reaction time as a measure of the meaningfulness of traffic signs. As an example of the use of reaction time, Dewar, Ells, and Mundy (1976) presented subjects with individual slides of three types of traffic signs, namely, warning, regulatory, and information. The subjects were to respond "yes" as quickly as possible whenever a warning or regulatory sign was presented, but were to make no response to the information sign. This was a "classification" task. Under other experimental conditions they were to respond as rapidly as possible when they identified a warning or regulatory sign. This was an "identification" task. The results of this study corresponded reasonably well with measures of on-the-road legibility distances of the signs used, thus suggesting that a reaction-time measure derived in a laboratory setting may have reasonable relevance for deriving estimates of on-the-road legibility of signs. However, some words of caution are in order regarding the use of reaction time as a measure of the meaningfulness of symbols. Green and Pew used reaction time as one of their measures in the study referred to above, and concluded that it was influenced by individual differences and that it decreases significantly with learning. Although it has been presumed that reaction time to symbols is generally correlated with *associative strength* (the degree of conceptual association of symbols with their meaning), their results did not support that assumption.

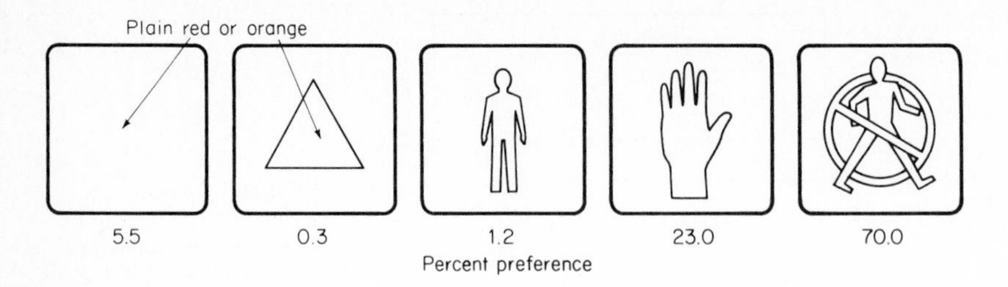

FIGURE 4-39
Percent of 330 pedestrians who responded that the symbol most clearly means *DON'T WALK*. All symbols were shown in orange and red, but with the first two only the color serves as the stimulus since there is no configuration involved. (*Source: Adapted from Robertson, 1977, fig. 3 and table 4, p. 40.*)

Expressed Preferences and Opinions Still another procedure that is used in evaluating symbolic displays is asking people for their preferences or opinions about the symbols in question. This procedure was used by Robertson (1977), for example, in the evaluation of street crossing lights for pedestrians. In one phase of his investigation 330 pedestrians in 10 cities were asked their opinions about the five designs of such signs shown in Figure 4-39. They were asked this question: Which of these symbols most clearly means DON'T WALK to you? The percents of responses for the five symbols are also shown in Figure 4-39. Rather clearly, the fifth symbol, the circle with a slash, was preferred by most pedestrians. In connection with other questions regarding color, red was (as expected) clearly preferred over orange to represent DON'T WALK, and green over white to represent WALK.

Other Procedures Certain other procedures and criteria have also been used in the evaluation of the adequacy of symbols in transmitting meaningful information to people, but they will not be discussed. It is relevant to emphasize the point that if there is any question about the relevance of symbols to be used, some evaluation of possible symbols should be carried out.

Perceptual Principles of Symbolic Design In discussing the use of coding symbols, Easterby (1967, 1970) makes the point that the effective use of such displays is predicated on perceptual processes, and postulates a number of principles that are rooted in perceptual research that generally would enhance the use of such displays. Certain of these principles will be summarized briefly and illustrated. The illustrations are in Figure 4-36. Although these particular examples are specifically applicable to machine displays (Easterby, 1970), the basic principles would be applicable in other contexts.

Figure/ground Clear and stable figure-to-ground articulation is essential, as illustrated in Figure 4-40*a*.

Figure boundaries A contrast boundary (essentially a solid shape) is preferable to a line boundary, as shown in Figure 4-40*b*. With different elements of a display to be depicted, the following practice should be followed: symbol dynamic–solid; moving or active part–outline; stationary or active part–solid.

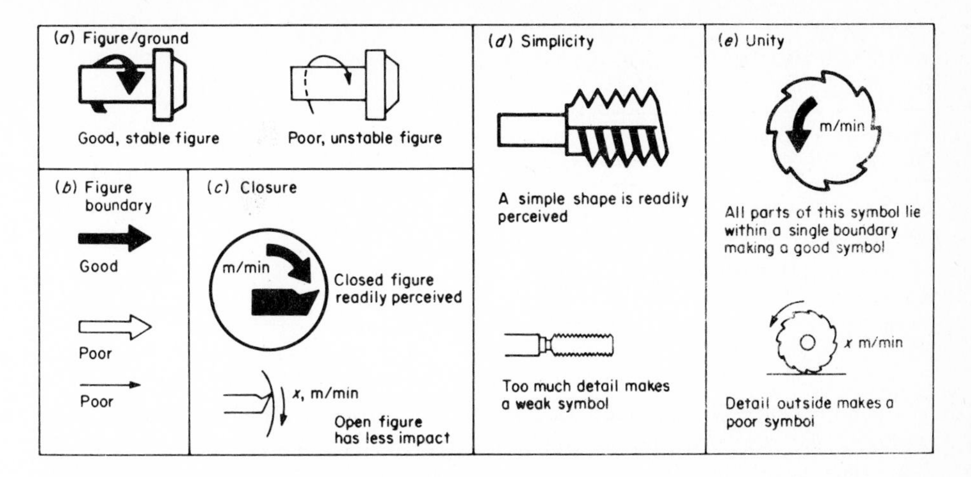

FIGURE 4-40
Examples of certain perceptual principles relevant to the design of visual code symbols. These particular examples relate to codes used with machines. (*Source: Adapted from Easterby, 1970.*)

Closure A closed figure as illustrated in Figure 4-40*c* enhances the perceptual process and should be used unless there is reason for the outline to be discontinuous.

Simplicity The symbols should be as simple as possible, consistent with the inclusion of features that are necessary, as illustrated in Figure 4-40*d*.

Unity Symbols should be as unified as possible. For example, when solid and outline figures occur together, the solid figure should be within the line outline figure, as shown in Figure 4-40*e*.

ROAD AND TRAFFIC SIGNS

Since road and traffic signs probably are the most common signs most of us are exposed to, and since more research has been carried out with such signs than with other types, we will include reference to a few such studies as examples of research in sign design.

Comparison of Symbolic and Verbal Signs

Of the various experiments that have been carried out with road signs, a number have dealt with comparative effectiveness of verbal and symbolic forms of presentation. In one such study by Ells and Dewar (1979), for example, subjects listened to a traffic sign message (such as "two-way traffic") and were then presented with a visual slide of a traffic sign. In the case of some trials verbal messages were presented on the slides, and in other trials symbolic representations were presented. The subjects were asked to say "yes" if the sign on the slide corresponded with the sign message that had been read orally to them, and to answer "no" if the two were different. A comparison of the mean reaction times for the two types of messages revealed a systematic superiority for the symbolic messages, as given in Table 4-10.

TABLE 4-10
REACTION TIME IN RESPONDING TO PRESENTATION OF SYMBOLIC AND VERBAL TRAFFIC SIGNS

Type of message	Type of sign	Reaction time (in s)	
		"Yes" response	"No" response
Symbolic	Warning	0.57	0.60
Verbal	Warning	0.64	0.70
Symbolic	Regulatory	0.59	0.68
Verbal	Regulatory	0.68	0.73

Source: Ells and Dewar, 1979.

In another phase of the study a similar comparison was made with signs under *nondegraded* and *degraded* viewing conditions, the results being summarized in Figure 4-41. Here, again, the advantage is in favor of the use of symbolic messages, especially under degraded viewing conditions.

This and other studies generally argue for the use of symbolic codes on road signs under most conditions, the primary explanation for such coding being that symbols tend to minimize the *recoding* of the stimuli presented. A second general principle in the design of road signs deals with the matter of standardization, at least across the geographical boundaries that are commonly traversed. The international road signs generally conform to both of these principles. Certain examples are shown in Figure 4-42. The international sign system has much higher marks on these principles than do the sign systems of the states of the United States, although the United States is moving in the direction of such principles.

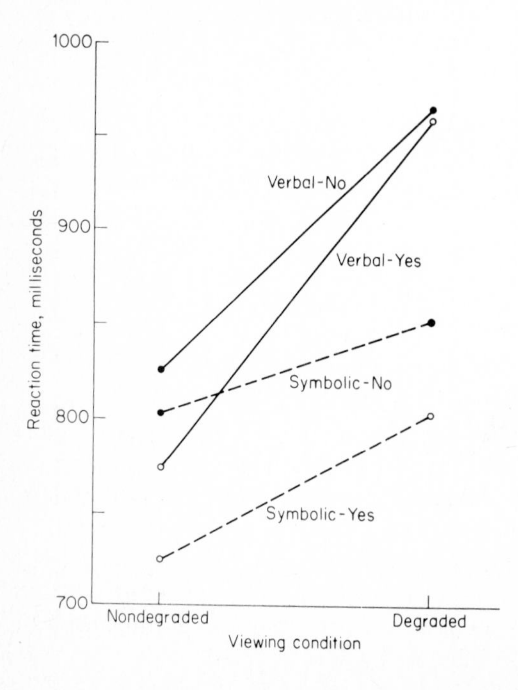

FIGURE 4-41
Mean reaction times of "yes" and "no" responses to symbolic and verbal traffic signs viewed under nondegraded and degraded viewing conditions. (*Source: Ells & Dewar, 1979, fig. 1, p. 168.*)

(*a*) Danger signs (*b*) Instruction signs (*c*) Information signs

FIGURE 4-42
Examples of a few International Road Signs. These signs conform to two useful design principles, namely, they are directly symbolic of their meaning, and they are standardized across countries.

Actual Use of Road Signs

Even the best-designed signs in the world, however, serve no purpose if they are not used. And there is (unfortunately) evidence that road signs frequently are not used by drivers. On the basis of a field study in Sweden, Johansson and Backland (1970) came to the dismal conclusion that " . . . the road sign system to a high degree does not achieve its purpose." Their study was carried out by placing any one of six signs (shown in Figure 4-43) on a main highway, each in such a position that drivers would have a clear view of it about 400 meters (400 m) away. Beyond the sign, around a slight rise and curve, was a police barrier, out of sight from the position where the sign could be viewed. The drivers were stopped by police at the barrier and asked certain questions, including: What was the last road sign you passed? The percentages of 2525 drivers who responded correctly to this question are given in Figure 4-43, these values ranging from 26 percent (sign *d*) to 76 percent (sign *a*). (These values can be considered as reflecting the human "reliability" associated with the signs.) Interestingly enough, there is a marked difference in the recall of the various signs despite the fact that they were very similar from the point of view of perceptual impressiveness. These differences suggest some sort of differential "selectivity" in giving attention to the signs, perhaps in terms of their potential relevance to the drivers. As an addend, it should be noted that a comparison of the responses of four different groups of drivers

FIGURE 4-43
Road signs used in a field study on a main highway in Sweden, along with the percent of drivers who recalled having seen each sign about a minute after passing it. (*Source: Adapted from Johansson & Backland, 1970.*)

(*a*)	(*b*)	(*c*)	(*d*)	(*e*)	(*f*)
50 300 m	Poliskontroll	Tjälskador 1km		300 m	
50 km/hr limit	Police control	Break in road	Other danger	Pedestrian	Wild animals
76 %	66 %	29 %	26 %	62 %	55 %

showed that those drivers who knew in advance of the experiment did much better (as one would expect) than those drivers who had known nothing about it.

The fact that many drivers presumably pass relevant road signs by with nary a glance is of course very disturbing. But it may be that with further research designs can be developed that will "get through" to drivers. A modest bit of encouragement on this score comes from an investigation by Hanscom (1975) of the effects of various designs and arrangements of icy bridge warning signs in reducing the speed of drivers. Figure 4-44 shows eight combinations of "advance location" and "at bridge" signs used in a study in West Virginia, along with the average reduction in speed associated with each of the eight combinations, under both dark and daylight conditions. Reductions in speed ranged from 6.0 mi/h to 0.9 mi/h for the various conditions, with reductions being generally greater in the dark (except for the eighth condition with no advance location sign). The use of alternately flashing beacons in the dark (conditions 1 and 3) seems to be useful, but the problem of activating such beacons would of course be an argument against their use.

But even the greatest reductions in speed are not overly impressive. Thus, the general problem of modifying the driving behavior of people must be retained on the human factors agenda as a piece of unfinished business.

DISPLAYS FOR SPECIAL SITUATIONS

There are of course scads of situations in which displays of one kind or another are (or could be) used. Although many display design problems can be resolved by the application of data or guidelines that have been based on other investigations, in some circumstances special attention must be given to the design problems, including in some instances the carrying out of experiments aimed at providing relevant data for the problem at hand. It is, of course, not feasible to discuss very many special display problems, but we will illustrate such problems by brief reference to the matter of vehicle conspicuity in traffic.

Vehicle Conspicuity

It stands to reason that the more adequately drivers are able to identify the presence and movements of vehicles in their traffic area, the more likely they are to avoid accidents. One of the most important factors influencing the detectability of an object in one's environment is the contrast between the object and the background. Although the colors of vehicles as contrasted with their backgrounds can be useful in this regard, the use of lights on vehicles can also provide desired contrast. In fact, Hörberg and Rumar (1979) cite evidence that the conspicuity of any car with low-beam running lights is equal or superior to the conspicuity of a car having the best color contrast for that background. Some years ago in the United States there was a flurry of interest in the use of a single front running light on automobiles, but this flurry (along with the sale of such lights) died out after a year or two. However, the demise of that practice may have been premature. Hörberg and Rumar summarize the results of a few studies that indicate that the use of daytime running lights raises the probability of detection

	Advance location	At bridge	Ambient condition	Speed reduction: Bridge entry	Speed reduction: On bridge
1	WATCH FOR ICE ON BRIDGE	ICE ON BRIDGE WHEN FLASHING	Dark	6.0*	5.9*
2	BRIDGE ICY AHEAD	WATCH FOR ICE	Dark	4.5*	4.7*
3	No sign	ICE ON BRIDGE WHEN FLASHING	Dark	4.4*	4.5*
4	BRIDGE ICY AHEAD	No sign	Daylight	4.2*	3.6*
5	WATCH FOR ICE ON BRIDGE	No sign	Daylight	3.0*	2.6*
6	BRIDGE ICY AHEAD	ICE ON BRIDGE WHEN FLASHING	Daylight	2.4*	2.4*
7	WATCH FOR ICE ON BRIDGE	WATCH FOR ICE	Daylight	0.9	1.5
8	No sign	WATCH FOR ICE	Dark	0.9	1.3

FIGURE 4-44
Eight combinations of "advance location" and "at bridge" warning signs and mean speed reductions associated with each under dark and daylight conditions. (*Source: Hanscom, 1975, fig. 1, p. 19.*)

of vehicles by drivers and a reduction of accidents, and report that because of such evidence daylight running lights are compulsory in Sweden and, to some extent, in Finland and Norway. (In these countries low-beam headlights are used as the running lights, instead of having "special" running lights.) Although vehicle conspicuity is affected by certain related variables such as viewing angle and intensity, we will not report the details of such effects. Rather, we want to emphasize the basic point that the conspicuity of vehicles is enhanced by the use of running lights, especially during the twilight hours, and that their use seems to contribute to the reduction of accidents.

As an illustration of another type of vehicular light that seems to offer promise of being useful in accident reduction, Voevodsky (1974) rigged up deceleration warning lights on 343 San Francisco taxicabs. Each light was mounted on the rear of the vehicle at the same height as existing stoplights. As the cab was caused to decelerate by the use of the brake, the warning light was caused to flash, the flash rate increasing exponentially with the rate of deceleration, meaning that the greater the deceleration the faster the flash rate. A comparison was made of the rear-end accidents in which these cabs were involved over a 10-month period (in which they traveled a total of 12.3 million miles) with another sample of cabs that were not so equipped. A summary of these accident data is given in Table 4-11.

It should be added that the drivers who were assigned the cabs with the warning lights were not any better drivers than the others, since there was no significant difference in other types of accidents between the two groups. In other words, the differences presumably were attributed to the fact that the drivers of the "following" cars responded more adequately to the warning lights than in the case of cabs without them. Voevodsky points out that the National Safety Council indicates that rear-end accidents account for about 2000 deaths per year, and states that if such warning lights would reduce rear-end accidents by about 60 percent (as in this study), there might be a saving of 1200 lives per year.

As Mortimer (1976) points out, there are many other aspects of vehicle lighting that offer promise of reducing vehicle accidents, but these examples will serve to point up the fact that the use of appropriately designed vehicle lights can aid in driving safety by providing relevant information to drivers about the presence and movements of other vehicles.

TABLE 4-11

ACCIDENT RATES IN MILLIONS OF MILES FOR TAXICABS WITH AND WITHOUT DECELERATION WARNING LIGHTS

Criterion	Cabs with warning lights	Cabs without warning lights
Rear-end collisions	3.51	8.91
Drivers injured	0.65	1.67
Cab damage ($)	398	1041

Source: Voevodsky, 1974.

GENERAL GUIDELINES IN DESIGNING VISUAL DISPLAYS

In this chapter we have discussed and illustrated a variety of visual displays. The sample of displays covered, however, is only suggestive of the wide range of visual displays that are in actual use. In the selection or design of visual displays for certain specific purposes, the basic type of display to use is sometimes virtually dictated by the nature of the information to be presented and the use to which it is to be put. In other circumstances, however, options may be available.

Although it is not possible to provide specific guidelines to follow in resolving every design problem, there are certain general guidelines and principles that can be followed in many situations. A few are given below, along with some words of caution where circumstances sometimes will justify deviations from these. These guidelines deal largely with some of the more conventionally used visual displays.

Quantitative Scales

- Digital or open-window displays are preferable if values remain long enough to read.
- Fixed-scale, moving-pointer designs are usually preferable to moving-scale, fixed-pointer designs.
- For long scales, a moving scale with tape on spools behind a panel, or a counter plus a circular scale, have practical advantages over a fixed scale.
- For values subject to continuous change, display all (or most) of the range used (as with circular or horizontal scale).
- If two or more items of *related* information are to be presented, consider integrated display.
- The smallest scale unit to be read should be represented on the scale by about 0.05 in or more.
- Preferably use a marker for each scale unit, unless the scale has to be very small.
- Use conventional progression systems of 1, 2, 3, 4, etc., unless there is reason to do otherwise, with major markers at 0, 10, 20, etc.

Qualitative Scales

- Preferably use a fixed scale with a moving pointer (to show trends).
- For groups, use circular scales, and arrange null positions systematically for ease of visual scanning, as at 9 o'clock or 12 o'clock positions.
- Preferably use extended pointers, and possibly extended lines between scales.

Status Indicators

- If basic data represent discrete, independent categories, or if basically quantitative data are always used in terms of such categories, use a display that represents each.

Signal and Warning Lights

- Minimum size used must be consistent with luminance and exposure time.
- With low signal-to-background contrast, red light is more visible.
- Flash rate of flashing lights of 1 to 10 per second presumably can be detected by people.

Representational Displays

- A moving element (such as an aircraft) should be depicted against a fixed background (as the horizon).
- Graphic displays that depict trends are read better if they are formed with lines than with bars.
- Pursuit displays usually are easier for people to use than compensatory displays.
- Cathode-ray-tube (CRT) displays are most effective when there are seven to nine or more scan lines per mm.
- In the design of displays of complex configurations (such as traffic routes and wiring diagrams), avoid unnecessary detail and use schematic representation if consistent with uses.

Alphanumeric Displays

- The typography of alphanumeric characters (design, size, contrast, etc.) is especially critical under adverse viewing conditions.
- Alphanumeric characters should be presented in groups of three or four for optimum short-term memory.
- Capital letters and numerals used in visual displays are read most accurately (*a*) when the ratio of stroke width to height is about 1:6 to 1:8 for black on white and somewhat higher (up to 1:10) for white on black, and (*b*) when the width is at least two-thirds the height.

Symbolic Displays

- Symbolic displays should be designed on the basis of the following perceptual principles: figure/ground; figure boundaries; closure; simplicity; and unity. In case the symbols do not clearly represent what they are supposed to represent they should be evaluated experimentally.

REFERENCES

Applied ergonomics handbook. Guilford, Surrey, England: IPC Science and Technology Press, Ltd., 1974.

Beringer, D. B., Williges, R. C., & Roscoe, N. The transition of experienced pilots to a frequency-separated aircraft attitude display. *Human Factors,* 1975, *17*(4), 401–414.

Broadbent, D. E. Language and ergonomics. *Ergonomics,* 1977, *8*(1), 15–18.

Burg, A. Visual acuity as measured by static and dynamic tests: A comparative evaluation. *Journal of Applied Psychology,* 1966, *50*(6), 460–466.

Cahill, M. C. Interpretability of graphic symbols as a function of context and experience factors. *Journal of Applied Psychology,* 1975, *60*(3), 376–380.

Carter, R. C., & Cahill, M. C. Regression models of search time for color-coded information displays. *Human Factors,* 1979, *21*(3), 293–302.

Chapanis, A. Words, words, words. *Human Factors,* February 1967, *7*(1), 1–17.

Christ, R. E. Review and analysis of color coding research for visual displays. *Human Factors,* 1975, *17*(6), 542–570.

Cohen, E., & Follert, R. L. Accuracy of interpolation between scale graduations. *Human Factors,* 1970, *12*(5), 481–483.

Conover, D. W., & Kraft, C. L. *The use of color in coding displays.* USAF, WADC, TR 55-471, October 1958.

Cornog, D. Y., & Rose, F. C. *Legibility of alphanumeric characters and other symbols: II. A reference handbook* (National Bureau of Standards, Miscellaneous 262-2). Washington, D. C.: U.S. Government Printing Office, February 1967.

Crawford, A. The perception of light signals: the effect of mixing flashing and steady irrelevant lights. *Ergonomics,* 1963, *6*, 287–294.

Dashevsky, S. G. Check-reading accuracy as a function of pointer alignment, patterning, and viewing angle. *Journal of Applied Psychology,* 1964, *48,* 344–347.

Dewar, R. E., Ells, J. G., & Mundy, G. Reaction time as an index of traffic sign perception. *Human Factors,* 1976, *18*(4), 381–392.

Dreyfus, H. *Symbol sourcebook,* New York: McGraw-Hill, 1972.

Easterby, R. S. Perceptual organization in static displays for man/machine systems. *Ergonomics,* 1967, *10*, 195–205.

Easterby, R. S. The perception of symbols for machine displays. *Ergonomics,* 1970, *13*(1), 149–158.

Elkin, E. H. *Effect of scale shape, exposure time and display complexity on scale reading efficiency* (TR 58-472). U.S. Air Force, WADC, Wright-Patterson Air Force Base, Ohio, February 1959.

Ells, J. G., & Dewar, R. E. Rapid comprehension of verbal and symbolic traffic sign messages. *Human Factors,* 1979, *21*(2), 161–168.

Feallock, J. B., Southard, J. F., Kobayashi, M., & Howell, W. C. Absolute judgments of colors in the Federal Standards System. *Journal of Applied Psychology,* 1966, *50*, 266–272.

Foster, J., and Coles, P. An experimental study of typographic cueing in printed text. *Ergonomics,* 1977, *20*(1), 57–66.

Green, P., and Pew, R. W. Evaluating pictographic symbols: An automotive application. *Human Factors,* 1978, *20*(1), 103–114.

Grether, W. F. Instrument reading: I. The design of long-scale indicators for speed and accuracy of quantitative readings. *Journal of Applied Psychology,* 1949, *33,* 363–372.

Grether, W. F., & Baker, C. A. Visual presentation of information. In H. A. Van Cott & R. G. Kinkade (Eds.), *Human engineering guide to equipment design* (Rev. ed.). Washington, D. C.: U.S. Government Printing Office, 1972.

Hanscom, F. R. An evaluation of icy bridge warning signs. *Traffic Engineering,* September 1975, *45*(5), 17–20.

Heglin, H. J. *NAVSHIPS Display Illumination Design Guide: II. Human Factors* (NELC/TD223). San Diego, Calif.: Naval Electronics Laboratory Center, July 1973.

Hemingway, J. C., & Erickson, R. A. Relative effects of raster scan lines and image subtense on symbol legibility on television. *Human Factors,* 1969, *11*(4), 331–338.

Hitt, W. D. An evaluation of five different coding methods. *Human Factors,* July 1961, *3*(2), 120–130.

Hörberg, U., & Rumar, K. The effect of running lights on vehicle conspicuity in daylight and twilight. *Ergonomics,* 1979, *22*(2), 165–173.

Hull, A. J. Human performance with homogeneous, patterned, and random alphanumeric displays. *Ergonomics,* 1976, *79*(6), 741–750.

Johansson, G., & Bäckland, F. Drivers and road signs. *Ergonomics,* 1970, *13*(6), 741–759.

Johnson, S. L., & Roscoe, S. N. What moves, the airplane or the world? *Human Factors*, 1972, *14*(2), 107-129.

Jones, M. R. Color coding. *Human Factors*, 1962, *4*, 355-365.

Jones, S. Symbolic representation of abstract concepts. *Ergonomics*, 1978, *21*(4), 573-577.

Klemmer, E. T., & Stocker, L. P. Effects of grouping of printed digits on forced-paced manual entry performance. *Journal of Applied Psychology*, 1974, *59*(6), 675-678.

Kurke, M. I. Evaluation of a display incorporating quantitative and check-reading characteristics. *Journal of Applied Psychology*, 1956, *40*, 233-236.

Markowitz, J. Optimal flash rate and duty cycle for flashing visual indicators. *Human Factors*, 1971, *13*(5), 427-433.

Mortimer, R. G. Motor vehicle exterior lighting. *Human Factors*, 1976, *18*(3), 259-272.

Muller, P. F., Jr., Sidorsky, R. C., Slivinske, A. J., Alluisi, E. A., & Fitts, P. M. *The symbolic coding of information on cathode ray tubes and similar displays.* USAF, WADC, TR 55-375, October 1955.

Munsell book of color. Baltimore: Munsell Color Co., 1959.

New Federal Standard on Colors. *Journal of the Optical Society of America*, 1957, *47*, 330-334.

Oatman, L. C. Check-reading accuracy using an extended-pointer dial display. *Journal of Engineering Psychology*, 1964, *3*, 123-131.

Peters, G. A., & Adams, B. B. These three criteria for readable panel markings. *Product Engineering*, May 25, 1959, *30*(21), 55-57.

Petersen, H. E., & Dugas, D. J. The relative importance of contrast and motion in visual detection. *Human Factors*, 1972, *14*(3), 207-216.

Plath, D. W. The readability of segmented and conventional numerals. *Human Factors*, 1970, *12*(5), 493-497.

Poulton, E. C. *Effects of printing types and formats on the comprehension of scientific journals* (APU 346). Cambridge, England: Applied Psychology Research Unit, 1959.

Poulton, E. C. Skimming lists of food ingredients printed in different sizes. *Journal of Applied Psychology*, 1969, *53*(1), 55-58.

Poulton, E. C. Size, style, and vertical spacing in the legibility of small typefaces. *Journal of Applied Psychology*, 1972, *56*(2), 156-161.

Reynolds, R. E., White, R. M., Jr., & Hilgendorf, R. L. Detection and recognition of colored signal lights. *Human Factors*, 1972, *14*(3), 227-236.

Robertson, H. D. Pedestrian preferences for symbolic signal displays. *Traffic Engineering*, June 1977, *47*(6), 38-42.

Roscoe, S. N. Airborne displays for flight and navigation. *Human Factors*, 1968, *10*(4), 321-332.

Sabeh, R., Jorve, W. R., & Vanderplas, J. M. *Shape coding of aircraft instrument zone markings* (Technical Note 57-260). USAF, WADC, Wright-Patterson Air Force Base, Ohio, March 1958.

Saenz, N. E., & Riche, C. V., Jr. Shape and color as dimensions of a visual redundant code. *Human Factors*, 1974, *16*(3), 308-313.

Schutz, H. G. An evaluation of formats for graphic trend displays—Experiment II. *Human Factors*, 1961, *3*, 99-107.

Shontz, W. D., Trumm, G. A., & Williams, L. G. Color coding for information location. *Human Factors*, 1971, *13*(3), 237-246.

Simon, C. W., & Roscoe, S. N. *Altimetry studies: II. A comparison of integrated versus separated, linear versus circular, and spatial versus numerical displays* (Technical Memorandum 435). Culver City, Calif.: Hughes Aircraft Company, May 1956.

Sinclair, H. J. Digital versus conventional clocks–A review. *Applied Ergonomics,* September 1971, *2*(3), 178–181.

Smith, S. L. Letter size and legibility. *Human Factors,* 1979, *21*(6), 661–670.

Smith, S. L. *Display color coding for visual separability* (MITRE Report MTS-10). Bedford, Mass.: MITRE Corp., August 1963.

Smith, S. L., & Thomas, D. W. Color versus shape coding in information displays. *Journal of Applied Psychology,* 1964, *48*, 137–146.

Teichner, W. H., & Krebs, M. J. Estimating the detectability of target luminances. *Human Factors,* 1972, *14*(6), 511–519.

Tinker, M. A. Prolonged reading tasks in visual research. *Journal of Applied Psychology,* 1955, *39*, 444–446.

Vartabedian, A. G. The effects of letter size, case, and generation method on CRT display search time. *Human Factors,* 1971, *13*(4), 363–368.

Vartabedian, A. G. Developing a graphic set for developing cathode-ray-tube display using a 7 × 9 dot matrix. *Applied Ergonomics,* 1973, *4*(1), 11–16.

Voevodsky, J. Evaluation of a deceleration warning light for reducing rear-end automobile collisions. *Journal of Applied Psychology,* 1974, *59*(3), 270–273.

Wong, K. W., & Yacoumelos, N. G. Identification of cartographic symbols from TV displays. *Human Factors,* 1973, *15*(1), 21–31.

Woodson, W. E., & Conover, D. W. *Human engineering guide for equipment designers* (2d ed.). Berkeley: University of California Press, 1964.

Zeff, C. Comparison of conventional and digital time displays. *Ergonomics,* 1965, *8*(3), 339–345.

CHAPTER 5

AUDITORY, TACTUAL, AND OLFACTORY DISPLAYS

We all depend upon our auditory, tactual, and olfactory senses in many aspects of our lives, including hearing our children cry or the doorbell ring, feeling the smooth finish on fine furniture, or smelling a cantaloupe to determine if it is ripe. As discussed in Chapter 3, information can come to us directly or indirectly. A child's natural cry is an example of direct information, while a doorbell's ring would be indirect information that someone is at the door. It is becoming increasingly possible to convert stimuli that are intrinsically, or directly, associated with one sensory modality into stimuli associated (indirectly) with another modality. Such technological developments have resulted in increased use of the auditory, tactual, and—to a lesser extent—olfactory senses, such as using buzzers to warn of fire or to warn blind people of physical objects in their path.

In this chapter we will discuss the use of the auditory, tactual, and olfactory senses as means of communication. Our focus will be on indirect information sources rather than direct, and the topic of speech will be deferred until Chapter 6.

HEARING

In discussing the hearing process, let us first describe the physical stimuli to which the ear is sensitive, namely, sound vibrations, and then we will discuss the anatomy of the ear.

The Nature and Measurement of Sound

Sound is originated by vibrations from some source. While such vibrations can be transmitted through various media, our primary concern is with those transmitted through the atmosphere to the ear. Two primary attributes of sound are *frequency* and *intensity* (or amplitude).

Frequency of Sound Waves The frequency of sound waves can be visualized if we think of a simple sound-generating source such as a tuning fork. When it is struck, the tuning fork is caused to vibrate at its "natural" frequency. In so doing, it causes the air molecules to be moved back and forth. This alternation creates corresponding increases and decreases in the air pressure.

The vibrations of a simple sound-generating source, such as a tuning fork, form *sinusoidal*, or *sine*, waves that can be represented as the projection of the movement of a point around a circle that is revolving at a constant rate, as shown in Figure 5-1. As point P revolves around its center O, its vertical amplitude, as a function of time, will be that represented by the sine wave. The height of the wave above the midline, at any given point in time, represents the amount of above-normal air pressure at that point in time. Positions below the midline, in turn, represent the reduction in air pressure below normal. One complete revolution of the circle represents one cycle, as shown in Figure 5-1. The number of cycles per second is the *frequency* of the sound. Frequency

FIGURE 5-1
Reproduction of sinusoidal, or sine, wave. The magnitude of the alternating changes in air pressure caused by a sound-generating source with a given frequency can be represented by a sine wave. A sine wave is represented as the projection of a point on the circumference of a circle as that point rotates about its center at a constant speed. The lower part of the figure depicts the changes in the density of molecules of the air caused by the vibrating source.

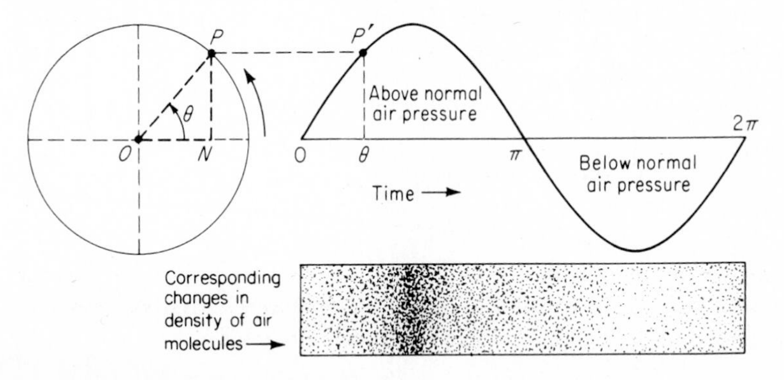

is expressed in hertz (Hz), which is equivalent to cycles per second. On the musical scale, middle C has a frequency of 256 Hz. Any given octave has double the frequency of the one below it, so one octave above middle C has a frequency of 512 Hz. In general terms, the human ear is sensitive to frequencies in the range of 20 to 20,000 Hz, although it is not equally sensitive to all frequencies. In addition, there are marked differences in sensitivity among individuals.

The frequency of a physical sound is associated with the human sensation of *pitch*. *Pitch* is the name given to the highness or lowness of a tone. Since high frequencies yield high-pitched tones, and low frequencies yield low-pitched tones, we tend to think of pitch and frequency as synonymous. Actually there are other factors that influence our perception of pitch besides frequency. For example, the intensity of a tone influences perceived pitch. When increased in intensity, low-frequency tones (less than about 1000 Hz) become lower in pitch, while high-frequency tones (greater than about 3000 Hz) become higher in pitch. Tones between 1000 and 3000 Hz, on the other hand, are relatively insensitive to intensity induced pitch changes. Incidentally, complex sounds such as those produced by musical instruments are very stable in perceived pitch regardless of whether the instruments are played loudly or softly—happily for us music lovers!

Intensity of Sound Sound intensity is associated with the human sensation of *loudness*. As in the case of frequency and pitch, there are several factors besides intensity which influence our perception of loudness. We, however, have deferred the discussion of loudness to Chapter 15 where we will present information illustrating the interplay of frequency and intensity in the perception of loudness.

Sound intensity is defined in terms of power per unit area, for example, watts per square meter (W/m^2). Because the range of power values for common sounds is so tremendous, it is convenient to use a logarithmic scale to characterize sound intensity. The *bel* (B) (named after Alexander Graham Bell) is the basic unit of measurement used. The number of bels is the logarithm (to the base 10) of the ratio of two sound intensities. Actually, the most convenient and most commonly used measure of sound intensity is the *decibel* (dB). A decibel is $^1/_{10}$ of a bel.

Unfortunately, there are no instruments available for directly measuring the sound power of a source. However, since sounds are pressure waves that vary above and below normal air pressure, these variations in air pressure can be directly measured. Luckily for us, sound power is directly proportional to the *square* of the sound pressure. The sound-pressure level (SPL), in decibels, is therefore defined as:

$$\text{SPL (dB)} = 10 \log \frac{P_1{}^2}{P_0{}^2}$$

where $P_1{}^2$ is the sound pressure squared of the sound one wishes to measure, and $P_0{}^2$ is the standard reference sound pressure squared that represents zero decibels. With a little algebraic magic, we can simplify the above equation so that we deal with sound pressure rather than sound pressure squared.

The equation for SPL becomes:

$$\text{SPL (dB)} = 20 \log \frac{P_1}{P_0}$$

Note, P_0 cannot equal zero, because the ratio P_1/P_0 would be infinity. Therefore, a sound pressure greater than zero must be used to represent zero decibels. Then, when P_1 equals P_0, the ratio will equal 1.0, and the log of 1.0 is equal to zero, i.e., SPL will equal zero dB. The most common sound-pressure reference value used to represent zero dB is 20 micronewtons per square meter (20 μN/m^2). The μN stands for 0.000001 newton (0.000001 N).[1] This sound pressure is roughly equivalent to the lowest intensity, 1000 Hz pure tone, which a healthy adult can just barely hear under ideal conditions. It is possible, therefore, to have sound-pressure levels that are actually less than zero dB, that is, they have sound pressures less than 20 μN/m^2.

The decibel scale is a logarithmic scale, therefore an increase of 10 dB represents a tenfold increase in sound power and a hundredfold increase in sound pressure (remember that power is related to the square of pressure). In like manner, it can be shown that doubling the sound power will raise the SPL by 3 dB.

Sound pressure is measured by sound-level meters. The American National Standards Institute (ANSI) and the American Standards Association, have established a standard to which sound-level meters should conform. This standard requires that three different *weighting* networks (designated *A*, *B*, and *C*) be built into such instruments. Each network responds differently to low or high frequencies according to standard *frequency-response* curves. We will have more to say about this in Chapter 15 when we discuss noise. For now, we need to be aware that when reporting a sound intensity, the specific weighting network used for the measure must be specified, for example, "The *A*-weighted sound level is 45 dB," "sound level (*A*) = 45 dB," "SL*A* = 45 dB," or "45 dB*A*." Figure 5-2 shows the decibel scale with examples of several sounds that fall at varying positions along the scale.

Complex Sounds There are very few sounds that are pure tones. Even tones from musical instruments are not pure, but rather consist of a fundamental frequency in combination with certain others (especially harmonic frequencies that are multiples of the fundamental). Most complex sounds, however, are nonharmonic. Complex sounds can be depicted in two ways. One of these is a waveform which is the composite of the waveforms of its individual component sounds. Figure 5-3 shows three component waveforms and the resultant composite waveform.

The other method of depicting complex sounds is by use of a sound spectrum which divides the sound into frequency bands and measures the intensity of the sound in each band. A frequency band analyzer is used for this purpose. The four curves of

[1] This can also be expressed as 0.0002 dyne per square centimeter (0.0002 dyn/cm^2) or 0.0002 microbar (0.0002 μbar).

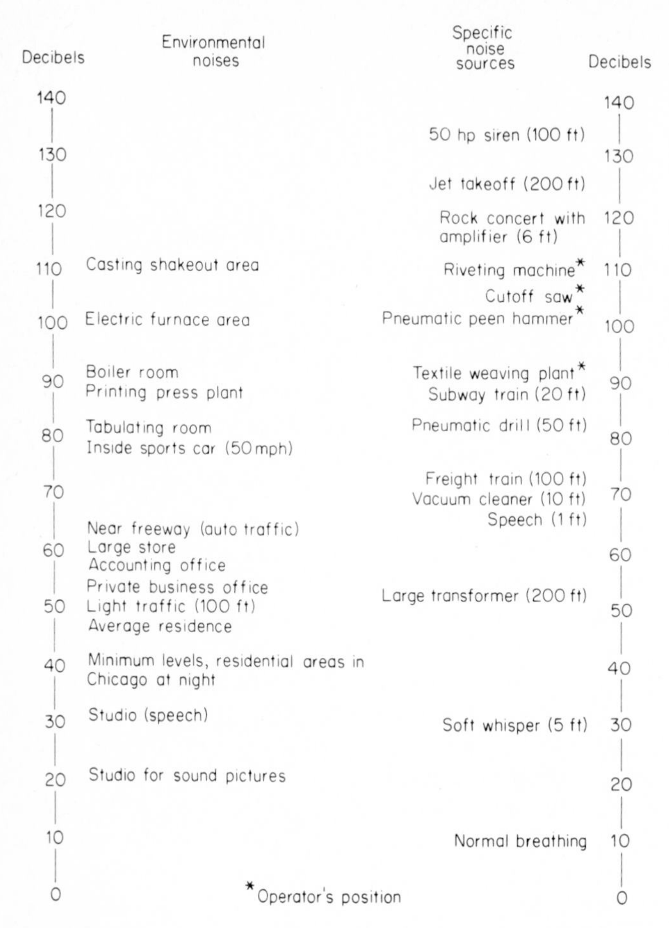

FIGURE 5-2
Decibel levels (dB) for various sounds. Decibel levels are *A*-weighted sound levels measured with a sound-level meter. (*Source: Peterson & Gross, 1972, fig. 2-1, p. 4.*)

Figure 5-4 illustrate spectral analyses of the noise from a rope-closing machine. Each curve was generated using a different sized frequency band (*bandwidth*), namely, an octave, a half octave, a third of an octave, and a thirty-fifth of an octave. The narrower the bandwidth, the greater the detail of the spectrum and the lower the sound level of each bandwidth. There have been various practices in the division of the sound spectrum into octaves, but the current preferred practice as set forth by the American National Standards Institute (ANSI) is that of dividing the audible range into 10 bands, with the *center* frequencies being 31.5, 63, 125, 250, 500, 1000, 2000, 4000, 8000, and 16,000 Hz. (Many existing sets of sound and hearing data are presented using the previous practice of defining octaves in terms of the *ends* of the class intervals instead of their *centers*.)

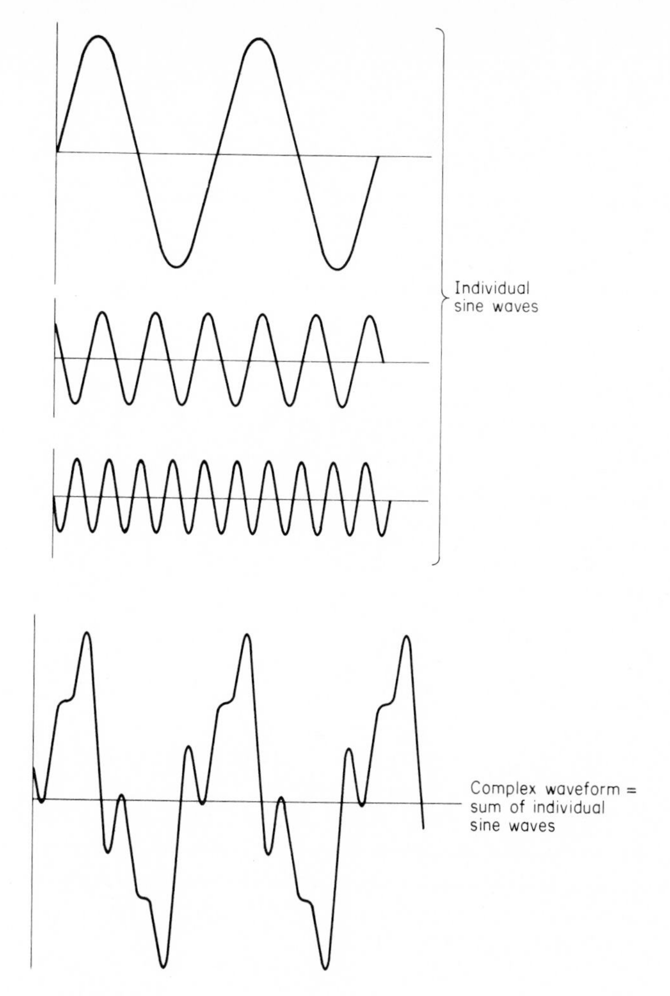

FIGURE 5-3
Waveform of a complex sound formed from the three individual sine waves. (*Source: Minnix, 1978, fig. 1-11.*)

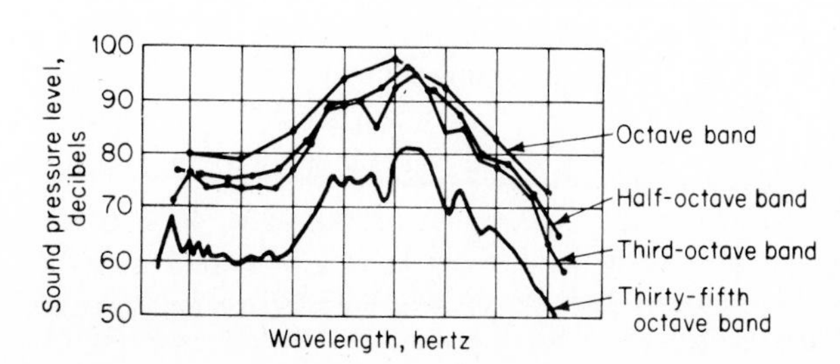

FIGURE 5-4
Spectral analyses of noise of a rope-closing machine, using analyzers of varying bandwidths. The narrower the bandwidth, the greater the detail and the lower the level of any single bandwidth. (*Source: Adapted from* Industrial noise manual, *1966, p. 25.*)

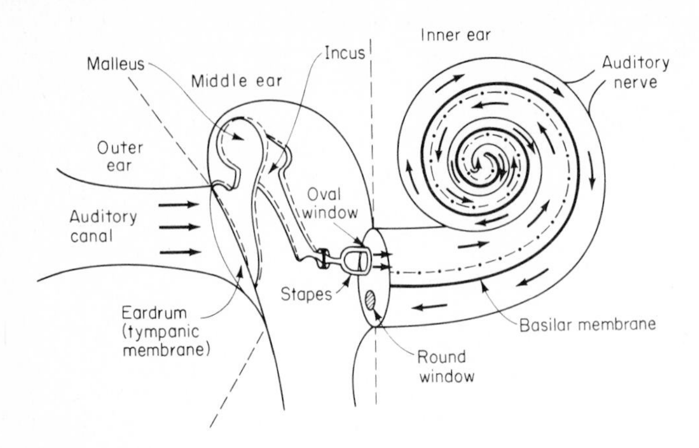

FIGURE 5-5
Schematic drawing of the ear, showing auditory canal through which sound waves travel to the tympanic membrane, which in turn vibrates the ossicles of the middle ear. This vibration is transmitted through the membrane of the oval window to the cochlea, where the vibrations are transmitted by liquid through membranes to sensitive hair cells, which send nerve impulses to the brain.

The Anatomy of the Ear

The ear has three primary anatomical divisions, namely, the *outer ear*, the *middle ear*, and the *inner ear*. These are shown schematically in Figure 5-5.

The Outer Ear The outer ear, which collects sound energy, consists of the external part (called the *pinna* or *concha*), the auditory canal (the *meatus*), which is a bayonet-shaped tube about an inch long that leads inward from the external part, and the eardrum (the *tympanic membrane*) at the end of the auditory canal.

The Middle Ear The middle ear is separated from the outer ear by the *tympanic membrane*. The middle ear includes a chain of three small bones called *ossicles*, the *malleus*, the *incus*, and the *stapes*. These three ossicles, by their interconnections, transmit vibrations from the eardrum to the oval window of the inner ear. The stapes acts something like a piston on the oval window, its action transmitting the changes in sound pressure to the fluid of the inner ear, on the other side of the oval-window membrane. The tympanic membrane has an area approximately 22 times that of the oval window. Therefore 22 times as much sound energy is collected there as could be collected by the oval window alone. All of this is transmitted by the ossicles. Thus, the pressure of the foot of the stapes against the oval window is amplified to about 22 times that which could be effected by applying sound waves directly to the oval window (Guyton, 1969, p. 300).

The middle ear also contains two muscles attached to the ossicles. The *tensor tympani muscle* attaches to the malleus, and the *stapedius muscle* attaches to the stapes. These muscles represent a protection for the inner ear against very intense sounds. When the muscles tighten, they reduce sound transmitted to the inner ear.

The Inner Ear The inner ear or *cochlea* is a spiral-shaped affair that resembles a snail. If uncoiled it would be about 30 millimeters (30 mm) long, with its widest section (near the oval window) being about 5 or 6 mm. The inner ear is filled with a fluid. The stapes of the middle ear acts on this fluid like a piston, driving it back and forth in response to changes in the sound pressure. These movements of the fluid force into vibration a thin membrane called the *basilar membrane*, which in turn transmits the vibrations to the *organ of Corti*, which contains hair cells and nerve endings that are sensitive to very slight changes in pressure. The neural impulses picked up by these nerve endings are transmitted to the brain via the *auditory nerve*.

The Conversion of Sound Waves into Sensations

While the mechanical processes that are involved in the ear have been known for some time, the procedures by which sound vibrations are "heard" and differentiated are still not entirely known. In this regard, the numerous theories of hearing fall generally into two classes. The *place* (or resonance) theories are postulated on the notion that the fibers at various positions (or places) along the basilar membrane act, as Geldard (1972) puts it, like harp or piano strings. Since these fibers vary in length, they are differentially sensitive to different frequencies and thus give rise to sensations of pitch. In turn, overtones stimulate a series of fibers spaced down the membrane. On the other hand, the *frequency* (or telephone) theories are based on the tenet that the basilar membrane vibrates as a whole, thus functioning much like the diaphragm of a telephone or microphone.

As Geldard (1972) comments, it is not now in the cards to be able to accept one basic theory to the exclusion of any others as explaining all auditory phenomena. Actually, Geldard indicates that the current state of knowledge would prejudice one toward putting reliance in a place theory as related to high tones and in a frequency theory as related to low tones, one principle thus giving way to the other in the middle range of frequencies. As with theories in various areas of life, this notion of the two theories being complementary to each other still needs to be regarded as tentative pending further support or rejection from additional research.

One consequence of all this is that the ear is not equally sensitive to all frequencies of sound. Although we will discuss this more fully in Chapter 15, it is important to know that, in general, the ear is less sensitive to low-frequency sounds (20 to approximately 500 Hz) and more sensitive to higher frequencies (1000 to approximately 5000 Hz). That is, a 4000-Hz tone, at a given sound pressure level, will seem to be louder than, say, a 200-Hz tone at the same sound pressure level.

Masking

Masking is a condition in which one component of the sound environment reduces the sensitivity of the ear to another component. Operationally defined, *masking* is the amount by which the threshold of audibility of a sound (the masked sound) is raised by the presence of another (masking) sound. In studying the effects of masking, an experimenter typically measures the absolute threshold (the minimum audible level) of a sound (the sound to be masked) when presented by itself and then measures its threshold in the presence of the masking sound. The difference is attributed to the masking effect. The concept of masking is central to discussions of auditory displays. In select-

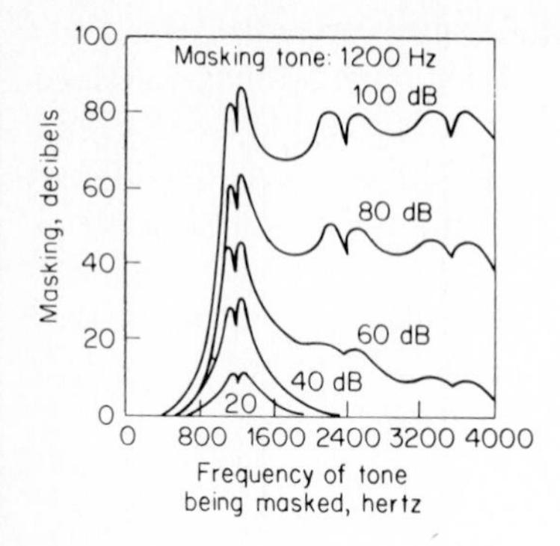

FIGURE 5-6
The effects of masking pure tones by a pure tone of 1200 Hz (100, 80, 60, 40, and 20 dB). (*Source: Wegel & Lane, 1924.*)

ing a particular auditory signal for use in a particular environment, we must consider the masking effect of any noise on the reception of that signal.

The effects of masking vary with the type of masking sound and with the masked sound itself–whether pure tones, complex sound, white noise, speech, etc. Figure 5-6 illustrates the masking of a pure tone by another pure tone. A few general principles can be gleaned from this. The greatest masking effect occurs near the frequency of the masking tone and its harmonic overtones. As we would expect, the higher the intensity of the masking tone, the greater the masking effect. Notice, however, that with low-intensity masking tones (20 to 40 dB) the masking effect is somewhat confined to the frequencies around that of the masking tone, but with higher-intensity masking tones (60 to 100 dB), the masking effect spreads to higher frequencies. The masking of pure tones by narrow band noise (i.e., noise concentrated in a narrow band of frequencies) is very similar to that shown in Figure 5-6.

In the masking of pure tones by wide band noise, the primary concern is with the intensity of the masking noise in a "critical band" around the frequency of the masked tone. The size of the critical band is a function of the center frequency, with larger critical bands being associated with higher frequencies (Deatherage, 1972, p. 139).[2]

The nature of masking effects depends very much upon the nature of the two sounds in question, as discussed by Geldard (1972, pp. 215-220) and by Deatherage (1972). Although these complex effects will not be discussed here, the effects of masking in our everyday lives are of considerable consequence, as, for example, the noise of a hair dryer drowning out the sound of the telephone ringing.

AUDITORY DISPLAYS

The nature of the auditory sensory modality offers certain unique advantages for presenting information as contrasted with the visual modality, which has its own advantages. One set of comparisons of these two senses was given in Table 3-3, Chapter 3. On the basis of such comparisons and of other cues, it is possible to crystallize certain

[2] The critical bandwidth is technically defined as that frequency band of sound which contains sound power just equal to that of a pure tone centered in the critical band when the tone is just audible at its masked threshold.

types of circumstances in which auditory displays usually would be preferable to visual displays. Some of these circumstances are given below:

- When the origin of the signal is itself a sound
- When the message is simple and short
- When the message will not be referred to later
- When the message deals with events in time
- When sending warnings or when the message calls for immediate action
- When presenting continuously changing information of some type, such as aircraft, radio range, or flight-path information
- When the visual system is overburdened
- When speech channels are fully employed (in which case auditory signals such as tones should be clearly detectable from the speech)
- When illumination limits use of vision
- When the receiver moves from one place to another

Obviously the application of the above set of guidelines should be tempered with judgment rather than being followed rigidly. There are, of course, other circumstances in which auditory displays would be preferable. In the above guidelines particular mention should be made of the desirability of restricting auditory messages to those that are short and simple (except in the case of speech), since people do not do well in short-term storage of complex messages.

In a sense there are three types of human functions involved in the reception of auditory signals, these depending on the nature of the task in question, as follows: (1) *detection* (determining if a given signal is or is not present, such as a warning signal); (2) *relative discrimination* (differentiating between two or more signals when presented close together); and (3) *absolute identification* (identifying a particular signal of some class, when only the one is presented). Relative discrimination and absolute identification can be made on the basis of any of several stimulus dimensions, such as intensity, frequency, duration, and direction (the difference in intensity of signals transmitted to the two ears).

Detection of Signals

The detection of auditory signals can be viewed in the frame of reference of signal detection theory (as discussed in Chapter 3), whether the signals occur in "peaceful" surroundings or in environments that are permeated by some ambient noise. As indicated in the discussion of signal detection theory, if at all possible the signal plus noise (SN) should be distinct from the noise (N) itself. When there is some overlap, the signal cannot invariably be detected from the noise; this confusion is, in part, a function of masking. When the signal (the masked sound) occurs in the presence of noise (i.e., the masking sound), the threshold of detectability of the signal is elevated–and it is this "elevated" threshold that should be exceeded by the signal if it is to be detected accurately.

In quiet surroundings (when one doesn't have to worry about clacking typewriters, whining machines, or screeching tires), a sound of about 40 to 50 dB above absolute threshold normally would be sufficient to be detected. However, such detectability would vary somewhat with the frequency of the signal (Pollack, 1952) and its duration. With respect to duration, the ear does not respond instantaneously to sound. For pure tones it takes about 200 to 300 milliseconds (ms) to "build up" (Munson, 1947) and about 140 ms to decay (Stevens & Davis, 1938) although wide-band sounds build up and decay more rapidly. Because of these lags, signals of less than about 200 to 500 ms in duration do not sound as loud as those of longer duration. Thus, auditory signals (especially pure tones) should be at least 300 ms in duration; if they have to be shorter than that, their intensity should be increased to compensate for reduced audibility. Although detectability increases with signal duration, there is not much additional effect beyond a few seconds in duration.

In rather noisy conditions, the signal intensity has to be set at a level far enough above that of the noise to ensure detectability. In this regard Deatherage (1972) has suggested a rule of thumb for specifying the optimum signal level, as follows: that the signal intensity (at the entrance of the ear) be about midway between the masked threshold of the signal, in the presence of noise, and 110 dB.

Fidell, Pearsons, and Bennett (1974) have developed a method for predicting the detectability of complex signals (complex in terms of frequency) in noisy backgrounds. Although we will not present the details of the procedure, it is based on the observation that human performance is a function of the signal-to-noise ratio in the single one-third octave band to which human sensitivity is highest. The basic data employed in the proposed method are therefore the one-third octave band spectra of the signal and background noise.

The Use of Filters Under some circumstances it is possible to enhance the detectability of signals by filtering out some of the noise. This is most feasible when the predominant frequencies of the noise are different from those of the signal. In such an instance some of the noise can be filtered out and the intensity of the remaining sound raised (signal plus the nonfiltered noise). This has the effect of increasing the signal-to-noise ratio, thus making the signal more audible.

Relative Discrimination of Auditory Signals

The relative discrimination of signals on the basis of intensity and frequency (which are the most commonly used dimensions) depends in part on interactions between these two dimensions.

Discrimination of Intensity Differences An impression of human abilities to discriminate intensity differences is given in Figure 5-7 (Deatherage, 1972). This reflects the just-noticeable differences (JNDs) for certain pure tones and for wide-band noise of various sound-pressure levels. It is clear that the smallest differences can be detected with signals of higher intensities, such as at least 60 dB above the absolute threshold. The JNDs of signals above 60 dB are minimal for intermediate frequencies (1000 and 4000 Hz) signals, these differences of course having implications if the auditory signals are to be discriminated by intensity.

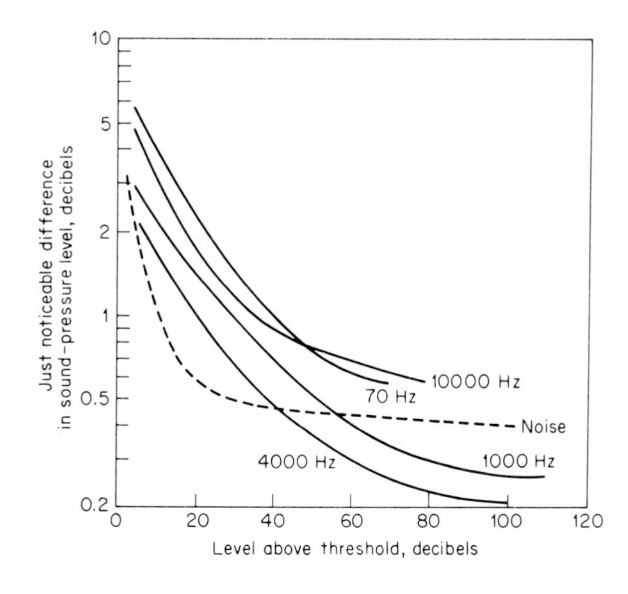

FIGURE 5-7
The just-noticeable differences (JNDs) in sound intensity for pure tones of selected frequencies and for wide-band noise. (*Source: Deatherage, 1972, p. 147, as based on data from Riesz, 1928, and Miller, 1947.*)

Discrimination of Frequency Differences Some indication of the ability of people to tell the difference between pure tones of different frequencies is reflected by the results of an early study by Shower and Biddulph (1931) as shown in Figure 5-8. This figure shows the JNDs for pure tones of various frequencies at levels above threshold. The JNDs are smaller for frequencies below about 1000 Hz (especially for high intensities) but increase rather sharply for frequencies above that. Thus, if signals are to be

FIGURE 5-8
Just-noticeable differences (JNDs) for pure tones at various levels above threshold. (*Source: Shower & Biddulph, 1931.*)

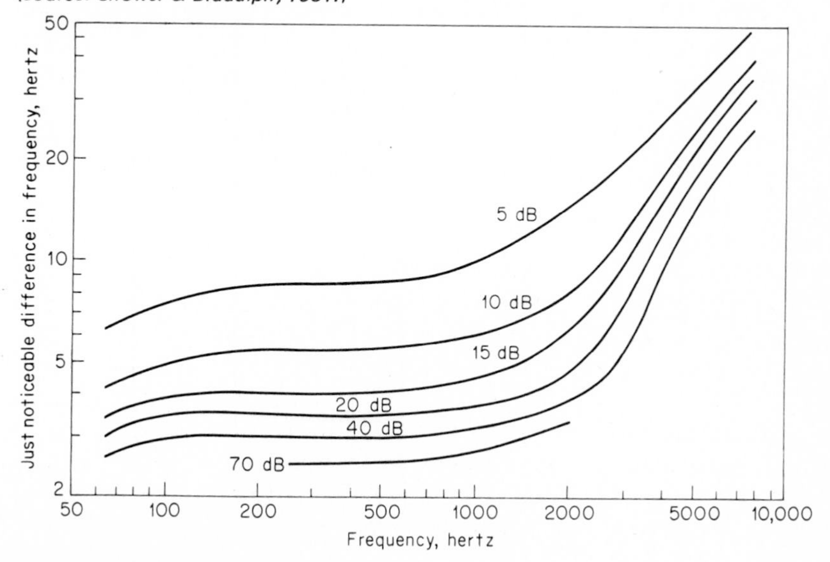

discriminated on the basis of frequency, it usually would be desirable to use signals of lower frequencies. This practice may run into a snag, however, if there is much ambient noise (which usually tends to consist of lower frequencies) that might mask the signals. A possible compromise is that of using signals in the 500- to 1000-Hz range (Deatherage, 1972). It is also obvious from Figure 5-8 that the JNDs are smaller for signals of high intensity than for those of low intensity, thus suggesting that signals probably should be at least 30 dB above absolute threshold. Duration of the signal is also important in discrimination of frequency differences. Gales (1979) reports that discrimination is best when stimulus duration is in excess of 0.1 s.

Absolute Identification of Auditory Signals

The JND for a given stimulus dimension reflects the minimum difference that can be discriminated on a relative basis. But in many circumstances it is necessary to make absolute identification of an individual stimulus (such as the frequency of a single tone) presented by itself. The number of "levels" along a continuum that can be so identified usually is quite small, as mentioned in the discussion of information theory in Chapter 3. Table 5-1 gives the number of levels that can be identified for certain common auditory dimensions.

Multidimensional Coding If the amount of information to be transmitted by auditory codes is substantial (meaning that discriminations among numerous signals need to be made), it is possible to use a multidimensional code system. For example, Pollack and Ficks (1954) used various combinations of several dimensions, such as direction (right ear versus left ear), frequency, intensity, repetition rate, on-and-off time fraction, and duration. Such a system obviously imposes the requirement for training the receiver regarding the "meaning" of each unique combination. In using such multidimensional codes, however, it is generally better to use more dimensions with fewer "steps" or "levels" of each (such as eight dimensions with two steps of each dimension) than to use fewer dimensions and more levels of each (such as four dimensions with four steps of each). It is on the basis of the many different facets (dimensions) of sounds that we can identify the voices of individuals, a dripping water faucet, or a squeaking hinge.

TABLE 5-1
LEVELS OF AUDITORY DIMENSIONS IDENTIFIABLE ON AN ABSOLUTE BASIS

Dimension	Levels
Intensity (pure tones)	4 to 5
Frequency	4 to 7
Duration	2 to 3
Intensity and frequency	9

Source: Deatherage, 1972; Van Cott & Warrick, 1972.

Principles of Auditory Display

As in other areas of human factors engineering, most guidelines and principles have to be accepted with a few grains of salt, since specific circumstances, trade-off values, etc., may argue for their violation. With such reservations in mind, a few guidelines for the use of auditory displays are given below. These generally stem from research and experience. Some are drawn in part from Mudd (1961) and Licklider (1961).

I General principles

A *Compatibility:* Where feasible, the selection of signal dimensions and their encoding should exploit learned or natural relationships of the users, such as high frequencies associated with up or high, and wailing signals with emergency.

B *Approximation:* Two-stage signals should be considered when complex information is to be presented, these stages to consist of:

1 Attention-demanding signal: to attract attention and identify a general category of information.

2 Designation signal: to follow the attention-demanding signal and designate the precise information within the general class indicated above.

C *Dissociability:* Auditory signals should be easily discernible from any ongoing audio input (be it either meaningful input or noise). For example, if a person is to listen concurrently to two or more channels, the frequencies of the channels should be different if it is possible to make them so.

D *Parsimony:* Input signals to the operator should not provide more information than is necessary.

E *Invariance:* The same signal should designate the same information at all times.

II Principles of presentation

F *Avoid extremes of auditory dimensions:* High-intensity signals, for example, can cause a startle response and actually disrupt performance.

G *Establish intensity relative to ambient noise level:* This is simply saying that the intensity level should be set so that it is not masked by the ambient noise level.

H *Use interrupted or variable signals:* Where feasible, avoid steady-state signals and, rather, use interrupted or variable signals. This will tend to minimize perceptual adaptation.

I *Don't overload the auditory channel:* Only a few displays should be used in any given situation. Too many displays can be confusing and will overload the operator. (For example, during the Three Mile Island nuclear crisis, over 60 different auditory warning displays were activated.)

III Principles of installation of auditory displays

J *Test signals to be used:* Such tests should be made with a representative sample of the potential user population, to be sure the signals can be detected by them.

K *Avoid conflict with previously used signals:* Any newly installed signals should not be contradictory in meaning to any somewhat similar signals used in existing or earlier systems.

L *Facilitate changeover from previous display:* Where auditory signals replace some other mode of presentation (e.g., visual), preferably continue both modes for a while, to help people become accustomed to new auditory signals.

Auditory Displays for Specific Purposes

For illustrative purposes, a few types of auditory displays for specific purposes will be discussed below.

Warning and Alarm Signals The unique features of the auditory system lend auditory displays to special use for signaling warnings and alarms. For this purpose the various types of available devices have their individual characteristics and corresponding advantages and limitations. A summary of such characteristics and features is given in Table 5-2 (Deatherage, 1972, p. 126).

As one example of research in this area, Adams and Trucks (1976) tested reaction time to eight different warning signals ("wail," "yelp," "whoop," "yeow," intermittent horns, etc.) in five different ambient noise environments. Across all five environments, the most effective signals were the "yeow" (descending change in frequency from 800 to 100 Hz every 1.4 s) and the "beep" (intermittent horn, 425 Hz, on for 0.7 s, off for 0.6 s). The least effective signal was the "wail" (slow increase in frequency from 400 to 925 Hz over 3.8 s and then a slow decrease back to 400 Hz). Figure 5-9 shows the mean reaction time to these three signals as a function of overall signal intensity in one of the ambient noise environments tested. Reaction time decreased with

TABLE 5-2
THE CHARACTERISTICS AND FEATURES OF CERTAIN TYPES OF AUDIO ALARMS

Alarm	Intensity	Frequency	Attention-getting ability	Noise-penetration ability
Diaphone (foghorn)	Very high	Very low	Good	Poor in low-frequency noise
Horn	High	Low to high	Good	Good
Whistle	High	Low to high	Good if intermittent	Good if frequency is properly chosen
Siren	High	Low to high	Very good if pitch rises and falls	Very good with rising and falling frequency
Bell	Medium	Medium to high	Good	Good in low-frequency noise
Buzzer	Low to medium	Low to medium	Good	Fair if spectrum is suited to background noise
Chimes and gong	Low to medium	Low to medium	Fair	Fair if spectrum is suited to background noise
Oscillator	Low to high	Medium to high	Good if intermittent	Good if frequency is properly chosen

Source: Deatherage (1972, table 4-2).

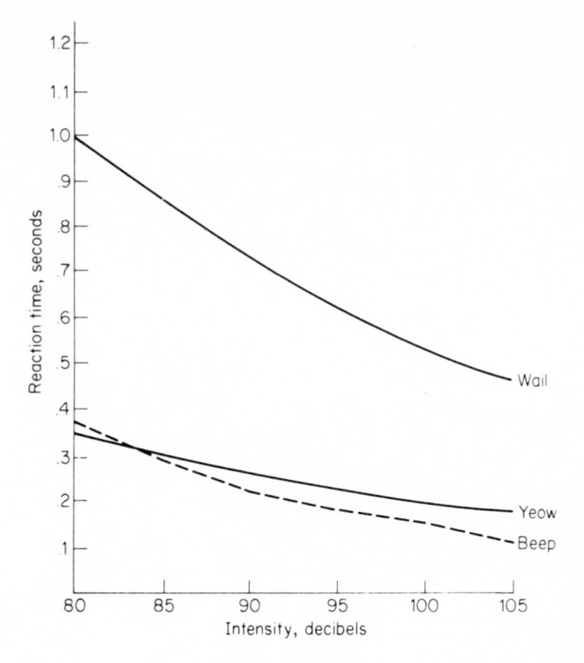

FIGURE 5-9
Mean reaction time to three auditory warning systems in an ambient noise environment. Reaction time decreases with increased signal intensity. (*Source: Adapted from Adams & Trucks, 1976, fig. 4.*)

increased signal intensity. The difference in reaction time between the most effective and the least effective signals also decreased somewhat as intensity increased.

This relationship between reaction time and signal intensity may only hold for simple reactions, that is, where the same response is to be made for all signals. Where a different response is to be made to different signals (choice reaction time), the relationship between reaction time and signal intensity is not so simple. Van der Molen and Keuss (1979) used two signals (1000 Hz and 3000 Hz) and either required the subjects to make the same response to both (simple reaction time), or a different response to each (choice reaction time). Figure 5-10 illustrates their results. For simple reaction time, increases in signal intensity resulted in faster reaction times. For choice reaction time, however, the fastest reaction times were for moderate-intensity signals, rather than high-intensity signals. What appears to be happening is that the high-intensity signals elicit a startle reflex, which may be helpful for efficient task performance only if the same response is demanded by both signals. But when either of two responses must be provided, the startle reflex acts to degrade performance. The intensity level of warning devices should, therefore, be chosen with concern for the response requirements placed on the operator.

In the selection or design of warning and alarm signals, the following general design recommendations have been proposed by Deatherage (1972, pp. 125, 126) and by Mudd (1961), these being in slightly modified form.

- Use frequencies between 200 and 5000 Hz, and preferably between 500 and 3000 Hz, because the ear is most sensitive to this middle range.

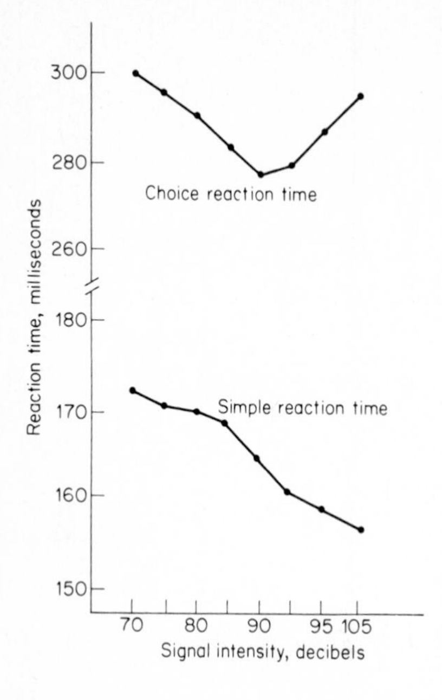

FIGURE 5-10
Relationship between reaction time and signal intensity for choice reaction time and simple reaction time tasks. (*Source: Adapted from Van der Molen & Keuss, 1979, fig. 1.*)

- Use frequencies below 1000 Hz when signals have to travel long distances (over 1000 ft), because high frequencies do not travel as far.
- Use frequencies below 500 Hz when signals have to "bend around" major obstacles or pass through partitions.
- Use a modulated signal (1 to 8 beeps per second or warbling sounds varying from 1 to 3 times per second), since it is different enough from normal sounds to demand attention.
- Use signals with frequencies different from those that dominate any background noise, to minimize masking.
- If different warning signals are used to represent different conditions requiring different responses, each should be discriminable from the others, and moderate intensity signals should be used.
- Where feasible, use a separate communication system for warnings, such as loudspeakers, horns, or other devices not used for other purposes.

Radio Range Signals An example of a situation that requires discrimination between sounds is the reception of radio range signals in aircraft, in which the A (dot-dash) or the N (dash-dot) is heard if the pilot is to the left or to the right, respectively, of the center beam. The signal of one beam is on when that of the other is off. When the beams are of equal strength, a continuous signal is heard, meaning that the pilot is on the center beam. Under adverse noise conditions the difference between the A and N signals may not be properly identified, and pilots may think they are to the right of the beam when actually they are to the left, or vice versa.

Some of the aspects of this particular problem have been investigated under simulated signal and noise conditions that were somewhat like those encountered in an airplane (Flynn, Goffard, Truscott, & Forbes, 1945). Without discussing the details of this investigation, the results probably have important implications for various auditory displays. In particular, the signal-to-noise ratio was found to be a much more critical factor in effective auditory discriminations than was the intensity of the signal itself.

Mobility Aids for the Blind With advances in technology, several devices have been built which use auditory information to aid the blind in moving about in their environment. Kay, Bui, Brabyn, and Strelow (1977) distinguish two types of mobility aids, clear path indicators (where the aim is to provide essentially a "go"-"no-go" signal indicating whether it is safe to proceed along the line of travel) and environmental sensors (which attempt to allow users to build up a picture of their environment). Examples of clear path indicators would include: *Pathsounder* (Russell, 1969), which emits an audible buzzing sound when there is something 6 ft ahead, and a distinctively different high-pitched beeping sound when an object is within 3 ft of the user; and *Single Object Sensor* (Kay et al., 1977), which emits repetitive clicks, coded in terms of repetition rate to the distance of objects. A *binaural* (using both ears) form of display also allows perception of left-right location of objects.

An example of an environmental sensor is the *Sonicguide* (Kay, 1973), built into an eyeglass frame (as is the Single Object Sensor). Sonicguide is one of the more complex mobility aids. It gives the user three types of information about an object: frequency differences to determine distance, directional information, and information about the surface characteristics of an object.

THE CUTANEOUS SENSES

In everyday life people depend upon their cutaneous (or somesthetic or skin) senses much more than they realize. There is, however, an ongoing question as to how many cutaneous senses there are, this confusion existing in part because of the basis on which the senses can be classified. As Geldard (1972, pp. 258, 259) points out, we can classify them *qualitatively* (on the basis of their observed similarity—that is, the sensations generated); in terms of the *stimulus* (i.e., the form of energy that triggers the sensation, as thermal, mechanical, chemical, or electrical); or *anatomically* (in accordance with the nature of the sense organs or tissues involved). With respect to the anatomical structures involved, it is still not clear how many distinct types of nerve endings there are, but for convenience Geldard considers the skin as housing three more or less separate systems of sensitivity, one for pressure reception, one for pain, and one responsive to temperature changes. Although it has been suggested that any particular sensation is triggered by the stimulation of a specific corresponding type of nerve ending or receptor, this notion seems not to be warranted. Rather, it is now generally believed that the various receptors are specialized in their functions through the operation of what Geldard refers to as the principle of *patterned response*. Some of the cutaneous receptors are responsive to more than one form of energy (such as mechanical pressure and thermal changes) or to certain ranges of energy. Through complex interactions among the various types of nerve endings as they are stimulated by various forms and

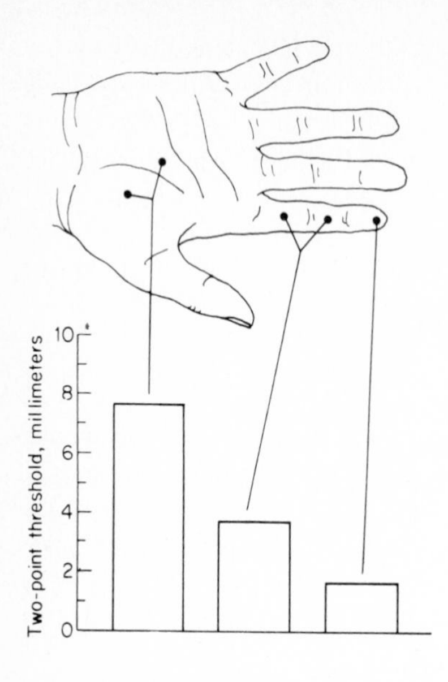

FIGURE 5-11
Median two-point threshold for fingertip, finger, and palm. (*Source: Vallbo & Johansson, 1978, fig. 6.*)

amounts of energy, we experience a wide "variety" of specific sensations that we endow with such labels as *touch, contact, tickle,* and *pressure* (Geldard, 1972).

For the most part, tactual displays have utilized the hand and fingers as the principal receptors of information. Craft workers, for centuries, have used their sense of touch to detect irregularities and surface roughness in their work. Interestingly, Lederman (1978) reports that surface irregularities can be detected more accurately when the person moves an intermediate piece of paper or thin cloth over the surface than when the bare fingers are used alone. Inspection of the coachwork of Aston Martin sports cars, for example, is done by rubbing over the surface with a cotton glove on the hand.

Not all parts of the hand are equally sensitive to touch. One common measure of touch sensitivity is the *two-point threshold*, the smallest distance between two pressure points at which the points are perceived as separate. Figure 5-11 shows the median two-point threshold for fingertip, finger, and palm (Vallbo and Johansson, 1978). The sensitivity increases (i.e., the two-point threshold becomes smaller) from the palm to the fingertips. Thus, tactual displays that require fine discriminations are best designed for fingertip reception. Tactual sensitivity is also degraded by low temperatures; therefore, tactual displays should be used with extreme caution in low temperatures. With this bit of background information let us now take a look at (or more appropriately feel our way through) some specific types of tactual displays.

TACTUAL DISPLAYS

Although we depend upon our cutaneous senses very much in everyday living, these senses have been used only to a limited degree as the basis for the "intentional" transmission of information to people by the use of tactual displays. The primary uses of the cutaneous senses for tactual displays to date have been as substitutes for hearing, especially as aids to the deaf, and as substitutes for seeing, especially as aids to the blind.

The most frequent types of stimuli used for tactual displays have been mechanical vibration or electrical impulses. Information can be transmitted by mechanical vibration based on such physical parameters as location of vibrators, frequency, intensity, or duration. Electrical impulses can be coded in terms of electrode location, pulse rate, duration, intensity, or phase.

Substitutes for Hearing

Tactual displays have generally found three applications as substitutes for hearing: reception of coded messages, perception of speech, and localization of sound.

Reception of Coded Messages Both mechanical and electrical energy forms have been used to transmit coded messages. In an early example, using mechanical vibrators, Geldard (1957) developed a coding system using 5 chest locations, 3 levels of intensity, and 3 durations, which provided 45 unique patterns (5 X 3 X 3), which in turn were used to code 26 letters, 10 numerals, and 4 frequently used words. One subject could receive messages at a rate of 38 words per minute, somewhat above the military requirements for Morse code reception of 24 words per minute.

Electrical energy has some advantages over mechanical vibrators, as noted by Hawkes (1962). Electrodes are easily mounted on the body surface, and have less bulk and a lower power requirement than mechanical vibrators. Sensitivity to mechanical vibration is dependent upon skin temperature, whereas the amount of electrical current required to reach threshold apparently is independent of skin temperature. A disadvantage of electrical energy is that it can elicit pain.

Perception of Speech Kirman (1973) reviews numerous attempts to transmit speech to the skin through mechanical and electrical stimulation. All attempts to build successful tactile displays of speech have been disappointing at best. One reason given for the poor performance of such devices is that the resolving power of the skin, both temporally and spatially, is too limited to deal with the complexities of speech. Kirman (1973) and Richardson and Frost (1977), however, believe that improvements in tactual displays in the future will someday permit tactile comprehension of speech at rapid rates.

Localization of Sounds Richardson and his colleagues (Frost & Richardson, 1976; Richardson & Frost, 1979; Richardson, Wuillemin, & Saunders, 1978) report that subjects are able to localize a sound source by use of a simple tactile display with an accuracy level comparable to normal audition. The sound intensity is picked up by microphones placed over each ear. The outputs of these microphones are amplified and fed to two vibrators, upon which the subjects rest their index fingers. The intensity of vibration on each finger is then proportional to the intensity of the sound reaching the ear. It is this difference in intensity, especially when one moves one's head, that allows localization of sound with normal audition.

Substitutes for Seeing

Tactual displays have been most extensively used as substitutes for seeing–as an extension of our eyes. The most exciting possibilities in the use of tactual displays lie in this realm, especially with the advances being made in low-cost microelectronics technology. Dobelle (1977), for example, has developed and tested an experimental unit that uses direct electrical stimulation of the brain. It produces genuine sensations of vision–indeed volunteers who have been blind for years complain of eyestrain. A television camera picks up a black-and-white picture which is converted into a matrix of 60 or so electrodes implanted in the primary visual cortex of the brain. Each electrode, when stimulated, produces a single point of light in the visual field. Several subjects have had the implant and have been able to recognize white horizontal and vertical lines on black backgrounds. There are, of course, problems with such a technique, and the best that can be expected in the foreseeable future, according to Dobelle, is low-resolution black-and-white slow-scan images.

We will now review some applications of tactual displays as substitutes for seeing, ranging from the relatively mundane to the more sophisticated.

Identification of Controls Another use of the tactual sense is with respect to the design of control knobs and related devices. Although these are not *displays* in the conventional sense of the term, the need to correctly identify such devices may be viewed within the general framework of displays. The coding of such devices for tactual identification includes their shape, texture, and size. Figure 5-12, for example, shows two sets of shape-coded knobs such that the knobs within each group are rarely confused with each other (Jenkins, 1947). In Chapter 9, we will discuss in more detail methods of tactually coding controls.

Reading Printed Material Probably one of the most widely known tactual displays for reading printed material is braille printing. Braille print for the blind consists of raised "dots" formed by the use of all the possible combinations of six dots numbered and arranged thus:

1 . . 4
2 . . 5
3 . . 6

A particular combination of these represents each letter, numeral, or common word. The critical features of these dots are position, distance between dots, and dimension (diameter and height), all of which have to be discriminable to the touch.

Because of the cost of braille printing there are only limited amounts of braille reading material available to blind persons. The combination of research relating to the sensitivity of the skin and technological developments now permit us to convert optical images of letters and numbers into tactile vibrations that can be interpeted by blind persons. The Optacon (for *op*tical-to-*tac*tile *con*verter), developed at the Stanford University Electronics Laboratories, consists of essentially two components. One

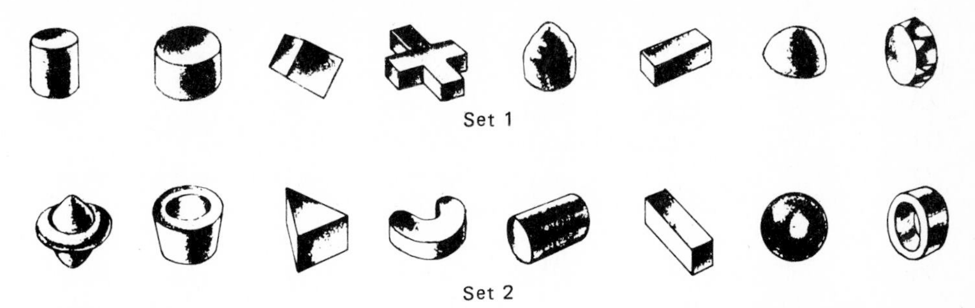

FIGURE 5-12
Two sets of knobs for levers that are distinguishable by touch alone. The shapes in each set are rarely confused with each other. (*Source: Jenkins, 1947.*)

of these is an optical pickup "camera," about the size of a pocketknife, and a "slave" vibrating tactile stimulator that reproduces images that are scanned by the camera (Bliss, Katcher, Rogers, & Shepard, 1970; Linvill, 1973). A photograph of the Optacon is shown in Figure 5-13. The 144 phototransistors of the camera in a 24 by 6 array serve to control the activation of 144 corresponding vibrating pins that "reproduce" the visual image. The vibrating pins can be sensed by one fingertip. The particular features of the vibrating pins and of their vibrations were based on extensive previous psychophysical research relating to the sensitivity of the skin to vibrations. [For what interest they may be, the specifications of the vibrating pins were: size of pin, 10 mils;

FIGURE 5-13
Photograph of Optacon for use in converting visual images into tactile vibrations. (*Photograph, Stanford Electronics Laboratories. The Optacon is produced by Telesensory Systems, Inc.*)

space between 6 rows, 100 mils; space between 24 pins in each row, 50 mils; frequency of vibrations, 230 Hz; depth of skin indentation, 2.6 mils (Bliss et al., 1970)].

The letter *O*, for example, produces a circular pattern of vibration that moves from right to left across the user's fingertip as the camera is moved across the letter.

Learning to "read" with the Optacon takes time. For example, one subject took about 160 hours (160 h) of practice and training to achieve a reading rate of about 50 words per min. Most users, however, only achieve rates of 10 to 15 words per min. This rate is well below that of most readers of braille, some of whom achieve a rate of 200 words per min. At the same time, such a device does offer the possibility for blind persons of being able to read conventional printing at least at a moderate rate.

Navigation Aids To navigate in an environment, we need a spatial map so that we can "get around." In recent years there has been a flurry of research directed at the development of tactual maps for blind persons. Tactual maps for the blind are very similar to conventional printed maps, except the density of information must be reduced because symbols must be large and well spaced to be identified by the sense of touch.

Environmental features can be classified as *point* features (e.g., a bus stop), *linear* features (e.g., a road) and *areal* features (e.g., a park). Much research effort has been expended to develop sets of discriminable symbols for maps. James and Armstrong (1976) have suggested a standard set of point and line symbols, as shown in Figure 5-14. These have been used in the United Kingdom and other parts of the world

FIGURE 5-14
A standardized set of line and point symbols for tactual maps. (*Source: James & Armstrong, 1976, as shown in Armstrong, 1978, fig. 6.*)

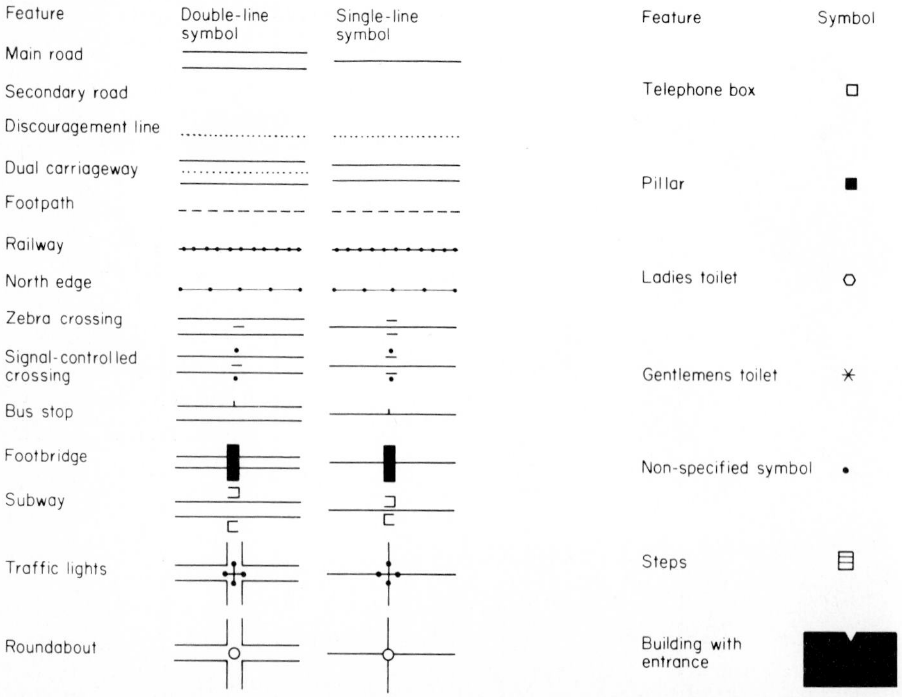

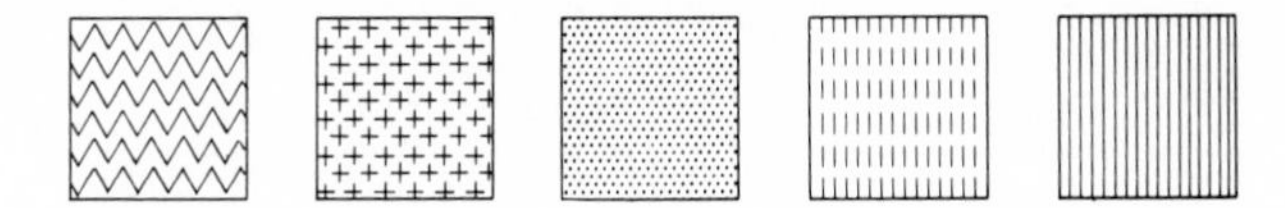

FIGURE 5-15
Discriminable areal symbols for use on tactual maps. (*Source: James & Gill, 1975, adapted from Armstrong, 1978, fig. 1.*)

(Armstrong, 1978). James and Gill (1975) investigated areal symbols and identified five discriminable symbols as shown in Figure 5-15.[3]

Although there has been a fair amount of research done on symbol discriminability, there has been relatively little research studying the ability of blind persons to use spatial maps. What few studies there have been, however, seem to indicate that such maps can be used effectively to aid navigation in unfamiliar environments (James & Armstrong, 1975).

Mobility Aids Two of the mobility aids we discussed under auditory displays also have built-in tactual stimulators, i.e., the Pathsounder (Russell, 1969) and the Sonicguide (Kay, 1973); these devices mainly sense obstructions in the field of view and warn the user. Another device, developed at the Smith-Kettlewell Institute of Visual Sciences and the Department of Visual Sciences of the University of the Pacific, is called the *Tactile Vision Substitution System* (TVSS) (Collins & Bachy-y-Rita, 1973). This system, illustrated in Figure 5-16, senses objects with a pair of "seeing eyeglasses" and transmits the images via a flexible-fiber optic bundle to a lightweight television camera, which in turn converts the visual image into a corresponding pattern of electronic impulses. These impulses stimulate a matrix of electrodes which are worn around the waist.

The system is still in the development stage. Despite some success in using the TVSS to identify and localize simple objects in an uncluttered setting, in its present form it is not suitable for general use by the blind (Jansson, 1978).

Tracking-Task Displays Although the task of tracking will be discussed in Chapter 8, we want to illustrate here the use of tactual displays for such tasks. Hofmann and Heimstra (1972), for example, used electrodes applied to the right and left sides of the neck to transmit directional information to subjects. The magnitude of the error was represented by the intensity of the stimulation. The study also involved the use of visual and auditory displays. Although the three types of displays resulted in some differences in the results as reflected by four different criteria used, the average percents of "efficiency" across all four measures were as follows:

Vision:	56%
Audition:	67%
Electrocutaneous:	64%

[3] See Lederman and Kinch (1979) for a review of areal symbol research and Hampshire (1979) for a review of methods for producing tactile graphic material.

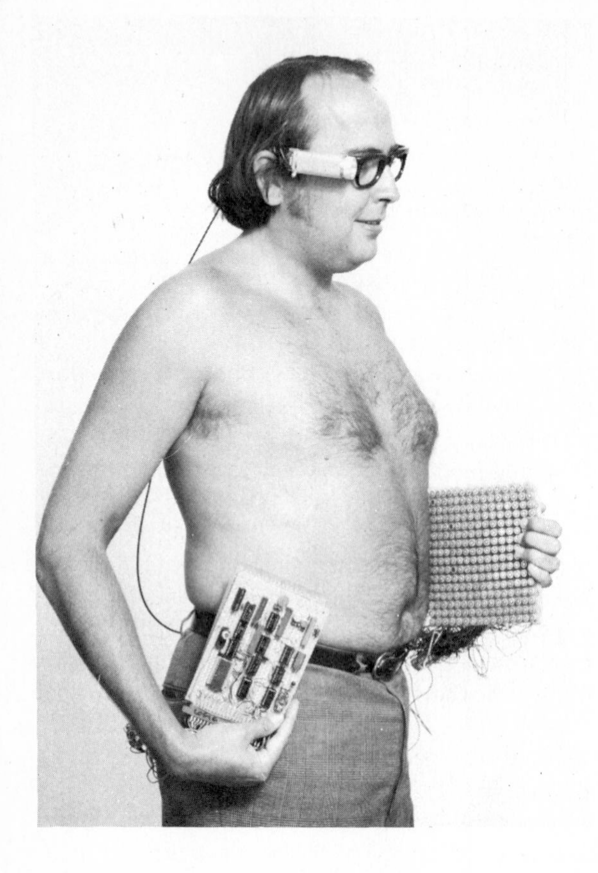

FIGURE 5-16
A portable electronic image projection seeing-aid system for the blind [the Tactile Vision Substitution System (TVSS)]. The image sensed by the "seeing" eyeglasses is transmitted by fiber optics to a small TV camera that converts the image into corresponding electronic pulses by an array of small electrodes shown here in the left hand. When in use, the components shown here are worn around the waist, the array of electrodes being in contact with the skin. *(Photograph, Smith-Kettlewell Institute of Visual Sciences, University of the Pacific.)*

The fact that the electrocutaneous display resulted in nearly as efficient a tracking performance as auditory signals offers reasonable promise of the potential use of such displays in at least certain tracking tasks.

Jagacinski, Miller, and Gilson (1979) used the tactual display shown in Figure 5-17 to transmit directional and magnitude information in a tracking task. Although somewhat poorer than visual tracking, under certain conditions tactual tracking yielded results approaching that of visual tracking.

Thus, tactual displays may have potential for use in tracking, especially if the task is relatively simple and the demands on the subject are not great.

Discussion

Relative to the other senses, the cutaneous senses seem generally suitable for transmitting only a limited number of discrete stimuli. Although we would then not expect tactual displays to become commonplace, there are at least two types of circumstances for which future developments of such displays would seem to be appropriate. In the first place, such displays offer some promise of being useful in certain special circumstances in which the visual and auditory senses are overburdened; in such instances tac-

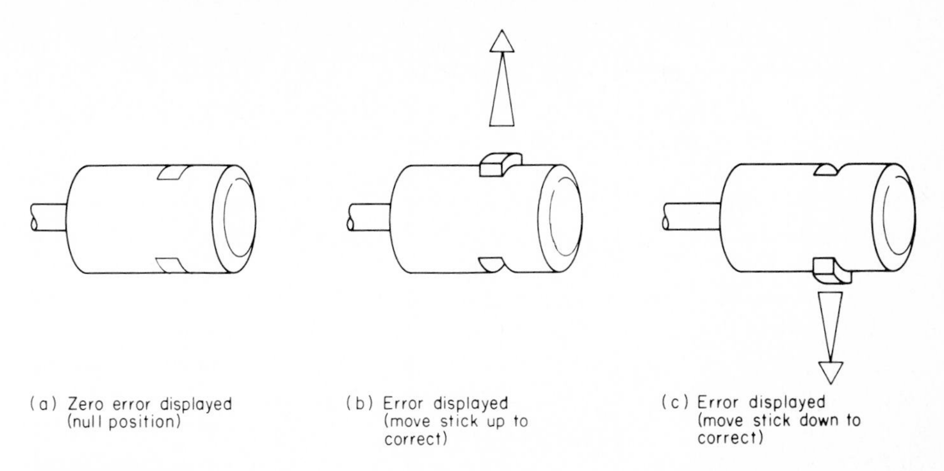

FIGURE 5-17
The tactual display used to present directional information to subjects in a tracking task. The correct response to condition *b* was to move the control stick up; to condition *c*, to move the stick down. (*Source: Jagacinski, Miller, & Gilson, 1979, fig. 1.*)

tual displays might be used, such as for warning purposes. And in the second place, as we have seen, they have definite potential as aids to blind persons.

THE OLFACTORY SENSE

In everyday life we depend on our sense of smell to give us information about things that otherwise is not easily obtainable—for example, the fragrance of a flower, the smell of fresh-brewed coffee, and the odor of sour milk. The olfactory sense is simple in its construction, but its workings remain somewhat of a mystery. The sense organ for smell (called the *olfactory epithelium*) is a small, 4 to 6 square centimeters (cm^2), patch of cells located in the upper part of each nostril. These *olfactory cells* contain *olfactory hairs* which actually do the detecting of different odors. The cells are connected directly to the olfactory areas of the brain.

There is little doubt that the sensation of odor is caused by airborne molecules of a volatile substance entering the nose. Research workers, however, are still not in agreement on just what property of these molecules is responsible for the kind of odor perceived, or whether the sense of smell is composed of distinct primary odors, and if so, what they are (Amoore, 1970).

Many people think that the sense of smell is very good—in one way it is and in other ways it isn't. The nose is apparently a very sensitive instrument for detecting the presence of odors; this sensitivity depends on the particular substance and the individual doing the sniffing. Isobutyl isobutyrate (a fruity odor), for example, can be detected in concentrations of about three parts per million in water (Amoore, 1970).

Surprisingly, however, our sense of smell is not outstanding when it comes to making absolute identifications of specific odors (we do far better comparing odors on a relative basis). The number of different odors we can identify depends on a number of factors including the types of odors to be identified and the amount of training. Desor and Beauchamp (1974), for example, found that untrained subjects could identify approximately 15 to 32 common stimuli (coffee, paint, banana, tuna fish, etc.) by their odors alone. With training, however, some subjects could identify up to 60 stimuli without error. This may seem high when compared with the visual or auditory sense, but it is not. These odor stimuli were complex "whole odors" as opposed to simple chemical odors. The analogy with the visual sense is the difference between simple color patches and complex pictures or faces. We can identify literally hundreds of different faces visually.

When it comes to identifying odors that differ only in intensity, we can identify about four different intensities (Engen & Pfaffmann, 1959). Thus, overall, the sense of smell is probably not to be counted on for identifying a specific stimulus from among many, but it can be effective in detecting the presence of an odor.

OLFACTORY DISPLAYS

Olfactory displays, needless to say, have not found widespread application. Part of the reason for this is that they cannot be depended upon as a reliable source of information because people differ greatly with respect to their sensitivity to various odors; a stuffy nose can reduce sensitivity; people adapt quickly to odors so that the presence of an odor is not sensed after a little while of exposure; the dispersion of an odor is hard to control; and some odors make people feel sick.

Despite all these problems, olfactory displays do have some useful applications—primarily as warning devices. The gas company, for example, adds an odorant to natural gas so that we can detect gas leaks in our homes. This is no less an "information display" than if a blinking light were used to warn us of the leak.

The author has seen an unusual sign in a computer room which reads:

> IF THE RED LIGHT IS BLINKING OR YOU SMELL WINTERGREEN
> –EVACUATE THE BUILDING.

The fire prevention system in the building releases carbon dioxide when a fire is detected. When the system was initially activated, the people in the building did not detect the presence of the carbon dioxide, and several people lost consciousness from the lack of oxygen. A wintergreen odor was added to the gas as a warning.

In the prior examples odor was added to a gas to act as a warning that the gas was being released, intentionally in the case of carbon dioxide, and unintentionally in the case of natural gas. Another example of an olfactory display uses odor to signal an emergency not associated with the gas itself. Several underground metal mines in the United States use a "stench" system to signal workers to evacuate the mine in an emergency. The odor is released into the mine's ventilation system and is quickly carried throughout the entire mine. This illustrates one general advantage of olfactory displays; they can penetrate vast areas which might not be economically reached by visual or auditory displays. Note, however, this is somewhat of a unique situation, consisting of vast areas (often hundreds of miles of tunnels) with no visual or auditory access, and a closed ventilation system for controlling the dispersion of the odorant.

Olfactory displays will probably never become widespread in application, but they do represent a unique form of information display which could be creatively integrated into very special situations to supplement more traditional forms of displays.

REFERENCES

Adams, S., & Trucks, L. A procedure for evaluating auditory warning signals. *Proceedings of the 6th Congress of the International Ergonomics Association and technical program for the 20th Annual Meeting of the Human Factors Society,* Santa Monica, Calif.: Human Factors Society, 1976.

American National Standards Institute (ANSI). *Standard specification for octave, half-octave, and third-octave band filter sets* (ANSI S1.11-1966). New York: American National Standards Institute, 1966.

Amoore, J. *Molecular basis of odor.* Springfield, Ill.: Charles C Thomas, 1970.

Armstrong, J. The development of tactual maps for the visually handicapped. In G. Gordon (Ed.), *Active touch.* Elmsford, N.Y.: Pergamon, 1978.

Bliss, J. C., Katcher, M. H., Rogers, C. H., & Shepard, R. P. Optical-to-tactile image conversion for the blind. *IEEE Transactions on Man-Machine Systems,* March 1970, *11*(1), 58–65.

Collins, C. C., Bachy-y-Rita, P. Transmission of pictorial information through the skin. *Advances in biological and medical physics,* 1973, *14*, 285–315.

Deatherage, B. H. Auditory and other sensory forms of information presentation. In H. P. Van Cott and R. G. Kinkade (Eds.), *Human engineering guide to equipment design.* Washington, D.C.: U.S. Government Printing Office, 1972.

Desor, J., & Beauchamp, G. The human capacity to transmit olfactory information. *Perception and Psychophysics,* 1974, *16*, 551–556.

Dobelle, W. Current status of research on providing sight to the blind by electrical stimulation of the brain. *Visual Impairment and Blindness,* September 1977, pp. 290–297.

Engen, T., & Pfaffmann, C. Absolute judgments of odor intensity. *Journal of Experimental Psychology,* 1959, *58,* 234–237.

Fidell S., Pearsons, K., & Bennett, R. Prediction of aural detectability of noise signals. *Human Factors,* 1974, *16,* 373–383.

Flynn, J. P., Goffard, S. J., Truscott, I. P., & Forbes, T. W. *Auditory factors in the discrimination of radio range signals: Collected informal reports* (OSRD Report 6292). Cambridge, Mass.: Harvard University, Pyscho-acoustic Laboratory, Dec. 31, 1945.

Frost, B., & Richardson, B. Tactile localization of sounds: Acuity, tracking moving sources, and selective attention. *Journal of the Acoustical Society of America,* 1976, *59,* 907–914.

Gales, R. Hearing characteristics. In C. M. Harris (Ed.), *Handbook of noise control* (2d ed.). New York: McGraw-Hill, 1979.

Geldard, F. A. Adventures in tactile literacy. *American Psychologist,* 1957, *12,* 115–124.

Geldard, F. A. *The human senses* (2d ed.). New York: Wiley, 1972.

Guyton, A. *Function of the human body* (3d ed.). Philadelphia: Saunders, 1969.

Hampshire, B. The design and production of tactile graphic materials for the visually impaired. *Applied Ergonomics,* June 1979, pp. 87–96.

Hawkes, G. *Tactile communication* (Report 62-11). Oklahoma City, Okla.: Federal Aviation Agency, Aviation Medical Service Research Division, Civil Aeromedical Research Institute, 1962.

Hofmann, M. A., & Heimstra, N. W. Tracking performance with visual, auditory, or electrocutaneous displays. *Human Factors,* 1972, *14*(2), 131–138.

Industrial noise manual (2d ed.). Detroit: American Industrial Hygiene Association, 1966.

Jagacinski, R., Miller, D., & Gilson, R. A comparison of kinesthetic-tactual and visual displays via a critical tracking task. *Human Factors,* 1979, *21,* 79-86.

James, G., & Armstrong, J. An evaluation of a shopping center map for the visually handicapped. *Journal of Occupational Psychology,* 1975, *48,* 125-128.

James, G., & Armstrong, J. Handbook on mobility maps. *Mobility Monograph,* University of Nottingham, 1976, No. 2.

James, G., & Gill, J. A pilot study on the discriminability of tactile areal and line symbols. *American Foundation for the Blind Research Bulletin,* 1975, *29,* 23-33.

Jansson, G. Human locomotion guided by a matrix of tactile point stimuli. In G. Gordon (Ed.), *Active touch.* Elmsford, N.Y.: Pergamon, 1978.

Jenkins, W. O. The tactual discrimination of shapes for coding aircraft-type controls. In P. M. Fitts (Ed.), *Psychological research on equipment design* (Research Report 19). Army Air Force, Aviation Psychology Program, 1947.

Kay, L. Sonic glasses for the blind: A progress report. *AFB Research Bulletin,* 1973, *25,* 25-28.

Kay, L., Bui, S., Brabyn, J., & Strelow, E. A single object sensor: A simplified binaural mobility aid. *Visual Impairment and Blindness,* May 1977, pp. 210-213.

Kirman, J. Tactile communication of speech: A review and an analysis. *Psychological Bulletin,* 1973, *80,* 54-74.

Lederman, S. Heightening tactile impressions of surface texture. In G. Gordon (Ed.), *Active touch.* Elmsford, N.Y.: Pergamon, 1978.

Lederman, S. & Kinch, S. Texture in tactual maps and graphics for the visually handicapped. *Visual Impairment and Blindness,* June 1979, pp. 217-227.

Licklider, J. C. R. *Audio warning signals for Air Force weapon systems* (TR 60-814). U.S. Air Force, Wright Air Development Division, Wright Patterson Air Force Base, March 1961.

Linvill, J. G. *Research and development of tactile facsimile reading aid for the blind (the Optacon).* Stanford, Calif.: Stanford Electronics Laboratories, Stanford University, March 1973.

Miller, G. A. Sensitivity to changes in the intensity of white noise and its relation to masking and loudness. *Journal of the Acoustical Society of America,* 1947, *19,* 609-619.

Minnix, R. The nature of sound. In D. M. Lipscomb and A. C. Taylor, Jr. (Eds.), *Noise control: Handbook of principles and practices.* New York: Van Nostrand Reinhold, 1978.

Mudd, S. A. *The scaling and experimental investigation of four dimensions of pure tone and their use in an audio-visual monitoring problem.* Unpublished Ph.D. thesis, Lafayette, Ind.: Purdue University, 1961.

Munson, W. A. The growth of auditory sensitivity. *Journal of the Acoustical Society of America,* 1947, *19,* 584.

Peterson, A. P. G., & Gross, E. E., Jr. *Handbook of noise measurement* (7th ed.). New Concord, Mass.: General Radio Co., 1972.

Pollack, I. Comfortable listening levels for pure tones in quiet and noise. *Journal of the Acoustical Society of America,* 1952, *24,* 158.

Pollack, I., & Ficks, L. Information of elementary multidimensional auditory displays. *Journal of the Acoustical Society of America,* 1954, *26,* 155-158.

Richardson, B., & Frost, B. Sensory substitution and the design of an artificial ear. *The Journal of Psychology,* 1977, *96,* 259-285.

Richardson, B., & Frost, B. Tactile localization of the direction and distance of sounds. *Perception and Psychophysics,* 1979, *25,* 336-344.

Richardson, B., Wuillemin, D., & Saunders, F. Tactile discrimination of competing sounds. *Perception and Psychophysics,* 1978, *24*, 546–550.

Riesz, R. R. Differential intensity sensitivity of the ear for pure tones. *Physiological Review,* 1928, *31*, 867–875.

Russell, L. *Pathsounder instructor's handbook.* Cambridge, Mass.: Massachusetts Institute of Technology, Sensory Aids Evaluation and Development Center, 1969.

Shower, E. G., & Biddulph, R. Differential pitch sensitivity of the ear. *Journal of the Acoustical Society of America,* 1931, *3*, 275–287.

Stevens, S. S., & Davis, H. *Hearing, its psychology and physiology.* New York: Wiley, 1938.

Vallbo, A., & Johansson, R. The tactile sensory innervation of the glabrous skill of the human hand. In G. Gordon (Ed.), *Active touch,* Elmsford, N.Y.: Pergamon, 1978.

Van Cott, H. P., & Warrick, M. J. Man as a system component. In H. P. Van Cott & R. G. Kinkade (Eds.), *Human engineering guide to equipment design* (Rev. ed.). Washington, D.C.: Superintendent of Documents, 1972.

Van der Molen, M., & Keuss, P. The relationship between reaction time and intensity in discrete auditory tasks. *Quarterly Journal of Experimental Psychology,* 1979, *31*, 95–102.

Wegel, R. L., & Lane, C. E. The auditory masking of one pure tone by another and its probable relation to the dynamics of the inner ear. *Physiological Review,* 1924, *23*, 266–285.

CHAPTER 6

SPEECH COMMUNICATIONS

In a human factors framework *speech* is both an *output* (from a speaker) and an *input* (to a hearer). Realizing this dual aspect, however, it seems most appropriate to discuss it here in connection with other inputs.

In many circumstances in which speech is used there is no particular problem in its transmission and reception. However, speech transmission can be affected adversely by noise, by the communication system (telephone, intercommunication system, radio, etc.), and by the hearing abilities of the hearer, and it is in such circumstances that speech and its transmission take on human factors implications. Such concern is especially relevant when the communications are particularly critical, such as in airport control tower operations. In the process of designing a speech communication system one needs to establish the criteria or standards that the system should meet in terms of the reception of speech, such as the relative importance of intelligibility (as in an airport control tower system), or perhaps naturalness or quality (as in a home telephone system—where it is important to be able to recognize voices).

With any such criteria or standards nailed down, the designer can consider the implications of the various "components" of the speech communication system as they might influence the fulfillment of these criteria, these components including the following: the message to be transmitted; the speaker; the transmission system itself (i.e., telephone, radio, intercommunication), including whatever noise may be present; and the hearer.

It is not feasible in this chapter to deal extensively with the many facets of speech communications. For our purposes, then, we will first discuss briefly the nature of speech, with particular reference to the processes by which speech sounds are formed and the characteristics of speech sounds. We will then discuss speech intelligibility and its measurement, and certain aspects of the components of speech communication systems that can affect intelligibility such as the message itself, the speaker, the transmission system (including certain types of speech distortion), noise and the hearer.

THE NATURE OF SPEECH

The ability of people to speak is of course integrally tied into, and is dependent upon, the respiratory process. As we breathe we inhale air into, and exhale air out of, the lungs, this process being accompanied by alternating increase and decrease in the space occupied by the chest. Although the breathing process normally occurs through the nose, we can also breathe through the mouth.

The Speech Organs and their Functions

The various organs involved in speech include the lungs, the larynx (with its vocal cords), the pharynx (the channel between the larynx and the mouth), the mouth or oral cavity (with tongue, teeth, lips, and velum), and the nasal cavity. The vocal cords consist of folds that can open and close very rapidly, thus causing frequent stoppage of the breath stream. (The opening between the folds is called the *glottis*.) Speech sounds are generated by the sound waves that are created by the breath stream as it is exhaled from the lungs and modified by the speech organs. In speech, the rate of opening and closing of the vocal folds controls the basic frequency of the resulting speech sounds. As the sound wave is transmitted from the vocal cords it is further modified in three resonators, the pharynx, the oral cavity, and the nasal cavity. By manipulating the mandible (in the back of the oral cavity) and the articulators (the tongue, lips, and velum) we are able to so modify the sound wave in order to utter various speech sounds. In addition, by controlling the velum and the pharyngeal musculature we can influence the nasal quality of the articulation.

Types of Speech Sounds

The processes involved in speech make it possible to produce a very large number of *phonemes* (or speech sounds). These fall into two basic groups, namely the vowels and consonants. The vowels are produced by a relatively unobstructed air passage, with variations in the vowel sounds being controlled largely by laryngeal mechanisms and by the positioning of the tongue and lips. The consonants, in turn, are produced with more restricted air passage and more varied manipulation of the articulators. There are subclasses of both vowels and consonants, as follows:

Vowels

Pure vowels, as in *pet, hit, put, nut, hot,* and *pat.*

Diphthongs, vowel sounds that involve some lingual movement (or what is sometimes called a *slide*), as in *boy, how,* and *sigh.*

Consonants

Stops, produced by completely stopping the passage of air through the oral cavity. There are three types of stops: oral stops (such as *b, p, d, t, g,* and *k*); nasal stops (such as *m* and *n*); and a class of affricated stops that are characterized by stop closure followed by a *delayed* release (such as the *ch* in *ch*ur*ch* and the *j* and *ge* in *j*ud*ge*).

Fricatives or *spirants,* produced by forming a narrow slit or groove for air passage, as *th* in *th*eir.

Laterals, formed by closing the middle line of the mouth and leaving an air passage around one or both sides, as *l* in *l*et.

Trills, caused by rapid vibration of an articulator, as the trilled *r* in certain European languages. There are also taps as in bi*tt*er and flaps as in mur*d*er.

Retroflex sounds, as in the American English *r*ead or bee*r*.

The English language includes about 16 different vowel sounds, the specific number depending on whether certain closely related sounds are counted as the same or different; and it includes about 22 consonant sounds, making a total of about 38 phonemes. Other languages employ some phonemes that we do not use, and vice versa. Often two languages will have the same phoneme, but it will be "patterned" differently, as the *p* in Spanish and English.

The variations in air pressure associated with any given sound (the speech wave) can be represented graphically in various ways. One such representation is the *waveform*, which shows the variations in air pressure over time. Another representation is the *spectrum*, which depicts the combinations of frequency and intensity (speech power) of the speech sound. The combinations of many individual speech sounds in turn can be represented by an "overall" or "continuous" spectrum which shows the combinations of frequencies and intensities of the many individual sounds; further, it is possible to derive an estimate of the overall intensity level across all frequencies.

Intensity of Speech

The average intensity, or speech power, of individual speech sounds varies tremendously, with the vowels generally having much greater speech power than the consonants. For example, the *a* as pronounced in *talk* has roughly 680 times the speech power of *th* as pronounced in *then*. This is a difference of about 28 dB.

The overall intensity of speech, of course, varies from one person to another and also for the same individual. For example, when one talks almost as softly as possible, speech has an intensity of about 46 dB, and when one talks almost as loudly as possible, it has a decibel level of about 86. The results of a survey by Fletcher (1953, p. 77) of the telephone-speech levels of a good-sized sample of people showed a range from about 50 to about 75 dB, with a mean of about 66 dB. There was, however, a heavy concentration of cases from about 60 to 69 dB, with about 40 percent falling within 3 dB above or below the mean.

Spectrum of Speech

As indicated above, each speech sound has its own unique spectrum of sound frequencies, although there are, of course, differences in these among people, and individuals can (and do) shift their spectra up and down the register depending in part on the cir-

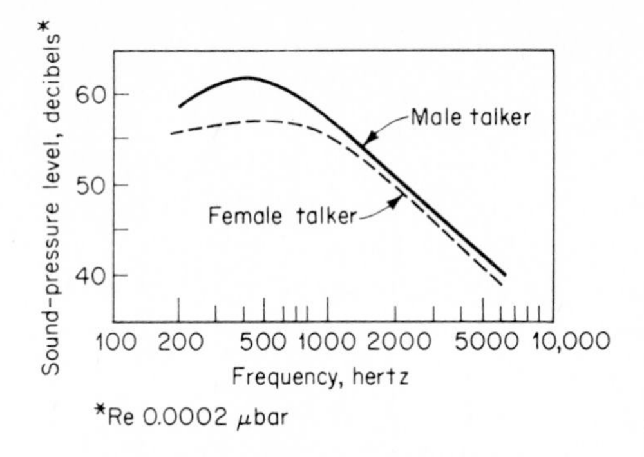

FIGURE 6-1
The octave-band long-time average spectrum of speech of samples of males and females. (*Source: Adapted from Kryter, 1972, fig. 5-2, p. 164.*)

cumstances, such as in talking in a quiet conversation (when frequency and intensity are relatively low), or talking to a group (when the frequency and intensity tend to be higher), or screeching at the children to stay out of trouble (when the frequency and intensity typically are near their peak). In general, the overall speech level for men is higher than for women, but in addition the spectrum for men has more dominant components in the lower frequencies than in the case of women. These differences are shown in Figure 6-1, which shows the overall level of speech of samples of men and women. This shows up the fact that the speech of men tends to be characterized by higher levels in the lower frequencies than the speech of women. The overall level for women is about 3 dB lower than that for men.

INTELLIGIBILITY OF SPEECH

In many communication situations the most relevant criterion of voice communications is that of intelligibility, that is, the extent to which the transmitted message is understood by the listener. Some quantitative indication of the intelligibility of communications can serve various purposes, such as speech research, evaluating speech communication systems, evaluating the effects of noise and other conditions on intelligibility, and measuring the speech effectiveness and hearing abilities of individuals.

Tests of Speech Intelligibility

The most straightforward method of measuring speech intelligibility is by the use of some standardized test. Such tests usually involve the transmission of speech material to individuals who are asked to repeat what they hear, or to make some other response. The speech material may be presented directly by trained speakers or it may be presented by the use of tapes or records. There are various types of speech intelligibility tests, and although this is not the place to treat them in detail, brief descriptions will be given of a few types or examples.

- Nonsense-syllable tests. Subjects are presented with a number of nonsense syllables and are asked to repeat them. They are scored on the basis of the accuracy of their responses.

• Phonetically balanced (PB) word lists. The words in these lists are proportionately representative of various types of speech sounds used in everyday speech. Subjects are scored on the basis of the accuracy with which they repeat the words.

• Diagnostic Rhyme Test (DRT). This test, reported by Voiers (1977), consists of pairs of words with similar "rhymes" (such as *goat-coat*) but with other speech elements that are different. In the usual administration of the test the listener has a printed list of pairs of words, and when one of the words is presented orally the listener marks off the word he judges to have been uttered. The scoring is based on the number of correct responses.

• Sentence tests. There are various types of tests that involve the use of sentences. One type, for example, includes sentences that consist of questions to which there is only one answer; subjects are scored in terms of the correctness of their answers. One particular test has been developed to measure speech intelligibility in noise (the SPIN test reported by Kalikow, Stevens, & Elliott, 1976). The test items are sentences that are presented in babble-type noise, and the listener's response is the final word in the sentence (the key word) which is always a monosyllabic noun. There are two types of sentences. One consists of high-predictability items (for which the key word is somewhat predictable from the context), and the other consists of low-predictability items (for which the key word cannot be predicted from the context). The scoring is based on the number of correct responses.

The Articulation Index (AI)

The use of appropriate speech intelligibility tests can well serve the various purposes referred to above, whether the objective be that of evaluating the speaking or hearing abilities of individuals, or of evaluating the adequacy of communication systems and environmental variables such as noise. However, because these tests are time-consuming and costly, efforts have been made to develop procedures for deriving indirect indices of the intelligibility of speech that would avoid the necessity of using such tests. The articulation index (AI) is the result of one such effort. Although there are several methods of deriving an articulation index, they all involve about the same basic approach. The procedures of one such approach are summarized below, in particular, the one-third-octave-band method (ANSI, 1969; Kryter, 1972). (Other methods provide for the use of data based on 20 bands or octave bands.)

1 For each one-third octave band, plot on a worksheet, such as in Figure 6-2, the band level of the speech peak reaching the listener's ear. The specific procedures for deriving such band levels are given by Kryter, but an "idealized" spectrum for males is presented in Figure 6-2 as the basis for illustrating the derivation of the articulation index.

2 Plot on the worksheet the band levels of steady-state noise reaching the ear of the listener. An example of such a spectrum is presented in Figure 6-2.

3 Determine at the center frequency of each of the bands on the worksheet the difference in decibels between the level of speech and that of the noise. When the noise level exceeds that of speech, assign a zero value. When the speech level exceeds that of

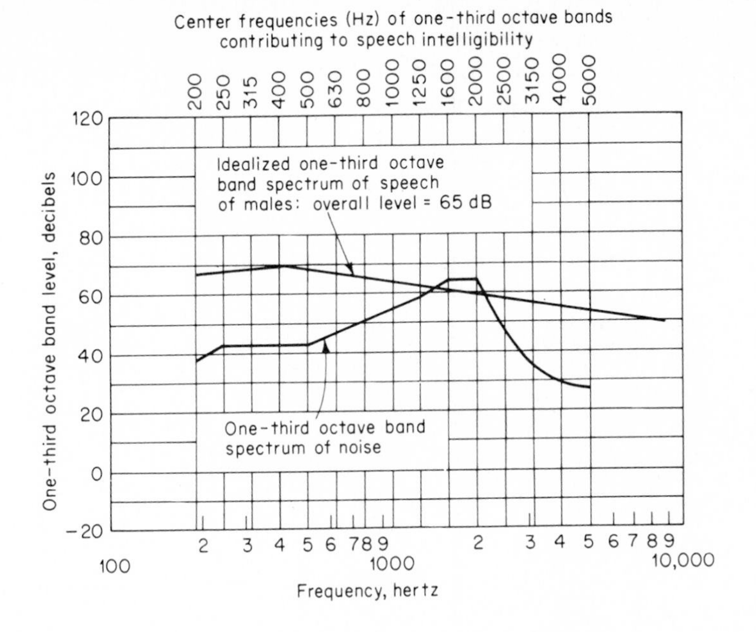

1. Band	2. Speech peaks minus noise, dB	3. Weight	4. Column 2 x 3
200	30	0.0004	0.0120
250	26	0.0010	0.0260
315	27	0.0010	0.0270
400	28	0.0014	0.0392
500	26	0.0014	0.0364
630	22	0.0020	0.0440
800	16	0.0020	0.0320
1000	8	0.0024	0.0192
1250	3	0.0030	0.0090
1600	0	0.0037	0.0000
2000	0	0.0038	0.0000
2500	12	0.0034	0.0408
3150	22	0.0034	0.0758
4000	26	0.0024	0.0624
5000	25	0.0020	0.0500
		AI =	0.4738

FIGURE 6-2
Example of the calculation of an articulation index (AI) by the one-third-octave-band method. In any given situation the difference (in dB) between the level of speech and the level of noise is determined for each band. These differences are multiplied by their weights. The sum of these is the AI. (*Source: Adapted from Kryter, 1972.*)

noise by more than 30 dB, assign a difference of 30 dB. Record this value in column 2 of the table given as part of Figure 6-2.

4 Multiply the value for each band derived by step 3 above by the weight for that band as given in column 3 of the table given with Figure 6-2, and enter that value in column 4.

5 Add the values in column 4. This sum is the AI.

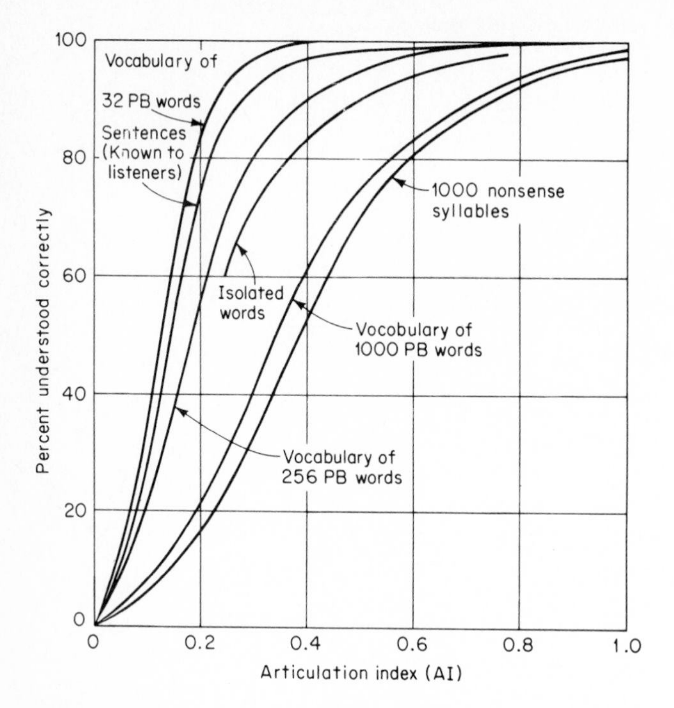

FIGURE 6-3
Relationship between the articulation index (AI) and the intelligibility of various types of speech-test materials. (*Source: Adapted from Kryter, 1972, and French & Steinberg, 1947.*)

The AI, in a sense, is a summated index of the differences–across the octave bands–between a typical speech spectrum and the spectrum of the background noise. Although it is not a particularly meaningful index by itself, its value lies in the fact that it can be used to derive an estimate of speech intelligibility of various types of speech material, specifically in the form of an intelligibility score.

Intelligibility Score

An intelligibility score is simply the percentage of spoken material of any particular type that is understood by the listener. The relationships between AI values and intelligibility scores for various types of speech material are shown in Figure 6-3. We will refer back to this figure later. The relationships shown in that figure are approximations in that they depend upon such factors as the speaking skills of the speakers and the hearing skills of listeners. The figure shows that intelligibility depends very much on the nature of the speech material. However, it has been suggested that if the AI of a speech communication system is less than about 0.3, the system would be inadequate for transmission of conventional speech material (Beranek, 1947).

COMPONENTS OF SPEECH COMMUNICATIONS

If we need to do something about improving the intelligibility of speech communication in some system, we need to do so in terms of the individual components involved, such as the message itself, the speaker, the transmission system, and the hearer.

The Message

Under adverse communication conditions such as noise, some speech messages or message units are more susceptible to degradation than others. Under such conditions, then, it behooves one to construct messages in such a way as to increase the probability of their getting through.

The Vocabulary Used If we picked a word randomly from the dictionary, the probability of your guessing the correct one would be infinitesimally small. (P. S. Would you have guessed *pedantic*?) But if we arbitrarily restrict our language to two words (for example, *pedantic* and *jostle*), your probability of guessing the right one would be 50-50. Under extremely noisy conditions, when it is difficult to make out the speech sounds, the total number of possible words that *might* be used has a marked influence on the correct recognition of the words—the smaller the possible vocabulary, the greater the probability of recognition. This general principle has been confirmed in experiments such as the one by Miller, Heise, and Lichten (1951). In this experiment subjects were presented with words from vocabularies of various sizes (2, 4, 8, 16, 32, and 256 words, and unselected monosyllables). These were presented under noise conditions with signal-to-noise ratios ranging from −18 to 9 dB. The results, shown in Figure 6-4, show clearly that the percent of words correctly recognized was very strongly correlated with the size of the vocabulary that was used.

Phonetic Characteristics of Message Components Within limits it is sometimes possible in speech communications to select and use those speech components with phonetic characteristics that aid in differentiating between and among the various components that are used. The International Civil Aviation Organization (ICAO), for example, has developed a word list that represents the letters of the alphabet: *Alfa, Bravo,*

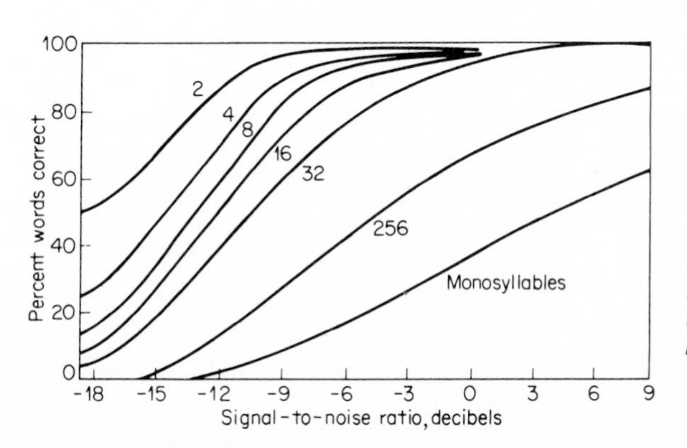

FIGURE 6-4
Intelligibility of words from vocabularies of different sizes under varying noise conditions. The numbers 2, 4, 8, etc., refer to the number of words in the vocabulary used. (*Source: Miller, Heise, & Lichten, 1951. Copyright © 1951 by the American Psychological Association and reproduced by permission.*)

Charlie, Delta, etc. The words in this list generally have two characteristics: each word has considerable speech power (and therefore can "get through" a fair amount of noise); and further, these words are not readily confused with each other.

In connection with the use of letters and digits in speech communications, Hull (1976) has demonstrated that certain characters are more easily confused with each other than are other characters. For example, the letters within each of the following groups frequently are confused with each other: *DVPBGCETQU; FXSH; KJA; MN;* and *YIR.* The upshot of such experimental findings is that when auditory transmission is difficult, and if letters are used in the messages by themselves (or in words in which the interpretation of the letters is critical), one should try to avoid (or at least minimize) the use of letters within those subsets that are easily confused with each other.

The Context of the Message Another factor that influences the intelligibility of speech is the context of the message or message components. This is essentially a problem of expectancy. If we were to hear the expression "A rolling ______ gathers no moss," but failed to distinguish the third word, we could readily supply it in the context of the expression. But it would be hard for us to fill in this blank: "On Wednesday he ______ ." The effect of context on intelligibility was illustrated in Figure 6-3. In that figure it can be seen that, for any given AI (as derived by the method discussed earlier in this chapter), sentences are more intelligible than isolated words, and isolated words, in turn, are more intelligible than separate syllables. Thus, components in the context of a meaningful conceptual message stand a better chance of being picked up against a noisy background than do those that have no such contextual backdrop.

Clearly what we are talking about here is redundancy and "information" load and how it correlates with intelligibility. The first context above makes the missing element *stone* highly redundant. It therefore carries little information with it. Intelligibility is, therefore, much increased where an element is redundant. The second context above ("On Wednesday he ______ "), on the other hand, provides no relevant cues about (and no redundancy regarding) the missing item. Knowing what the missing item would be therefore carries a high degree of information; but when there are no cues from the context to suggest what that missing item might be, the intelligibility of the statement would be low if that item could not be understood. Data based on the SPIN test referred to before tend to emphasize the relevance of the context of messages as very important in influencing intelligibility.

The Speaker

As we all know, the intelligibility of speech depends in part on the character of the speaker's voice. Although, in common parlance, we refer to such features of speech as enunciation, research has made it possible to trace down certain specific features of speech that affect its intelligibility. In a study by Bilger, Hanley, and Steer (1955), for example, it was found that the speech of "superior" speakers (as contrasted with less intelligible speakers) had a longer "syllable duration," and that they spoke with greater intensity, utilized more of the total time with speech sounds (and less with pauses), and varied their speech more in terms of fundamental vocal frequencies.

The differences in intelligibility of speakers generally are due to the structure of their articulators (the organs that are used in generating speech) and the speech habits people have learned. Although neither of these factors can be modified very much, it has been found that appropriate speech training usually can bring about moderate improvements in the intelligibility of speakers.

The Transmission System

Speech transmission systems (such as telephones and radios) can produce a variety of forms of distortion, such as frequency distortion, filtering, amplitude distortion, and modifications of the time scale. If intelligibility (and not fidelity) is important in a system, it should be noted that certain types of distortion (especially amplitude distortion) still can allow intelligibility. Since high-fidelity systems are very expensive, it is useful to know what effects various forms of distortion have on intelligibility to be able (it is hoped) to make better decisions about the design or selection of communication equipment. For illustrative purposes, we shall discuss the effects of a couple of types of distortion, namely, filtering and amplitude distortion.

Effects of Filtering on Speech The filtering of speech consists basically in blocking out certain frequencies and permitting only the remaining frequencies to be transmitted. Filtering may be the fortuitous, unintentional consequence, or in some instances the intentional consequence, of the design of a component. Most filters eliminate frequencies *above* some level (a *low-pass* filter) or frequencies *below* some level (a *high-pass* filter). Typically, however, the cutoff, even if intentional, usually is not precisely at a specific frequency, but rather tapers off over a range of adjacent frequencies. A given filtering affects the intelligibility of certain sounds more than others. Fletcher (1953, pp. 418–419), for example, points out that long *e* is recognized correctly about 98 percent of the time if the frequencies either above or below 1700 Hz are eliminated. But while *s* is affected only slightly by eliminating frequencies below 1500, its intelligibility is practically destroyed by eliminating frequencies above 4000. Most short vowels have important sound components below 1000 Hz, and 20 percent error in their recognition occurs when frequencies below that level are eliminated; but the elimination of frequencies above 2000 has little effect on their intelligibility.

However, the intelligibility of normal speech does not depend entirely upon intelligibility of each and every speech sound. For example, filters that eliminate all frequencies above, or all frequencies below, 2000 Hz will *in the quiet* still transmit speech quite intelligibly, although the speech does not sound natural. The distortion effects on speech of filtering out certain frequencies are summarized in Figure 6-5. This shows, for high-pass filters and for low-pass filters with various cutoffs, the percentage of intelligibility of the speech that is transmitted. It can be seen, for example, that the filtering out of frequencies above 4000 Hz or below about 600 Hz has relatively little effect on intelligibility. But look at the effect of filtering out frequencies above 1000 Hz or below 3000 Hz! Such data as those given in Figure 6-5 can provide the designer of communications equipment with some guidelines to follow when trying to decide how much filtering can be tolerated in the system.

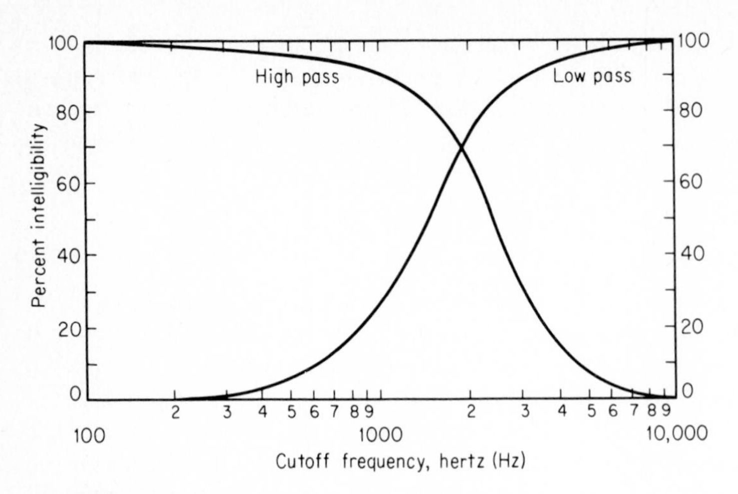

FIGURE 6-5
Effect on intelligibility of elimination of frequencies by the use of filters. A low-pass filter permits frequencies below a given cutoff to pass through the communication system, and eliminates frequencies above that cutoff. A high-pass filter, in turn, permits frequencies above the cutoff to pass and eliminates those below. (*Source: Adapted from French & Steinberg, 1947.*)

Effects of Amplitude Distortion on Speech *Amplitude distortion* is the deformation which results when a signal passes through a nonlinear circuit. One form of such distortion is *peak clipping*, in which the peaks of the sound waves are clipped off and only the center part of the waves left. Although peak clipping is produced by electronic circuits for experiments, some communication equipment has insufficient amplitude-handling capability to pass the peaks of the speech waves and at the same time provide adequate intensity for the lower-intensity speech components, thus reducing intelligibility. Since peak clipping impairs the quality of speech and music, it is not used in regular broadcasting, but it is sometimes used in military and commercial communication equipment. In such cases premodulation clippers are built into the transmitters, thus reducing the peaks, but the available power is then used for transmitting the remainder of the speech waves. *Center clipping*, on the other hand, eliminates the amplitudes below a given value and leaves the peaks of the waves. Clipping is more of an experimental procedure than one used extensively in practice. The amount of clipping can be controlled through electronic circuits. Figure 6-6 illustrates the speech waves that would result from both forms of clipping.

The effects of these two forms of clipping on speech intelligibility are very different, as shown in Figure 6-7. It can be seen that peak clipping does not cause major degradation of intelligibility even when the amount of clipping (in decibels) is reasonably high. On the other hand, even a small amount of center clipping results in rather thorough garbling of the message. The reason for this difference in the effects of peak and center clipping is the difference in phonetic characteristics of the vowels (which generally have more speech power) and of the consonants (which generally have lower speech power). Thus, when the peaks are lopped off, we reduce the power of the

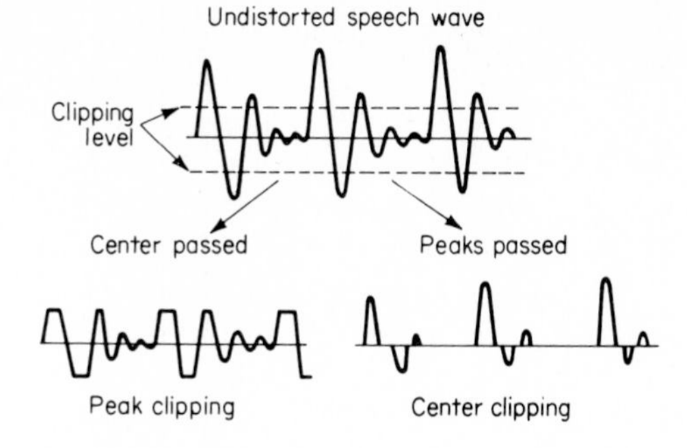

FIGURE 6-6
Undistorted speech wave and the speech waves that would result from peak clipping and center clipping. (*Source: Miller, 1963, p. 72.*)

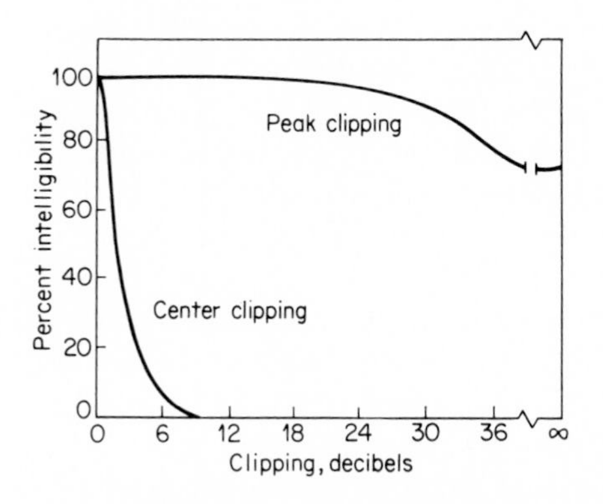

FIGURE 6-7
The effects on speech intelligibility of various amounts of peak clipping and of center clipping. (*Source: Licklider & Miller, 1951.*)

vowels, which are less critical in intelligibility, and leave the consonants essentially unscathed. But when we cut out the center amplitudes, the consonants fall by the wayside, thus leaving essentially the high peaks of the vowels. Since intelligibility is relatively insensitive to peak clipping, the communications engineers can shear off the peaks and repackage the waveforms, using the available power to amplify the weaker but more important consonants.

Noise

Some noise is ever with us, but in large doses it is a potential bugaboo as far as speech communications are concerned, whether it be ambient noise (noise in the environment) or in-line noise in a communication system. Various approaches have been carried out with the view toward developing systematic procedures for evaluating the effects of noise on speech. The indices that resulted from a couple of these approaches will be described briefly, namely the Preferred-Octave Speech Interference Level (PSIL) and the Noise Criteria (NC) curves.

Preferred-Octave Speech Interference Level (PSIL) This index, reported by Peterson and Gross (1978) can be used as a gross basis for comparing the relative effectiveness of speech transmission under different environments of reception. For any given situation, it is actually the simple numerical average of the decibel levels of noise in three octave bands, namely, those with centers at 500, 1000, and 2000 Hz. Thus, if the decibel levels of noise in the three octave bands are 70, 80, and 75 dB, respectively, the PSIL would be their average, 75 dB. The PSIL is useful as a rough index for estimating the effects of noise on speech intelligibility, especially if the noise spectrum is relatively flat. It loses some of its value if the noise has intense low-frequency components, has an irregular spectrum, or consists primarily of pure tones. The speech interference level (SIL) (without the "preferred") usually refers to the arithmetic average of the levels in an earlier-used set of three octave bands: 600 to 1200; 1200 to 2400; and 2400 to 4800 Hz.

Some indication of the relationship between PSIL (and SIL) values and speech intelligibility under different circumstances, as presented by Webster (1969, pp. 49–73), is shown in Figure 6-8. In particular this shows the effects of noise in face-to-face communications when the speaker and the hearer are at various distances from each other when the background noise is characterized by various PSIL and SIL values. For the distance and noise level combinations in the white area of the figure, a normal voice would be satisfactory, but for those combinations above the "maximum vocal effort line" it would probably be best to get out the boy scout wigwag flags or send up Indian smoke signals.

The subjective reactions of people to noise levels in private offices and in large offices (secretarial, drafting, business machine offices, etc.) were elicited by Beranek and Newman (1950) by questionnaires. The results of this survey were used to develop a rating chart for office noises, as shown in Figure 6-9. The line for each group represents PSILs (baselines) that were judged to exist at certain subjective ratings (vertical scale). The dot on each curve represents the judged upper limit for intelligibility. The judged limit for private offices (normal voice at 9 ft) was slightly above 45 dB, and for larger offices (slightly raised voice at 3 ft) was 60 dB.

FIGURE 6-8
Voice level and distance between talker and listener for satisfactory face-to-face communication as limited by ambient noise level (expressed in PSIL and SIL). For any given noise level (such as a PSIL of 70) and distance (such as 8 ft) it is possible to determine the speech level that would be required (in this case a "shout"). (*Source: Adapted from Webster, 1969, fig. 19, p. 69.*)

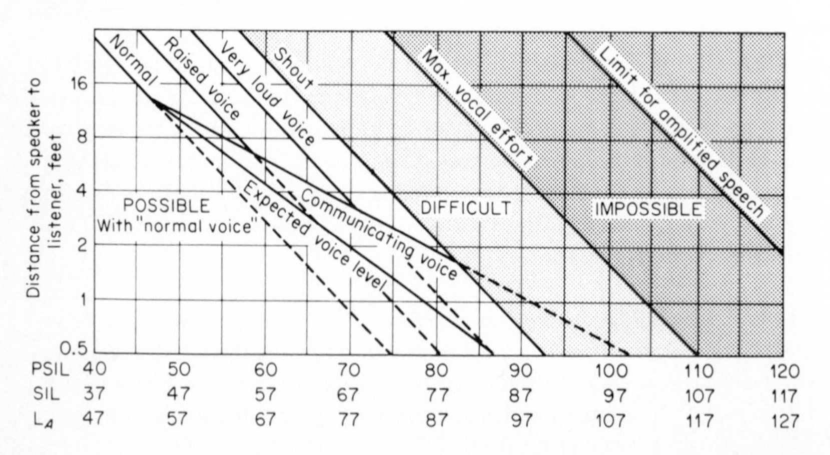

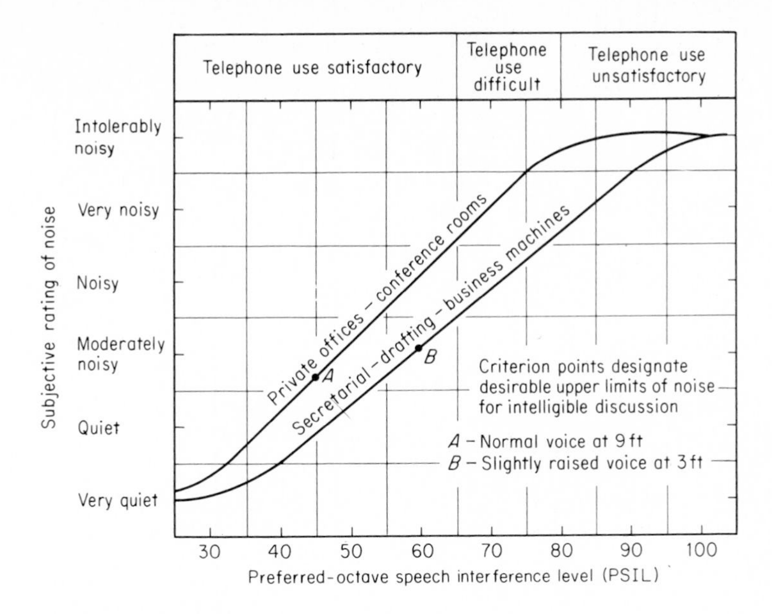

FIGURE 6-9
Rating chart for office noises. (*Source: Beranek & Newman, 1950, as modified by Peterson & Gross, 1978, to reflect the current practice of using octave bands with centers at 500, 1000, and 2000 Hz.*)

The judgments with regard to telephone use are given in Table 6-1. These judgments were made for long-distance or suburban calls. For calls within a single exchange, about 5 dB can be added to each of the listed levels, since there is usually better transmission within a local exchange.

Criteria for control of background noise in various communication situations have been set forth by Peterson and Gross based on earlier standards by Beranek and Newman. These, expressed as PSILs, are given in Table 6-2.

TABLE 6-1
EFFECTS OF PREFERRED-OCTAVE SPEECH INTERFERENCE LEVELS (PSIL) ON TELEPHONE USE

PSIL, dB	Telephone use
Less than 60	Satisfactory
60 to 75	Difficult
Above 80	Impossible

Source: Peterson & Gross, 1978, p. 38.

TABLE 6-2
MAXIMUM PERMISSIBLE PREFERRED-OCTAVE SPEECH INTERFERENCE LEVELS (PSIL) FOR CERTAIN TYPES OF ROOMS AND SPACES

Type of room	Maximum permissible PSIL (measured when room is not in use)
Secretarial offices, typing	60
Coliseum for sports only (amplification)	55
Small private office	45
Conference room for 20	35
Movie theater	35
Conference room for 50	30
Theaters for drama, 500 seats (no amplification)	30
Homes, sleeping areas	30
Assembly halls (no amplification)	30
Schoolrooms	30
Concert halls (no amplification)	25

Source: Peterson & Gross, 1978, table 3-5, p. 39.

Noise Criteria (NC) Curves These curves, originally developed by Beranek (1957), are particularly useful for evaluating background noise inside office buildings and in rooms and halls in which speech communications are important. A set of NC curves is shown in Figure 6-10. In use, the noise spectrum of the area is plotted on the chart, and each octave-band level is then compared with the NC curves to find the one that penetrates to the highest NC level. The corresponding value of the NC curve is the NC rating of the noise. As an example, the noise level of an office is shown on the figure in encircled crosses. This noise would have a rating of 38 since that is the value of the octave band of the noise that penetrates to the highest NC level.

Reverberation

Reverberation is the effect of noise bouncing back and forth from the walls, ceiling, and floor of an enclosed room. As we know from experience, in some rooms or auditoriums this reverberation seems to obliterate speech or important segments of it. Figure 6-11 shows approximately the reduction in intelligibility that is caused by varying degrees of reverberation (specifically, the time in seconds that it takes the noise to die down). This relationship essentially is a straight-line one.

Effects of Earplugs on Speech In a sense, earplugs are part of the transmission system, since they intervene between the environment and the receiver. Although their purpose is to prevent or minimize hearing loss, one might expect that they would also reduce the intelligibility of speech. In fact, using earplugs may actually increase the intelligibility of speech under high noise levels. Under low noise levels, however, the use of earplugs may impair speech intelligibility somewhat. It is in high-noise-level situations, however, that there would be a greater likelihood that earplugs would be worn,

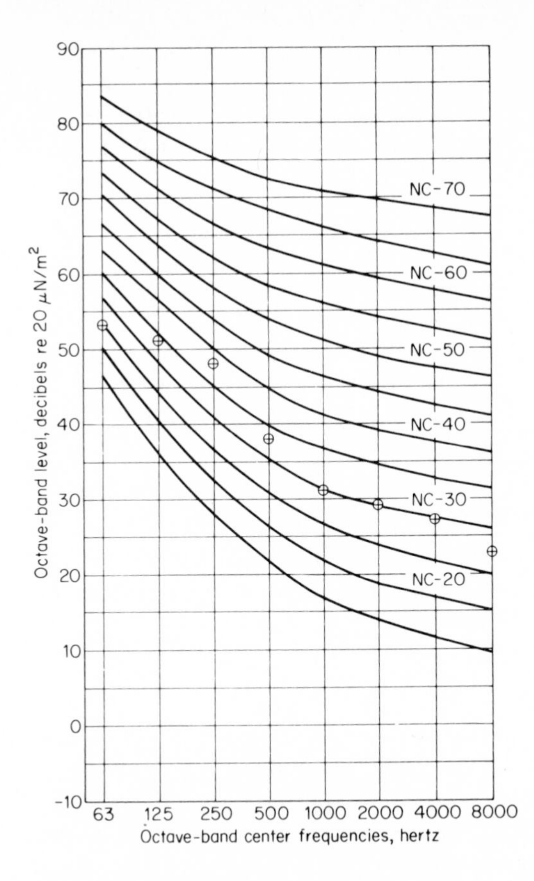

FIGURE 6-10
Noise criteria (NC) curves for use in evaluating noise levels of rooms or spaces in which speech communications are important. The noise spectrum of any given type of area should not exceed the recommended level for that type of area (such as shown in terms of PSIL in Table 6-2). The spectrum for an office is shown in encircled crosses; the rating of that office is 38. (*Source: Schultz, 1968, fig. 1, p. 637, as presented by Peterson & Gross, 1978, fig. 3-6.*)

and in such conditions they can be effective. The explanation for this is that at high noise levels a point is reached where additional intensity cannot be discriminated; at such levels the difference between the intensity of the signal (in this case speech) and of its background noise cannot be discriminated. The value of earplugs under such circumstances is to bring the levels of both the signal and the background noise down to the point where the difference between them can be discriminated.

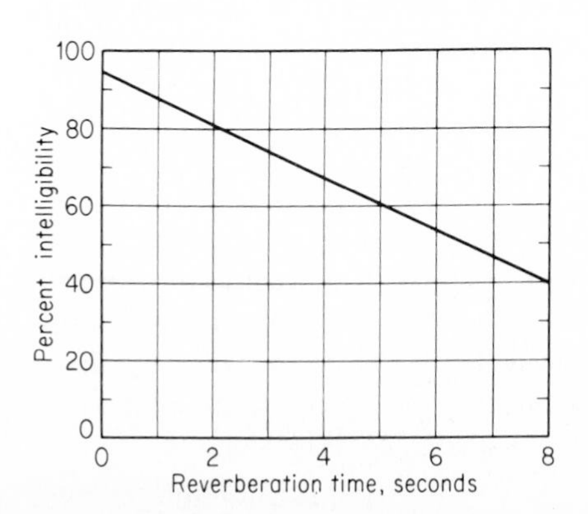

FIGURE 6-11
Intelligibility of speech in relation to reverberation time. The longer the reverberation of noise in a room, the lower the intelligibility of speech. (*Source: Adapted from Fletcher, 1953.*)

In connection with the use of earplugs, however, there have been some recent inklings that some people, when wearing earplugs, may tend to lower their voices, thereby reducing the intelligibility of their speech. To the extent that this might occur, such individuals should be trained to speak more loudly than they otherwise would, or, conversely, to remove their earplugs when speaking.

The Hearer

The hearer or listener is the last link in the communication chain. For receiving speech messages under noise conditions, the hearer should have normal hearing, should be trained in the types of communications to be received, should be reasonably durable in withstanding the stresses of the situation, and should be able to concentrate on one of several conflicting stimuli.

RECENT INNOVATIONS IN SPEECH COMMUNICATIONS

In recent years there have been certain interesting developments in speech communications that offer some promise of being useful in specific types of circumstances. We will refer briefly to a couple of these developments.

Computerized Voice Recognition

One such development is the use of computers for the recognition of voices. Such recognition could be relevant for ensuring authorized access to computers or in connection with security matters (as in identifying persons who are authorized to have access to certain facilities or locations). One technique that has been developed for this purpose is reported in an article on voice operated computer systems in *Applied Ergonomics*. Each individual in question repeats each of a specified list of 150 words into a noise-canceling microphone. These repetitions, in turn, enable the system to obtain an average voice pattern from the slight variations which occur each time the speaker pronounces a given word. This pattern is then stored in the memory of the computer. When any such word is spoken by an individual the pattern of the spoken voice can be compared by the computer with the already recorded pattern in order to determine if the two are the same or different. In this particular situation the "recognition" of spoken words serves as the basis for instructions to the computer. In the application of this technique to security circumstances the comparison would be made to confirm (or disconfirm) the identity of an individual.

Voice-Operated Control

In the situation described above the recognition of spoken words serves to give instructions to a computer. However, there are various types of circumstances in which the human voice might be used (via a computer) for the control of some process or operation. Further reference to the use of the voice for such control will be made in Chapter 9.

DISCUSSION

When feasible, of course, speech communications should be carried out under favorable conditions, uncluttered with noise. However, in many circumstances it is not possible to reduce noise at its source; (one cannot stop the rolling mills of a steel mill for people to communicate with others). Under these and other circumstances it is necessary to look to other elements of the total communication system, rather than to the noise source itself, for possibilities of improving the intelligibility of speech. On the engineering design side of the coin, the possibilities to consider are those of minimizing the transmission of noise if possible (through acoustical treatment and other means), improving the design of the communication equipment, and modifying the nature of the messages to be used; and on the personnel side of the coin, the possibilities are those of selection and training of speakers and hearers, where these are feasible.

REFERENCES

American National Standards Institute (ANSI). *Methods for the calculation of the articulation index* (Report 535). New York: American National Standards Institute, 1969.

Beranek, L. L. The design of speech communication systems. *Proceedings of the Institute of Radio Engineers,* New York: 1947, *35,* pp. 880–890.

Beranek, L. L. Revised criteria for noise in buildings. *Noise Control,* 1957, *3*(1), 19–27.

Beranek, L. L. & Newman, R. B. Speech interference levels as criteria for rating background noise in offices. *Journal of the Acoustical Society of America,* 1950, *22,* 671.

Bilger, R. C., Hanley, T. D., & Steer, M. D. *A further investigation of the relationships between voice variables and speech intelligibility in high level noise.* TR for SDC, 104-2-26, Project 20-F-8, Contract N6ori-104, Lafayette, Ind.: Purdue University (mimeographed), 1955.

Fletcher, H. *Speech and hearing in communication.* Princeton, N. J.: Van Nostrand Reinhold, 1953.

French, N. R., & Steinberg, J. C. Factors governing the intelligibility of speech sounds. *Journal of the Acoustical Society of America,* 1947, *19,* 90–119.

Hull, A. J. Reducing sources of human error in transmission of alphanumeric codes. *Applied Ergonomics,* 1976, 7(2), 75–78.

Kalikow, D. N., Stevens, K. N., & Elliott, L. L. Test of speech intelligibility in noise using sentences with controlled word predictability. *Journal of the Acoustical Society of America,* 1976, *60,* Suppl. No. 1, p. S28.

Kryter, K. D. Speech communication. In H. P. Van Cott and R. G. Kinkade (Eds.), *Human engineering guide to equipment design.* Washington, D.C.: U.S. Government Printing Office, 1972.

Licklider, J. C. R., & Miller, G. A. The perception of speech. In S. S. Stevens (Ed.), *Handbook of experimental psychology.* New York: Wiley, 1951.

Miller, G. A. *Language and communication* (paperback ed.). New York: McGraw-Hill, 1963.

Miller, G. A., Heise, G. A., & Lichten, W. The intelligibility of speech as a function of the context of the test materials. *Journal of Experimental Psychology,* 1951, *41,* 329–335.

Peterson, A. P. G., & Gross, E. E., Jr. *Handbook of noise measurement* (8th ed.). New Concord, Mass.: Gen Rad Co., 1978.

Schultz, T. J. Noise criterion curves for use with the USASI preferred frequencies. *Journal of the Acoustical Society of America,* 1968, *43*(3), 637, 638.

Voice operated computer systems. *Applied Ergonomics,* 1975, *6*(1), 46, 47.

Voiers, W. D. Diagnostic evaluation of speech intelligibility, paper 34 in M. E. Hawley (Ed.), *Speech intelligibility and speaker recognition.* Stroudsburg, Pennsylvania: Dowden, Hutchinson & Ross, Inc., 1977.

Webster, J. C. *Effects of noise on speech intelligibility* (ASHA Reports 4). Washington, D.C.: The American Speech and Hearing Association, 1969, pp. 49–73.

PART THREE

HUMAN OUTPUT AND CONTROL

CHAPTER 7

HUMAN ACTIVITIES: THEIR NATURE AND EFFECTS

Any form of human work (whether chopping wood for a fireplace, assembling doorknobs, driving a taxi, programming a computer, or figuring one's income tax) requires some combination of activities—sensory, perceptual, mental, and physical in nature. The performance of these activities is of course accompanied by related physiological changes and sometimes psychological changes. Further, the adjustment of people to various physical and social environments also may be accompanied by physiological and psychological changes.

Certain work activities or environmental conditions may be such that the accompanying physiological or psychological effects are within reasonable or acceptable limits. In other circumstances, however, there may be undesirable physiological or psychological consequences. In such circumstances the concepts of *stress* and *strain* arise. In general terms *stress* refers to some undesirable condition, circumstance, task, or other factor that impinges upon the individual, and *strain* refers to the effects of the stress. Some of the possible sources of stress are given in Figure 7-1, these being based on the work, the environment, and the circadian (diurnal) rhythm, the more specific sources being subdivided into physiological and psychological. (This particular figure, from Singleton, 1971, does not include social and interpersonal environmental sources of stress, which can have both physiological and psychological effects.) Stress can have either external or internal origins, or some combination of these. The external origins can include the work activity itself, the physical or social aspects of the work situation, pressures from supervisors or peers, and other aspects of the work environment. In turn, Jex and Clement (1979) point out that certain internal states of individuals, such as concern for consequences of failure, can serve as internal sources of stress.

Granting that certain work-related factors would serve as external sources of stress

Source of stress	Physiological	Psychological
Work	Heavy work Immobilization	Information overload Vigilance
Environment	Atmospheric Noise/vibration Heat/cold	Danger Confinement
Circadian	Sleep loss	Sleep loss

Measures of strain				
Chemical	Electrical	Physical	Activity	Attitudes
Blood content Urine content Oxygen consumption Oxygen deficit Oxygen recovery curves Calories	EEG (electroencephalogram) EKG (electrocardiogram) EMG (electromyograph) EOG (electrooculogram) GSR (galvanic skin response)	Blood pressure Heart rate Sinus arrhythmia Pulse volume Pulse deficit Temp. of body Respiratory rate	Work rate Errors Blink rate	Boredom Other attitudinal factors

FIGURE 7-1
Primary sources of stress and primary measures of strain as induced by stress. (*Source: Adapted from Singleton, 1971, fig. 4.*)

to virtually anyone, it must be recognized that some such factors might be sources of stress for some individuals but not for others. These differences can arise from various individual characteristics such as those of a physical, mental, attitudinal, or emotional nature. Individual differences in such characteristics might be reflected, for example, in differences in capacities to adjust or adapt to different external conditions. In addition, a person's perception of his or her ability to deal with the job and its external features could cause some people to experience stress but not others. As another example, the recognition of the importance of the consequences of one's work might trigger a stressful reaction on the part of some individuals.

In at least some circumstances, sources of possible stress can be minimized or eliminated, as by modifying the work equipment or procedures, changing work schedules, modifying the environment, or other means. Since such modification can influence certain physiological processes and physical activities of people, this chapter includes an overview of the various physical structures that are involved in human motor activity, and a discussion of some of the "concomitants" of work, that is, some of the physiological, subjective or psychological, and performance variables associated with work. In addition the chapter includes a discussion of certain work- and environmental-related matters of energy expenditure, and some coverage of certain aspects of the physical movements and activities that people can perform.

BIOMECHANICS OF MOTION

Biomechanics deals with the various aspects of physical movements of the body and body members.[1] The abilities of people to perform various types of motor activities

[1] For a more extensive treatment of biomechanics, especially as related to equipment design, the reader is referred to Damon, Stoudt, & McFarland (1966), and to Tichauer (1978). A glossary of terms used in biomechanics is included in *American National Standard Industrial Engineering Terminology: Biomechanics*, ANSI Z94.1-1972, The American Society of Mechanical Engineers, New York.

depend essentially on the physical structure of the body (the skeleton), the skeletal muscles, the nervous system, and the metabolic processes.

The Skeletal Structure

The basic structure of the body consists of the skeleton, there being 206 bones that form the skeleton. Certain bone structures serve primarily the purpose of housing and protecting essential organs of the body, such as the skull (which protects the brain) and the rib cage (which protects the heart, lungs, and other internal organs). The other (skeletal) bones—those of the upper and lower extremities and the articulated bones of the spine—are concerned primarily with the execution of physical activities, and it is these that are particularly relevant to our subject. The skeletal bones are connected at body joints, there being two general types of joints that are principally used in physical activities, namely *synovial* joints and *cartilaginous* joints. The synovial joints include hinge joints (such as the fingers and knees), pivot joints (such as the elbow, which is also a hinge joint), and ball-and-socket joints (such as the shoulder and hip). The primary examples of cartilaginous joints are those of the vertebrae of the spine, which make possible, collectively, considerable rotation and forward bending of the body.

The Skeletal Muscle System

The bones of the body are held together at their joints by ligaments. The skeletal muscles (also called *striated* or *voluntary* muscles) consist of bundles of muscle fibers that have the property of contractility; the muscle fibers serve to convert chemical energy into mechanical work. The two ends of each muscle blend into tendons, which, in turn, are connected to different skeletal bones in such a manner that when the muscles are activated, they apply some form of mechanical leverage.

Neural Control of Muscular Activity

The nerves entering a muscle are of two classes, namely, *sensory* nerves and *motor* nerves. Some of the sensory nerves are associated with the cutaneous senses (touch, heat and cold, pain, etc.). The other sensory nerves are the proprioceptors, which are distributed through the muscles, the tendons, and the covering of the bones and which provide kinesthetic feedback to aid in muscular control. The motor nerves actually control the actions of the muscles.

The execution of physical activities depends upon learning which can be viewed in a two-stage, hierarchical frame of reference. To borrow terms from computer lingo, the first stage, the *executive program*, deals with the overall purpose or plan of the act and is essentially under conscious control by the central nervous system. The second stage, the *subroutine*, deals with the control of the specific movements that are required to complete the act. Through practice the subroutines typically become so learned that they are performed automatically once the "executive plan" is put into action. The "learning" of the subroutines is based on the establishment of neural connections at the neural center of the motor nerve pathways. Thus, if at the executive program level one decides to go upstairs, the subroutines of doing so typically are executed without conscious control.

Muscle Metabolism

Metabolism is the collective chemical process of the conversion of foodstuffs into two forms, namely mechanical work and heat. Some of the mechanical work is of course used internally in the processes of respiration and digestion. Other mechanical work is used externally, as in walking and performing physical tasks. In either case, heat is generated, usually in amounts that are excessive to body needs; this surplus heat must be dissipated by the body. The contraction of a muscle requires energy, the basic source of this energy being *glycogen*, which can be thought of as a large number of glucose molecules bound together to form one large molecule. The conversion of glycogen into energy consists of a chemical reaction that ends in the production of lactic acid. However, the lactic acid needs to be dissipated by being broken down into water and carbon dioxide. The first stage (the breakdown of glycogen into lactic acid) does not require oxygen, and is said to be *anaerobic.* The second stage (the breakdown of lactic acid into water and carbon dioxide) is said to be *aerobic.*

At the initiation of physical activity the muscles can utilize the glycogen which is already available. But the amount of glycogen and hence glucose available is small; so if the activity is continued, the body needs to replenish these nutrients from the blood, along with a supply of oxygen, which is required in the second stage. When an adequate supply of oxygen is provided, little or no lactic acid accumulates. If the level of activity requires more oxygen than is provided by the normal rate of blood flow through the cardiovascular system, the system adjusts itself to fulfill the increased demands, in particular by increasing the breathing rate to bring additional oxygen into the lungs, and by increasing the heart rate to pump more blood through the "pipes" of the cardiovascular system. From the heart the blood is pumped through the lungs, where it picks up a supply of oxygen, which is then carted by the blood to the muscles where the oxygen is needed. With at least moderate rates of work, the heart rate and breathing rate are normally increased to the level that provides enough oxygen to perform the physical activities over a continuing period of time. However, when the amount of oxygen delivered to the muscles fails to meet the requirements (as when the level of physical activity is high), lactic acid tends to accumulate in the blood. If the rate or duration of physical activity results in continued accumulation of lactic acid, the muscles will ultimately cease to respond. If the rate of removal of lactic acid does not keep pace with its formation, additional oxygen must be supplied after cessation of the activity to remove the remaining lactic acid. This is referred to as the *oxygen debt.* Since this debt has to be paid back, the heart rate and breathing rate do not immediately settle back to prework levels when work ceases, but rather slow down gradually until the borrowed oxygen is replaced.

Basal Metabolism The *basal metabolic rate* is that which is required simply to maintain the body in an inactive state. Although it varies from individual to individual, the average for adults usually ranges from about 1500 to 1800 kilocalories per day (kcal/d) (Schottelius & Schottelius, 1978).[2] Considering the basal level plus the energy required for a relatively sedentary existence, Passmore (1956) estimates that about 500 kcal are required for 8 h in bed plus about 1400 kcal for nonworking time, adding up to a total of 1900 kcal/day; in turn, Lehmann (1958) estimates the corresponding requirements (basal metabolism and leisure) at around 2300 kcal/d, and Schottelius and Schottelius estimate that the typical adult who lives a fairly sedentary life utilizes about 2400 kcal/day. Thus, various estimates of the total nonworking calorie requirements range from about 1900 to about 2400 kcal/day. (The physiological costs of work will be mentioned later.)

Movement of Body Members

The study of human movements as a function of the construction of the musculoskeletal system is referred to as *kinesiology* (Tichauer, 1978). In this regard the body members operate as levers. As illustrated by Tichauer, for example, the forearm has its fulcrum at the elbow, and one of the muscles of the upper arm (the *brachialis*) provides the activating force just in front of the elbow to bend the forearm. There are three different types of levers, and various body movements represent all three types.

Types of Movements Certain of the basic types of movements that are performed by the body members are described below, along with their associated jargon in kinesiology:

- *Flexion:* bending, or decreasing the angle between the parts of the body
- *Extension:* straightening, or increasing the angle between the parts of the body
- *Adduction:* moving toward the midline of the body
- *Abduction:* moving away from the midline of the body
- *Medial rotation:* turning toward the midline of the body
- *Lateral rotation:* turning away from the midline of the body
- *Pronation:* rotating the forearm so that the palm faces downward
- *Supination:* rotating the forearm so that the palm faces upward

Some of these basic movements are illustrated in Figure 7-2, along with the following values for each (as based on a sample of 39 men selected to represent the major physical types in the military services): mean angle (in degrees) and 5th and 95th percentile angles (computed from the standard deviations for the sample). In this, as in other aspects of biomechanics, there are the ever-present individual differences, including the effects of physical condition and the ravages of age.

[2] The energy unit generally used in physiology is the kilocalorie (abbreviated kcal) or Calorie (with a capital C to distinguish it from the gram-calorie). The *kilocalorie* is the amount of heat required to raise the temperature of a kilogram of water from 15 to 16 degrees Celsius (°C). The relation of the kilocalorie to certain other units of energy measurement is

1 kcal = 426.85 kilogram-meters (426.85 kg-m)
1 kcal = 3087.4 foot pounds (3087.4 ft-lb)
1 kcal = 1000 calories (1000 cal = 1 C)

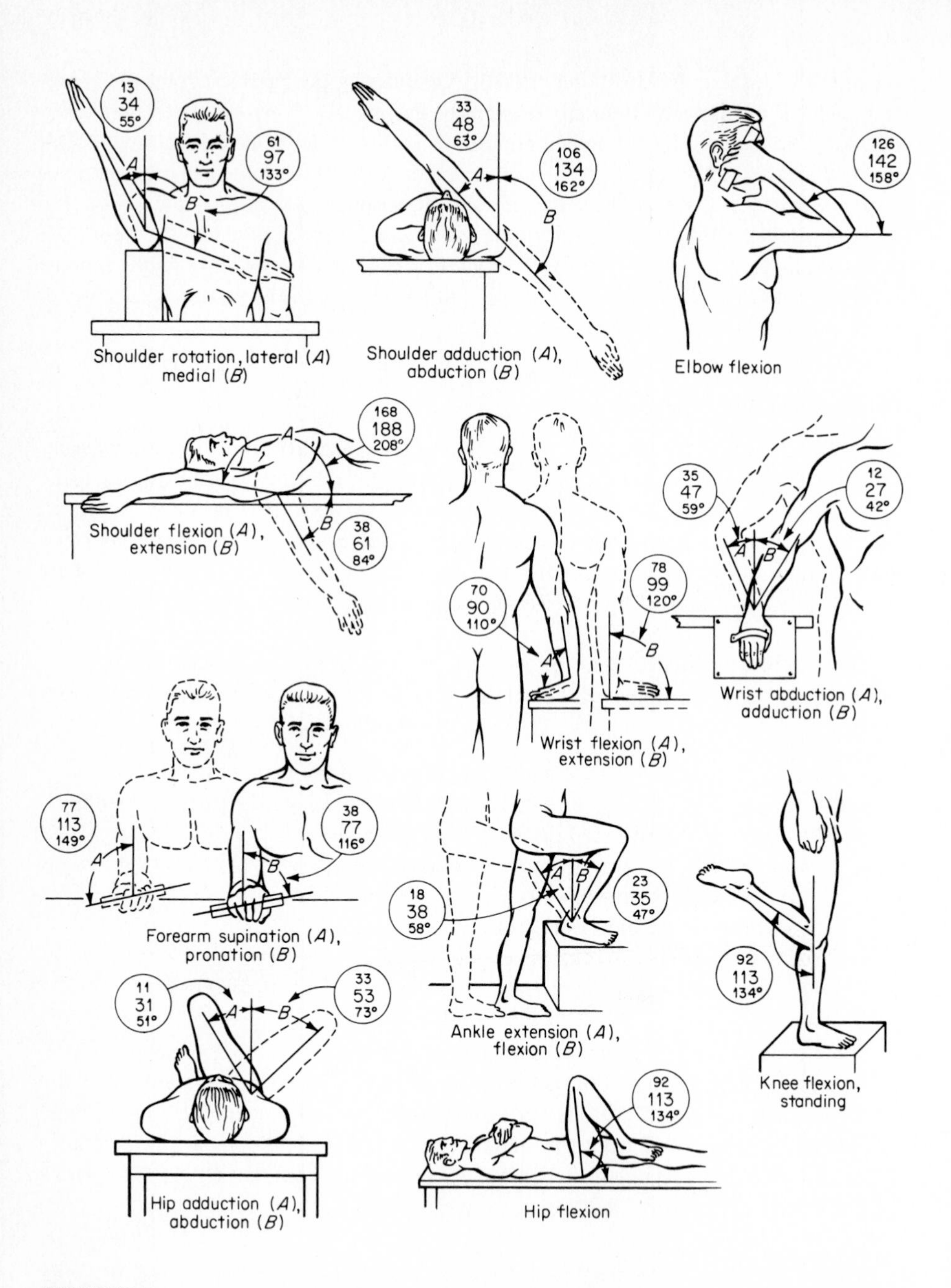

FIGURE 7-2
Range of certain movements of the upper and lower extremities, based on a sample of 39 men selected to represent the major physical types in the military services. The three values (in degrees) given for each angle are the 5th percentile, the mean, and the 95th percentile, respectively, of voluntary (not forced) movements. (***Source:* Dempster, 1955, as reanalyzed by Barter, Emanuel, & Truett, 1957.**)

Although specific movements of body members can be described in terms of these basic movements, in describing work activities it usually is preferable to do so in more operational terms. There are different ways in which movements can be so classified, one of them being given below:

- *Positioning* movements are those in which the hand or foot moves from one specific position to another, as in reaching for a control knob.
- *Continuous* movements are those which require muscular control adjustments of some type during the movement, as in operating the steering wheel of a car or guiding a piece of wood through a band saw.
- *Manipulative* movements involve the handling of parts, tools, control mechanisms, etc., typically with the fingers or hands.
- *Repetitive* movements are those in which the same movement is repeated, as in hammering, operating a screwdriver, and turning a handwheel.
- *Sequential* movements are several relatively separate, independent movements in a sequence.
- A *static* adjustment is the absence of a movement, consisting of maintaining a specific position of a body member for a period of time.

Various types of movements may be combined in sequence so that they blend one into another. For example, placing the foot on a brake pedal is a positioning movement, but this may be followed by a continuous movement of adjusting the amount of brake pressure to the conditions of the situation. Similarly, a continuous movement may include holding a position (a static adjustment) for a short time.

Measurement of Movements There are different aspects of physical movements that can be measured, these including *range* of movements, *force* applied during the activity (i.e., strength), *endurance, speed,* and *accuracy.* For measuring these, various kinds of gadgetry are used, such as timing devices, motion pictures, strain gauges, and dynamometers. One rather interesting device used for this purpose is a *force platform.* This is a small platform on which a subject stands when carrying out some physical activity. By the use of some sensing elements below the platform (such as piezoelectric crystals) it is possible to sense and then automatically record the forces generated by the subject in each of three planes, namely, vertical, frontal, and transverse. The original force platform was developed by Lauru (1954); other platforms have been used experimentally by Barany (1963) and by Greene, Morris, and Wiebers (1959). Such devices are sensitive to slight differences in physical movements and can thus lend themselves to use in comparing the three-dimensional forces in different activities. Recordings of the forces in the operation of manual and of electric typewriters are shown in Figure 7-3 for comparison.

It has been proposed that such force-time recordings, as possible indices of energy expenditure, are nearly as accurate as metabolic measurements and can thus be used as a measure of physiological cost of a given motion (Brouha, 1960, p. 103). In fact, Brouha presents data for oxygen cost of certain work activities that show high correlations (ranging from .83 to .96) with data from the force platform.

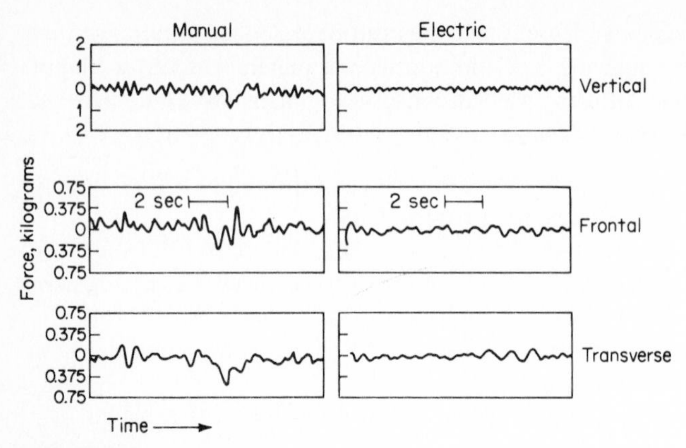

FIGURE 7-3
Forces in three dimensions (vertical, frontal, and transverse) in the operation of a manual and an electric typewriter, recorded with a force platform. (*Source: Brouha, 1960, p. 106.*)

THE CONCOMITANTS OF HUMAN ACTIVITY

Human activity (or exposure to environmental conditions) usually is accompanied by any of several types of effects or consequences, these generally having to do with physiological conditions of the individual, subjective reactions of the individual, or the performance of the individual. These "effects" can be either desirable or undesirable. Certain undesirable effects may be indicative of strain. Both for research and operational purposes measurements of these may be relevant.

Measures of Physiological Functions

Numerous types of physiological functions or conditions vary as the consequence of work or environmental exposure. Under desirable conditions measures of these functions may be at or near their normal levels. Under conditions of stress, however (such as those referred to in Figure 7-1), certain forms of physiological strain may occur, such as those for which measures are also referred to in Figure 7-1. Wierwille (1979) refers to certain other possible indices of physiological strain (that he calls physiological work-load measures), these being flicker-fusion frequency or FFF (which is the lowest frequency at which a flickering "on-off" light appears to be constant); evoked cortical potential or ECP (which is somewhat like the EEG except that the electrical potentials are measured on the scalp or upper neck); body fluid analysis (a complex chemical analysis of body fluids); eye and eyelid movements; and muscle tension.

Certain measures are indicative, in their own way, of the general level of physiological strain, others tend to reflect muscular conditions, and still others tend to be more indicative of mental or perceptual activity.

Measures of General Physiological Status In exposure to physical stress, the physiological strain is manifested in cardiorespiratory functions. There are many measures of these functions, such as heart rate, blood pressure, respiratory rate, oxygen consumption, and stroke (pulse) volume. (*Stroke volume* is the volume of blood pumped through the arteries for every heartbeat.) Heart rate and oxygen consumption probably are the most commonly used. These two measures tend to be reasonably well correlated with each other across varying levels of general dynamic muscular work, but neither is a very sensitive index of static muscular work, "local" dynamic muscular work (that of specific muscles or muscle groups), or mental work (Burger, 1969). Heart rate is reasonably indicative of the effects of heat stress and emotional stress, but is also related to individual factors (constitution, physical condition, sex, etc.), and is therefore less suitable as an absolute index of the load imposed by various types of work than is oxygen consumption.

There are certain derivatives of oxygen consumption and heart rate that are sometimes used. The *oxygen debt*, for example, is the amount of oxygen that is required by the muscles after the beginning of work, over and above that which is supplied to them by the circulatory system during their activity. This debt needs to be "repaid" after the cessation of work, and is reflected in the elevated rate (i.e., above resting level) of oxygen consumption in the recovery process, as illustrated in Figure 7-4. This figure represents the "theoretical" pattern as it would be expected to occur under a steady-state condition of work. However, as Bălănescu (1979) has demonstrated, the actual patterns may differ from this model, especially when the work activity varies during the work period. In any event, however, the oxygen that is originally "borrowed" at the beginning of the work period, or during it, is ultimately "repaid."

Another measure based on heart rate is the heart-rate recovery curve as used by Brouha (1960), this being a curve of the heart rate measured at certain intervals after work (such as 1, 2, and 3 min). In general terms, the more strenuous the work activity, the longer it takes the heart rate to settle down to its prework level.

Since no single physiological measure is a completely adequate index of physiological strain, various attempts have been made to derive composite indices that would more adequately reflect the level of such strain. Burger, for example, proposes the use of an index of circulatory load. An approximation of such an index is the product of heart rate, mean blood pressure, and stroke volume. In turn, Dukes-Dobos, Wright, Carlson, and Cohen (1976) experimented with a combination of 18 cardiorespiratory measures and found a combination of 9 of them to be most sensitive to changes in the level of physical activity. The combination of these was used in the derivation of a cardiorespiratory variance score (CVS). However, such composite measures are still not completely adequate, and in addition the problems of instrumentation in their measurement impose serious constraints on their use in many practical situations. Thus, it is still the most common practice to use single parameters such as heart rate or oxygen

FIGURE 7-4
Illustration of the oxygen debt under conditions of steady-state work activity.

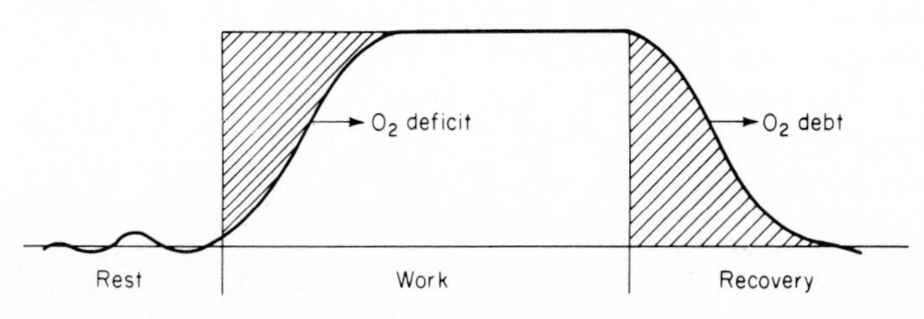

FIGURE 7-5
Apparatus for monitoring heart rate and obtaining sample oxygen consumption levels of a person while working on the job. (*Photograph courtesy of Human Factors Department of Eli Lilly and Company.*)

consumption as measures of physiological strain. As measures of the energy cost of work (the source of stress) oxygen consumption or calories frequently are used, usually expressed in units per minute or per hour.

Certain physiological measures can be obtained while people are performing regular work activities, such as shown in Figure 7-5.

Measures of Local Muscular Condition One of the available measures of the physiological condition of individual muscles or muscle groups is electromyographic (EMG) recordings, which are also called *myograms*. These are inked tracings of the electrical impulses that occur during work, and provide estimates of the magnitude of voluntary muscular activity. Examples of such recordings are shown in Figure 7-6 for four muscles of one subject when applying a constant torque of 60 ft-lb and of 15 ft-lb on a steel socket. There have been several problems associated with the recording, quantification, and interpretation of such recordings. In this regard Khalil (1973) has developed procedures for summating the action potentials of several muscles that are monitored simultaneously. The values given in Figure 7-6 show the individual and summated values for the four muscles. Although Khalil believes that his method of summating the recordings is suitable for measuring both static (i.e., isometric) and dynamic muscular exertion, Tichauer (1978) makes the point that the interpretation of myograms of dynamic tasks is much more complex than that of interpreting those for static work. He expresses the opinion that it is useless in most dynamic situations to attempt the numeric quantification of such recordings, but indicates that their "qualitative" interpretation by experienced investigators can be very useful.

In this regard Örtengren, Andersson, Broman, Magnusson, and Petersén (1975) used the electromyographic technique to assess the amount of localized muscle fatigue as-

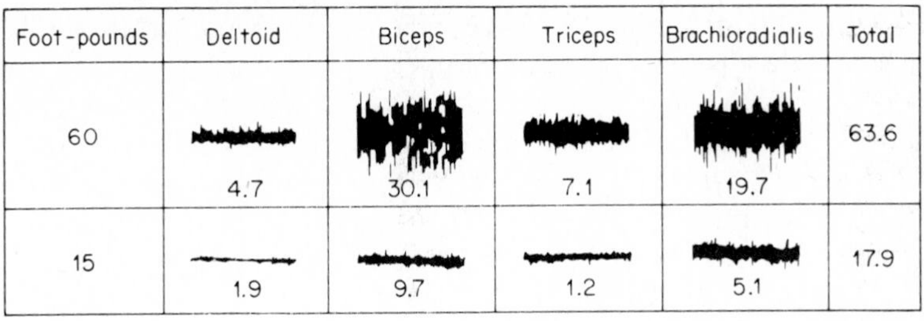

Foot-pounds	Deltoid	Biceps	Triceps	Brachioradialis	Total
60	4.7	30.1	7.1	19.7	63.6
15	1.9	9.7	1.2	5.1	17.9

FIGURE 7-6
Electromyograms recorded for four muscles of a subject maintaining a constant torque of 60 ft-lb and of 15 ft-lb. The sum of the four values is an index of the total amount of energy expended. (*Source: Adapted from Khalil, 1973, fig. 3.*)

sociated with certain manual tasks in the Volvo automobile factory in Göteborg, Sweden. In particular they were interested in the analysis of EMG data that would reveal "incidents" of muscular fatigue for various types of work activities. The incidents were identified by significant changes in the spectrum of the EMG recordings. Certain of the results are summarized in Table 7-1, in particular the mean number of such incidents per car.

The magnitude of these differences suggests that EMG recordings may have utility in identifying work activities that are conducive to high levels of muscular fatigue, with the thought that certain modifications in the work would reduce the level of such fatigue.

Measures of Mental Activity Although mental activities involve physiological processes, efforts to obtain measurements of such processes that could serve as indices of mental strain have not been particularly rewarding. Some current attention is being given to the use of sinus arrhythmia (SA) (also called heart rate variability, HRV). This is essentially a measure of the irregularity of the heart rate. There are various techniques that have been used in measuring sinus arrhythmia, some of which involve the spectral analysis of the differences between heartbeats. Although some technical problems still remain regarding the measurement of sinus arrhythmia, our interests are in its potential use in measuring mental activity. To date the results at best reflect something of a mixed bag. Kalsbeek (1971) concluded that sinus arrhythmia, as he

TABLE 7-1
INCIDENTS OF MUSCULAR FATIGUE FOR THREE TYPES OF TASKS

Type of work	Mean number of incidents
Assembly	0.08
Wet-rubbing: light	0.04
Wet-rubbing: heavy	0.36

measured it, decreases with an increase in mental load. This tendency was somewhat supported by Boyce (1974), who reported a mean SA score for an easy task of 58 and for a difficult task of 52. However, the results of a study by Hyndman and Gregory (1975) only partially confirmed such findings, but they did report a decrease in such scores during a perceptual task that required a physical response. In turn, Mulder (1979, pp. 327-343) reports some changes in the distribution of differences in heartbeats for paced versus unpaced tasks and for tasks of varying difficulty versus a resting condition.

Although Luczak (1979) agrees with those who claim that HRV tends to decrease with mental load, he points out that it is also influenced by other variables, such as by speech on the part of the subject, physical activity, emotional condition, and thermal conditions. These and other constraints seem to argue against its use as a measure of mental load except under very carefully controlled conditions.

Some probing efforts with electroencephalogram (EEG) measures as related to performance on a vigilance type of task are reported by Milošević (1978). His reported correlations between certain EEG measures (especially the theta and alpha rhythms) with certain criteria of detection of signals during the vigilance task suggest that the EEG may be found to be a useful index of mental strain in such tasks. In summarizing much of the research relating to what Wierwille (1979) refers to as physiological work-load measures, he makes the dismal statement that few if any of them are at present proven to the extent that they can be applied, at least to mental work-load problems of air crews. Of those he did discuss, the most promising were pupil dilation, evoked cortical potentials (ECP), and body fluid analysis.

Although certain physiological measures may ultimately be found to be useful for measuring mental activities, Ursin and Ursin (1979) refer to a problem that could serve as a stumbling block for such use. In particular they refer to the serious difficulties in discriminating between possible physiological effects of emotional factors and the information work load associated with work activities. Thus, we probably must agree with Wierwille that no single technique can now be recommended as the definitive behavioral measure of what he calls operator (mental) work load. In turn, he indicates that the strongest research support exists for using subjective opinions of people and certain task-analytic (i.e., performance) measures.

Subjective Reactions

Measures of a wide variety of subjective reactions can be used in connection with analysis of work- and environmental-related variables, usually these reactions being measured by the use of questionnaires or structured interviews. These reactions tend to fall into two broad, but overlapping, classes. On the one hand are judgments, opinions, or other reactions that generally deal with sources of possible stress, such as reactions about the work activity (judgments about its difficulty, complexity, etc.), or about the environment (judgments regarding heat, cold, noise, distractions, etc.), or even about one's emotional or "affective" state (fear of or interest in the consequences of failure, or interest in or motivation toward the work). There is considerable evidence that people's judgments about the physical effort required in various types of work are relatively reliable and valid (Hogan, Ogden, Gebhardt, & Fleishman, 1980; Stamford, 1976; Wardle, 1978). Hogan et al. (1980), for example, report correlations of .75 and .72 between ratings of physical effort of various tasks and two indices of metabolic costs of the tasks, when the raters were simply given lists of the tasks to be rated in terms of effort. In addition, one sample of 20 subjects performed 24 manual materials-handling tasks whose work costs were actually measured. They then rated these tasks in terms of the physical effort required. These ratings had a correlation of .88 with the actual work costs of the tasks. Thus, it is quite evident that people's judgments of physical effort of work tasks correspond reasonably well with the actual effort required.

The second type of subjective opinion deals more with the reactions to, or attitudes about, the effects of work or environment, these generally dealing with possible measures of strain. Such reactions tend to be associated with various attitudinal dimensions, such as satisfaction versus dissatisfaction, interest versus boredom, the degree of "pressure" caused by work, etc. In the case of both classes of variables the reactions can of course range from those of a positive, affirmative nature to those of a negative nature.

Although responses of people to questionnaires or interviews are far from being perfectly reliable or valid, and although there are marked individual differences in subjective reactions, research with such measures indicates that the results can be used with reasonable confidence. Borg (1978), for example, reports that data from his and other studies show that people can make reasonably adequate estimates of work difficulty.

Discussion There are probably two factors that argue for the use of measures of subjective reactions of people in the analysis of human factors associated with work activities and environments. In the first place is the important fact of individual differences (differences in reactions to the work or situational variables as such, and to the "effects" thereof). And in the second place, although physiological measures have important uses, they have their limitations, as in the measurement of mental or emotional strain resulting from work or environmental variables. For these (and perhaps other) reasons there is very strong current support for the use of measures of subjective reactions in the analysis of work- and environmental-related variables, as reflected by the opinions of, for example, Curry, Jex, Levison, and Stassen (1979), Johansson, Aronsson, and Lindström (1978), Sanders (1979), and Sheridan and Stassen (1979). In this regard Borg placed considerable emphasis on the "perceived" difficulty of work, implying that such perceptions may be more important than the actual level of physiological strain and work load.

Performance Measures

In some circumstances measures of the actual performance of people on tasks (either in a laboratory or on the job) are used as indices of the effects of work or environmental conditions. Degradation in performance (from what is otherwise considered as normal) usually is considered as an indication of stress.

Aside from the measurement of performance of whatever task or job is in question, it is sometimes the practice to use a secondary task in the assessment of work load. With this method the subjects perform two tasks concurrently, one usually being considered as the primary task and the other as a secondary task. Soede (1979, pp. 445–467) points out that the secondary task can be used in two different ways. In one of these a *paced* secondary task, when performed concurrently with the primary task, can affect performance on the primary task, the reduction being some indication of the possible overloading as related to the reserve capacity of the subject. With the other method, a *self-paced* secondary task is used concurrently with the primary task (recall that a "self-paced" task is one whose pace is controlled by the worker). With this method performance on the primary task usually is not affected, and the level of performance on the secondary task presumably reflects the "spare capacity" of the individual when performing the primary task. In some instances the difference in performance on the secondary task when performed by itself, versus when performed concurrently with the primary task, usually is considered as an index of the spare capacity when the primary task is performed by itself.

The dual-task method, however, has serious constraints. In the discussion of time sharing in Chapter 3, for example, it was pointed out that performance on certain types of tasks used in combination is affected more by time sharing than is performance on combinations of other types of tasks. In addition there is the possibility of unexpected interactions between certain tasks. Thus, as Ogden, Levine, and Eisner (1979) point out, the choice of a secondary task can pose a problem. Because of these and other limitations of this method it should be used with caution.

ENERGY EXPENDITURE IN PHYSICAL ACTIVITIES

Although human beings are not now used as sources of energy nearly as much as in days gone by, some occupations still require substantial physical effort, at least at certain times or as accumulated over the workday, and in some countries the use of human beings as major sources of energy is almost dictated by economic considerations. When human physical activity in work is potentially dangerous to health and safety, some modification of the work is in order, whether by appropriate redesign of the equipment and work space, by modification of methods, or by reduction of work periods or work pace.

Energy Expenditures of Gross Body Activities

In order to give some "feel" for the numerical values of energy expenditures for different kinds of physical activities, it may be useful to present the physiological costs of certain everyday activities; the following examples are given in kilocalories per minute (kcal/min) (Edholm, 1967): sleeping, 1.3; sitting, 1.6; and standing, 2.25. In connection with rate of body movement (as in walking and running), the price per unit of work (in physiological costs) goes up with increasing rate. This is shown quite clearly

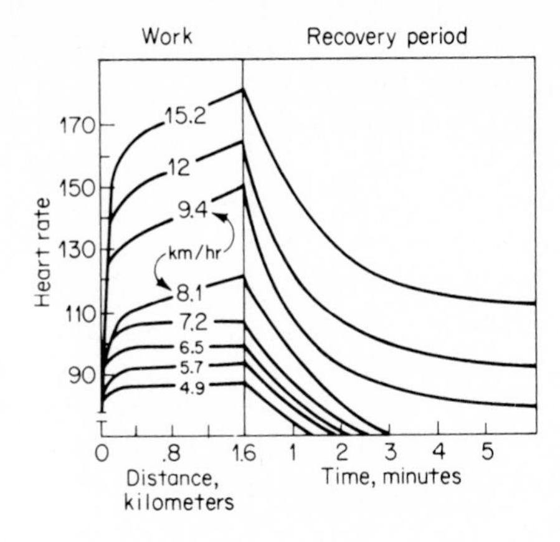

FIGURE 7-7
Heart rate during and after a march of 1.6 km (1 mi) at various speeds, in kilometers per hour. (*Source: LeBlanc, 1957, as presented by Monod, 1967.*)

in Figure 7-7, which shows the heart rate during and after marching a 1600-m (about a 1-mi) course, when the 18 subjects marched at speeds ranging from 4.9 to 15.2 kilometers per hour (km/h) (about 3.0 to 9.4 mi/h). The increasing energy cost scoots up rather sharply at speeds of 8.1 km/h and above. In addition, recovery time also increases markedly. The energy costs of body exercises of different intensities are shown further in Figure 7-8. The work loads were the consequence of running on a treadmill at certain combinations of speed and inclines. The oxygen-consumption curves are obviously steeper the higher the work load; the lightest work led to exhaustion in about 3 min, the heaviest in about 30 s. These and other examples clearly indicate the trade-offs in human work, in particular that the physiological price of work–per unit of work–is greater at higher rates of work than at more moderate rates.

FIGURE 7-8
Oxygen consumption from the onset to the end of running on a treadmill with certain work loads that were the consequence of specified speeds and inclines; the energy requirements of the work loads are themselves expressed in terms of an independently derived oxygen-consumption index. (*Source: Margaria, Mangili, Cuttica, & Cerretelli, 1965.*)

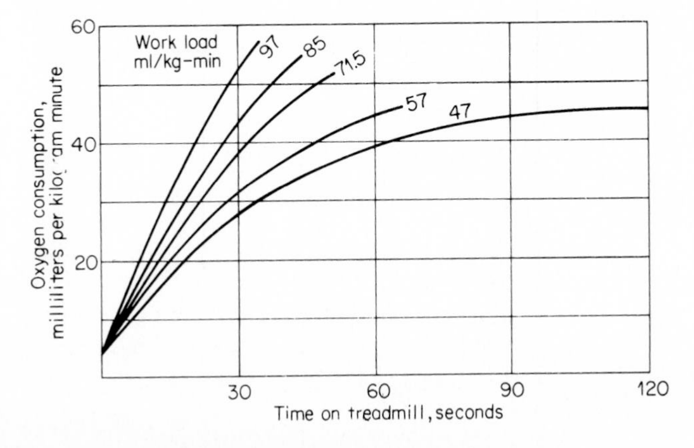

Energy Expenditures of Specific Activities

The discussion above dealt simply with the energy expenditures of lugging the body around, as, for example, at different paces, inclines, etc. The energy expenditures of various types of activity vary somewhat for individuals, but estimates of the approximate energy costs for certain specific types of work are given in Figure 7-9. The energy costs for these range from 1.6 to 16.2 kcal/min.

However, the energy cost for certain types of work can vary with the manner in which the work is carried out. The differential costs of methods of performing an activity are illustrated by the several methods of carrying a load as used in various cultures. Seven such methods were compared by Datta and Ramanathan (1971) on the basis of oxygen requirements; these methods and the results are shown in Figure 7-10. The requirement of the most efficient method (the double pack) is used as an arbitrary

FIGURE 7-9
Examples of energy costs of various types of human activity. Energy costs are given in kilocalaries per minute. (*Source: Passmore & Durnin, 1955, as adapted and presented by Gordon, 1957.*)

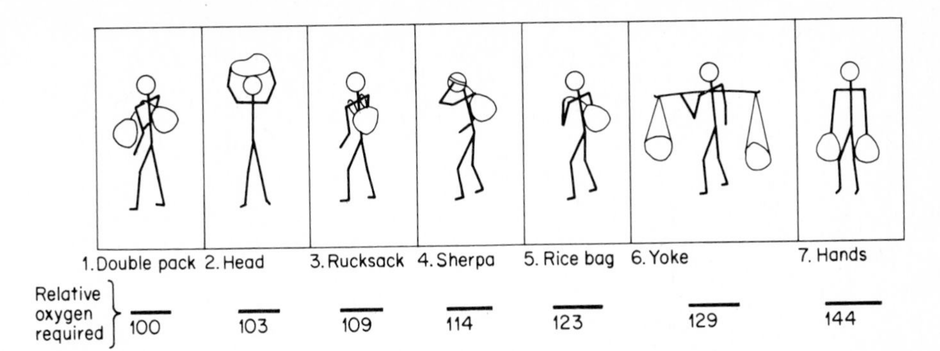

FIGURE 7-10
Relative oxygen consumption of seven methods of carrying a load, with the double-pack method used as a base of 100 percent. This illustrates that the manner in which an activity is carried out can influence the energy requirements. (*Source: Adapted from Datta & Ramanathan, 1971.*)

base of 100 percent. There are advantages and disadvantages to the various methods over and above their oxygen requirements, but the common denominator of the most efficient methods reported in this and other studies is that of maintenance of good postural balance, one that affects the body's center of gravity the least.

Energy Expenditures of Different Postures

The posture of workers when performing some tasks is another factor that can influence energy expenditure. In this regard certain agricultural tasks in particular have to be carried out at or near ground level, as in picking strawberries. When performing such work, however, any of several postures can be assumed. The energy costs of certain such postures were measured in a study by Vos (1973) in which he used a task of picking up metal tags placed in a standard pattern on the floor. A comparison of the energy expenditures of five different postures is given in Figure 7-11. This figure shows

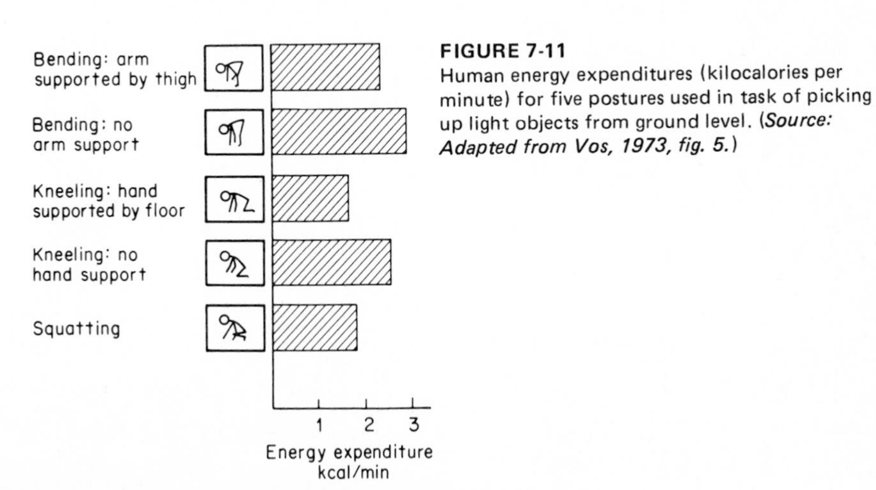

FIGURE 7-11
Human energy expenditures (kilocalories per minute) for five postures used in task of picking up light objects from ground level. (*Source: Adapted from Vos, 1973, fig. 5.*)

that a kneeling posture with hand support and a squatting posture required less energy than the other postures. (The kneeling posture, however, precludes the use of one hand and might cause knee discomfort over a period of time.) On the basis of another phase of the study it was shown that a sitting posture (with a low stool) is a bit better than squatting, but a sitting posture is not feasible if the task requires moving from place to place. Although this particular analysis dealt with postures used on ground-level tasks, differences in postures used in certain other tasks also can have differential energy costs.

Energy Expenditure and Rate of Activity

Energy costs are related to the rate or pace of activity, as well as to the type of activity. One indication of this is shown in Figure 7-12, which shows the relative efficiency of stair climbing at different speeds. However, there is evidence that the optimum varies for different age groups (Salvendy & Pilitsis, 1971) and for individuals, and that individuals seem to be able to determine the pace that is most "natural" for themselves, which tends to be the one that involves the minimum energy expenditure for each cycle (Corlett & Mahadeva, 1970).

Related to the notion of an optimum pace for any given activity is the question as to whether, with certain types of work (such as assembly operations), the work pace should be set for the worker or controlled by the worker (i.e., paced or self-paced).

Keeping Energy Expenditures within Bounds

If those who are concerned with the nature of human work activities (design engineers, industrial engineers, supervisors, administrators, etc.) want to keep energy costs within reasonable bounds, it is necessary for them to know both what those bounds should be and what the costs are (or would be) for specific activities (such as those shown in Figure 7-9).

Energy Costs of Grades of Work As a starter toward this, the definitions of different grades of work shown in Table 7-2 may be helpful.

In discussing energy expenditures over the period of the conventional working day,

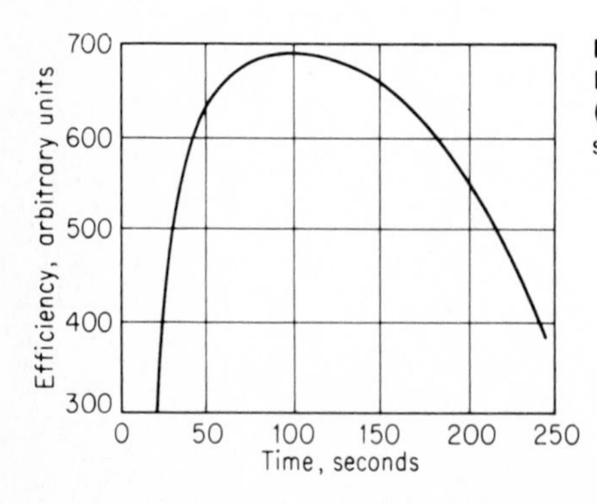

FIGURE 7-12
Efficiency of stair climbing at different speeds (speed is given by time in seconds to climb the stairs). (*Source: Schottelius & Schottelius, 1978.*)

TABLE 7-2
ENERGY COSTS OF VARIOUS GRADES OF WORK

Grade of work	Energy expenditure		Approximate oxygen consumption, liters/min
	kcal/min	kcal/8 h	
Unduly heavy	over 12.5	over 6000	over 2.5
Very heavy	10.0-12.5	4800-6000	2.0-2.5
Heavy	7.5-10.0	3600-4800	1.5-2.0
Moderate	5.0-7.5	2400-3600	1.0-1.5
Light	2.5-5.0	1200-2400	0.5-1.0
Very light	under 2.5	under 1200	under 0.5

Source: Christensen, 1953, pp. 93-108.

Lehmann (1958) estimates that the maximum energetic output a normal man can afford in the long run is about 4800 kcal/d; subtracting his estimate of basal and leisure requirements of 2300 kcal/d leaves a maximum of about 2500 kcal/d available for the working day. This figures out to be about 5 kcal/min. But although he proposes this as a maximum, he suggests about 2000 kcal/d as a more normal load, this averaging out to be about 4.2 kcal/min. Edholm (1967, p. 91) proposes a somewhat more conservative value, suggesting that the 2000 kcal/d expenditure should be considered as a maximum and that work levels preferably should be kept somewhat below this. Granting some modest differences between these and other physiological standards, we nonetheless get an impression of the general level of physiological costs that should not be exceeded.

Work and Rest If we accept some ceiling (such as 4 or 5 kcal/min) as a desirable upper limit of the average energy cost of work (exclusive of basal requirements), it is manifest that if a particular activity per se exceeds that limit, there must be rest to compensate for the excess. In this connection Murrell (1965, p. 376) presents a formula for estimating the total amount of rest (scheduled or not scheduled) required for any given work activity, depending on its average energy cost. This formula (with different notations) is

$$R = \frac{T(K-S)}{K-1.5}$$

in which R is rest required in minutes; T is total working time; K is average kilocalories per minute of work; and S is kilocalories per minute adopted as standard. The value of 1.5 in the denominator is an approximation of the resting level in kilocalories per minute. If we adopt as S a value of 4 kcal/min and want to figure R for a 1-h period ($T =$ 60 min), our formula becomes

$$R = \frac{60(K-4)}{K-1.5}$$

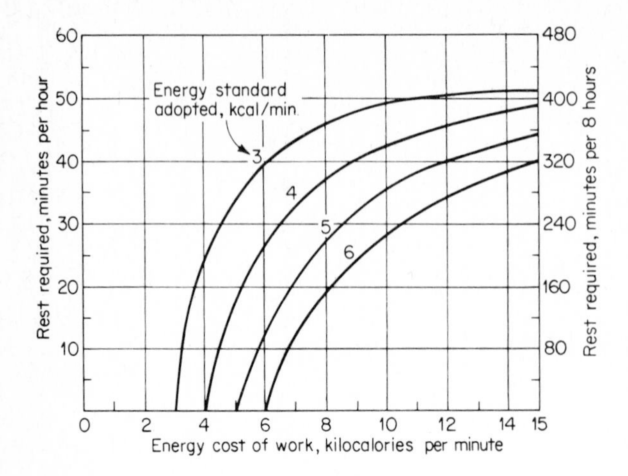

FIGURE 7-13
Total rest requirements for work activities of varying energy costs, for energy-expenditure standards (ceilings) of 3, 4, 5, and 6 kcal/min; a generally accepted standard is 4 to 5 kcal/min. The rest requirements for maintenance of the adopted standard are given per hour (left) and per 8-h day (right). (*Source: Based on a formulation of Murrell, 1965, p. 376.*)

Applying this to a series of values of K, we can obtain R values shown in the next to the top curve of Figure 7-13 ($S = 4$). The other curves (for values of $S = 3, 5$, and 6) are given for comparison when lower ($S = 3$ kcal/min) or higher ($S = 5$ or 6 kcal/ min) standards of energy expenditure might seem appropriate. The lowest curve (for a value of $S = 6$), however, undoubtedly represents a level of activity that probably could not be maintained very long, except possibly by the hardiest among us. This general formulation needs to be accepted with a fair sprinkling of salt, in part because of individual differences in physical condition. Further, although the curves in Figure 7-13 swing down to the zero-rest-required line, we should keep in mind that this formulation deals only with the physiological costs of work. Because of *other* consideration, some rest must be provided for virtually any kind of continuous work, even though its physiological costs are nominal.

Work Limits of Local Muscle Groups Although the overall energy cost of an activity might be within reasonable limits, it is of course possible for individual muscles or muscle groups to be worn to a frazzle with excessive use. If the rate of contraction of a muscle or muscle group is low enough, it can function almost indefinitely, but at higher rates it can become completely fatigued and cease to function at all. This is shown in Table 7-3. We can see that the rate of 1 contraction of the muscle per 10 s did not produce complete fatigue and permitted almost indefinite continuation of work; faster rates (1 contraction per 4, 2, or 1 s) produced such stress that the muscle ceased to function after 31, 18, or 14 contractions, respectively.

Discussion If a given type of physical activity is within the reasonable physical ability limits of an individual (and individuals obviously vary in such limits), the outer bounds of the ability of the individual to *continue* the activity are prescribed by the total energy costs, by the recovery rates of individual muscles or muscle groups, or by both. These possible constraints in effect should dictate the work rate that should not be exceeded for continued effectiveness of the individual in performing the activity.

TABLE 7-3
MUSCLE FATIGUE AS RELATED TO RATE OF MUSCLE CONTRACTION

Number of contractions	Fatigued by	Work done
1 per 1 s	14 contractions	0.912 kg-m
1 per 2 s	18 contractions	1.080 kg-m
1 per 4 s	31 contractions	1.842 kg-m
1 per 10 s	No fatigue (no stress)	Almost indefinite

Source: Tuttle and Schottelius, table 6-4, p. 110.

STRENGTH AND ENDURANCE

Strength is the maximal force muscles can exert isometrically in a single voluntary effort, that is, the muscular capacity to exert force under static conditions, and is usually measured by the use of an external device such as a hand dynamometer or a device for measuring the force exerted against some object. The measurement of such force, however, depends not only on the intrinsic muscle strength but also on the subject's motivation, the experimenter's instructions, and even the measurement index used (e.g., whether use of a peak value or an average of two or three efforts). Since most human activities consist of dynamic efforts rather than static efforts, Kroemer (1970) raises serious questions about the use of measures of strength per se, pointing out that it is mechanically difficult, if not impossible, to predict an individual's performance on a dynamic task (for example, turning a crank) from a measurement of static force capacity (for instance, holding a weight).

Arm Strength

With the above cautions in mind, let us illustrate studies of maximum strength with some data from Hunsicker (1955), who tested the arm strength of 55 subjects who made movements in each of several directions, with the upper part of the arm in each of five positions, as illustrated in Figure 7-14*a*. Some of the results are shown in Figure 7-14*b*, in particular the maximum strength of the 5th percentile of 55 males. It is frequently the practice, in dealing with data relating to strength, to use the 5th percentile value as the maximum force to be overcome by users of equipment being designed, since this would in general ensure that 95 percent of the individuals in question would have that strength level or more.

We can see that pull and push movements are clearly strongest, but that these are noticeably influenced by the position of the hand, with the strongest positions being at angles of 150 and 180°. The differences among the other movements are not great, but what patterns do emerge are the consequence of the mechanical advantages of such movements, considering the angles involved and the effectiveness of the muscle contractions in applying leverage to the body members. Although left-hand data are not shown, the strength of left-hand movements is roughly 10 percent below that of movements of the right hand.

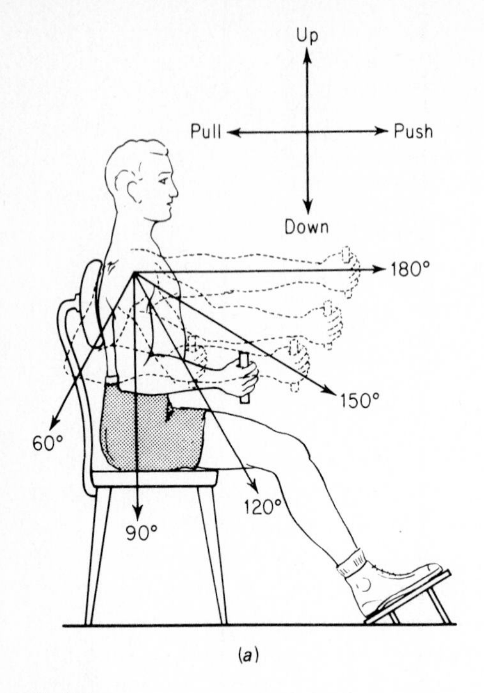

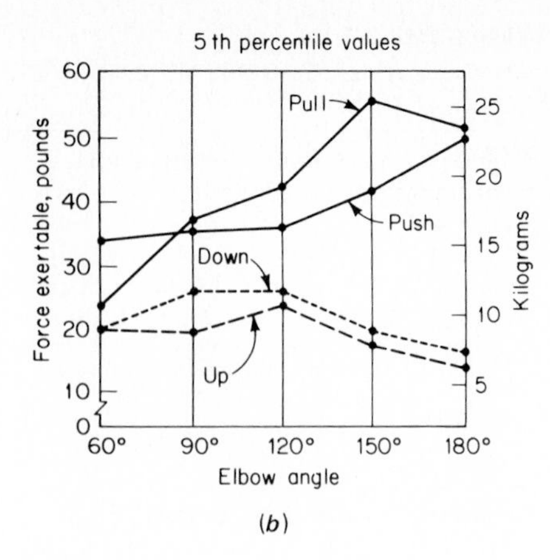

FIGURE 7-14
(*a*) Side view of subjects being tested for strength in executing push, pull, up, and down movements at each of five arm positions. (*b*) Maximum arm strength of 5th percentile of 55 male subjects. (*Source: Based on data from Hunsicker, 1955.*)

Endurance

If one considers the endurance of people to maintain a given muscular force, we can all attest from our own experience that such ability is related to the magnitude of the force. This is shown dramatically in Figure 7-15, which depicts the general pattern of endurance time as a function of force requirements of the task. It is obvious that people can maintain their maximum effort very briefly, whereas they can maintain a force of around 25 percent or less of their own maximum for a somewhat extended period (10 min or more). (Note that this relationship is based on each individual's *own* measures of strength.) The implication of this relationship is fairly obvious—that if it is necessary to require individuals to maintain force over a period of time, the force required should be well below each individual's own static force capacity.

It should be added that one commonsense concept of *endurance* refers to the abil-

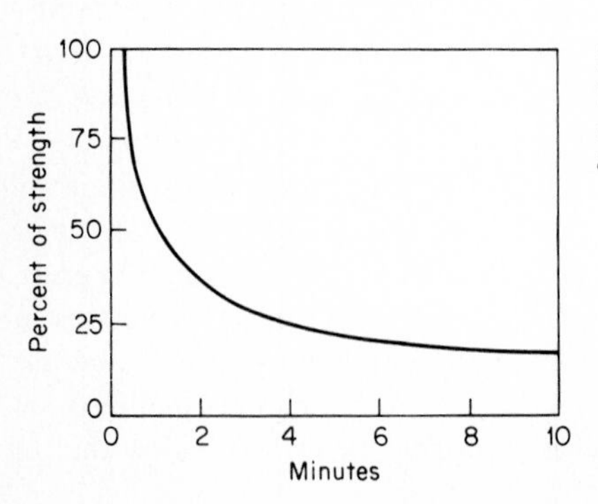

FIGURE 7-15
Endurance time as a function of force requirements. (*Source: Kroemer, 1970, fig. 4, as adapted from various sources.*)

ity to keep up some general body activity over a period of time (rather than the exertion of a given muscle or muscle group). In this frame of reference the *endurance* of individuals would be a function of the total energy cost of the activity and the energy expenditure the individuals can reasonably maintain over time. If the energy costs exceed reasonable limits, rest should be provided to keep total energy requirements within bounds.

Discussion

As indicated earlier, there are certain human variables that are related to muscle strength and endurance. Observations about a few of these, based largely on the discussion by Damon, Stoudt, and McFarland (1966), are given below:

- *Age:* Strength reaches a maximum by the middle to late twenties and declines slowly but continuously from then on, until at about age 65 strength is about 75 percent of that exerted in youth. Despite such reduction in strength, however, there are indications that continuous-work capacity does not decrease with age, up to about 60, at least in moderate environments (Snook, 1978). Snook suggests, however, that this finding may not apply in hot environments or for intermittent work that requires short durations of high energy expenditure.
- *Sex:* On the average, women's strength is about two-thirds that of men.
- *Body build:* Although body build is related to strength and endurance, the relationships are complicated; for example, athletic-looking individuals generally are stronger than others, but less powerfully built persons may be more efficient; and for rapidly fatiguing, severe exercise, slender subjects are best, with obese subjects worst; and for moderate exercise, those with normal build are best.
- *Exercise:* Exercise can increase strength and endurance within limits, these increases frequently being in the range of 30 to 50 percent above beginning levels. Incidentally, there are some indications that at least moderate continuous exercise over the years sometimes can stave off some of the typical decline in physical condition with increasing age.

SPEED AND ACCURACY OF MOVEMENTS

Speed generally is the primary requirement in executing movements that are otherwise not difficult or demanding, such as in applying the brake pedal of an automobile or reaching for parts to be assembled. In turn, accuracy is the primary requirement in executing such movements as those in tracking (in which continuous control is required), in certain positioning actions that require precision and control, and in certain manipulative activities. However, in some circumstances both speed and accuracy may be required.

Response Time

Many movements are triggered by some external stimulus such as a changing traffic light or auditory warning signal. The time to make a movement following such a stimulus actually consists of a combination of delays; the nature of these delays and the range of typical times in milliseconds (ms) required for them have been summarized by Wargo (1967) as follows: receptor delays, 1 to 38; neural transmission to the cortex, 2 to 100; central-process delays, 70 to 300; neural transmission to muscle, 10 to 20; and muscle latency and activation time, 30 to 70. These add up to a total ranging from 113 to 528 ms. The total time to make a response following a stimulus frequently is referred to as *reaction time*. However, one can differentiate between the time to initiate a movement (this being a more restrictive definition of reaction time) and the time to make the movement (sometimes called *movement time*).

Simple and Choice Reaction Time *Simple reaction time* is the time to make a specific response when only one particular stimulus can occur, usually when an individual is anticipating the stimulus (as in conventional laboratory experiments). Reaction time is usually shortest in such circumstances, typically ranging from about 150 to 200 ms (0.15 to 0.20 s), with 200 ms being a fairly representative value; the value may be higher or lower depending on the stimulus modality and the nature of the stimulus (including its intensity and duration), as well as on the subject's age and other individual differences. In the case of *choice reaction time* (when there are two or more stimuli and two or more possible responses), reaction time typically increases, the increase being due to such factors as the time for identification of the particular stimulus, the need for "recoding" the stimulus, the time to make a decision, and of course the number of stimuli and corresponding responses. Some indication of the influence of number of choices is given below (summarized from various sources by Damon et al., 1966, p. 239):

Number of choices:	1	2	3	4	5	6	7	8	9	10
Approximate reaction time, s:	0.20	0.35	0.40	0.45	0.50	0.55	0.60	0.60	0.65	0.65

Expectancy Most data on simple and choice reaction times come from laboratories in which the subject is anticipating a stimulus. (And in some industrial circumstances people actually are waiting for a stimulus.) However, when stimuli occur infrequently or when they are not expected, the ante is raised. This was illustrated, for example, in a study by Warrick, Kibler, and Topmiller (1965), in which typists at their regular jobs were asked to press a button whenever a buzzer sounded, the buzzer going off only once or twice a week over a period of 6 months. The reaction time (actually the total response time) to the "unexpected" signals averaged about 100 ms above that when the subjects had received an advance warning.

As another example to illustrate the point, Johansson and Rumar (1971) collected data in Sweden on the response time of 321 automobile drivers in applying the brake pedal following an auditory signal. This was done under two conditions, one in which the drivers were anticipating a signal within the next 10 kilometers (10 km) (6 mi), and the other under a "surprise" condition, with no advance warning. The mean response times for these two conditions were:

Condition	Mean response time
"Anticipation"	0.54 s
"Surprise"	0.73 s

The response times for the "surprise" conditions for some subjects were over 2 s, and the investigators estimated that over half the drivers would take over 0.9 s (900 ms) to respond. In general, then, there is strong evidence to support the contention that reaction time is much longer when people are not "expecting" to have to make a response than when they are anticipating some signal or cue.

Other Factors that Influence Reaction Time There are numerous other factors that can influence reaction time. For example, practice typically reduces reaction time. In addition, if the response to be made is compatible with one's "expectations" the time will be shorter than if it is not compatible.

Movement Time The time to effect a movement following a signal would of course vary with the type and distance of movement, but it has been estimated (Wargo) that a minimum of about 300 ms (0.30 s) can be expected for most control activities. Adding this value to an estimated reaction time of 200 ms would result in a total response time of about 500 ms. However, the nature and distance and location of the response mechanism can influence the total time. This is illustrated, for example, by the result of a study by Pattie (1973), who investigated the time to activate four types of possible emergency power cutoff devices as they might be used on agricultural tractors. The mean times to activate the devices in response to a buzzer are given in Table 7-4.

Discussion We can see that the time required to make certain responses can be influenced by a number of variables, such as the nature of the stimulus, the number of choices, the degree of expectancy, and the device used. In some instances the total time required can be of substantial consequence. For example, the response time of the pilot of a supersonic aircraft on a collision course can be as long as 1.7 s, this being the simple addition of 0.3 s for visual acquisition of the other aircraft, 0.6 s for recognition of the impending danger, 0.5 s for selection of a course of action, and 0.3 s for initiation of the desired control response. Add the response time of the aircraft itself, and it would be futile to take any action if the planes were closer than about 4 mi.

When time is, as they say, of the essence, we should not throw up our hands in complete despair of the time lag in human responses. There are, indeed, ways of aiding and abetting people in responding rapidly to stimuli. Speed requirements, for example, can be reduced by taking actions such as using sensory modalities with shortest reaction time, presenting stimuli in a clear and unambiguous manner, minimizing the num-

TABLE 7-4
MEAN TIMES TO ACTIVATE FOUR POWER-CUTOFF DEVICES ON A TRACTOR

	Mean time, ms
Clutch	613
Toggle switch (underneath steering wheel)	498
Horn rim (on steering wheel)	412
Rim blow (on under edge of steering wheel)	337

ber of alternatives from which to choose, giving advance warning of stimuli if possible, using body members that are close to the cortex to reduce neural transmission time, using control mechanisms that minimize response time, and training the individuals. In more exotic circumstances, one can even bypass the human physical response by the direct use of electrical muscle-action potentials for effecting control responses (Wargo).

Positioning Movements

Positioning movements are made when a person reaches for something or moves something to another location, usually by hand; they are then *travel* movements of the body member. The time and accuracy of such movements can be influenced by such factors as the nature of the stimulus that triggers the movement, the distance and direction of the movement, and visual versus nonvisual (i.e., "blind") control. Certain positioning movements can be dissected into two or three relatively distinct components, namely, reaction time (the time to initiate a response following the stimulus that triggers it), primary or gross travel time (to bring the body member near the terminal), and a secondary or corrective type of motion to bring the body member to the precise position desired. Where there is an automatic fixed terminal (such as on a typewriter carriage), the secondary, or corrective, component virtually drops out of the picture.

Time and Distance of Movements In the execution of positioning movements, reaction time is almost a constant value, unrelated to the distance of movement. This is shown, for example, in Figure 7-16, which is based on a pair of related studies in which the subjects moved a sliding device to a marked position when a buzzer was sounded (Brown & Slater-Hammel, 1949; Brown, Wieben, & Norris, 1948). Three different distances were used, namely, 2.5, 10, and 40 cm, the movements being left to right, and inward and outward. This figure also shows that movement time is related to distance but is not proportionate to distance. This lack of linear relationship between distance and time probably can be attributed to the time required for acceleration to the maximum speed, and (except where there is a mechanical terminal) the secondary, or corrective, movement in bringing the body member to the precise terminal.

Direction of Positioning Movements Because of the nature of our physical structures, motions in certain directions can be made more rapidly than those in other directions. We will use as an example data from Schmidtke and Stier (1960), who had subjects make positioning movements with the right hand in eight directions in a horizontal plane from a center starting point. Their results, shown in Figure 7-17, show the average times to make the movements in the eight directions. The pattern suggests that, in biomechanical terms, controlled arm movements that are primarily based on a pivoting of the elbow (as toward the lower left or upper right) take less time than those that require a greater degree of upper-arm and shoulder action (as toward the lower right or upper left). It should be added that evidence from other sources indicates that such movements are also more accurate.

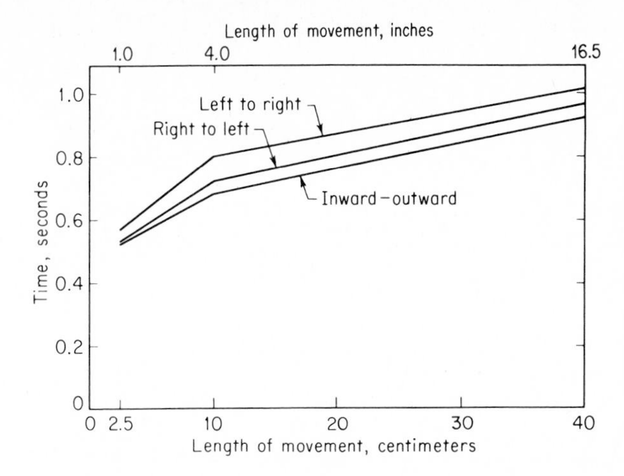

FIGURE 7-16
Times required for horizontal positioning movements of different lengths. Times given are from the sounding of a buzzer to completion of movement. Reaction time, which was essentially the same for all movements (about 0.25 s), is included in time values. (*Source: Adapted from Brown & Slater-Hammel, 1949, and Brown, Wieben, & Norris, 1948.*)

Blind Positioning Movements When visual control of movements is not feasible, the individual needs to depend on kinesthetic sense for feedback. Probably the most usual type of blind positioning movement is one in which the individual moves a hand (or foot) in free space from one location to another, as in reaching for a control device when the eyes are otherwise occupied. The very well-known study by Fitts (1947) probably provides the best available data relating to the accuracy of the *direction* of such movements in free space. He used an arrangement with targets positioned around the subject at 0, 45, 90, and 135° angles left and right, in three tiers, namely, a center (reference) tier, and tiers 45° above and below the center tier. The blindfolded subjects were given a marker with a sharp point, which they pressed against each target

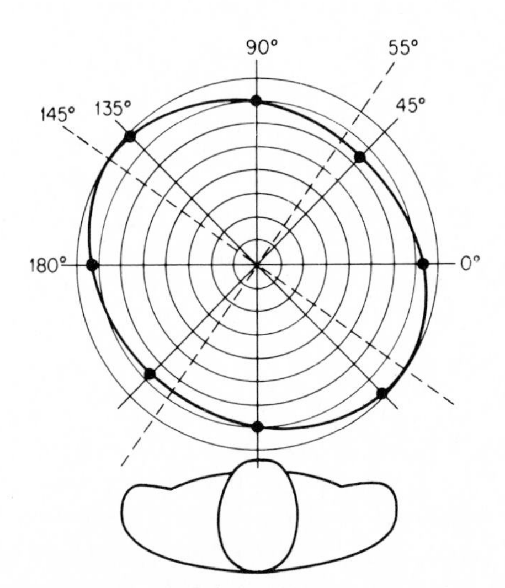

FIGURE 7-17
Average times of hand movements made in various directions. Data were available for the points indicated by black dots; the oval was drawn from these points and represents assumed, rather than actual, values between the recorded points. The concentric circles represent equal increments of time to provide a reference for the average movement times depicted by the oval. (*Source: Adapted from Schmidtke & Stier, 1960.*)

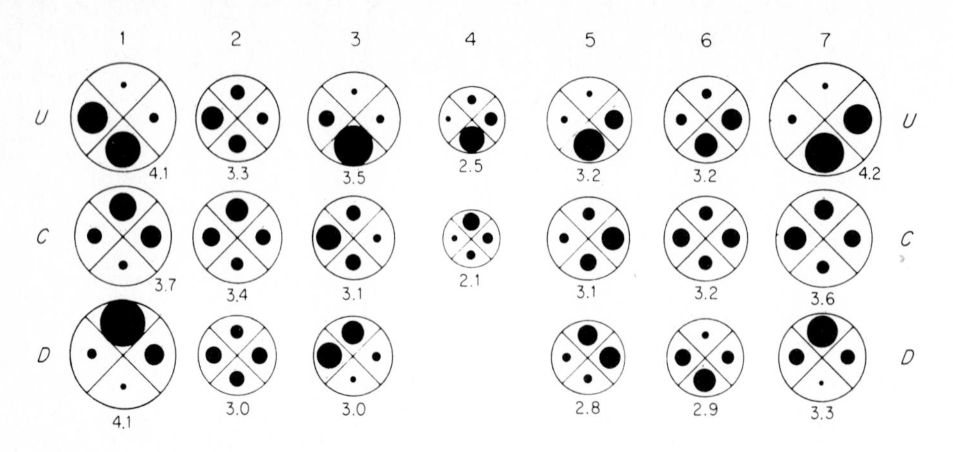

FIGURE 7-18
Relative accuracy scores for different areas in blind-positioning study by Fitts (1947). The position of the circles represents the location of targets, which ranged from 135° left (number 1) to 135° right (number 7), number 4 being straight ahead. The three tiers represent those up, center, and down. The size of each circle represents the relative number of errors, so small circles indicate greater accuracy. The relative size of the four dark circles within each light one is proportional to the errors in each quadrant of the target. (*Source: USAF, AFHRL, Human Engineering Division.*)

when they tried to reach to it. A bull's-eye was scored zero, and marks in subsequent circles were scored from 1 to 5, marks outside the circles being scored 6.

Figure 7-18 shows the results. Each circle in this figure represents the subjects' accuracy in hitting the target in the corresponding position. The size of the circle is proportional to the average accuracy score for that target; the smaller the size, the better the accuracy. The circles within circles (those in the four quadrants) indicate, relatively, the proportionate number of marks that were made in each quadrant. From this figure it can be seen that blind positioning movements can be made with greatest accuracy in the dead-ahead positions and with least accuracy in the side positions. With regard to the level of the targets, the accuracy is greatest for the lowest tier, average for the middle tier, and poorest for the upper tier. Further, right-hand targets can be reached with a bit more accuracy than left-hand ones.

Thus, in general, in the positioning of control devices or other gadgets that are to be reached for blindly, positions closer to the center and below shoulder height usually can be expected to be reached more accurately than those farther off to the sides or higher up.

Continuous Movements

Continuous movements are those that require accurate control over the span of the movement. Deviations from the desired path are produced by tremor of the body member. An interesting approach to the study of tremor during continuous movements was that carried out by Mead and Sampson (1972), in which they had subjects move a stylus (15 in long with a 4-in, 90° bend at the tip) along a narrow groove in any one of the four positions shown in Figure 7-19. As the subject moved the stylus in

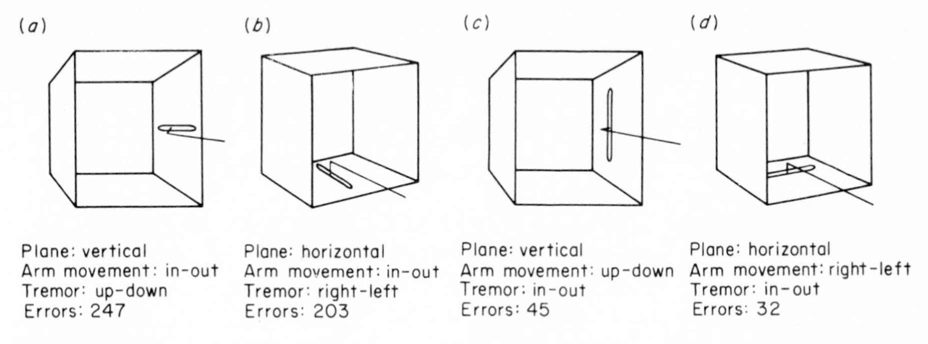

FIGURE 7-19
Directions and planes of arm movements with stylus as used in study of hand steadiness, with direction of hand tremor and number of "errors" (number of times the stylus touched the side of the groove) for each condition. (*Source: Adapted from Mead & Sampson, 1972, fig. 1 and table 1.*)

the directions indicated in that figure, any time the stylus touched the side of the groove an "error" was electronically recorded. These errors, which can be viewed as measures of tremor, are shown in Figure 7-19, and indicate that tremor was greatest during an in-out arm movement in the vertical plane (in which the tremor was up-down), and that it was least during a right-left arm movement in the horizontal plane (in which the tremor was in-out).

It is interesting to note that the results of this study are reasonably in line with the study of positioning movements as shown in Figure 7-17 since there was less tremor for movement *d* that was based on an essentially lateral component (left to right, with the arm moving at the elbow as a pivot) than movements *a* and *b* that consisted more of an in-out direction (involving more upper-arm and shoulder movement). Movement *c* (which resulted in relatively little tremor) probably benefited from the fact that the motion was based on forearm movement pivoted at the elbow.

Continuous control movements such as those involved in tracking tasks will be discussed further in Chapter 8.

Manipulative Movements

Most *manipulative movements* involve the use of the hand, the fingers, or both, as in handling items, in assembling parts, or in using hand tools or control devices. (However, one of the authors has seen in Pakistan and India amazingly facile meat cutters who ply their trade with the knives held between their toes!) Because of the varied nature of manipulative movements, it is not feasible to present any generalizable discussion of them. However, Chapter 8 will include some discussion of the use of control devices.

Repetitive Movements

Any given type of *repetitive movement* consists of successive performance of the same action. Such movements may be either self-paced or paced (self-paced tasks being paced by the worker and paced tasks being controlled by some external factor such as the action of a machine or the movement of a conveyor belt). Although there are differences in the specific manner in which individuals perform the same task, the actions of an individual reflect a "subroutine" of the muscular responses that the individual has acquired that are automatically executed in accordance with the "executive program" mentioned before.

Probably the critical aspect in the performance of repetitive tasks is the requirement for work pauses from time to time, especially in the case of paced work. However, as Corlett (1977) points out, it is difficult to specify the nature of the pause (active or passive) and when it should be taken, since this would depend on the nature of the work and in part on differences among individuals. In the absence of specific guidelines for providing such pauses, those responsible for the supervision of such work need to take cognizance of whatever relevant cues there may be that would be useful in determining when to give such relief pauses, with particular attention to the reactions of the individuals involved.

Sequential Movements

In most instances *sequential movements* are of the same general kind, varying in some differentiating feature, as in operating a keyboard. In some instances, however, a potpourri of types of movements may occur in sequence, such as those in starting a car on a rainy night, which might include turning on the ignition, turning on the lights, and turning on the windshield wipers. Most of the research on sequential movements relates to the first type (movements of the same type), especially in the use of keyboards. A discussion of keyboards will follow in Chapter 9.

In connection with sequential movements in which the hand moves from one place to another, one particular point should be made, namely, that there are indications that the time required for shifting from one position to another (travel time) can be affected by the nature of the manipulation performed at each position. A review of research relating to this interaction has led Schappe (1965) to conclude that travel time of a body member (such as a hand) is in fact influenced by the manipulation activities of the terminals of the travel movement, and that both of these are influenced by perceptual factors. Such interactions raise questions about the *additivity* of the times "allowed" for various elemental motions in the use of predetermined time systems for estimating time allowances for complete operations. However, the extent to which the estimate of total times might be adversely affected by such interactions probably is not known. However, some predetermined time systems do, to a certain degree, take such interaction into account in deriving total time allowances for sequences of movements.

Static Reactions

In *static reactions*, certain sets of muscles typically operate in opposition to each other to maintain equilibrium of the body or of certain portions of it. Thus, if a body member, such as the hand, is being held in a fixed position, the various muscles controlling hand movement are in a balance that permits no net movement one way or the other.

The tensions set up in the muscles to bring about this balance, however, require continued effort, as most of us who have attempted to maintain an immobile state for any length of time can testify. In fact, it has been stated that maintaining a static position produces more wear and tear on people than some kind of adjustive posture.

Deviations from static postures are of two types: those called *tremor* (small vibrations of the body member) and those characterized by a gross drifting of the body or body member from its original position.

Tremor in Maintaining Static Position Tremor is of particular importance in work activities in which a body member must be maintained in a precise and immovable position (as in holding an electrode in place when welding). An interesting aspect of tremor, incidentally, is that the more a person tries to control it, the worse it usually is. The following are four conditions that help to reduce tremor:

1 Use of visual reference.

2 Support of body in general (as when seated) and of body member involved in static reaction (as hand or arm).

3 Hand position. (There is less hand tremor if the hand is within 8 in above or below the heart level.)

4 Friction. (Contrary to most situations, mechanical friction in the devices used can reduce tremor by adding enough resistance to movement to counteract in part the energy of the vibrations of the body member.)

MANUAL HANDLING TASKS

Many jobs and activities in other aspects of life require the manual handling or movement of objects. The physical movements and associated demands involved in such activities are so varied that we can only touch on certain illustrative aspects. And it should be kept in mind that such factors as individual differences, physical condition, sex, etc. markedly influence the abilities of individuals to perform such activities. In connection with manual handling tasks, Snook (1978) reports that such tasks represent the principal source of compensable work injuries in the United States, and that 79 percent of the manual handling injuries affect the lower back. The use of appropriate handling procedures and the designing of jobs to fit the workers can reduce such injuries significantly.

Lifting Tasks

Lifting tasks of course make up a large proportion of the manual handling tasks in industry.

Methods of Lifting With regard to methods of lifting loads from the floor (or near floor level) the evidence argues for the following technique (adapted from Davies, 1972): (1) keep feet far enough apart for balanced distribution of weight; (2) keep knees and hips bent, back reasonably straight; (3) keep arms as near to body as possible, with load as close to body as possible; (4) wherever possible, use whole hand, not just fingers; (5) lift smoothly with no jerks. The actual lifting is performed largely by an extension of the legs. A posture somewhat akin to this is shown in Figure 7-20 (example I), along with another lifting technique (example II) as used in a comparative study of these two methods by Tichauer (1971) in which he obtained electromyograph recordings for two muscles (the gluteus maximus and the sacrospinalis). Examples of these recordings show less muscle activity for technique I than for technique II, thus tending to support the method recommended by Davies.

Work Pace and Energy Cost in Lifting When the work activity consists of frequent or virtually continuous lifting, the efficiency of the work—that is, the energy cost per unit of work—is influenced by both the range of the lifting and the work pace. With regard to the range, for example, the energy cost of lifting objects from the floor to about 20 in (51 cm) is about half again as much as lifting the same weight from about 20 to 40 in (51 to 102 cm) (Davies). This is because of the additional effort in lowering and raising the body. Some indication of the relative efficiency of lifting various weights over certain vertical ranges is shown in Figure 7-21 (Frederick, 1959, as presented by Davies). This shows, for each of four lift ranges, the most efficient weight to be lifted in terms of energy cost per unit of work. The fact that the lift range of 40 to 60 in (102 to 152 cm) was the most efficient suggests that workplaces preferably should be designed so that the primary lifting is within that range.

As a further elaboration of the initial study, Frederick developed a formula for estimating the energy cost for any given number of lifts per hour for any given weight and lift range.[3] Alternatively, one can estimate the number of lifts per hour that could be executed and still keep the total energy expenditure within some specified desirable limit, as 200 kcal/h.

Frequency of Lifting Aside from the possibility of estimating the energy costs in lifting weights at various paces, it is in order to consider the judgments of people regarding what they consider to be "acceptable" weights to be lifted in relation to the frequency with which they are to be lifted. In this regard, as part of a series of studies dealing with manual handling tasks, Snook had subjects lift tote boxes with handles at various frequency rates, the subjects adjusting the weights to those that they considered "acceptable". The boxes were lifted three specified distances (9.8, 20.1, and 29.1 in, or 25, 58, and 76 cm) at three levels (floor level to knuckle height, knuckle height to shoulder height, and shoulder height to upper-arm reach). The widths of the tote boxes were also varied. (It should be added that in this and other studies the subjects were industrial workers and served as subjects for several days.)

Since it is not feasible to present all of the data from the lifting study, illustrative

[3] The formula for developing such estimates is:

$$\text{kcal/h} = \frac{f \times a \times w \times c}{1000}$$

in which f = number of lifts per hour; a = lifting height in ft; w = weight in lb; and c = energy in gram-calories per foot pound taken from Figure 7-20.

As an example, assuming a 200 kcal/h limit, with $a = 2$ ft, $w = 40$ lb, and $c = 4$ gram-calories per foot pound, f would be 625 lifts per hour (Davies, 1972).

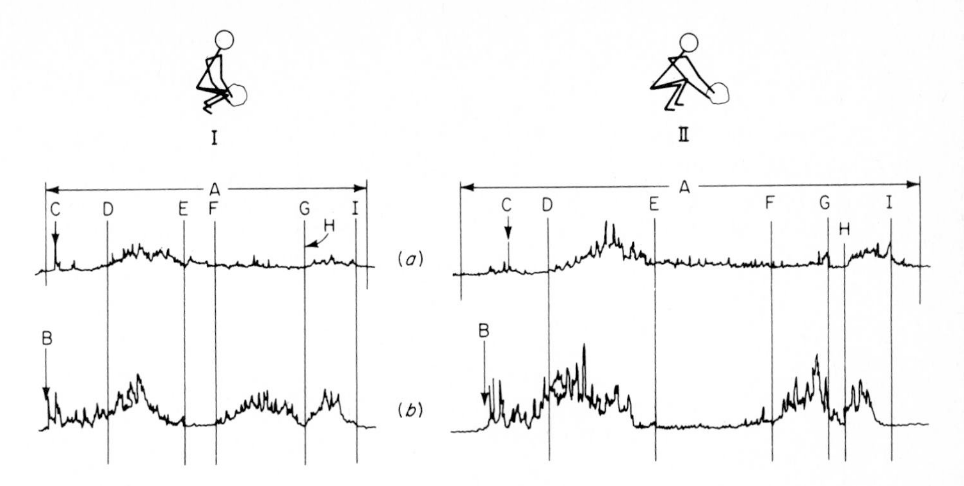

FIGURE 7-20
Electromyograph recordings of two muscles (*a*, gluteus maximus, and *b*, sacrospinalis) made for one subject using two lifting techniques. Technique I incurred less muscle activity than technique II. (The letters and lines represent recordings of corresponding segments of the movements made by the two methods.) (*Source: Adapted from Tichauer, 1971, fig. 9.*)

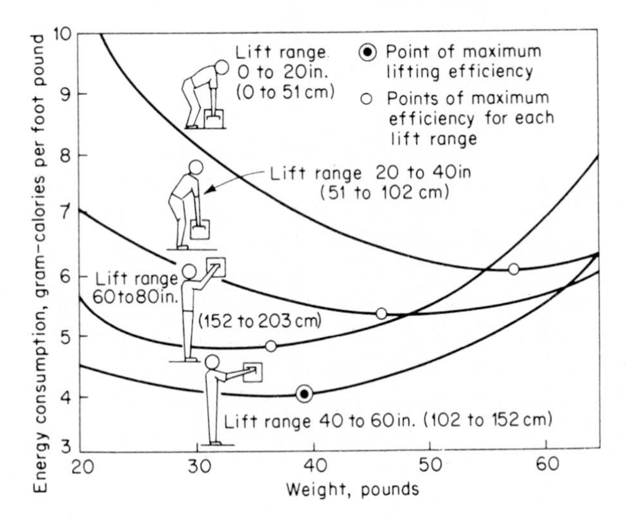

FIGURE 7-21
Energy consumption in lifting per unit of work for various weights and specified lift ranges. The most efficient lifting is with a weight of about 40 lb for a lift range from about 40 to 60 in. (*Source: Frederick, 1959, as presented by Davies, 1972, fig. 3.*)

data are given in Figure 7-22. This shows data on the "acceptable" weights chosen by male subjects for various frequencies of lift for the three vertical distances of the floor-to-knuckle-height lifts, for a tote box 14.2 in (36 cm) in width. In particular this shows the weights chosen by the 90th percentile subjects, which really means that 90 percent of the subjects chose heavier weights, and 10 percent chose lighter weights. This value then represents weights that were acceptable to 90 percent of the subjects. The very marked differences in weights that were considered as acceptable for the different frequencies argues for the adjustment of such weights if the items are to be lifted frequently, or for reducing the frequency if the weights involved are heavy.

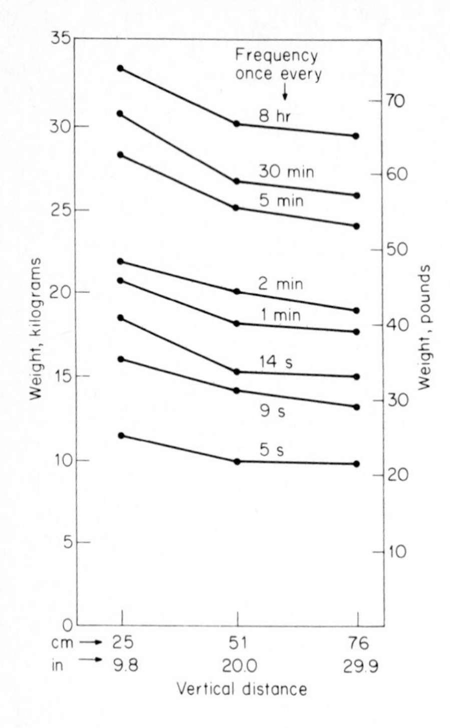

FIGURE 7-22
Maximum weights chosen as acceptable by 90 percent of a sample of male industrial workers for a lifting task performed at different frequencies, for three vertical distances of lift using a tote box 14.2 in (36 cm) wide. (*Source: Snook, 1978, table 1, p. 971.*)

A comment is in order regarding the matter of age, since age influences the physical abilities of people. The International Occupational Safety and Health Information Centre has set forth a set of recommended weight limits for occasional lifting by any method for persons of various age groups. Those recommendations are given in Table 7-5. Snook makes the point that his results compare rather favorably with these recommendations. In considering the lifting activities involved in jobs, however, one should keep in mind the fact that various factors can influence the acceptable loads, as pointed out by Konz and Coetzee (1978) and by Snook. The extensive tables in Snook's paper take at least some variables into account, such as width of containers, distances of lift, and sex.

TABLE 7-5
RECOMMENDED WEIGHT LIMITS FOR LIFTING TASKS AS GIVEN BY THE INTERNATIONAL OCCUPATIONAL SAFETY AND HEALTH CENTRE

	Weight, lb, for specified age groups (approximate kilogram equivalents in parentheses)					
	14–16	**16–18**	**18–20**	**20–35**	**35–50**	**Over 50**
Male	33 (15)	42 (19)	51 (23)	55 (25)	45 (20)	36 (16)
Female	22 (10)	26 (12)	31 (14)	33 (15)	29 (13)	22 (10)

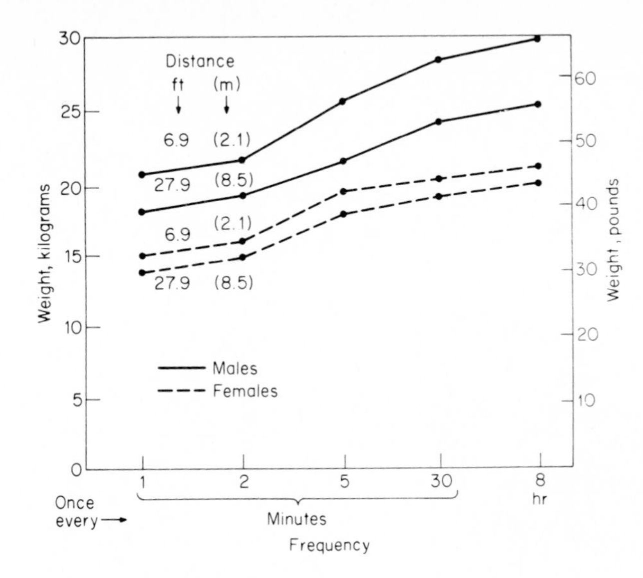

FIGURE 7-23
Maximum weights chosen as acceptable by 90 percent of a sample of male and female industrial workers for carrying loads at knuckle height, at different frequencies, for two distances. (*Source: Snook, 1978, table 10, p. 980.*)

Carrying Tasks

As with lifting tasks, there are marked differences in the weights people find acceptable, depending on the frequency with which the activity is carried out. Aside from individual differences, there are systematic differences between males and females. As an example, Figure 7-23 presents data from Snook for males and females on the maximum weights reported as acceptable by 90 percent of the subjects, for two distances of carry. The effects of frequency are particularly noticeable, with carrying distance having some additional effect on the level of acceptable loads. That figure presents data for carrying at knuckle height. When carrying at elbow height, the acceptable weight levels are somewhat lower.

Pushing Tasks

Additional data on manual handling tasks from Snook deal with acceptable levels of weights in pushing tasks. The data shown in Figure 7-24 also reveal systematic differences in the maximum acceptable loads depending on the frequency with which the tasks are performed, with appreciable differences in distances pushed, but only slight sex differences (in fact, for the shortest distance there was no difference by sex). This example represents data for pushing at average shoulder height. Pushing at elbow or knuckle height appears not to influence the levels reported as acceptable.

WORK LOAD

The discussion earlier in the chapter referred briefly to the notions of stress and work load, there being implications that certain kinds or levels of work and of environmental conditions can bring about changes in various indexes of strain, or result in work degradation.

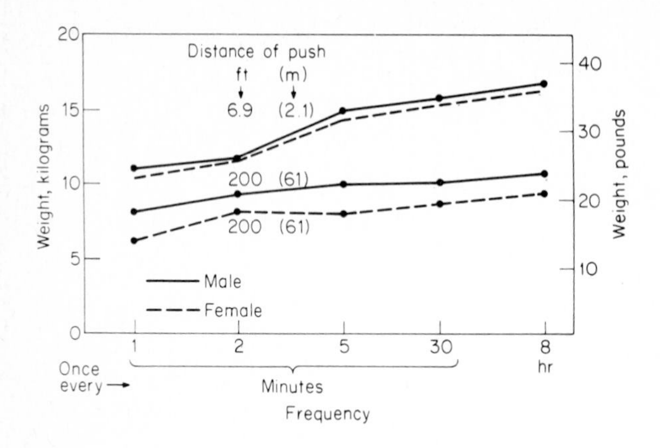

FIGURE 7-24
Maximum weights chosen as acceptable by 90 percent of a sample of male and female industrial workers for a pushing task at shoulder height, at different frequencies, for two distances. (*Source: Snook, 1978, tables 5, 6, 7, 8, pp. 975-978.*)

Concept of Work Load

As Leplat (1978) and Welford (1978) imply, the concept of *work load* is rather elusive. Leplat, for example, calls attention to two views of the concept, as follows: (1) the load as characteristic of the task, hence the obligations and compulsions it imposes on the worker (he prefers to refer to this as the requirements of the job); and (2) the load or consequence for the worker in performing the task (he reserves the term *load* for this concept). In line with this, Welford expresses the opinion that the basic idea of work load is most easily understood in terms of muscular effort on the part of the individual, of which two types can be distinguished, namely the maximum instantaneous force executed in a particular task, and the amount of work done over a given period of time. The notion of work load in the case of mental activities is a bit more elusive, but one can envision in general terms the same two types, the maximum instantaneous mental effort and the amount of mental work done over a period of time.

Work Load and Capacity

Work load, however, cannot readily be separated from the consideration of the capacity of individuals, in particular maximum capacity. The relationship between capacity and demands of the task is shown in Figure 7-25*a*. So long as capacity exceeds demand, performance is limited by demand. But when demands of the task exceed the capacity of the individual, performance is of course limited by capacity. (The capacity of an individual, incidentally, is subject to variation over time.)

When demand does exceed capacity (with the implication of restricted performance), it may be, as Welford points out, that a change in strategy can bring the demands down to a level that is within the capacity limits of the individual. This is illustrated in Figure 7-25*b*. This shows how strategy x (as contrasted with strategy y) could bring the demands within the capacity limits of the individual. Depending on the nature of the task, the changes in strategy could consist of the use of a different design of machine or tool, the substitute of a powered device for a manual device, or a change in the work pace. The objectives of many human factors efforts are toward tak-

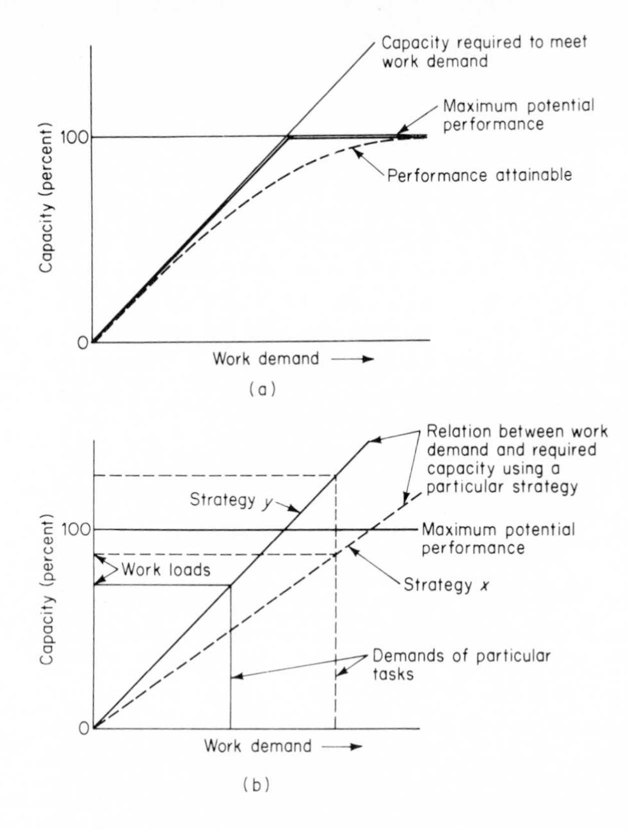

FIGURE 7-25
(*a*) Basic relationship between the capacity of individuals and the demands of a task.
(*b*) Illustration of how a change in strategy can modify the demands of a task to bring the demands within the capacity of individuals. (*Source: Adapted from Welford, 1978, figs. 1 and 3, pp. 152 and 156.*)

ing actions to keep the work loads of human activities within the bounds of human capacities.

Although work loads should be kept within reasonable upper bounds, Snook and Irvine (1969), referring to the intensity of work, make the point that low levels of work intensity increase the likelihood of boredom. It should be noted, however, that boredom is a reaction of an individual and not an attribute of the work activity. Granting this matter of individual differences, Snook and Irvine argue for the development of work activities at something of an "optimum" level, between the upper bounds that would induce exhaustion (that would require performance near one's capacity) and the lower bounds that would increase the likelihood of boredom. Where feasible, human factors efforts should be directed toward achieving this optimum.

Work Load and Stress

The work load imposed by a task or environmental condition may be well within an individual's capacity. But if it is not, the circumstance can serve as a source of stress.

Effects of Stress As indicated before, the effects of stress can be manifested in some form of strain, such effects frequently being physiological or psychological, or being manifested in changes in performance. However, there are some indications that continuous stress may be accompanied by effects on the health and well-being of individuals. This possible effect is illustrated by the results of an interesting investigation of various such manifestations as reported by Johansson et al. (1978) in comparing two groups of sawmill workers in Sweden. One group, a "high-risk" group, consisted of sawyers, edgers, and graders, whose work was characterized by short work cycles, a forced work tempo, and the requirement for continuous attention and alertness. The other group, a "low-risk" group, included stickers, repairmen, maintenance workers, etc., whose work was considered to be much less demanding.

One comparison of these two groups consisted of the frequency of various health symptoms as based on the combination of self-reports and clinical tests. The average numbers of symptoms for the two groups are given below:

Group	Frequency of symptoms
High risk	3.7
Low risk	2.3

There were fairly systematic differences for a number of symptoms, with particularly noticeable differences for headaches (36 percent versus 0 percent for the two groups) and "slight nervous disturbance" (36 percent versus 0 percent).

Job Factors Associated with Stress Reference has been made in this chapter to certain possible sources of stress, such as the nature of the work, work demands, work methods, rate of work, and environmental factors. (Later chapters deal more specifically with some of these factors.) However, some interesting insight regarding job demands that came from the study by Johansson et al. (1978) are relevant to our present discussion.

As one phase of their study they made a comparison between the high-risk group and the low-risk group of the "psychological job content" as measured by responses to a questionnaire, with the low-risk group generally responding to the various items of the questionnaire much more favorably than the high-risk group. Some of the items for which the low-risk group responded more favorably were: longer work cycles; more variation between tasks; more autonomous work; fewer demands for continuous attention; and more freedom to choose order of task items.

Other attitudinal differences were found regarding the working conditions, as based on another questionnaire, with the relative frequency of responses to certain items being given in Table 7-6 for the two groups.

Although only a few of the results are given in Table 7-6, the general pattern of results indicates that the high-risk group experienced a substantially higher incidence of undesirable health symptoms and "psychological" stress, the psychological stress tending to be associated with machine-paced work, short-cycle operations, demands for continuous attention, less variability in the job, and restricted social contacts at work, along with other job factors and working conditions.

As an interesting sideline to these results, Broadbent (1979) reports that, in one industrial situation, although most workers reported dissatisfaction in repetitive jobs, the

TABLE 7-6
ATTITUDES OF SAWMILL WORKERS ON "HIGH-RISK" AND "LOW-RISK" JOBS TO CERTAIN WORKING CONDITIONS

Working condition	High-risk group, %	Low-risk group, %
Consider job monotonous	100	50
Bored by job	57	0
Attribute job stress to lack of social contact at work	71	30
Feel distress or uneasiness on going to work	50	0

Source: Johansson et al., 1978.

level of "anxiety" (as based on psychiatric symptoms) was much higher for those on paced jobs (such as on a paced assembly line) than on an equally repetitive job on which the work pace was under the control of the worker. Thus it seems (as Broadbent points out) that although repetitive work may generate dissatisfaction on the part of many workers, lack of control over the job (such as with jobs that are paced) may be a dominant factor in connection with anxiety symptoms.

DISCUSSION

Many of the sketches, blueprints, and plans for the things people use or for the work they are to do predetermine the nature of the activities people are to perform on their jobs or elsewhere, and of the environments in which people will work and live. Some timely consideration given to the design of things and environments can pay handsome dividends in minimizing the possibility of stress and its accompanying consequences. Such consideration basically depends upon understanding of the nature of human beings with particular reference to their physical and physiological characteristics and their related abilities and limitations.

REFERENCES

Bălănescu, B. F. Some aspects concerning the dynamic evaluation of oxygen consumption in exercise tests. *Ergonomics*, 1979, *22*(12), 1337–1342.

Barany, J. W. The nature of individual differences in bodily forces exerted during a simple motor task. *Journal of Industrial Engineering*, 1963, *14*(6), 332–341.

Barter, J. T., Emanuel, I., & Truett, B. *A statistical evaluation of joint range data* (Technical Note 57-311). U.S. Air Force, WADC, 1957.

Borg, G. Subjective aspects of physical and mental load. *Ergonomics*, 1978, *21*(3), 215–220.

Boyce, P. R. Sinus arrhythmia as a measure of mental load. *Ergonomics*, 1974, *17*(2), 177–183.

Broadbent, D. E. Is a fatigue test now possible? *Ergonomics,* 1979, *22*(12), 1277-1290.

Brouha, L. *Physiology in industry.* New York: Pergamon, 1960.

Brown, J. S., & Slater-Hammel, A. T. Discrete movements in the horizontal plane as a function of their length and direction. *Journal of Experimental Psychology,* 1949, *39*, 84-95.

Brown, J. S., Wieben, E. W., & Norris, E. B. *Discrete movements toward and away from the body in a horizontal plane* (Contract N5ori-57, Report 6). USN, ONR, SDC, September 1948.

Burger, G. C. E. Heart rate and the concept of circulatory load. *Ergonomics,* 1969, *12*(6), 857-864.

Christensen, E. H. Physiological valuation of work in the Nykroppa iron works. In W. F. Floyd & A. T. Welford (Eds.), *Ergonomics Society Symposium on Fatigue.* London: Lewis, 1953.

Corlett, E. N. A study of a light repetitive task. *Applied Ergonomics,* 1977, *8*(2), 103-109.

Corlett, E. N., & Mahadeva, K. A relationship between a freely chosen working pace and energy consumption curves. *Ergonomics,* 1970, *13*(4), 517-524.

Curry, R., Jex, H., Levison, W., & Stassen, H. Final report (revised) of control engineering group. In N. Moray, *Mental workload: Its theory and measurement,* 1979, pp. 235-252.

Damon, A., Stoudt, H. W., & McFarland, R. A. *The human body in equipment design.* Cambridge, Mass.: Harvard, 1966.

Datta, S. R., & Ramanathan, N. L. Ergonomics comparison of seven modes of carrying loads on the horizontal plane. *Ergonomics,* 1971, *14*(2), 269-278.

Davies, B. T. Moving loads manually. *Applied Ergonomics,* 1972, *3*(4), 190-194.

Dempster, W. T. The anthropometry of body action. *Annals of the New York Academy of Sciences,* 1955, *63*, 559-585.

Dukes-Dobos, F. N., Wright, G., Carlson, W. S., & Cohen, H. H. Cardio-pulmonary correlates of subjective fatigue. Technical Program for the 20th Annual Meeting of the Human Factors Society, July 11-16, 1976, pp. 24-27.

Edholm, O. G. *The biology of work.* World University Library, New York: McGraw-Hill, 1967.

Fitts, P. M. A study of location discrimination ability. In P. M. Fitts (Ed.), *Psychological research on equipment design* (Research Report 19). Army Air Force, Aviation Psychology Program, 1947.

Frederick, W. S. Human energy in manual lifting. *Modern Materials Handling,* 1959, *14*(3), 74-76.

Gordon, E. E. The use of energy costs in regulating physical activity in chronic disease. *A.M.A. Archives of Industrial Health,* November 1957, *16*, 437-441.

Greene, J. H., Morris, W. H. M., & Wiebers, J. E. A method for measuring physiological cost of work. *Journal of Industrial Engineering,* May-June 1959, *10*(3).

Hogan, J. C., Ogden, G. D., Gebhardt, D. L., & Fleishman, E. A. Reliability and validity of methods for evaluating perceived physical effort. *Journal of Applied Psychology,* 1980, *65*(6), 672-679.

Hunsicker, P. A. *Arm strength at selected degrees of elbow flexion* (Technical Report 54-548). U.S. Air Force, WADC, August 1955.

Hyndman, B. W., & Gregory, J. R. Spectral analysis of sinus arrhythmia during mental loading. *Ergonomics,* 1975, *18*(3), 255-270.

International Occupational Safety and Health Organization Centre (CIS Information Sheet 3). Geneva, 1962.

Jex, H. R., & Clement, W. F. Defining and measuring perceptual-motor workload in manual control tasks. In N. Moray, *Mental Workload: Its theory and measurement,* 1979, pp. 125–177.

Johansson, G., Aronsson, G., & Lindstrom, B. O. Social psychological and neuroendocrine stress relations in highly mechanized work. *Ergonomics,* 1978, *21*(8), 583–599.

Johansson, G., & Rumar, K. Drivers' brake-reaction times. *Human Factors,* 1971, *13* (1), 23–27.

Kalsbeek, J. W. H. Sinus arrhythmia and the dual task of measuring mental load. In W. T. Singleton, J. G. Fox, & D. Whitfield (Eds.), *Measurement of man at work.* London: Taylor and Francis, 1971, pp. 101–113.

Khalil, T. M. An electromyographic methodology for the evaluation of industrial design. *Human Factors,* 1973, *15*(3), 257–264.

Konz, S. A., & Coetzee, J. Prediction of ratings of lifting difficulty from individual and task variables. *Human Factors,* 1978, *20*(4), 481–497.

Kroemer, K. H. E. Human strength: terminology, measurement, and interpretation of data. *Human Factors,* 1970, *12*(3), 297–313.

Lauru, L. The measurement of fatigue. *The Manager,* 1954, *22*, 299–303 and 369–375.

LeBlanc, J. A. Use of heart rate as an index of work output. *Journal of Applied Physiology,* 1957, *10*, 275–280.

Lehmann, G. Physiological measurements as a basis of work organization in industry. *Ergonomics,* 1958, *1,* 328–344.

Leplat, J. Factors determining work-load. *Ergonomics,* 1978, *21*(3), 143–149.

Luczak, H. Fractioned heart rate variability. *Ergonomics,* 1979, *22*(12), 1315–1323.

Margaria, R., Mangili, F., Cuttica, F., & Cerretelli, P. The kinetics of the oxygen consumption at the onset of muscular exercise in man. *Ergonomics,* 1965, *8*(1), 49–54.

Mead, P. G., & Sampson, P. B. Hand steadiness during unrestricted linear arm movements. *Human Factors,* 1972, *14*(1), 45–50.

Milošević, S. Vigilance performance and the amplitude of EEG activity, *Ergonomics,* 1978, *21*(11), 887–894.

Monod, P. H. La validité des mesures de fréquence cardiaque en ergonomie. *Ergonomics,* 1967, *10*(5), 485–537.

Moray, N. (Ed.). *Mental workload: Its theory and measurement.* New York: Plenum, 1979.

Mulder, G. Sinus arrhythmia and mental workload. In N. Moray, *Mental Workload: Its theory and measurement,* 1979.

Murrell, K. F. H. *Human performance in industry.* New York: Reinhold, 1965.

Ogden, G. D., Levine, J. M., & Eisner, E. J. Measurement of workload by secondary tasks. *Human Factors,* 1979, *21*(5), 529–548.

Örtengren, R., Andersson, G., Broman, H., Magnusson, R., & Petersén, I. Electromyography: Studies of localized muscle fatigue at the assembly line. *Ergonomics,* 1975, *18*(2), 157–174.

Passmore, R. Daily energy expenditure by man. *Proceedings of the Nutrition Society,* 1956, *15*, 83–89.

Passmore, R., & Durnin, J. V. G. A. Human energy expenditure. *Physiological Reviews,* 1955, *35,* 801-875.

Pattie, C. *Simulated tractor overturnings: A study of human responses in an emergency situation.* Ph.D. thesis. Lafayette, Ind.: Purdue University, May 1973.

Salvendy, G., & Pilitsis, J. Psychophysiological aspects of paced and unpaced performance as influenced by age. *Ergonomics,* 1971, *14*(6), 703-711.

Sanders, A. F. Some remarks on mental load. In N. Moray, *Mental workload: Its theory and measurement,* 1979, pp. 41-77.

Schappe, R. H. Motion element synthesis: an assessment. *Perceptual and Motor Skills,* 1965, *20,* 103-106.

Schmidtke, H., & Stier, F. Der aufbau komplexer bewegungsabläufe aus elementarbewegungen. *Forschungsberichte des landes Nordrhein-Westfalen,* 1960, No. 822, 13-32.

Schottelius, B. A., & Schottelius, D. D. *Textbook of physiology* (18th ed.). St. Louis: Mosby, 1978.

Sheridan, T. B., & Stassen, H. G. Definitions, models and measures of human workload. In N. Moray, *Mental workload: Its theory and measurement,* 1979, pp. 219-233.

Singleton, W. T. The measurement of man at work with particular reference to arousal. In W. J. Singleton, J. G. Fox, & D. Whitfield (Eds.), *Measurement of man at work.* London: Taylor and Francis, 1971, pp. 17-25.

Snook, S. H. The design of manual handling tasks. *Ergonomics,* 1978, *21*(12), 963-985.

Snook, S. H., & Irvine, C. H. Psychophysical studies of physiological fatigue criteria. *Human Factors,* 1969, *11*(3), 291-300.

Soede, M. On mental load and reduced mental capacity. In N. Moray, *Mental workload: Its theory and measurement,* 1979.

Stamford, B. A. Validity and reliability of subjective ratings of perceived exertion during work. *Ergonomics,* 1976, *19*(1), 53-60.

Tichauer, E. R. *The biomechanical basis of ergonomics.* New York: Wiley, 1978.

Tichauer, E. R. A pilot study of the biomechanics of lifting in simulated industrial work situations. *Journal of Safety Research,* September 1971, *3*(3), 98-115.

Tuttle, W. W., & Schottelius, A. A. *Textbook of physiology* (16th ed.). St. Louis: Mosby, 1969.

Ursin, H., & Ursin, R. Physiological indicators of mental workload. In N. Moray, *Mental workload: Its theory and measurement,* 1979, pp. 349-365.

Vos, H. W. Physical workload in different body postures, while working near to, or below ground level. *Ergonomics,* 1973, *16*(6), 817-828.

Wardle, M. G. A psychophysical approach to estimating endurance in performing physically demanding work. *Human Factors,* 1978, *20*(6), 745-747.

Wargo, M. J. Human operator response speed, frequency, and flexibility: a review and analysis. *Human Factors,* 1967, *9*(3), 221-238.

Warrick, M. J., Kibler, A. W., & Topmiller, D. A. Response time to unexpected stimuli. *Human Factors,* 1965, *7*(1), 81-86.

Welford, A. T. Mental work load as a function of demand, capacity, strategy and skill. *Ergonomics,* 1978, *21*(3), 157-167.

Wierwille, W. W. Physiological measures of aircrew mental workload. *Human Factors,* 1979, *21*(5), 575-593.

CHAPTER 8

HUMAN CONTROL OF SYSTEMS

The types of systems and mechanisms people control in their jobs and everyday lives vary tremendously, and of course influence the nature of the control devices people use and the human requirements for their use. By and large such control includes the following functions: activation of some machine or item of equipment (usually with an "on-off" switch); making discrete (i.e., separate, distinct) settings (as TV channel selector controls); making quantitative settings (as setting a thermostat); continuous control (as of an automobile); and data entry (as with a computer or typewriter). The control of some systems is of course extremely complex, as in controlling aircraft around a metropolitan airport or controlling the operation of a nuclear power plant. Whatever the nature of the system, the basic human functions involved are those referred to earlier, namely, information input, information processes leading to a decision, and action or response. The action taken serves as the control input to the system. In the case of some systems that require continuous control, there typically is some form of feedback to the operator.

HUMAN INPUT AND OUTPUT CHANNELS

In some circumstances the input to a system comes from the individual, such as a judgment that a particular action should be initiated (such as mowing the lawn). In the operation of most "typical" systems, however (as in production processes or the operation of vehicles) there are stimuli from the environment that serve to trigger responses on the part of an operator. In such instances the stimuli are of course received by one or another of the sensory mechanisms. The primary sensory modalities are vision and hearing. However, certain of the other senses serve as channels of useful input in some circumstances, such as the skin senses (pressure and temperature), the kinesthetic sense, the senses of body movement and equilibrium, and even the sense of smell. The inputs to the sensory receptors should of course consist of stimuli that are relevant to the control of the system in question. As illustrated in Figure 3-1 in Chapter 3, such stimuli may be received *directly* from the original (distal) source, or *indirectly* via some man-made display.

The primary output channels (i.e., the effectors) used in system control are the motor responses (especially with the hands and feet) and speech (as in aircraft-control-tower operations). Most typically the motor responses consist of the use of control devices such as levers, cranks, push buttons, and pedals. However, other responses are used in special circumstances, or offer some promise of greater use. These include certain of the physiological measures discussed in Chapter 7 (EEG, EMG, etc.). In addition, there have been some interesting developments in the use of eye movements in control processes. The choice of the output channel for any given situation would of course depend upon its appropriateness for the purpose at hand.

BETWEEN INPUT AND OUTPUT

The nature of the human involvement in control processes naturally depends on the type of system involved, the functions that human beings are intended to perform, and the types of mechanisms provided for them in the control processes. In a simple case this could consist of the activation of a machine or the dialing of a telephone number. In the case of a complex tracking task the individual is supposed to match the response to a continually changing input signal. In the case of an electric power plant the operator has a complex array of displays and controls to attend to. Whatever the level of complexity of the operation, usually some form of information processing and decision is sandwiched in between the information input (as from displays) and the output (usually some physical response with a control mechanism). The decisions involved in various control tasks can range from the very simple (virtually a conditioned response in some circumstances) to the complex. In the case of some operations the work load may tend to peak at certain times, thereby running the risk of mental overload as discussed in Chapter 7. An example of work-load peaking is illustrated in Figure 8-1, this showing the number of aircraft movements during each half hour of two successive days at London's Heathrow airport. The peaks could reflect circumstances in which the air traffic controllers might experience mental overload as discussed in Chapter 7.

One of the possible consequences of overload in control operations is the fact that short-term memory may cause people to forget relevant information. This forgetting effect was demonstrated by the results of a study of a simulated airport control situation involving communications between ground controllers and pilots. As one phase of this study the controllers were required to remember four-digit aircraft identification numbers (transponder codes). Figure 8-2 shows the forgetting curves of such information for "low-information" and "high-information" load conditions. The forgetting curve for the high-information load condition deteriorates more markedly.

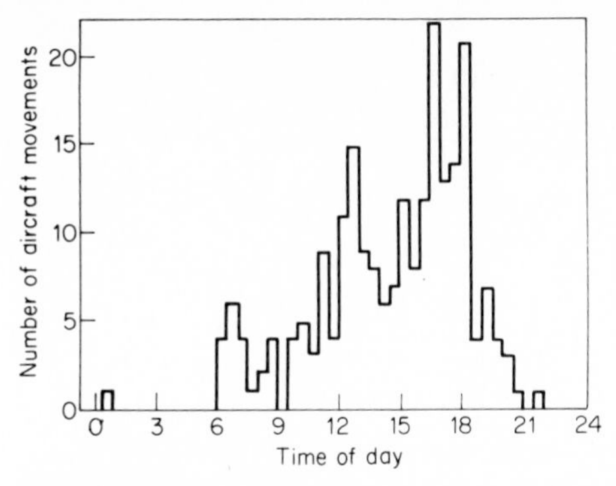

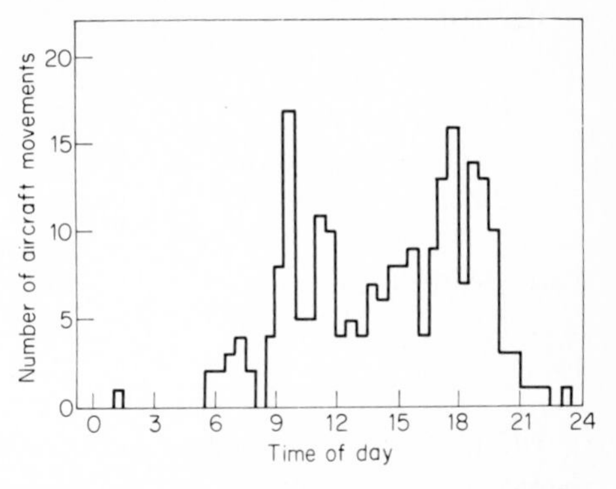

FIGURE 8-1
Number of aircraft movements during each half hour on two successive days at London's Heathrow Airport, reflecting the work load of the air traffic controllers. The peaks could represent circumstances in which the controllers might experience mental overload. (*Source: Brigham, 1976, fig. 6, p. 23.* This figure appeared as fig. 6 in Vol. 7, No. 1, p. 23 of *Applied ergonomics*, published by IPC Science and Technology Press, Ltd., Guilford, Surrey, U.K.)

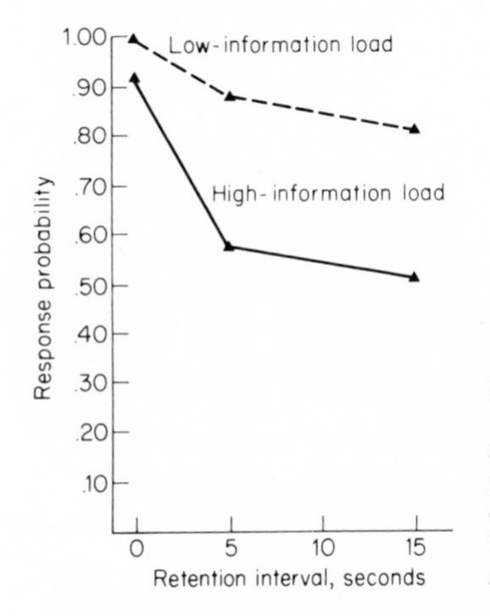

FIGURE 8-2
Forgetting curves of ground controllers in a simulated aircraft control experiment, the curves showing the percent of four-digit identification numbers (transponder codes) of aircraft for "low-information" and "high-information" load conditions. (*Source: Adapted from Loftus, Dark, & Williams, 1979, fig. 2, p. 176.*)

Factors that Can Influence Human Control

In the case of some control operations there are no particular problems in effecting appropriate control or any reasonable likelihood of undesirable consequences to the individuals involved. Operating a dishwasher, a TV set, or many simple machines in industrial situations seldom poses serious problems. However, the human control of systems or mechanisms can have important implications under certain conditions, as when it is critical that performance be at a satisfactory level, when there might be some undesirable effects upon the individuals (such as injury), and particularly when the operational demands are difficult. In this regard Drury and Baum (1976), referring particularly to the factors that might have some effect on task performance or strategy, differentiate between task variables and individual variables as follows:

Task variables

Input to the controller (especially its complexity)
Display complexity
Control complexity
Process complexity

Operator variables

Operator skills
Operator motivation

The task variables of course influence the nature and complexity of the decisions that a controller must make and the control actions that are based on such decisions.

Input-Output Relationships

In connection with the decisions and control responses of a controller, Drury and Baum emphasize the importance, during the design of a system, of the analysis of the relationships between system inputs and outputs. As one stage of such an analysis they propose the development of an input-output matrix such as shown in Figure 8-3. Such a matrix is developed in part on the basis of interviews with experienced personnel who are asked what system inputs affect each output, to describe how the output is affected, and to rate the importance of the effect, using a 10-point scale. Those ratings can then be used in the further development of the system, especially the design displays that would be used to present any relevant input and output information and the design and arrangement of controls that would be used in the operation. Chapters 4 and 5 deal with the design of displays and Chapter 9 deals with the design of controls. The next section of this chapter deals with the concept of compatibility that in part concerns the relationship between displays and controls.

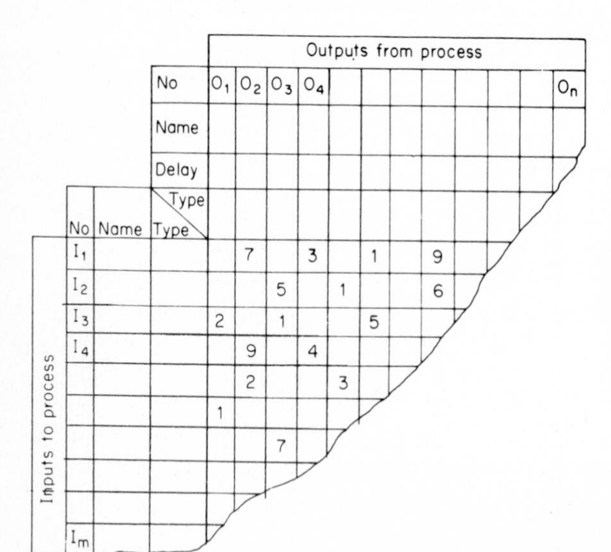

FIGURE 8-3
Illustration of an input-output matrix for use in system development. The values in the cells are ratings of the importance of the relationships and can aid in the design of the system, especially the displays and controls. (*Source: Drury & Baum, 1976, fig. 2, p. 7.* This figure appeared as a portion of fig. 2 in Vol. 7, No. 1, p. 7 of *Applied ergonomics*, published by IPC Science and Technology Press, Ltd., Guilford, Surrey, U.K.)

COMPATIBILITY

As discussed in Chapter 3, *compatibility* refers to the spatial, movement, or conceptual relationships of stimuli and responses, individually or in combination, that are consistent with human expectations.

Bases of Compatibility Relationships

There are three general bases for the formation of people's expectations. In the first place some compatibility relationships are clearly manifest or intrinsic to the context. This is especially the case with the physical relationships between displays and their corresponding controls, such as having four displays in a row with their respective controls beneath them. In the second place some compatibility relationships arise from the culture and are learned, such as red meaning "stop" and green meaning "go." And in the third place there are "response tendencies" that are characteristic of at least some (but not necessarily all) people, such as the tendency to turn a control knob clockwise to increase whatever value is associated with it. Some such tendencies are stronger than others in the sense that they are characteristic of larger proportions of the population. Actually these bases of compatibility blend with each other rather than represent clear-cut categories. Those compatibility relationships that involve movements of some type frequently are referred to as *population stereotypes.*

Whatever the origin of compatibility relationships, the concept has very definite human factors implications in the design of things people use, especially displays and controls, on the grounds that the use of compatibility principles generally will enhance the use of the item or system in question. However, we should keep in mind two possible limitations about compatibility relationships: (1) some are stronger than others (as being more obvious or more commonly recognized by people); and (2) there are circumstances in which it may be necessary to violate one compatibility relationship in order to take advantage of another one in the design of some system.

Spatial Compatibility

There are many variations of the theme of spatial compatibility, most of them spanning the gamut of physical similarities in displays and corresponding controls and their arrangement, and the arrangement of any given set of either displays or controls.

Physical Similarity of Displays and Controls Sometimes there exists the opportunity to design related displays and controls so there is reasonable correspondence of their physical features, and perhaps also of their modes of operation. Such a case was well illustrated years ago by Fitts and Seeger (1953). In this study three different displays and three different controls were used in all possible combinations. The displays consisted of lights in various arrangements. As a light would go on, the subject was to move a stylus along a corresponding channel to its end; when that location was reached, an electric contact would turn the light off. The three displays and controls are illustrated in Figure 8-4. The experimental procedures will not be described, but in general, different groups of subjects used each combination of the three stimulus

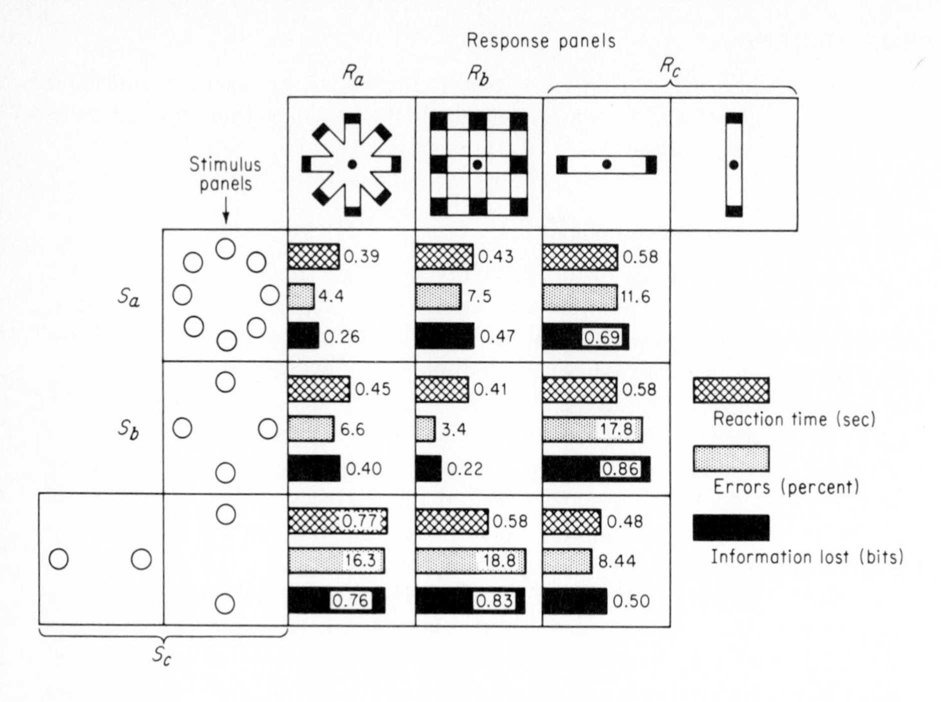

FIGURE 8-4
Illustrations of signal (stimulus) panels and response panels used in study by Fitts and Seeger. The values in any one of the nine squares are the average performance measures for the combination of stimulus panel and response panel in question. The compatible combinations are S_a-R_a, S_b-R_b, and S_c-R_c, for which results are shown in the diagonal cells. (*Source: Fitts & Seeger, 1953.*)

panels with the three response panels, their performance being measured in reaction-time errors and information lost. The results, also given in Figure 8-4, show that performance with any given stimulus or response panel was better when it was used in combination with its *corresponding* response or stimulus panel (S_a-R_a, S_b-R_b, and S_c-R_c) than when used in combination with a different configuration.

Physical Arrangement of Displays and Controls Both experiments and rational considerations lead one to conclude that for optimum use, corresponding displays and controls should be arranged in corresponding patterns. This aspect of compatibility was well demonstrated by the down-to-earth example dealing with a gadget used morning, noon, and night (and sometimes in between)—a study dealing with the arrangement of burner controls on a four-burner stove (Chapanis & Lindenbaum, 1959). The burners, in a sense, can be thought of as displays. The four arrangements that were tried out are shown in Figure 8-5. Fifteen subjects each had 80 trials on each design. They were told what burner to turn on, and their reaction time was recorded along with any errors they made. The number of errors made on the four designs is given in Table 8-1. Design I was also best in reaction time (over the last 40 trials), with design II next best. Clearly, design I was the most compatible.

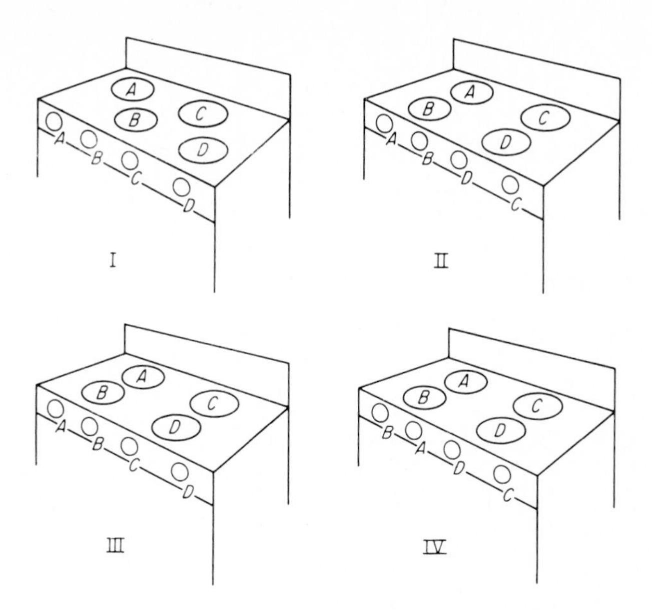

FIGURE 8-5
Control-burner arrangements of simulated stove used in experiment by Chapanis and Lindenbaum. (*Source: Chapanis & Lindenbaum, 1959.*)

In a further study dealing with controls of burners, Shinar and Acton (1978) asked subjects to indicate which of four unmarked controls they thought controlled each of four burners, the burners being arranged and labeled as those in example II of Figure 8-5. There was almost universal association of the two left controls with the two left burners, but there was considerable confusion as to which of the two controlled which burner. The same pattern existed with the right controls and burners. In other words,

TABLE 8-1
ERRORS IN EXPERIMENTAL USE OF BURNER CONTROLS OF STOVES FOR FOUR ARRANGEMENTS SHOWN IN FIGURE 8-5

Design	Number of errors, out of 1200 trials
I	0
II	76
III	116
IV	129

Source: Chapanis and Lindenbaum, 1959.

the "left versus right" stereotype was very strong, but there was no clear-cut stereotype for the two sets on the left or right. It will be noticed that in the Chapanis and Lindenbaum study the strongest stereotype was I, which has off-set burners that tend to lend them a linear, sequential pattern of relationships with the controls. Incidentally, Shinar and Acton found that of 49 models of stoves they found in stores *not one* was similar to type I of Chapanis and Lindenbaum.

Compatibility of Movement Relationships

There are several different types of circumstances in which movement relationships in systems can be viewed in a compatibility frame of reference. Following are some examples:

- Movement of a control device to *follow* the movement of a display (as moving a lever to the right to follow a right movement of a blip on a radarscope)
- Movement of a control device to *control* the movement of a display (as tuning a radio to a particular wavelength)
- Movement of a control device that produces a specific system response (as turning a steering wheel to the right to turn right)
- Movement of a display indication without any related response (as the clockwise turn of the hands of a clock)

For most movements involving a control action there usually is a particular pattern that is most compatible with people's expectations, that represents people's stereotypes. However, the compatibility of movement relationships is intertwined with various features of the displays, the controls, or both, and with their physical orientation to the user (such as whether they are in the same, or different, planes relative to the user). On the basis of research and experience certain principles of movement compatibility have emerged, some of which are discussed below. However, it should be added that some stereotypes are stronger than others, and that in many circumstances conflicts between and among certain principles are inevitable.

Rotary Controls and Rotary Displays in Same Plane In the case of fixed rotary scales with moving pointers the principle that has become firmly established is that a clockwise turn of the control is associated with a clockwise turn of the pointer, and that such rotation generally indicates an increase in the value in question. Conversely, a counterclockwise movement of the control and display pointer are associated with a decrease of the value. This general pattern is reflected in Figure 8-6.

For moving scales with fixed pointers Bradley (1954) postulated that the following principles generally would be desirable:

1 That the scale rotate in the same direction as its control knob (i.e., that a direct drive exist between control and display)
2 That the scale numbers increase from left to right
3 That the control turn clockwise to increase settings

However, it is not possible to incorporate all of these principles in all conventional assemblies. These incompatibilities occur because of the nature of the linkage (that is,

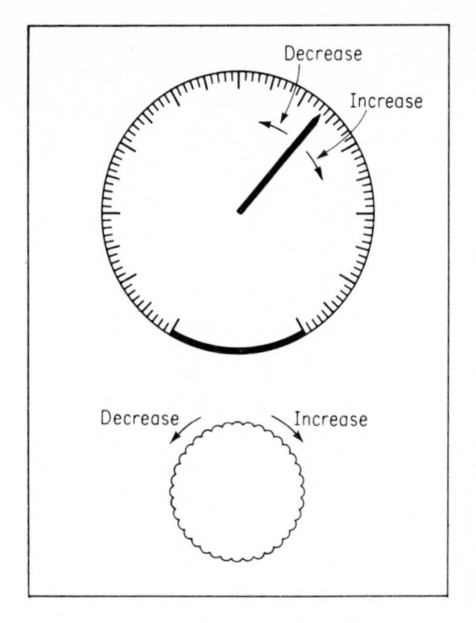

FIGURE 8-6
Illustration of the most compatible relationship between a fixed-scale display and a rotary control such as a knob.

the drive) between the control and display, direction of increase in scale numbers (left to right versus right to left), or both. As shown in Figure 8-7, only two combinations of these principles are possible with moving-scale displays. Since some principle must therefore be sacrificed, a question arises as to which combination of principles would be most effective in use. A study by Bradley reflects some light on this matter. He used various combinations of control and display relationships, including the four shown in Figure 8-7, with provision for obtaining the following types of criteria: (1) starting errors (an initial movement in the wrong direction); (2) setting errors (incorrect settings); and (3) rank-order preferences of the subject. Some of the results are given at the bottom of Figure 8-7 (the two or three sets of data for each criterion are from different groups of subjects used in different phases of the study in which only certain assemblies were used).

While assembly *A* incorporated the incompatible feature of a *counterclockwise* control turn bringing about an *increase* in scale value (rather than a decrease), nonetheless, this assembly was found throughout a series of subexperiments generally to be the best assembly, considering starting errors, setting errors, and preferences. In evaluating the apparent superiority of assembly *A* over certain others (for example, *C*), it seemed clear that a *reverse drive* (such as in *C* and *D*) tended to give rise to more starting errors than a direct drive.

While, in a sense, all this relates to a fairly incidental feature of display-control design, it probably does have broader implications. If within a situation there might be different *kinds* of compatibility (as, indeed, there are in this experiment), it may be pertinent to know *what* type of compatibility is the more (or most) critical in case of some possible conflict (for example, *direct-drive* compatibility seems to be more critical than the *clockwise-increase* principle with moving scales).

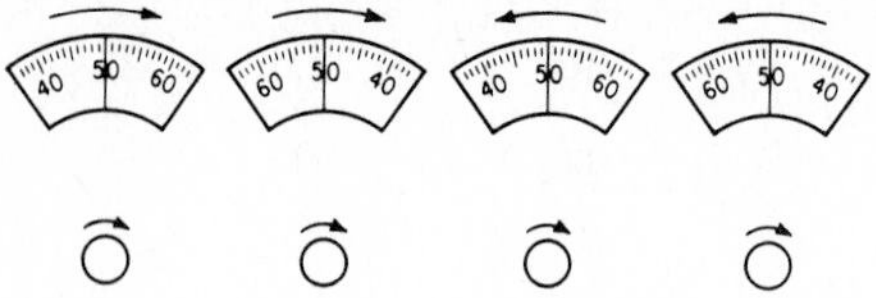

Assembly	A	B	C	D
Drive	Direct	Direct	Reversed	Reversed
Scale numbers increase	Left to right	Right to left	Left to right	Right to left
With clockwise knob movement setting will:	Decrease	Increase	Increase	Decrease

	A	B	C	D
Starting errors	13	11	87	106
	11		116	
Setting errors	0	9	1	8
	7	20		
Preference (number of times ranked "first")	31	22	17.5	1.5
	42		10	
	11	9		

FIGURE 8-7
Some of the moving-display and control-assembly types used in a study by Bradley. The various features of these related to three desirable characteristics are given below the diagrams; crosshatching indicates an undesirable feature. With the usual display orientation all three desirable features are not possible. Some data on three criteria are given for various groups of subjects at the bottom of the figure, indicating the general preferability of *A*. (*Source: Adapted from Bradley, 1954.*)

Rotary Controls and Linear Displays in Same Plane In the case of rotary controls and fixed linear displays, Warrick (1947) postulated the principle that the indicator of a linear display should move in the same direction as the nearest point on the control knob. This is illustrated in Figure 8-8.

FIGURE 8-8
Illustration of Warrick's principle of the compatible relationships between the rotation of knob controls and linear scales, with the indicator of the scale moving in the same direction as the closest point of the knob. With the knob at the end of the scale the relationship would be ambiguous. (*Source: Adapted from Warrick, 1947.*)

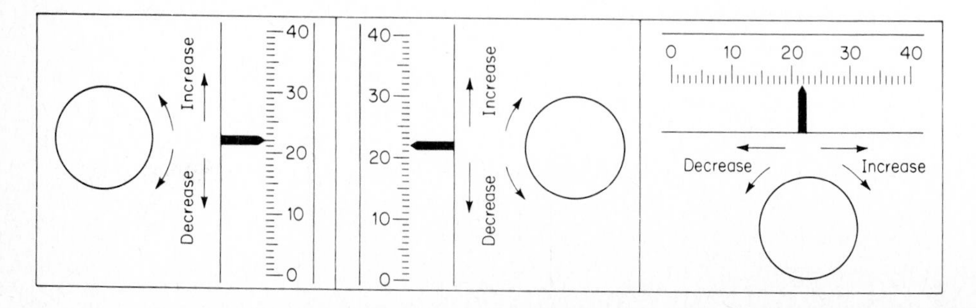

However, Brebner and Sandow (1976a and 1976b) point out that four separate principles might be involved in the use of rotary controls with linear displays (referring specifically to vertical displays), these being

1 Warrick's principle (discussed above)
2 *Clockwise for increase* (a clockwise turn of the knob being associated with an increase in value)
3 *Up for increase* on a vertical scale (a movement up is associated with an increase)
4 *Scale side* (the indicator will move in the same direction as that part of the control knob which is on the same side of the knob as the scalar markings, the directional part of the indicator, or both)

In their study Brebner and Sandow used illustrations of vertical linear scales and knob locations such as shown in Figure 8-9. The illustrations consisted of all possible combinations of the following four variables:

1 Indicator type (a *neutral* line as in examples 1, 6, and 8, and a *directional* pointer as in 2, 3, 4, 5, 7, and 8)
2 Scale direction (the scale increased *upward* as in 2, 3, 5, and 7, or *downward* as in 1, 4, 6, and 8)
3 Scale side (the scale markers were either *left* as in 3 and 4, or *right* as in 1, 2, 5, 6, 7, and 8)
4 Control position (eight possible positions as shown in the eight illustrations by solid circles)

The subjects of the experiment were shown slides of the 64 combinations and were asked to indicate whether they thought the control knob should be turned clockwise or counterclockwise to "move the indicator to 15." Certain of the principal results include the following: the scale direction (whether the scale increased up or down) had no appreciable effect on the results; the directional indicators (the pointers) resulted in more consistent responses than the neutral (line) indicators; Warrick's principle was generally supported, but the effect of scale side influenced the strength of this princi-

FIGURE 8-9
Examples of the 64 combinations of the following variables as used in study of rotary control knobs and vertical scales: indicator type; scale direction; scale side; and control position. (*Source: Adapted from Brebner & Sandow, 1976a, fig. 2.* This figure appeared as fig. 2 in Vol. 7, No. 1, p. 35 of *Applied ergonomics*, published by IPC Science and Technology Press Ltd., Guilford, Surrey, U.K.)

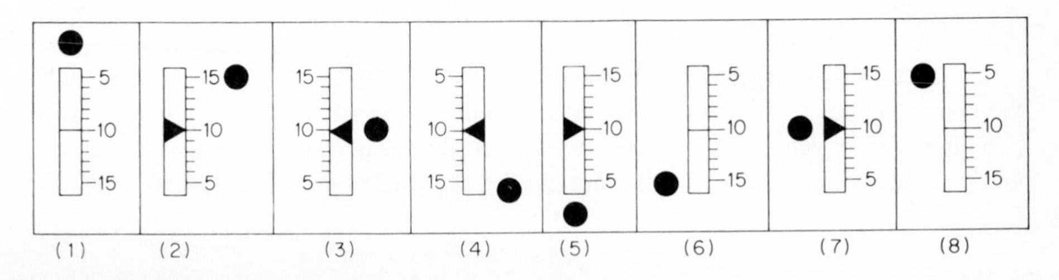

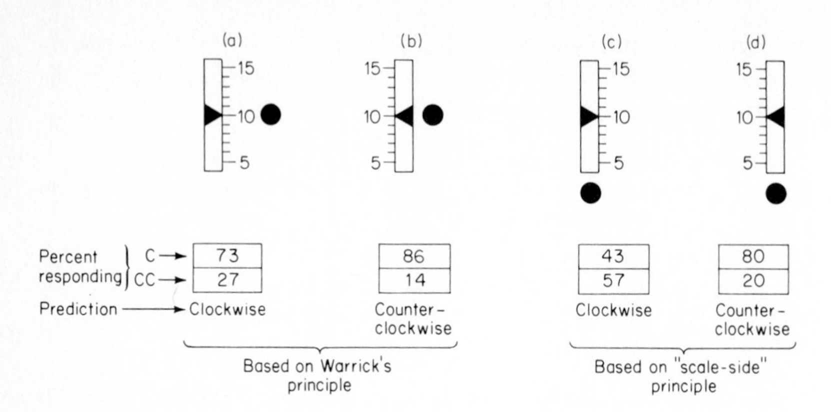

FIGURE 8-10
Percentages of "clockwise" (c) and "counterclockwise" (cc) responses to combinations of rotary controls and vertical scales when the "scale-side" principle contradicts (*a* and *c*) or reinforces (*b* and *d*) predicted response tendencies. (*Source: Adapted from Brebner & Sandow, 1976b, fig. 3.*)

ple (the results indicated that the stereotype based on Warrick's principle is stronger when the scale is on the side of the display *opposite* the rotary control than when it is on the same side). The effect of the scale-side variable is illustrated in Figure 8-10, which shows in *a* and *b* the proportions of clockwise (c) and counterclockwise (cc) responses that presumably were influenced by the side on which the scale was located, with greater consistency when the scale was on the side opposite the knob (as in *b*). The effect of scale side also carried over to the responses in circumstances in which the control knob was at the bottom of the scale rather than at the side, as illustrated by Figure 8-10*c* and *d*.

This rather extended discussion of this study is intended primarily to drive home the points that compatibility relationships in connection with movements are indeed complicated, and that there are possibly conflicting principles that serve to complicate the lives of control-display designers. However, it is clear from this and other studies that certain stereotypes are stronger than others, and that the best working arrangements will be those in which the various operating principles combine to facilitate and reinforce the same response tendency. The point should also be made that it is especially important that, when somewhat similar displays and controls are used in the same system or in different models of the same type of system, the same principles of design should always be followed.

Movement Relationships of Displays and Rotary Controls in Different Planes In the relationship between movement of display indicators and of control devices, some control devices may be in a different plane from the displays with which they are associated, such as the lateral cutting plane or orthogonal cutting plane. In one such study, the control knobs that were used caused a pointer to move along a straight-line scale (Holding, 1957). The knob and pointer were in different planes, as illustrated in Figure 8-11. More than 700 subjects were used in the study and were asked to "twist the

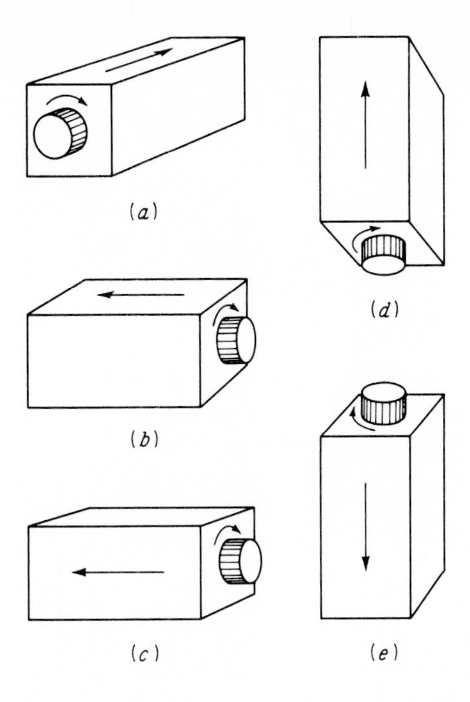

FIGURE 8-11
Illustration of some of the display-control relationships studied by Holding, showing the strongest relationships demonstrated by subjects. (*Source: Holding, 1957.*)

knob" in order to move the pointer. The strongest relationships displayed by the subjects between direction of control movement and direction of pointer movement are shown in Figure 8-11. The results led to the conclusion regarding human performance in such situations that people's responses tend to be one of two types: (1) a generalized clockwise tendency; and (2) a helical, or screwlike, tendency in which clockwise rotation is associated with movement away from (as with screws, bolts, etc.), and counterclockwise is associated with movement toward, the individual.

Movement Relationships of Rotary Vehicular Controls In the operation of most vehicles there is no "display" to reflect the "output" of the system; rather, there is a "response" of the vehicle. In such instances, if the wheel control is in a horizontal plane, an operator tends to assume an orientation toward the forward point of the control, as shown in Figure 8-12*a* (Chapanis & Kinkade, 1972). If the wheel is in a vertical plane, the operator tends to orient to the top of the control, as shown in Figure 8-12*b*.

A rather horrendous control problem one of the authors has seen is with a shuttle car for an underground coal mine that has a control wheel that controls left and right turns. This wheel is on the right-hand side of the car relative to the driver when going in one direction, but when going in the other direction the driver moves to the opposite side to face the direction of travel, so that the wheel is then on the driver's left. The control relationships of this are so complicated that new drivers have major problems in learning how to control the cars.

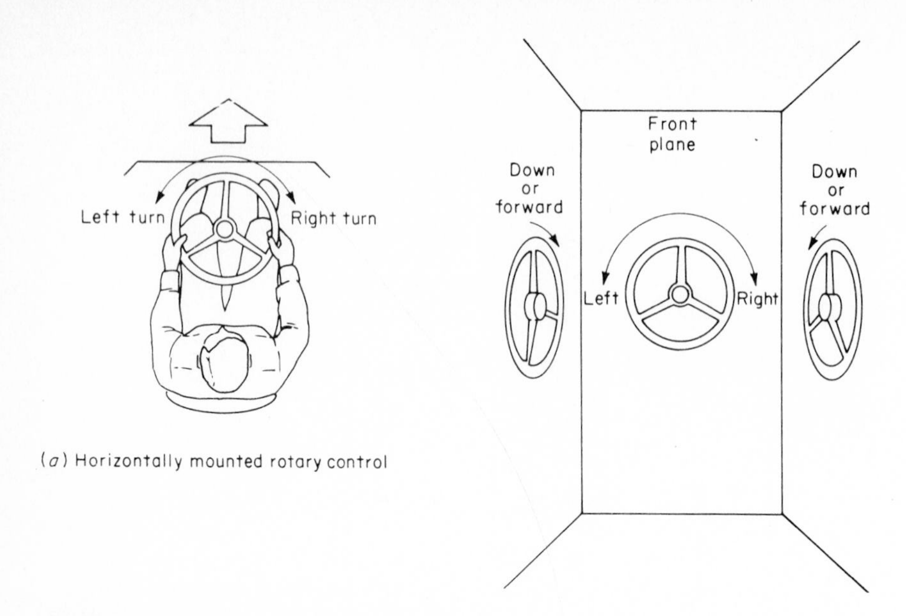

FIGURE 8-12
The most compatible relationships between the direction of movement of horizontally and vertically mounted rotary controls and the response of vehicles. (*Source: Adapted from Chapanis & Kinkade, 1972, figs. 8-6 and 8-7.*)

Movement Relationship of Stick-Type Controls The compatibility of stick-type controls and movements of associated display indicators was investigated in connection with a tracking task (Spragg, Finck, & Smith, 1959). Four combinations of control location and display movement were used, as illustrated in Figure 8-13. This figure also shows average tracking scores of subjects under these four conditions.

For a horizontally mounted stick (on the vertical display plane), the superiority of the *up-up* relationship (control movement up associated with display movement up) over the *up-down* relationship is evident. For a vertically mounted stick (on the horizontal lateral cutting plane) there was less difference between the *forward-up* and *forward-down* relationships, although the forward-up relationship was slightly superior (but not significantly so). The investigators concluded that for the kind of tracking used it was about equally effective to mount the stick in a horizontal or vertical position, provided one *avoids* the up-down relationship with a horizontally mounted stick.

Conceptual Compatibility

Probably the most common variety of conceptual compatibility relates to the associations in the use of coding systems, symbols, or other stimuli; these associations may be intrinsic (i.e., the use of visual symbols that represent things such as airplanes) or they may be culturally acquired. Certain illustrations were given in Chapter 3 and will not be discussed further here.

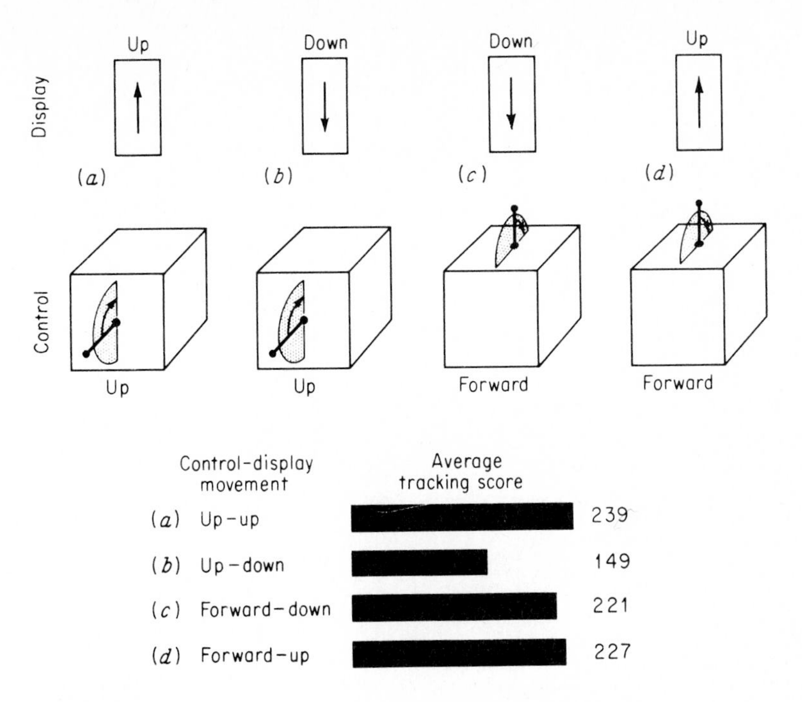

FIGURE 8-13
Tracking performance with horizontally mounted and vertically mounted stick controls and varying control-display relationships. (*Source: Adapted from Spragg, Finck, & Smith, 1959, data based on trials 9 to 16.*)

THE NATURE OF TRACKING OPERATIONS

Tracking operations present special problems in human control and so are singled out for some particular discussion in this chapter. *Tracking tasks* are those that require continuous control of something, as in driving an automobile, piloting a plane, following a racehorse with binoculars, or tracing a line on a sheet of paper. The basic requirement of a tracking task is to execute correct movements at correct times. In some instances the task is paced by the person doing it, as in driving an automobile. In other instances the task is externally paced, in that the individual has no control over the rate at which the task has to be performed, as in following the racehorse with binoculars.

The analysis of tracking operations can be extremely complex, and it is not appropriate in this text to deal extensively with such operations.[1] However, certain aspects of such operations are discussed briefly to provide some background regarding some of the concepts involved in tracking and some inklings of the many variables associated with tracking performance.

[1] For extensive treatment of tracking, the reader is referred to Kelley (1968), and to Poulton (1974). In addition various papers dealing with the applications of control theory in human factors are included in the August and October issues of *Human Factors*, 1977, *19*(4 and 5).

Inputs and Outputs in Tracking

In a tracking task some input specifies the desired output; this may be constant (e.g., steering a ship at a specified heading or flying a plane at an assigned altitude), or variable (e.g., following a winding road or tracking a maneuvering aircraft). Such input typically is received directly from the environment and sensed by mechanical sensors or by people. If it is sensed mechanically, it may be presented to operators in the form of signals on some display. The input signal is sometimes referred to as a *target* (and in certain situations it actually is a target), and its movement is called a *course*. Whether or not the input signal represents a real target with a course, or some other changing variable such as desired changes in temperature in a production process, it usually can be described mathematically and shown graphically. While in most instances the mathematical or graphic representations do not depict the real geometry of the input (and the input-output relations) in spatial terms, such representations do have utility in characterizing the input. The input, in effect, specifies the desired output of the "system," such as curves in a road specifying the desired path to be followed by an automobile.

There are different types of possible inputs. A *step* input is one that specifies a distinctly different desired output objective, such as an instruction from a control tower for an aircraft to change altitude, the need to change the temperature control of an oven at some point in time, or (as Poulton, 1974, suggests) even positioning a mark on a slide rule to a particular value. In some instances, the "track" may have many step inputs. A *ramp* input specifies that some particular output be kept at a specific value, such as velocity, rate of change of some variable, or acceleration. Examples include the requirements to maintain the speed (velocity) of a ship or airplane, to operate a crank at a constant rate, and to maintain a constant angular turn of an automobile as when making a turn in the form of an arc. Acceleration ramp inputs are used largely in laboratory situations. A *sine* input is characterized by a sinusoidal (i.e., sine) wave such as shown in the upper left corner of Figure 8-15 on page 235. A *complex* input is one that has little if any systematic character, such as the evasive movements of a military aircraft when attempting to avoid antiaircraft machine guns.

The output is usually brought about by a physical response with a control mechanism (if by an individual) or by the transmission of some form of energy (if by a mechanical element). In some systems the output is reflected by some indication on a display, sometimes called a *follower* or a *cursor*; in other systems it can be observed by the outward behavior of the system, such as the movement of an automobile. In either case it is frequently called the *controlled element.*

Pursuit and Compensatory Displays in Tracking

The input (the target) and the output (the controlled element) can be presented on either a *pursuit* display or a *compensatory* display, as illustrated in Figure 8-14. With a pursuit display both indications move, each showing its own location in space in relationship to the other. If the relative movement of the two elements is represented by a single value (a single dimension) they can be shown on a scale such as those in the upper part of Figure 8-14, whereas if this movement needs to be depicted in two dimen-

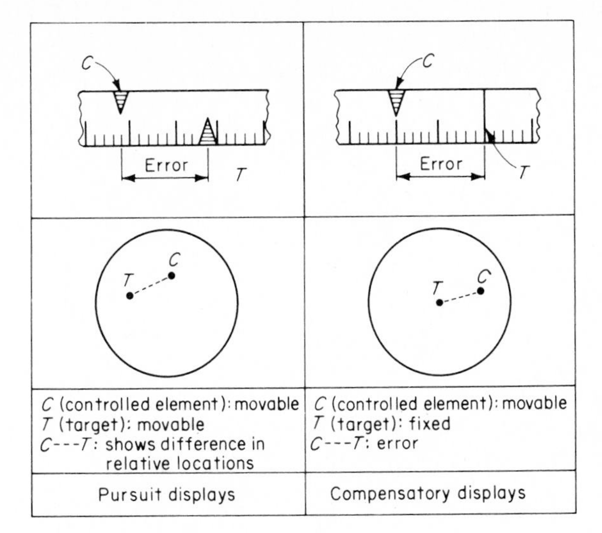

FIGURE 8-14
Illustrations of compensatory and pursuit tracking displays. A compensatory tracking display shows only the ***difference*** (error) between the target *T* and the controlled element *C*. A pursuit display shows the location (or other value represented) of both the target and the controlled element.

sions (as, say, in aiming a gun at a moving aircraft) the displays might be like those in the lower part of that figure. In a compensatory display (using either one or two dimensions) one of the two indications (the target or the controlled element) is fixed and the other moves. When the two are superimposed, the controlled element is *on target*; any difference represents an error, and the function of the operator is that of manipulating the controls to eliminate or minimize that error. The source of any difference (error), however, cannot be diagnosed; whether the target has moved or changed course or whether the controlled element has moved is not shown. Illustrations are shown in Figure 8-14 on the right side. With a pursuit display the operator is presented with information about the actual location of both elements, as shown on the left side of the figure. With a compensatory display the operator knows only the absolute error or difference. However, compensatory displays sometimes have a practical advantage in conserving space on an instrument panel since they do not need to represent the whole range of possible values or locations of the two elements.

Control Order of Systems

Control order refers to what we might call the hierarchy of control relationships between the movement of a control and the output it is intended to control. The nature of control orders and their implications for tracking tasks are very intricate, and we will not get too deep into this topic. To give some impression of control orders, however, let us begin with the concept of *position*. This refers to the output of the system, the objective of the tracking task being that of controlling the system so the output corresponds as closely as possible to the input. (The input of course specifies what the output should be.) Although we can readily envision the "position" of a vehicle in space, let us think of it in a broader context as a measure of the output. It might be shown as a moving pointer on a scale, a blip on a radar screen, an index of the revolutions per minute of a machine, or something else.

Position (Zero-Order) Control In a *position-control* tracking task the movement of the control device controls the output *directly*, such as moving a spotlight to keep it on an actor on a stage, following the movements of an athlete with a camera, or (in a laboratory task) following a moving curved line with a pen or other device. If the system involves a display, there is a direct relationship between the control movement and the display movement it produces.

Rate (First-Order) Control With a *rate-control* system the direct effect of the operator's movement is to control the rate at which the output is being changed. The accelerator of an automobile is a rate or first-order control device since it controls the speed (rate of change of position) of the automobile. In particular the distance by which the pedal is depressed controls the speed, although there is a time lag between depressing the pedal and the speed controlled by the pedal. In turn, the speed of the automobile controls its position along the road. The operation of certain machine guns is controlled by hand cranks that control the rate at which the gun changes direction, and would involve rate control.

Acceleration (Second-Order) Control *Acceleration* is the rate at which there is a change in the rate of movement of something. The operation of the steering wheel of an automobile is an example of acceleration control since the angle at which the wheel is turned controls the angle of the front wheels. In turn, the direction in which the automobile wheels point determines the rate at which the automobile turns. Thus, a given rotation of the steering wheel gives the automobile a corresponding acceleration toward its turning direction. Poulton (1974, p. 324) makes the point that the control of certain chemical plants or nuclear reactors may approximate an acceleration or second-order system.

Higher-Order Control Certain systems have control systems that can be considered as *higher-order* systems, such as third- or fourth-order control. A third-order control, for example, would involve the direct control of the rate of change of the acceleration (jerk), that then controls the rate, that finally controls the position of whatever is being controlled. The control of a large ship could approximate a third-order or even a fourth-order control because of the linkages between the person steering and the actual movement and position of the ship and the mass (that is, the weight) of the ship. In the control of a large ship, the *position* of the rudder control (which operates the rudder hydraulic system) produces a *rate of movement* of the rudder, and, in a chain-reaction manner, the position of the rudder results in the angular acceleration of the ship and, in turn, the *rate of change of lateral position* with respect to the desired course. In continuous-control processes that have a series of control linkages (such as a ship) the sequence of chain-reaction effects can be described in terms of mathematical functions, such as a change in the *position* of one variable changing the *velocity* (rate) of the next, the *acceleration* of the next, etc.

Control Responses with Various Control Orders

The operators of many systems are expected to make control responses that will bring about the desired operation of the system as implied by the input (such as a road to be followed, the flight path of an aircraft, or the appropriate temperature control over time of an industrial process). In the absence of any scheme for helping the operator, such control can be complicated with higher-level control orders. Examples of the appropriate responses for sine, step, and ramp inputs with position, rate, and acceleration control systems are shown in Figure 8-15. In each case the dotted line represents the response over time (along the horizontal) that would be required for satisfactory tracking of the input in question. In general, the higher the order of control, the greater is the number of controlled movements that need to be made by an operator in response

FIGURE 8-15
Tracking responses to sine, step, and ramp inputs which would be conducive to satisfactory tracking with position, rate, and acceleration control. The desired response, however, is not often achieved to perfection, and actual responses typically show variation from the ideal. In a positioning response to a step input, for example, a person usually overshoots and then hunts for the exact adjustment by overshooting in both directions, the magnitude diminishing until arriving at the correct adjustment.

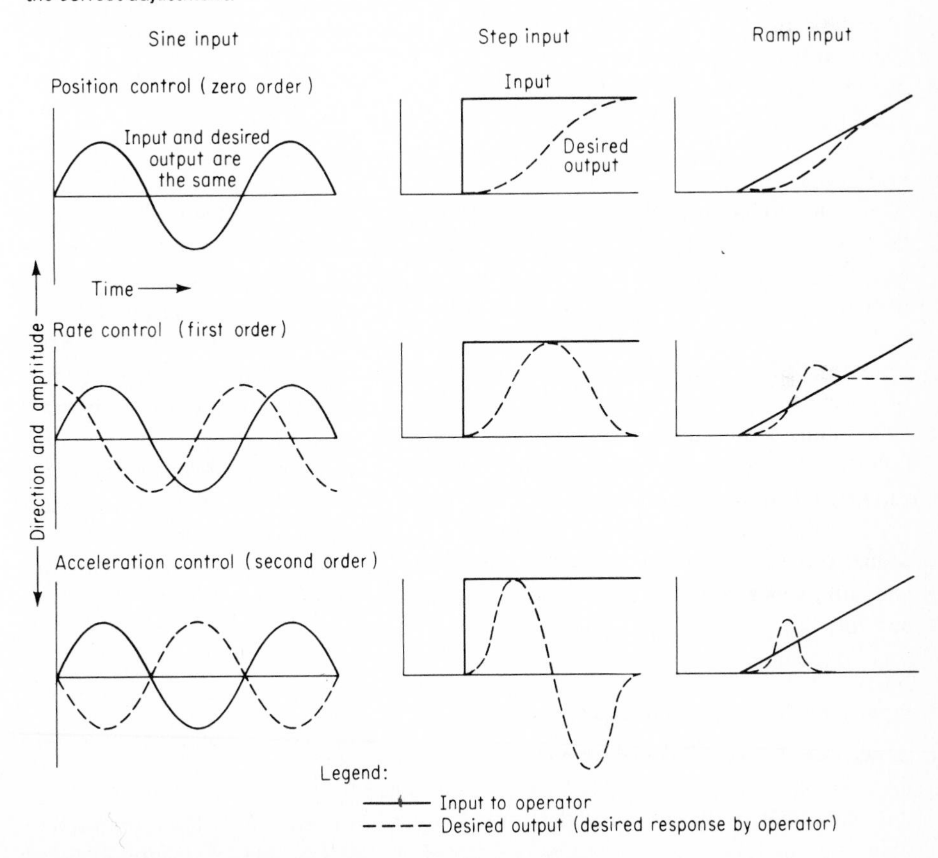

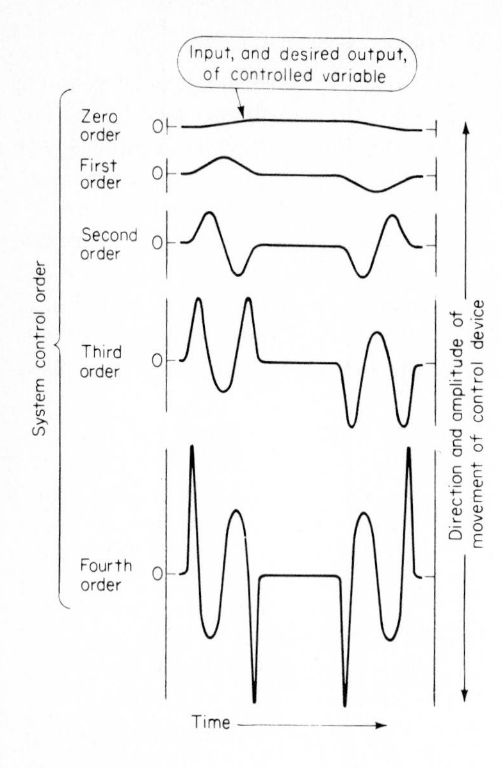

FIGURE 8-16
Control order illustrated by changes in a controlled variable and in its first four derivatives. Each line represents the changes over time that would have to be made with control systems of various control orders to make the controlled variable correspond to the input. (*Source: Adapted from Kelley, 1968, p. 31.*)

to any single change in the input, as illustrated in Figure 8-15. This is illustrated further in Figure 8-16. If the input, changing over time, follows the pattern (line) shown, and if the control system (whatever it may be) is zero order (i.e., position control), then the movement of the control device by an operator should correspond exactly with that line. But if, instead, the control system is a first-order system (i.e., rate control), then the operator needs to anticipate and make the response movements shown. In turn, the other lines represent the changes that the operator would need to make with a control device if, in fact, the control system were a second-, third-, or fourth-order system. Some of the higher-order systems are, in fact, especially characteristic of certain vehicles such as a ship when it is to make the slight deviation from, and return to, a straight-ahead course such as shown as the top (input) line of Figure 8-16. The rudder and flaps of aircraft serve to control the rate of change of heading and elevation and are second-order controls; and submarines typically have at least third-order control.

Procedures for Facilitating Tracking Tasks

As we can see from the previous discussion, the particular nature of the desired human response (output) in tracking tasks is predetermined by the type of input to the system (sine, step, ramp, etc.) and the control order of the system. In the case of a zero-order system (position control) as illustrated in the top line of Figure 8-16, the problem of deciding what response to make is relatively simple. In such a case the operator simply has to effect the amount of movement of the control that is required to bring about a specified degree of change in the controlled element (i.e., the system response, such as movement of a vehicle). The mental function in doing this is essentially an amplification process (in particular, the multiplication by a constant that represents the ratio of the input signal to the desired output response). In the case of higher-order systems, however, the mental functions of figuring out what movements to make involve more complex mental gymnastics, including those akin to mathematical differentiation, integration, and algebraic addition. In general, people do not do well in these kinds of mental operations; and when one combines two or more operations, such as differentiation, integration, multiplication, and addition, the task becomes virtually impossible to perform.

Thus, in the case of complex systems that require high-order controls (such as second-, third-, or fourth-order) some ways and means have to be devised for relieving the operator of the need to perform the mental functions that otherwise would be required, or of "compensating" for the disparity between response demands of the situation and the response capabilities of people. Certain such procedures are discussed below.

Aiding One such procedure is the use of aiding. Aiding was initially developed for use in gunnery tracking systems and is most applicable to tracking situations of this general type, in which the operator is following a moving target with some device. Its effect is to modify the output of the control in order to help the tracker. In *rate aiding* a single adjustment of the control affects both the rate and position components of the tracking system. Let us suppose we are trying to keep a high-powered telescope directed exactly on a high-flying aircraft by using a rate-aided system. When we fall behind the target, our control movement to catch up again would automatically speed up the *rate* of motion of our telescope (and thus, of course, its position). Similarly, if our telescope gets ahead of the target, a corrective motion would automatically slow down its rate (and influence its position accordingly). Such rate aiding would simplify the problem of quickly matching the rate of motion of the following device to that of the target and would thus improve tracking performance. In *acceleration aiding* the control movement controls three variables of the controlled element, namely acceleration, rate, and position.

The operational effect of aiding is to shift from the operator the mental operations that approximate those of differentiation, integration, and algebraic addition that are required in some tracking tasks, so the operator's chore is primarily that of amplification. (In effect this means the operator simply has to figure out the ratio of the movement of the control to that of the controlled element.) The operational effects of aiding depend upon a number of factors such as the nature of the input signal, the control order, and whether the system is a pursuit or compensatory type. Thus, aiding should be used selectively, in those control situations in which it is uniquely appropriate.

Quickening Quickening is particularly relevant for use in vehicular control systems that involve second-, third-, or higher-order control. The electronic instrumentation of quickened displays is very complex, but the effect of the use of such displays is generally to show the operator what control movement to make to bring about a particular output response. By its nature, a quickened system is most appropriate where the consequences of the operator's actions are not immediately reflected in the behavior of the system, but rather have a delayed effect, the delay frequently caused by the dynamics of the system, as in aircraft and submarines. Refer back again to Figure 8-16 and consider the responses that would be required for a second-order or third-order system (i.e., the lower curves of that figure); in quickening, the operator would still have to make those responses but would be shown what responses to make and would not have to go through the mental gymnastics of figuring what those complex movements should be (which, incidentally, would not be feasible).

Although quickening simplifies and improves some tracking tasks, it has certain possible disadvantages and limitations. For example, in a typical quickened system the operator is not provided with information regarding the current condition of the system, since the display shows primarily what control action to take. It should also be kept in mind that quickening does not have any appreciable advantage in very simple systems, or in systems where there is no delay in the system effect from the control action and where there is already immediate feedback of such system response.

Predictor Displays Still another procedure for simplifying the control of high-order systems (especially large vehicles such as submarines) is the use of predictor displays, as proposed primarily by Kelley (1962, 1968). In effect, predictor displays use a fast-time model of the system to predict the future excursion of the system (or controlled variable) and display this excursion to the operator on a scope or other device. The model repetitively computes predictions of the real system's future, based on one or more assumptions about what the operator will do with the control—e.g., return it to a neutral position, hold it where it is, or move it to one or another extreme. The predictions so generated are displayed to enable the operator to reduce the difference between the predicted and desired output of the system.

An example of a predictor display is shown in Figure 8-17, this showing the predicted depth error of a submarine as it would rise over a 10-s period. A computer-generated stylized representation of three aircraft-like predictor symbols is shown in Figure 8-18, this figure representing the predicted positions of an aircraft at three future points in time (7, 14, and 21 s) during a landing approach.

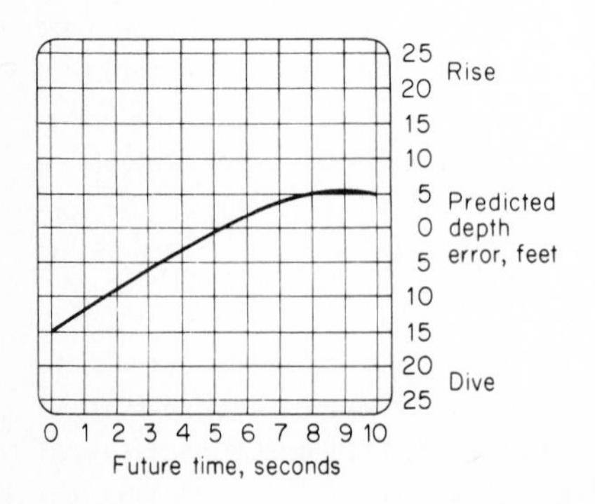

FIGURE 8-17
Example of a predictor display for a submarine. The display shows the predicted depth error, in feet, extrapolated to 10 s, assuming that the control device would be immediately returned to a neutral position. (*Source: Kelley, 1962.*)

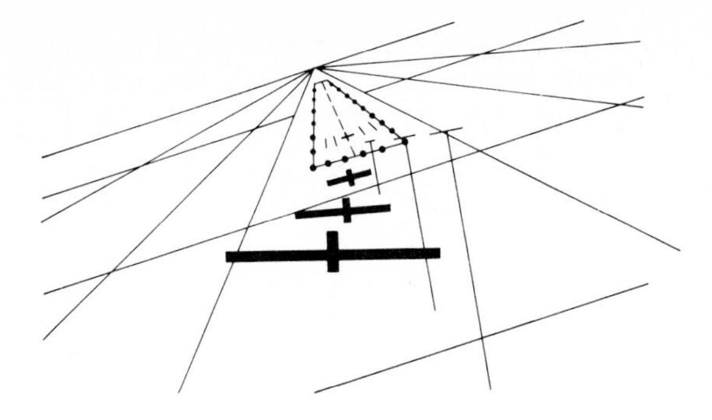

FIGURE 8-18
Stylized representation of aircraft symbols of a predictor display with runway, center line, touchdown zone, and gridlike ground texture cues. The symbols represent the predicted positions of the aircraft at three future points in time, 7, 14, and 21 s. (*Source: Gallaher, Hunt, & Williges, 1977, fig. 1, p. 550.*)

Predictor displays offer particular advantages for complex control systems in which the operator needs to anticipate several seconds in advance, such as with submarines, aircraft, and spacecraft. The advantages in such situations have been demonstrated by the results of experiments such as the one by Dey (1972) simulating the control of a VTOL (vertical takeoff and landing) aircraft. The values of one index of the deviations from a desired "course" with, and without, the use of a predictor display were 2.48 and 7.92, respectively. The evidence from this and other experiments shows a rather consistent enhancement of control performance with a predictor display.

Discussion The relative utility of the various possible methods of facilitating performance of tracking tasks becomes inextricably intertwined with the type of input (step, ramp, sine wave, etc.), with the control order (zero, first, second, etc.), and with the type of display (pursuit or compensatory). Aiding can be useful in certain circumstances but has limited general applicability. In connection with quickening, Poulton raised serious questions about the research strategies used in some studies and also referred to certain disadvantages of quickened displays (pp. 180–185). In general he concluded that "true motion" predictor displays are likely to be far easier and safer to use for control systems of high order than quickened displays (pp. ix, 189).

FACTORS THAT INFLUENCE TRACKING OPERATIONS

The effectiveness of human control of tracking operations is influenced by a wide variety of factors such as the nature of the displays and of the controls, and the features of the tracking system. A few such factors are discussed briefly below.

Type of Display: Pursuit and Compensatory

In reviewing the research relative to the possible merits of visual analog pursuit versus compensatory displays Poulton (1974, p. 166) concluded that, when there is a choice between the two, a conventional pursuit (true-motion) display would be preferable to a compensatory (relative-motion) display. His argument was based on the summary of numerous studies; see Table 8-2.

In the case of the seven studies in which compensatory tracking was reported to be best, Poulton argued that the results can be attributed to inappropriate experimental

TABLE 8-2
SUMMARY OF COMPARISON OF NUMEROUS STUDIES OF PURSUIT AND COMPENSATORY TRACKING

Control order	Number of studies	Difference in results reliably in favor of: Pursuit	Compensatory	Results inconclusive
Zero order	45	29	0	16
Higher order	34	14	7	13

Source: Poulton, 1974.

procedures, and that the results should be disregarded. In the case of three studies involving zero-order control included in the "inconclusive" column, the results did favor compensatory tracking, but Poulton points out that experimental variables probably influenced such results. (He cited only one instance in which a compensatory display was proved to be preferable, this being a circumstance in which two tracking tasks were carried out simultaneously using circular displays and crank controls.) Parts of the results from one study dealing with pursuit and compensatory displays are shown in Figure 8-19. The superiority of pursuit tracking is clearly evident in that figure.

Most of the experiments with pursuit and compensatory displays involved analog displays that present continuous tracking information. In an experiment with digital displays Kvälseth (1978b) found there was no difference in performance between the two types of displays. This experiment, then, may suggest that the general advantages of pursuit displays do not carry over completely in the case of digital (as opposed to analog) presentations.

Although pursuit displays generally result in better tracking performance than compensatory displays (at least with visual analog presentation) practical considerations may sometimes argue for the use of compensatory displays, especially because they may occupy less space on control panels.

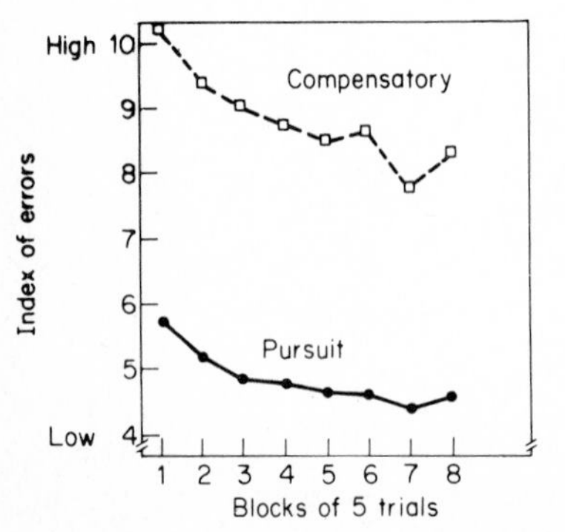

FIGURE 8-19
Comparison of errors of subjects when using a compensatory versus a pursuit display in a tracking task. (*Source: Adapted from Briggs & Rockway, 1966, fig. 2. Copyright © 1966 by the American Psychological Association. Reprinted by permission.*)

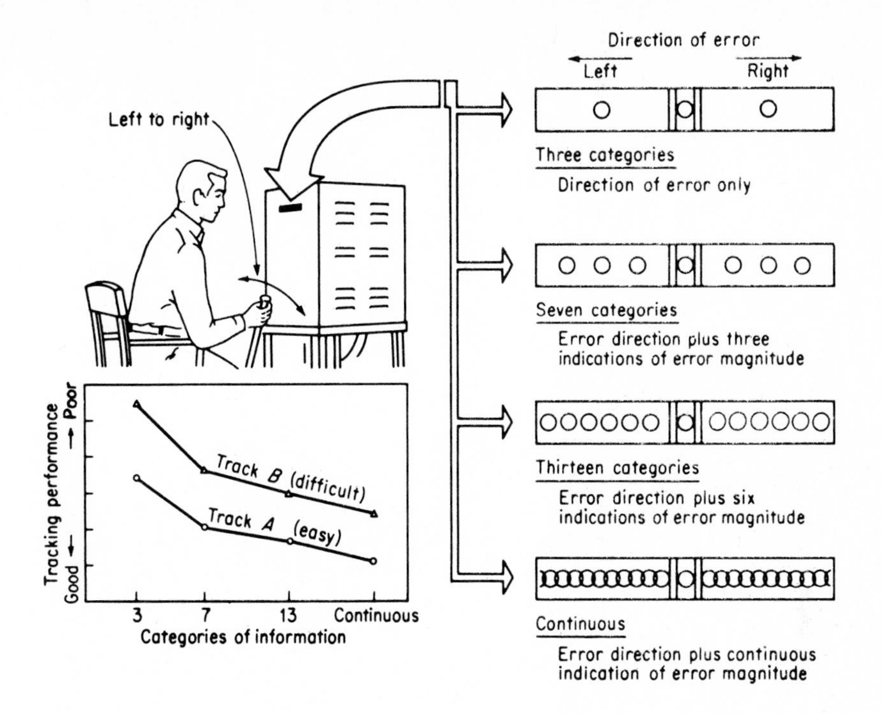

FIGURE 8-20
Compensatory displays used in study of the effects of specificity of feedback of error information, and tracking performance using such displays. The feedback of error was presented by the use of lights (3, 7, 13, and continuous). (*Source: Adapted from Hunt, 1961.*)

Specificity of Displayed Error in Tracking

In certain compensatory tracking systems the error (the difference between input and output) can be presented in varying degrees of *specificity*. Some such variations are shown in Figure 8-20 as they were used in a tracking experiment by Hunt (1961), these including: 3 categories of specificity (left, on target, and right); 7 categories; 13 categories; and continuous. The accuracy of tracking performance under these conditions (for two levels of task difficulty) indicates quite clearly that performance improved with the number of categories of information (greater specificity), this improvement taking a negatively accelerated form for tasks of both levels of difficulty. Although the results of other studies are not entirely consistent with these, the evidence suggests that in a tracking task, the mediating control functions are facilitated by the presentation of more specific, rather than less specific, display information.

Preview of Track Ahead

In some tracking situations an individual has some preview of the "track" ahead, as in driving an automobile on a winding road that can be seen ahead. (In a blinding snow storm or heavy fog, the individual does not have such a preview.) In general terms some preview of the input assists an operator in a tracking task. Although the nature of the tracking task presumably can influence the possible benefit of a preview, Kvälseth (1979) indicates that such preview is most beneficial if the preview shows that portion of the track that immediately precedes the "present" position, rather than showing a "lagged" preview, with a gap between what is previewed and the present position. The duration of the preview seems to be of less consequence than the opportunity to have at least some preview. Kvälseth (1978a) indicates that performance improves steadily as the preview span approaches approximately 0.5 s, beyond which the usefulness of the preview diminishes rapidly.

Anticipation

The *preview* of a tracking course input refers to the opportunity to see some segment of the input before having to make a control response. On the other hand, *anticipation* (or what is sometimes called *precognition*), refers to the operator's ability to predict what the future course will be without having any preview as such. Such anticipation is possible only if the previous input has had some systematic pattern that the operator can learn. In general terms, sine wave inputs are relatively easy to predict, especially with a pursuit display (as contrasted with a compensatory display). However, with practice some operators can learn the patterns of some relatively irregular inputs.

Paced versus Self-Paced Tracking

Most tracking tasks are self-paced, in that the person has control over the rate of the output, as in driving an automobile. In such an activity the driver usually can select the speed. With some tracking tasks, however, the pace is not under the individual's control. Poulton (p. 8), for example, refers to the fact that an airline pilot, when coming in to land, has to hold air speed within close limits as well as keeping the plane within a closely defined glide path. Tracking is easiest when the task is self-paced, and increases in difficulty as the degree of external pacing is increased.

Time Lags in Tracking

In certain tracking tasks there can be various types of lags. A *display time lag,* for example, consists of a delay in both the track input and the output of the system in question. A *response lag* is the time taken by the individual to make a response to an input. And a *control system lag* is the time between a control response of an individual and the response of the system under control. There are three basic types of control lag, these being illustrated in Figure 8-21 for a step input. Of these, *transmission time lag* simply delays the effect of a person's response; the output follows the control response by a constant time interval. An *exponential lag* refers to a situation in which the output is represented by an exponential function following a step input. And a *sigmoid time lag* is represented by the S-shaped curve of Figure 8-21.

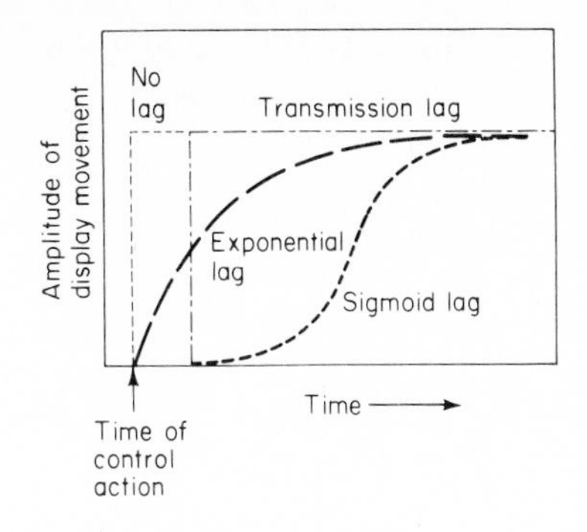

FIGURE 8-21
Illustration of three types of time lag following a step input. The dotted line represents the output when there is no lag, as (theoretically) might be the case when turning on a switch. The human response to a step input with most tracking systems tends to be somewhat similar to a sigmoid lag curve. (*Source: Poulton, 1974, fig. 11.9, p. 206.*)

The effects of various types of lags are intricately related to the various features of the tracking system and are not discussed here in detail. As an example of such complicated interrelationships, however, Kao and Smith (1978) present data that indicate that in a compensatory tracking task involving bimanual (two-handed) versus unimanual (one-handed) control, increasing time lags in the presentation of information on displays resulted in performance degradation for both forms of control with short and intermediate delays (up to 0.8 s), but that with a longer display lag (1.5 s) bimanual performance was more adversely affected than unimanual performance. The results are shown in Figure 8-22.

Aside from some of the intricacies of lags with various types of tracking systems, Poulton (p. 373), points out that all three types of control lag increase the error in tracking. Although such lags are generally undesirable, there sometimes are ways and means of minimizing these effects. For example, Poulton (p. 378) indicates that a design engineer may be able to reduce the effective order of a system from acceleration control to rate control by introducing an approximately exponential lag. As another illustration of methods of minimizing the effects of lag, Rockway (1954) has shown

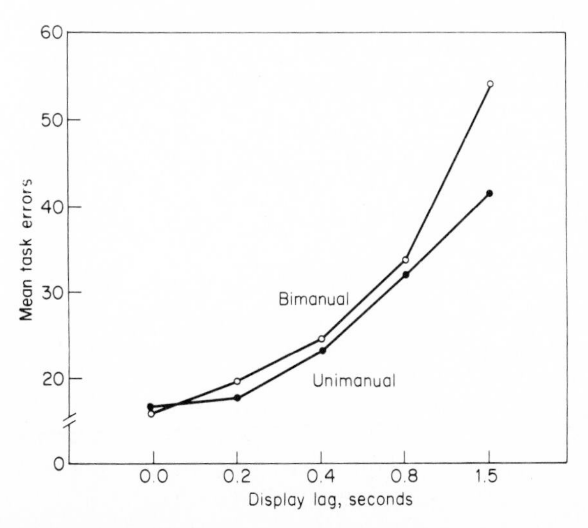

FIGURE 8-22
Illustration of the effects of various display lags in a compensatory tracking task performed with bimanual and unimanual control. The difference for the 1.5 s display lag is statistically significant. (*Source: Kao & Smith, 1978, fig. 2, p. 666.*)

that, with long (as opposed to short) delays, the control-response ratio (C/R ratio) can have some effect on tracking performance. The *C/D ratio* is the ratio of the amount of movement of the control device as related to the amount of corresponding movement of the display indication (see Chapter 9). In his study, Rockway found that high C/R ratios (1:3 and 1:6) resulted in performance degradation with long lags, whereas low C/R ratios (1:15 and 1:30) resulted in the maintenance or even improvement of performance with such lags. Thus, in some circumstances there are schemes that can be used to minimize the possible effects of time lags.

DISCUSSION

The basic functions of those who are to control a system or mechanism are rooted in the nature of the process or operation in question. However, specific design features of the system or mechanism predetermine the specific operational demands imposed on the controller, in particular the inputs that must be received, the types of decisions to be made, and the control actions to be taken. Considerations of human factors during the design stage can contribute to the design of systems and mechanisms that can be controlled adequately by human beings and that can contribute to the well-being of those who will be involved in subsequent control operations.

REFERENCES

Bradley. J. V. *Desirable control-display relationships for moving-scale instrument* (Technical Report 54-423). U.S. Air Force, WADC, September 1954.

Brebner, J., & Sandow, B. Direction-of-turn stereotypes—conflict and concord. *Applied Ergonomics*, 1976, *7*(1), 34-36. (a)

Brebner, J., & Sandow, B. The effect of scale side on population stereotype. *Ergonomics*, 1976, *19*(5), 571-580. (b)

Briggs, G. E., & Rockway, M. R. Learning and performance as a function of the percentage of pursuit tracking component in a tracking display. *Journal of Experimental Psychology*, 1966, *71*, 165-169.

Brigham, F. R. Decision-making problems in industrial control tasks. *Applied Ergonomics,* 1976, 7(1), 19-25.

Chapanis, A., & Kinkade, R. G. Design of controls. In H. P. Van Cott & R. G. Kinkade (Eds.), *Human engineering guide to equipment design.* Washington, D. C.: U.S. Government Printing Office, 1972.

Chapanis, A. & Lindenbaum, L. A reaction time study of four control-display linkages. *Human Factors,* November 1959, *1*(4), 1-7.

Dey, D. The influence of a prediction display on a quasi-linear describing function and remnant measured with an adaptive analog-pilot in a closed loop. *Proceedings of Seventh Annual Conference on Manual Control* (NASA SP-281). Washington, D. C.: National Aeronautics and Space Administration, 1972.

Drury, C. G., & Baum, A. S. Manual process control: A case study and a challenge. *Applied Ergonomics,* 1976, *17*(1), 3-9.

Fitts, P. M., & Seeger, C. M. *S-R* compatibility: spatial characteristics of stimulus and response codes. *Journal of Experimental Psychology,* 1953, *46*, 199-210.

Gallaher, P. D., Hunt, R. A., & Williges, R. C. A regression approach to generate aircraft predictor information. *Human Factors,* 1977, *19*(6), 549–555.

Holding, D. H. Direction of motion relationships between controls and displays in different planes. *Journal of Applied Psychology,* 1957, *41*, 93–97.

Hunt, D. P. The effect of the precision of informational feedback on human tracking performance. *Human Factors,* 1961, *3*, 77–85.

Kao, H. S. R., & Smith, K. U. Unimanual and bimanual control in a compensatory tracking task. *Ergonomics,* 1978, *21*(9), 661–669.

Kelley, C. R. Predictor instruments look to the future. *Control Engineering,* March 1962, 86f.

Kelley, C. R. *Manual and automatic control.* New York: Wiley, 1968.

Kvälseth, T. O. Effect of preview on digital pursuit control performance. *Human Factors,* 1978, *20*(3), 371–377.

Kvälseth, T. O. Human performance comparisons between digital pursuit and compensatory control. *Ergonomics,* 1978, *21*(6), 419–425.

Kvälseth, T. O. Digital man-machine control systems: The effects of preview lag. *Ergonomics,* 1979, *22*(1), 3–9.

Loftus, G. R., Dark, V. J., & Williams, D. Short term memory factors in ground controller/pilot communication. *Human Factors,* 1979, *21*(2), 169–181.

Poulton, E. C. *Tracking skill and manual control.* New York: Academic Press, 1974.

Rockway, M. R. *The effect of variations in control-display ratio and exponential time delay on tracking performance* (Technical Report 54-618). U. S. Air Force, WADC, December, 1954.

Shinar, D., & Acton, M. B. Control-display relationships on the four-burner range: Population stereotypes versus standards. *Human Factors,* 1978, *20*(1), 13–17.

Spragg, S. D. S., Finck, A, & Smith, S. Performance on a two-dimensional following tracking task with miniature stick control, as a function of control-display movement relationship. *Journal of Psychology,* 1959, *48*, 247–254.

Warrick, M. J. Direction of movement in the use of control knobs to position visual indicators. In P. M. Fitts (Ed.), *Psychological research on equipment design* (Research Report 19). Army Air Force, Aviation Psychology Program, 1947.

CHAPTER 9

CONTROLS

Human beings have demonstrated amazing ingenuity in designing machines for accomplishing things with less wear and tear upon themselves. Some of these machines perform more efficiently the functions that were once performed solely by hand, whereas others accomplish things that previously could only be dreamed of. Since most machines require human control, they must include control devices such as wheels, push buttons, or levers. It is through these control devices that people make their presence known to machines—that is, in addition to kicking the side of a candy machine. Control devices are not simply "extensions" of the human's extremities, but rather require various types of psychomotor action on the part of the operator. These devices must be designed to be suitable for the desired control actions in terms of sensory, psychomotor, and anthropometric characteristics of the intended users.

In this chapter we will examine some of the more common types of control devices and some of the factors which influence their use by humans. We will discuss hand controls, foot controls, data entry devices, and some other, more esoteric, control innovations. It will not be our intent to recite a comprehensive set of recommendations regarding specific design features of such devices, although a partial set of such recommendations is given in Appendix B.

FUNCTIONS OF CONTROLS

The primary function of a control is to transmit control information to some device, mechanism, or system. The type of information so transmitted can be divided into two broad classes, *discrete* information and *continuous* information. *Discrete* information is information that can only represent one of a limited number of conditions such as on-off; high-medium-low; boiler 1-boiler 2-boiler 3; or alphanumerics as A, B, and C, or 1, 2, and 3. *Continuous* information, on the other hand, can assume any value on a continuum, as for example speed (as 0 to 60 km/h); pressure [as 1 to 100 pounds per square inch (lb/in^2)]; position of a valve (as fully closed to fully open); or amount of electrical current [as 0 to 10 amperes (A)].

The information transmitted by a control may be presented in a display, or it may be manifested in the nature of the system response. Further, to complicate the situation a little, a secondary function of a control can be to serve, itself, as a display. An example here would be a rotary selector switch where the switch position indicates what information was input to the system.

Generic Types of Controls

There are numerous types of control devices available today. Certain types of controls are best suited for certain applications. One simple way to classify controls is based on the type of information they can most effectively transmit (discrete versus continuous) and the force normally required to manipulate them (large versus small). The amount of force required to manipulate a control is a function of the device being controlled, the mechanism of control, and the design of the control itself. Electrical and hydraulic systems typically require small forces to actuate controls, whereas direct mechanical linkage systems may require large forces. Table 9-1 lists some of the more common types of controls and the types of applications for which they tend to be most suited. Certain of these types of controls are illustrated in Figure 9-1.

TABLE 9-1
COMMON TYPES OF CONTROLS CLASSIFIED BY TYPE OF INFORMATION TRANSMITTED AND FORCE REQUIRED TO MANIPULATE

Force required to manipulate control	Type of information transmitted: Discrete	Continuous
Small	Push buttons (including keyboards)	Rotary knobs
	Toggle switches	Multirotational knobs
	Rotary selector switches	Thumb wheels
	Detent thumb wheels	Levers (or joysticks)
		Small cranks
Large	Detent levers	Handwheels
	Large hand push buttons	Foot pedals
	Foot push buttons	Large levers
		Large cranks

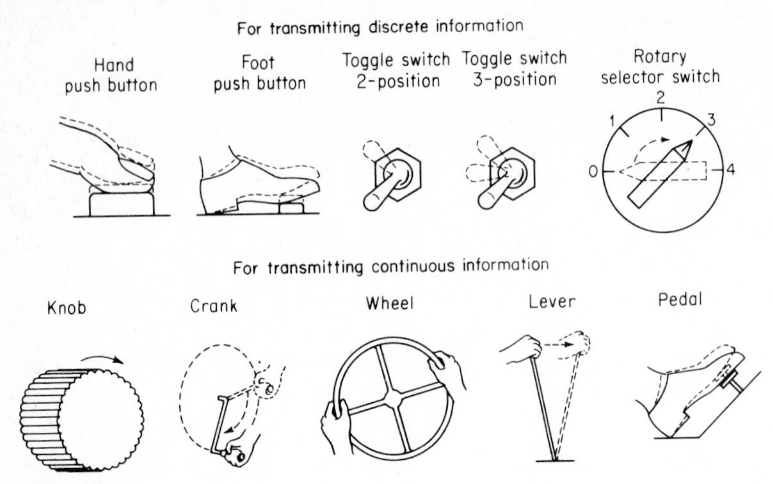

FIGURE 9-1
Examples of some types of control devices classified by the type of information they best transmit.

FACTORS IN CONTROL DESIGN

Although certain types of controls might be considered more appropriate for one application than for another, the overall utility of the control can be greatly influenced by such factors as ease of identification, size, control-response ratio, resistance, lag, backlash, deadspace, and location. We will discuss some of these factors here. Location of controls will be deferred to Chapter 11.

Identification of Controls

Although the correct identification of controls is not always critical in some circumstances (as in operating a pinball machine), there are some operating circumstances in which their correct and rapid identification is of major consequence—even of life and death. For example, McFarland (1946, pp. 605-608) cites cases and statistics relating to aircraft accidents that have been attributed to errors in identifying control devices. For example, confusion between landing gear and flap controls was reported to be the cause of over 400 Air Force accidents in a 22-month period during World War II. It is with such circumstances in mind that control identification becomes important.

Gamst (1975) reports that, in some railroad locomotives, engineers intending to turn off the signal lights grasp the wrong control and actually turn off the fuel pumps instead, thus killing all the diesel engines in the train.

The identification of controls is essentially a coding problem; the primary coding methods include shape, texture, size, location, operational method, color, and labels.

As discussed in Chapter 3, the choice of a coding method and the specific codes depends on the detectability, discriminability, compatibility, meaningfulness, and standardization of the codes selected. The utility of a coding method or specific set of codes typically is evaluated by such criteria as the number of discriminable differences that people can make (such as the number of shapes they can identify), bits of information, accuracy of use, and speed of use.

Shape Coding of Controls The discrimination of shape-coded controls is essentially one of tactual sensitivity. The procedure generally used in the selection of controls that are not confused with each other is illustrated by a study by Jenkins (1974b), in which he had 25 controls mounted on a rotating lazy Susan. Each subject, blindfolded, was presented with one knob and touched it for 1 s. The experimenter then rotated the turntable to a predesignated point from which the subject went from knob to knob, feeling each in turn, trying to identify the one previously touched. It was then possible to determine which knobs were confused with which other knobs. (Using this procedure, Jenkins identified two sets of eight knobs, such that the knobs within each group were rarely confused with each other. These two sets of knobs were presented in Figure 5-12.)

Following essentially the same tack as that mentioned above, the United States Air Force has developed 15 knob designs which are not often confused with each other. These designs are of three different types, each type being designed to serve a particular purpose (Hunt, 1953):

- *Class A: Multiple rotation.* These knobs are for use on continuous controls (1) which require twirling or spinning, (2) for which the adjustment range is one full turn or more, and (3) for which the knob position is not a critical item of information in the control operation.
- *Class B: Fractional rotation.* These knobs are for use on continuous controls (1) which do not require spinning or twirling, (2) for which the adjustment range usually is *less* than one full turn, and (3) for which the knob position is not a critical item of information in the control operation.
- *Class C: Detent positioning.* These knobs are for use on discrete setting controls, for which knob position can be an important item of information in the control operation.

The 15 knobs in these three classes are shown in Figure 9-2.[1]

If in addition to being individually discriminable by touch, the controls have shapes that are symbolically associated with their use, the learning of their use usually is simplified. In this connection, the United States Air Force has developed a series of 10 knobs that have been standardized for aircraft cockpits. These standard knob shapes, besides being distinguishable from each other by touch, include some that also have symbolic meaning. In Figure 9-3, which includes these shapes, it will be seen, for example, that the landing-gear knob is like a landing wheel, the flap control is shaped like a wing, and the fire-extinguishing control resembles the handle on some fire extinguishers.

[1] A few of these knobs may be confused with each other, and such combinations should not be used together if identification is critical. These combinations were *ab, co, cd#, do#, eg#, kp, ln, lo, np*, and *op#*. Those with a number sign (#) were confused only with gloves on and were not confused without gloves. In terms of size, Hunt (1953) suggests for the maximum dimension not less than ½ in (except for Class *C* where ¾ in is the minimum suggested) and not more than 4 in. In height, not less than ½ in and not more than 1 in is recommended.

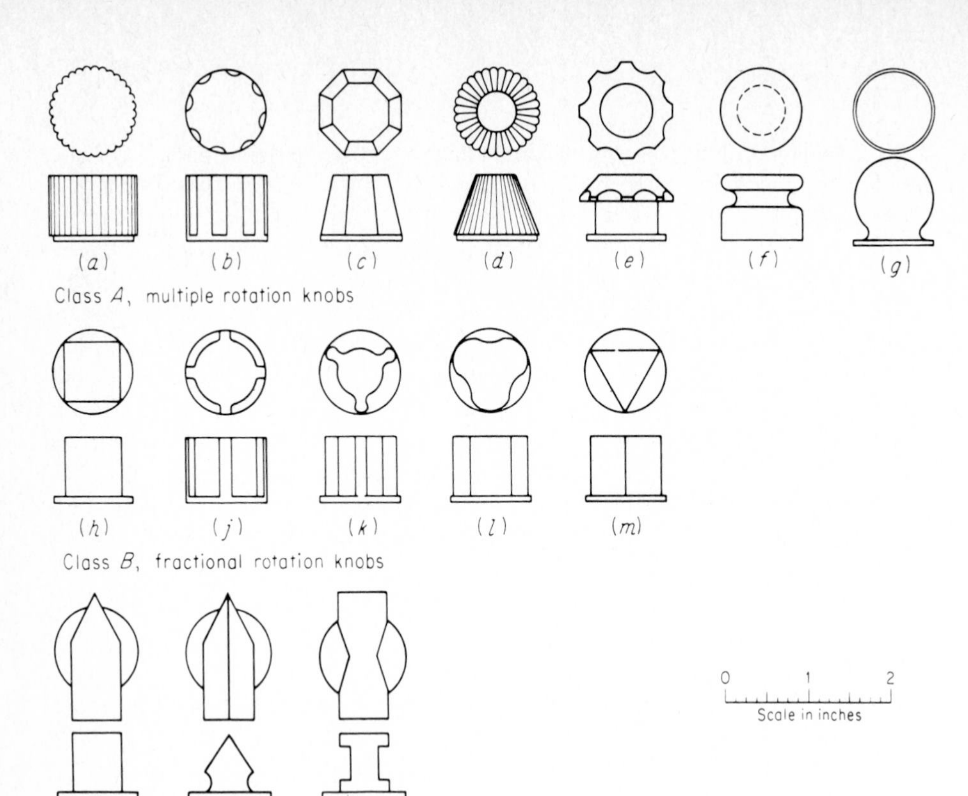

FIGURE 9-2
Knob designs of three classes that are seldom confused by touch. (*Source: Adapted from Hunt, 1953.*)

FIGURE 9-3
Standardized shape-coded knobs for United States Air Force aircraft. A number of these have symbolic associations with their functions, such as a wheel representing the landing-gear control. (*Source: USAF, Human factors engineering, AFSC design handbook.*)

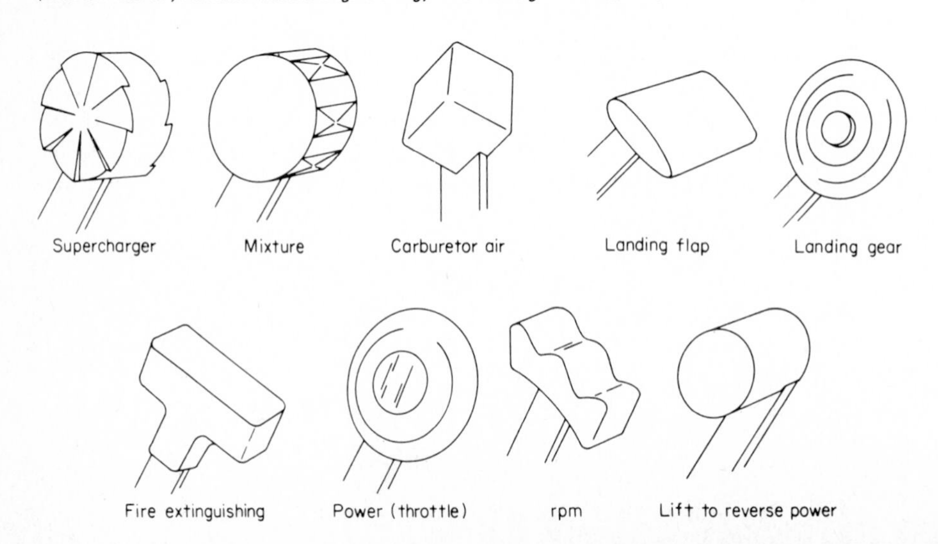

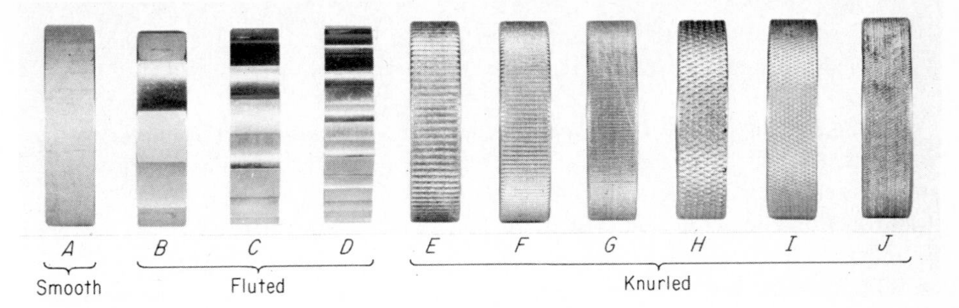

FIGURE 9-4
Illustration of some of the knob designs used in study of tactual discrimination of surface textures. Smooth: *A*; fluted: *B* (6 troughs), *C* (9), *D* (18); and knurled: *E* (full rectangular), *F* (half rectangular), *G* (quarter rectangular), *H* (full diamond), *I* (half diamond), and *J* (quarter diamond). (*Source: Bradley, 1967.*)

Texture Coding of Controls In addition to shape, control devices can be varied in their surface texture. This characteristic was studied (along with certain other variables) in a series of experiments with flat cylindrical knobs such as those shown in Figure 9-4 (Bradley, 1967). In one phase of the study, knobs of this type (2-in diameter) were used, and subjects were presented with individual knobs through a curtained aperture and were asked to identify the particular design they felt. The smooth knob was not confused with any other, and vice versa; the three fluted designs were confused with each other, but not with other types; and the knurled designs were confused with each other, but not with other designs. It should be added that with gloved hands and with smaller-sized knobs (in a later phase of the study) there was some cross-confusion among classes, but this was generally minimal. The investigator proposes that three surface characteristics can thus be used with reasonably accurate discrimination, namely, smooth, fluted, and knurled.

Size Coding of Controls Size coding of controls is not as useful for coding purposes as shape, but there may be some instances where it is appropriate. When such coding is used, the different sizes should of course be such that they are discriminable one from the others. Part of the study by Bradley reported above dealt with the discriminability of cylindrical knobs of varying diameters and thickness. It was found that knobs that differ by ½ in in diameter and by ⅜ in in thickness can be identified by touch very accurately, but that smaller differences between them sometimes result in confusion of knobs with each other. Incidentally, Bradley proposes that a combination of three surface textures (smooth, fluted, and knurled), three diameters (¾, 1¼, and 1¾ in), and two thicknesses (⅜ and ¾ in) could be used in all combinations to provide 18 tactually identifiable knobs.

Aside from the use of coding for individual control devices, size coding is part and parcel of ganged control knobs, where two or more knobs are mounted on concentric shafts with various sizes of knobs superimposed on each other like layers of a wedding cake. When this type of design is dictated by engineering considerations, the differences in the sizes of superimposed knobs need to be great enough to make them clearly distinguishable, as illustrated in Figure 9-11 on page 262 (Bradley, 1969).

Location Coding of Controls Whenever we shift our foot from the accelerator to the brake, feel for the light switch at night, or grasp for a machine control that we cannot see, we are responding to *location coding*. But if there are several similar controls from which to choose, the selection of the correct one may be difficult unless they are far enough apart so that our kinesthetic sense makes it possible for us to discriminate. Some indications about this come from a study by Fitts and Crannell as reported by Hunt (1953). In this study blindfolded subjects were asked to reach for designated toggle switches on vertical and horizontal panels, the switches being separated by 1 in (2.5 cm). The major results are summarized in Figure 9-5, which shows the percentage of reaches that were in error by specified amounts when the panels were in horizontal and vertical positions. The results have been simplified by combining several conditions. Thus, on the average approximately 40 percent of the reaches to horizontally arranged toggle switches were in error by more than 1 in (2.5 cm), while only about 15 percent were in error by the same amount when the switches were arranged vertically. The results indicate, quite clearly, that accuracy was greatest when toggle switches were arranged vertically. For vertically arranged switches, only a very small percentage of reaches were in error by more than 2½ in (6 cm); and since these errors are in both directions from the control, the controls should be separated by twice this distance (i.e., 5 in or 13 cm) so that the errors of one switch will not result in the activation of

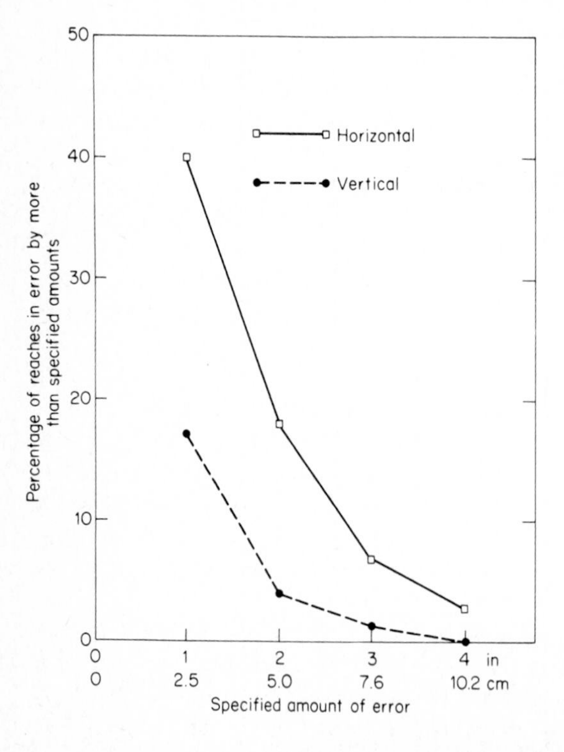

FIGURE 9-5
Accuracy of blind reaching to toggle switches (nine in a row on switch box) with switch box positioned horizontally and vertically. Each data point represents the mean of several conditions. (*Source: Adapted from Fitts & Crannell, as presented by Hunt, 1953.*)

an adjacent switch. For horizontally arranged switches, there should be at least 8 in (20 cm), and preferably more, between them if they are to be recognized by location without the aid of vision.

Operational Method of Coding Controls In the *operational method* of coding controls, each control has its own unique method for its operation. For example, one control might be of a push-pull variety, and another of a rotary variety. Each can be activated *only* by the movement that is unique to it. Thus, the control must be tried before it can be known if the correct control was selected. It is quite apparent that this scheme would be inappropriate if there were any premium on time in operating a control device and where operating errors were of considerable importance. When such a method is used, it is desirable that compatibility relationships be utilized, if feasible. By and large, this method of coding should be avoided except in those individual circumstances in which it seems to be uniquely appropriate.

Color Coding of Controls Color coding was discussed in Chapter 4 in reference to visual displays. Color can be useful for identifying controls as well. A moderate number of coding categories can be used, and colors can often be picked that are meaningful, for example red for an emergency stop control. One disadvantage, of course, is that the user must look at the control to identify it. Thus, color coding cannot be used in situations with poor illumination, or in which the control is likely to become dirty. Color, however, can be combined with some other coding method, such as size, to enhance discriminability or increase the number of coding categories available.

Label Coding of Controls Labels are probably the most common method of identifying controls and should be considered as the minimum coding requirement for any control. A large number of controls can be coded with labels and, if properly chosen, do not require much learning to comprehend. Extensive use of labels as the only means of coding controls is not desirable, however. Seminara, Gonzalez, and Parsons (1977) report that in most nuclear power plant control rooms, for example, there are literally walls of identical controls, distinguished only by labels. Workers often improvise their own coding methods including attaching uniquely shaped draft beer dispenser handles to the controls. Labels take time to read, and thus should not be the sole coding method where speed of operation is important. If labels are used they should be placed above the control (so that the operator's hand will not cover them when reaching for the control) and should be visible to the operator before reaching for the control.

Discussion of Coding Methods In the use of codes for identification of controls, two or more code systems can be used in combination. Actually, combinations can be used in two ways. In the first place, *unique combinations* of two or more codes can be used to identify separate control devices, such as the various combinations of texture, diameter, and thickness mentioned before (Bradley, 1967). And, in the second place, there can be completely *redundant codes*, such as identifying each control by a distinct shape *and* by a distinct color. Such a scheme probably would be particularly useful when accurate identification is especially critical. In discussing codes, we should be remiss if we failed to make a plug for standardization in the case of corresponding controls that are used in various models of the same type of equipment, such as automobiles and tractors. When individuals are likely to transfer from one model to another of the same general equipment type, the same system of coding should be used if at all possible. Otherwise, it is probable that marked "habit interference" will result and that people will revert to their previously learned modes of response.

There are some general considerations that affect the choice of a control code which should be kept in mind, such as whether the operators will have the luxury of being able to look at the control to identify it, whether they will have to continuously monitor a display, whether they will work in the dark, or whether they have a visual handicap. If vision is restricted, then only shape, size, texture, location, or operational method should be considered.

A second consideration is maintenance. When designing a system, including the controls, one must take into consideration the maintenance of that system. If shape, size, texture, or color coding is used, one must keep a stock of different spare controls and run the risk that the wrong control will be replaced if one is damaged. Thus, if there is a significant chance that controls will be damaged and have to be replaced, perhaps location or operational method coding should be considered.

We have been discussing coding methods mostly in terms of accuracy and speed of identification. Some coding methods, such as shape, size, and texture, may, however, influence the operation of the controls themselves. Carter (1978) compared the speed and accuracy of using knobs of different sizes and shapes. The study was aimed at designing underwater diving equipment so the tests took place in cold water with subjects wearing gloves, and the knobs were located on different parts of the body. The relative speed and accuracy of operating the knobs was somewhat dependent on the location of the knobs, but there were clear differences in performance using knobs that differed in shape, size, or both. It is important to select control codes that are discriminable, compatible, and meaningful, but it is also important to consider the possible effect of the specific codes on the speed and accuracy with which the control is operated. Incidentally, controls that require the user to grip them tightly should not use shape codes with sharp edges or protruding points.

Control-Response Ratio

In continuous control tasks or when a quantitative setting is to be made with a control device, a specified movement of the control will result in a system response. The system response may be represented on a display or it may not, as in turning the steering wheel of a car. The ratio of the movement of the control device to the movement of the system response we call the *control-response ratio* (C/R ratio). In the past, the C/R ratio has been called the *control-display ratio* (C/D ratio). It was felt that control-*display* was really not appropriate when no real display was present and that control-*response* is a more general term. Movement of the control, display, or system may be measured in linear distance (in the case of levers, vertical dials, etc.) or in angles or number of revolutions (in the case of knobs, wheels, circular displays, etc.). A very

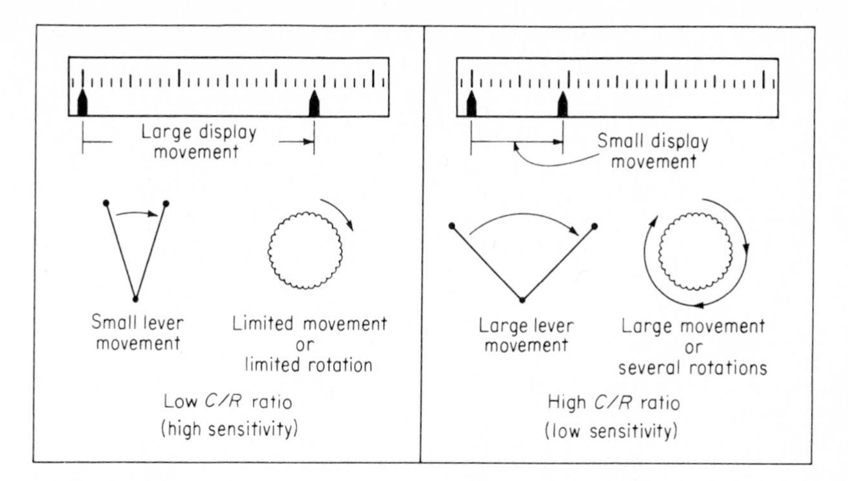

FIGURE 9-6
Generalized illustrations of low and of high control response ratios (C/R ratios) for lever and rotary controls. The C/R ratio is a function of the linkage between the control and display.

sensitive control is one which brings about a marked change in the controlled element (display) with a slight control movement; its C/R ratio would be low (a small control movement is associated with a large display movement). Examples of low and high C/R ratios are shown in Figure 9-6.

C/R Ratios and Control Operation The performance of human beings in the use of continuous or quantitative-setting control devices which have associated display movements is distinctly affected by the C/R ratio. This effect is not simple, but rather is a function of the nature of human motor activities when using such controls. In a sense, there are two types of human motions in such tasks. In the first place, there is essentially a *gross* adjustment movement (travel time or a sluing movement) in which the operator brings the controlled element (say, the display indicator) to the approximate desired position. This gross movement is followed by a *fine* adjustment movement, in which the operator makes an adjustment to bring the controlled element right to the desired location. (Actually, these two movements may not be individually identifiable, but there is typically some change in motor behavior as the desired position is approached.)

Optimum C/R Ratios The determination of an optimum C/R ratio for any continuous control or quantitative-setting control task needs to take into account these two components of human motions. Where the control in question is a knob, the general nature of the relationships is reflected by the results of studies by Jenkins and Connor (1949), and is illustrated in Figure 9-7. That figure illustrates the essential features of the relationships, namely, that travel time drops off sharply with decreasing C/R ratios and then tends to level off, and that adjustment time has the reverse pat-

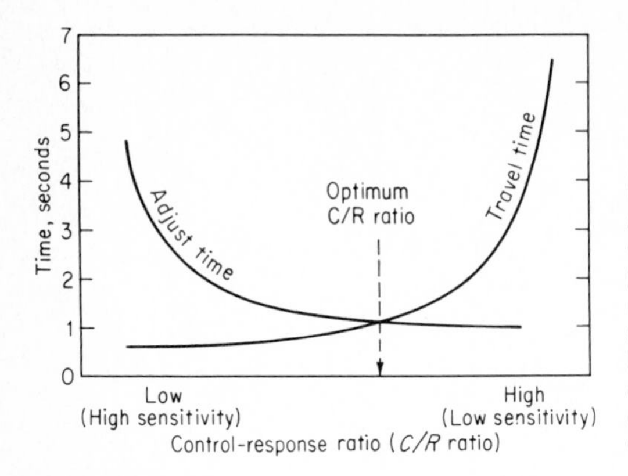

FIGURE 9-7
Relationship between C/R ratio and movement time (travel time and adjust time). While the data are from a study by Jenkins and Connor, the specific C/R ratios are not meaningful out of that context, so are omitted here. These data, however, depict very typically the nature of the relationships, especially for knob controls. (*Source: W. L. Jenkins & Connor, 1949.*)

tern. Thus, if you were trying to find a radio station using a control with a high C/R ratio (where it takes a couple of turns of the control to move the pointer on the dial a small distance), it would take you some time to get the pointer across the dial to the general location of the station (travel time). Once there, however, you could quickly "home in" on the station without overshooting it. With a low C/R ratio (where a small turn of the control sends the pointer across the dial), you can get the pointer near the station in no time, but it would take you a considerable time to "home in" because you would tend to overshoot the station. In determining the optimum C/R ratio, one must often trade off travel and adjustment time. The optimum ratio is somewhere around the point of intersection as shown in Figure 9-7. Within this general range, the combination of travel and adjustment time usually would be minimized. It should be pointed out that when a joystick or a lever is used instead of a control knob, the curve for travel time is almost a horizontal straight line. This is because it is almost as fast to move a stick through 180° as through a smaller angle. The optimum C/R ratio is then based on minimizing adjustment time only.

The numerical value for the optimum C/R ratio is a function of the type of control (knob, lever, crank, etc.), size of the display, tolerance permitted in setting the control, and other system parameters such as lag. Unfortunately, there are no formulas for determining what C/R ratio would be optimum for given circumstances. Rather, this ratio should be determined experimentally for the control and display being contemplated.

Resistance in Controls

In Chapter 8 we briefly mentioned control feedback as a factor that influences system control. Control feedback is intimately related to control resistance. The major part of the feedback or "feel" of a control is due to the various types of resistance associated with the control. Control manipulation takes principally two forms, the amount of displacement of the control and the amount of force applied to the control. The operator of the control can sense both movement and force through the proprioceptive and kinesthetic senses. Force and movement therefore are the primary sources of control feedback.

To use most controls requires both force to overcome resistance, and some displacement of the control (such as depressing a brake pedal or turning a steering wheel). Some controls, however, require only force or only displacement. A pure displacement control would have virtually no resistance to movement, and the only type of feedback would be the amount of movement of the body member. These are called *free-position* or *isotonic* controls. A *pure force* or *isometric* control, on the other hand, involves no displacement. The output is related to the amount of force or pressure that is applied to the control (these are sometimes called *stiff-sticks*). Pure force controls are becoming more common with advances being made in electronics and servo control mechanisms.

It is difficult to specify which type of feedback, force or displacement, is most useful. However, some research and experience, plus a bit of conventional wisdom, suggest that in many circumstances a combination of the two is useful, such as in the operation of a steering wheel. In special circumstances, however, one type or the other may have an advantage. For example, from the scant amount of literature available, Poulton (1974) concludes that pure force controls are superior to a combination force-displacement (spring-centered) control when used with higher-order tracking control systems (see Chapter 8), and a relatively fast-moving, gyrating target to track.

Types of Resistance All control devices have some resistance, although in the case of free-positioning (isotonic) controls this is virtually negligible. Resistance can of course serve as a source of feedback, but the feedback from some types of resistance can be useful to the operator while the feedback from other types of resistance can have a negative effect. In some circumstances the designer can design or select those controls that have resistance characteristics that will minimize possible negative effects and that possibly can enhance performance. This can be done in a number of ways, such as with servomechanisms, with hydraulic or other power, and with mechanical linkages. The primary types of resistance are elastic (or spring-loaded), static and coulomb friction, viscous damping, and inertia. We will discuss each of these in turn.

- *Elastic resistance:* Such resistance (as in spring-loaded controls) varies with the *displacement* of a control device (the greater the displacement, the greater the resistance). The relationship may be linear or nonlinear. A major advantage of elastic resistance is that it can serve as useful feedback, combining both force and displacement in a redundant manner. Poulton indicates that for tracking tasks, spring-loaded controls are best for compensatory tracking using a position control system with a slow ramp target to track (see Chapter 8). With elastic resistance, the control automatically returns to the null position when the operator releases it; hence it is ideal for "dead-man" switches. By so designing the control that there is a distinct gradient in resistance at critical positions (such as near the terminal), additional cues can be provided. In addition, such resistance permits sudden changes in control direction but reduces the likelihood of undesirable activation caused by accidental contact with the control.
- *Static and coulomb friction: Static friction,* the resistance to initial movement, is maximum at the initiation of a movement but drops off sharply. *Coulomb (sliding)*

friction continues as a resistance to movement, but this friction force is not related to either velocity or displacement. This can be illustrated by pressing a pencil eraser firmly against a flat horizontal surface with one hand while trying to slide the eraser slowly by pushing it horizontally with the other hand. You have to exert some force before you can get the pencil (or control) to move, i.e., you have to overcome the static friction. Once the pencil is moving, however, a smaller amount of force, to overcome the coulomb friction, will keep it moving no matter how fast it is moving or how far you move it. If you try and change directions you will notice a delay and some irregularity in the movements. With a couple of possible exceptions, static and coulomb friction tend to cause degradation in human performance. This is essentially a function of the fact that there is no systematic relation between such resistance and any aspect of the control movement such as displacement, speed, or acceleration; thus, it cannot produce any meaningful feedback to the user of the control movement. On the other hand, such friction can have the advantages of reducing the possibility of accidental activation and of helping to hold the control in place.

- *Viscous damping: Viscous damping* feels like moving a spoon through thick syrup. The amount of force required to move the spoon (or control) to overcome viscous damping is related to the velocity of the control movement. The faster you move the control, the more viscous damping you have to overcome. Viscous damping generally has the effect of resisting quick movement and of helping to execute smooth control, especially in maintaining a prescribed *rate* of movement. The feedback, however, being related to *velocity* and not displacement, probably is not readily interpretable by operators. Such resistance, however, does minimize accidental activation.
- *Inertia:* This is the resistance to movement (or change in direction of movement) caused by the mass (weight) of the mechanism involved. It varies in relation to *acceleration.* A force exerted on an inertia control will have little effect at first because it produces only an acceleration of the control (the greater the force, the greater the acceleration). The acceleration has to operate for a little while before the control will begin to move very much. Once the control begins to move, however, it is hard to stop. Inertia opposes quick control movements and thus probably would be a disadvantage in tracking. As Poulton points out, the human arm has enough inertia of its own. Inertia does, however, aid in smooth control and can improve the accuracy of turning a crank at a fixed speed (Poulton). It also reduces the possibility of accidental activation of the control.

Combining Resistances Almost all controls that move involve some forms of resistance, often more than one type. The effects of more than one type of resistance on control operation can be complex. Howland and Noble (1955), for example, show how time on target with a position control is affected by various combinations of control resistance. They used a knob control manipulated with the fingertips. Relatively small amounts of resistance were used. Table 9-2 summarizes the results. Elastic resistance alone resulted in the best performance, even better than no resistance at all. The addition of inertia by itself, or in combination with the other types of resistance, always resulted in a decrement in performance. The worst combination was elastic resistance plus inertia.

TABLE 9-2
EFFECTS OF VARIOUS COMBINATIONS OF CONTROL RESISTANCE ON TRACKING PERFORMANCE

	Type of control resistance present			
Condition	**Elastic***	**Viscous damping†**	**Inertia‡**	**Percent time on target**
1	Yes	No	No	60
2	Yes	Yes	No	52
3	No	Yes	No	44
4	No	No	No	41
5	No	No	Yes	35
6	No	Yes	Yes	35
7	Yes	Yes	Yes	34
8	Yes	No	Yes	27

*Elastic resistance: 0.02 in • lb/°
†Viscous damping: 0.02 (in • lb)/°/s
‡Inertia: 0.002 (in • lb)/°/s^2
Source: Howland & Noble, 1955.

Ability of People to Judge Resistance Regardless of what variety of resistance is intrinsic to a control mechanism, if it is to be used as a source of useful feedback, the meaningful differences in resistance have to be such that the corresponding pressure cues can be discriminated. Such discriminations were the focus of an investigation by Jenkins (1947a), with three kinds of pressure controls, namely, a stick, a wheel, and a pedal (like a rudder control in a plane). After some training and practice in reproducing specified forces, a series of trials was made by each subject. Measures of actual pressure exerted were then compared with the pressures that the subjects attempted to reproduce. The difference limens by pounds of pressure for the various types of controls are shown in Figure 9-8. The *difference limen* is the average difference that can just barely be detected; thus, two pressures have to differ by an amount greater than the limen to be detected as being different. The systematic drop-off of all the curves between 5 and 10 lb implies that if differences in pressure are to be used as feedback in operation of control devices of the types used, the pressures used preferably should be around or above these values (and perhaps more for pedals because of the weight of the foot). The experimenter suggested that if varying levels of pressure discrimination are to be made, the equipment should provide a wide range of pressures up to 30 or 40 lb. Beyond these pressures, the likelihood of fatigue increases, and also the likelihood of slower operation.

Deadspace

Deadspace in a control mechanism is the amount of control movement *around the null position* that results in no movement of the device being controlled. It is almost inevitable that some deadspace will exist in a control device. Deadspace of any consequence usually affects control performance, with the amount of effect being related to the sensitivity of the control system. This is indicated in Figure 9-9 (Rockway, 1957). It

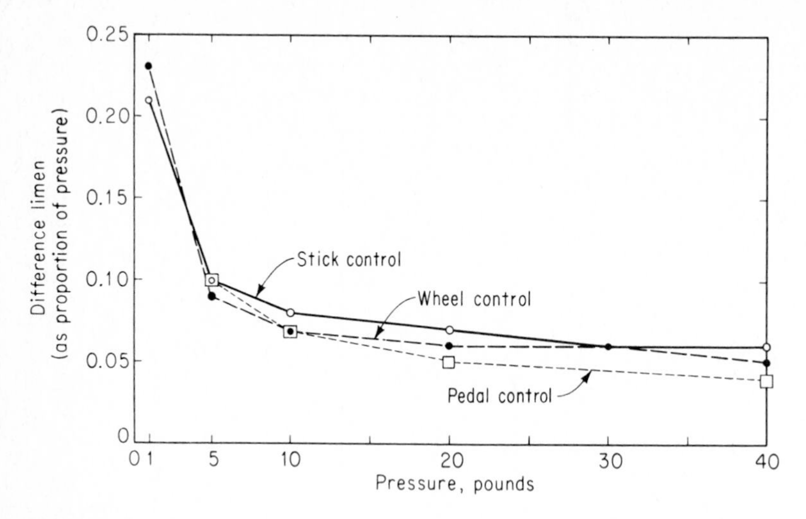

FIGURE 9-8
Difference limens for three control devices for various pressures that were to be reproduced. (*Limen* is the standard deviation divided by the standard pressure.) (*Source: W. O. Jenkins, 1947a.*)

can be observed that tracking performance deteriorated with increases in deadspace (in degrees of control movement that produced no movement of the controlled device). But the deterioration was less with the less-sensitive systems than with more sensitive systems. This, of course, suggests that deadspace can, in part, be compensated for by building in less-sensitive C/R relationships.

Rogers (1970) found that increasing the deadspace in a control resulted in an almost linear increase in the time needed to acquire a target. In the experiment a ball had to be spun at a minimum velocity to move the response marker. Deadspace corresponded to the minimum velocity required.

Poulton (1974) suggests that deadspace may be more detrimental in compensatory tracking tasks than in pursuit, but with higher-order control systems a small amount of deadspace may be helpful when the control is spring-centered. This is because the control may not always spring back to exactly the same position when it is released.

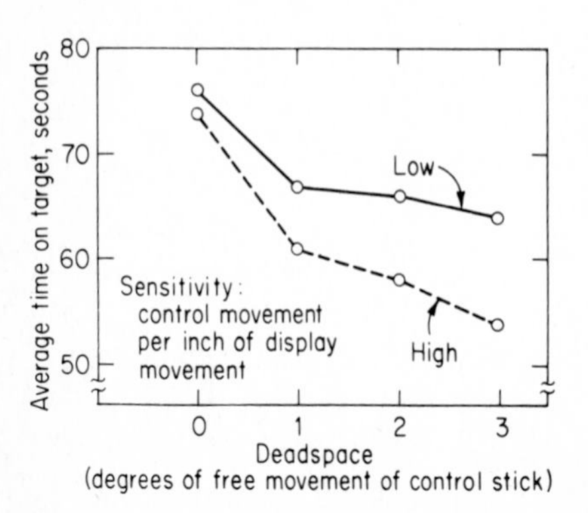

FIGURE 9-9
Relationship between deadspace in a control mechanism and tracking performance for various levels of control sensitivity. (High sensitivity corresponds to 0.56 percent per inch. Low sensitivity equals 3.33 percent per inch.) (*Source: Adapted from Rockway, 1957.*)

Backlash

The best way to think of *backlash* is to imagine operating a joystick or lever with a loose hollow cylinder fitted over it. When the cylinder is moved to the right, for example, the stick touches it on the left. If you are moving the cylinder to the right and then reverse direction, the stick does not start to return to the left until it comes up against the right side of the cylinder. Until the stick contacts the cylinder, the operator's control movements have no effect on the system. In essence, then, *backlash* is deadspace *at any control position.*

Typically, operators cannot cope very well with backlash. This effect was illustrated by the results of an investigation by Rockway and Franks (1959) using a control task under varying conditions of backlash and of display gain (which is essentially the reciprocal of C/R ratio). The results show that performance deteriorated with increasing backlash for all display gains, but was most accentuated for high gains. The implications of such results are that if a high display gain is strongly indicated (as in, say, high-speed aircraft), the backlash needs to be minimized in order to reduce system errors; or conversely, if it is not practical to minimize backlash, the display gain should be as low as possible—also to minimize errors from the operation of the system.

DESIGN OF SPECIFIC HAND-OPERATED CONTROLS

The factors discussed above relate to all types of controls, rather than to specific types of controls (although some of the factors are more relevant to tracking controls than to other types). We would like to now turn our attention to control design parameters of specific control types. It is neither relevant nor feasible for us to discuss or illustrate the many specific design parameters or the multitude of control mechanisms that ingenious people have concocted. It will be useful, however, to illustrate how specific design features have a direct bearing on the adequacy with which people can use a given control for its intended purpose. In this section we will discuss briefly some aspects of rotary switches, cranks and handwheels, and stick-type controls. We will discuss data entry devices and foot-operated controls later in this chapter.

Selector Switches

As a case in point, in one investigation a comparison was made of the use of four types of rotary selector switches for making settings of three-digit numbers, and, separately, for reading three-digit numbers already set into the four switches. One type of switch was a fixed-scale model with a moving pointer, as shown in Figure 9-10; the others were moving-scale models with fixed pointers with 10, 3, or 1 of the 10-scale positions visible. Some of the results of the study are given in Table 9-3.

In the task of making *settings* of specified three-digit numbers, there was little difference in accuracy among the four styles, but the fixed-scale resulted in lowest setting times and was first in preferences of the users. However, in the task of *reading* three-digit values at which the switches were set, the fixed-scale had the most errors. These differences, of course, need to be considered in the light of the *use* to which selector switches are to be put.

FIGURE 9-10
Illustration of fixed-scale, moving-pointer selector switches used in study by Kolesnik. See text for discussion. (*Source: Kolesnik, 1965. Reproduced by permission of Autonetics Division, North American Rockwell Corporation.*)

TABLE 9-3
PERFORMANCE USING VARIOUS SELECTOR SWITCHES

Type of switch	Average setting time, s	Preference, rank order	Reading errors
Fixed-scale	4.5	1	287
Moving-scale, 10 digits shown	5.4	3	135
Moving-scale, 3 digits shown	5.8	2	101
Moving-scale, 1 digit shown	6.3	4	92

Source: Kolesnik, 1965.

Concentrically Mounted Knobs

Space restrictions and other considerations sometimes argue for the use of concentrically mounted (or "ganged") knobs, such as those illustrated in Figure 9-11. Granting situational advantages for such controls, there are also some possible disadvantages associated with them, especially the possibility of inadvertent operation of adjacent knobs. If the knobs are too thin, the fingers may scrape against and operate the knob behind, and if the diameter distance is too small, the fingers may overlap and operate the knob in front. The problem, then, is one of identifying the optimum dimensions of such knobs, to minimize such possibilities. In a study dealing with this, Bradley (1969) used various combinations of such knobs, and various performance criteria (errors, reach time, and turning time). He found that the dimensions shown in Figure 9-11 were optimum.

FIGURE 9-11
Dimensions of concentrically mounted knobs that are desirable in order to allow human beings to differentiate knobs by touch. (*Source: Adapted from Bradley, 1969.*)

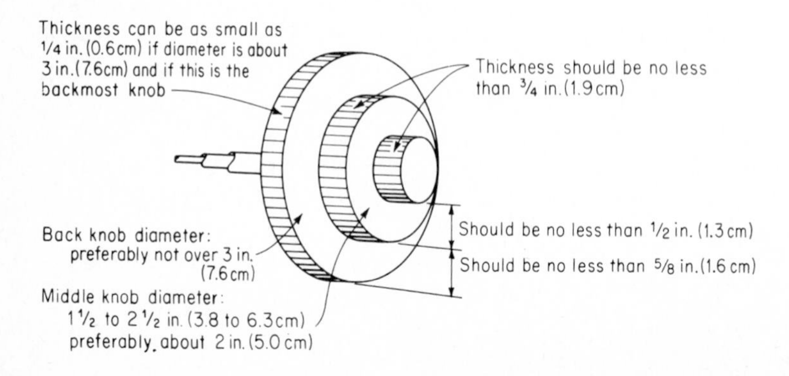

Cranks and Handwheels

Cranks and handwheels frequently are used as a means of applying force to perform various types of functions, such as moving a carriage or a cutting tool, or lifting objects. They can also be used as the control device for a tracking task. We will present the results of three studies which illustrate the effect of size, friction, and position on various aspects of operator performance. It should become clear that the optimum configuration depends on the aspect of operator performance one wishes to optimize.

Size of Cranks and Handwheels In a study by Davis (1949) several different sizes of cranks and handwheels were used to control the position of a pointer on a dial. In general, about one revolution of the crank was required in making the settings. The cranks and handwheels were mounted so that the plane of rotation was parallel to the frontal plane of the body. By using torques of 0, 20, 40, 60, and 90 in lb, it was possible to determine the time required to make settings under different friction-torque conditions that typically occur in the operation of such controls.

Figure 9-12 shows average times for certain torque levels by size of crank or handwheel. A couple of points are illustrated by this figure. In the first place, times for making settings with 0 torque were shortest for small sizes of cranks and wheels,

FIGURE 9-12
Average times for making settings with vertically mounted handwheels and cranks of various sizes under different torque conditions. (*Source: Adapted from Davis, 1949.*)

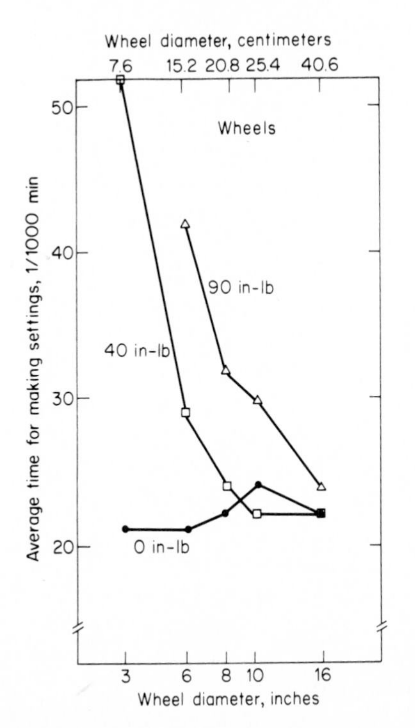

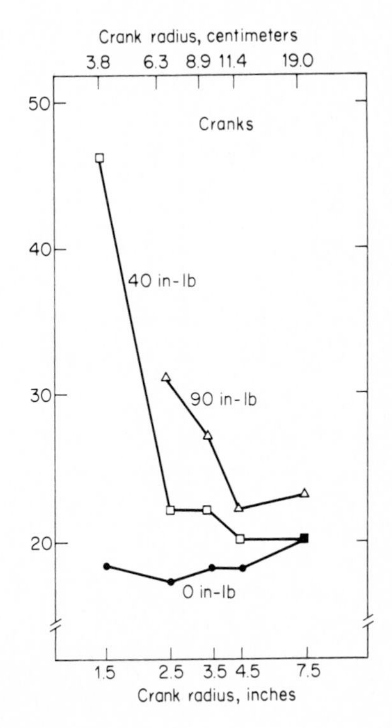

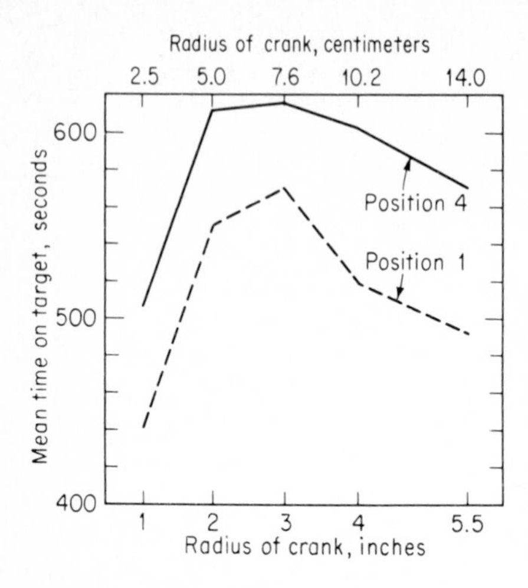

FIGURE 9-13
Accuracy of pursuit tracking with various sizes of cranks. The two curves show the relationship between radius of crank and time on target (TOT) for two combinations of crank positions (**1**, both cranks vertical and facing body; **4**, left crank same as 1 but right crank vertical and at right angles to body). (*Source: Adapted from Swartz, Norris, & Spragg, 1954, fig. 1.*)

whereas for torques of 40 and 90 in lb, the small cranks and wheels were definitely inferior to the larger ones. In the second place, the average times for cranks were, in general, lower than those for handwheels. In some cases, however, the time differences were not great, especially with the larger sizes.

Swartz, Norris, and Spragg (1954) investigated the relationship between size of crank and performance in a two-handed tracking task. The cranks were located in two positions (1) both cranks vertical and facing the body; and (2) the left crank vertical and facing the body, the right crank vertical and at a right angle to the body. Some of the results are shown in Figure 9-13 for time on target (TOT). The results indicate that TOT was greatest for the 2- and 3-in cranks (5.0- and 7.6-cm) and that the position of the cranks also influenced tracking performance.

Work Output with Different Cranks The work output of subjects using three sizes of cranks and five levels of resistance was investigated by Katchmer (1957) with a view to identifying fairly optimum combinations. The cranks were adjusted to waist height for each subject, and the subjects were instructed to turn the crank at a rapid rate until they felt they could no longer continue the chore (or until 10 min had passed). A summary of certain data from the study is presented in the four graphs of Figure 9-14. In total foot pounds of work (Figure 9-14*a*), the highest work levels occurred for the moderate torque loads (30 and 50 in lb) using the 5-in (13-cm) and 7-in (18-cm) radius cranks. Torque had no effect on the work output when using the small 4-in (10-cm) crank. Also, the larger the crank, the greater the total work output. Looking at the average time that subjects worked before quitting (Figure 9-14*b*), we see that, again, the larger the radius of the crank, the longer the subjects would continue to turn it, except under the higher torque conditions. But now, in terms of time, the optimum torque is 10 in lb, not 30 to 50 in lb as was the case with work output. The results for speed of cranking (Figure 9-14*c*) are not clear-cut, and it is difficult to reach general conclusions regarding optimum size and torque. Look now at Figure 9-14*d*, horsepower per minute. Here the optimum torque is 90 in lb; note, however, that the subjects could sustain that level only for a very short time (about 1 min).

These four graphs dramatically illustrate the point that a particular design may be optimum for one criterion, but not for another (in this case, the different criteria being tolerance time, work output, speed, and horsepower per minute). It is important, therefore, that the aspect or aspects of human performance one wants to maximize be specified before design parameters are chosen.

FIGURE 9-14
Average performance measures for subjects using 4-, 5-, and 7-in radius cranks under 10, 30, 50, 70, and 90 in lb torque resistance. (*Source: Adapted from Katchmer, 1957.*)

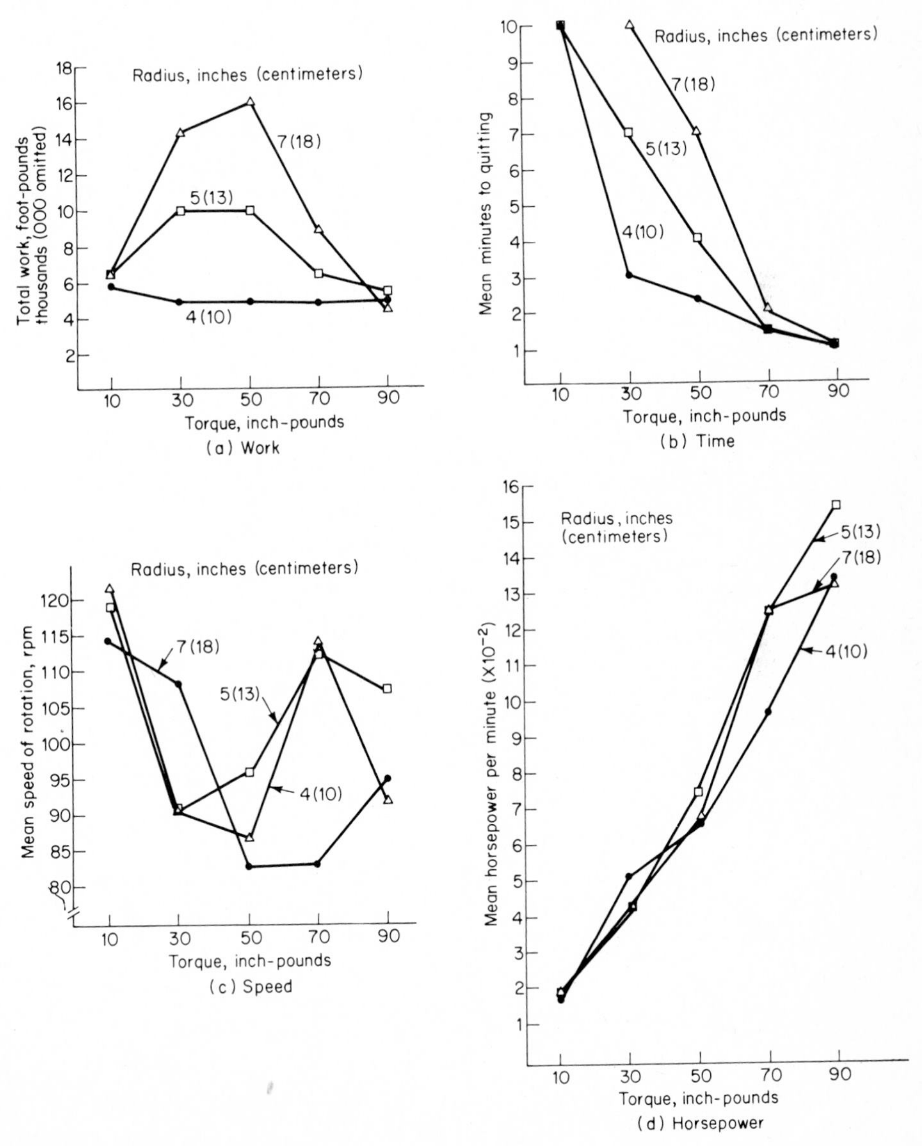

Stick-Type Controls

At first thought, one might think that the length of a joystick or lever would dramatically affect the speed or accuracy with which it could be used. This doesn't seem to be true, however, Jenkins and Karr (1954), for example, found that varying the length of a joystick from 12 to 30 in (30.5 to 76.2 cm) was relatively unimportant as long as the C/R ratio was around 2.5 or 3.0. Hartman (1956), using a tracking task, found only moderate effects on performance by varying joystick length from 6 to 27 in (15 to 69 cm). There was about a 10 percent advantage when using 18-in (46-cm) length sticks relative to other lengths.

Carrying this matter of stick length a step further, Hammerton and Tickner (1966) compared tracking performance using the thumb, hand, and forearm controls pictured in Figure 9-15. It was found that with high gain and a 2-s time lag, the hand control

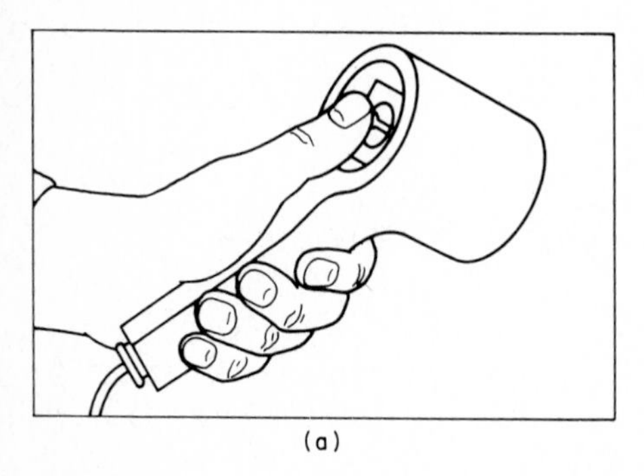
(a)

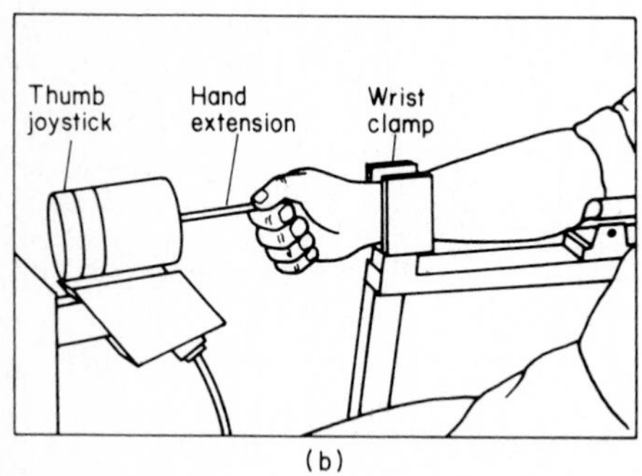

(b)

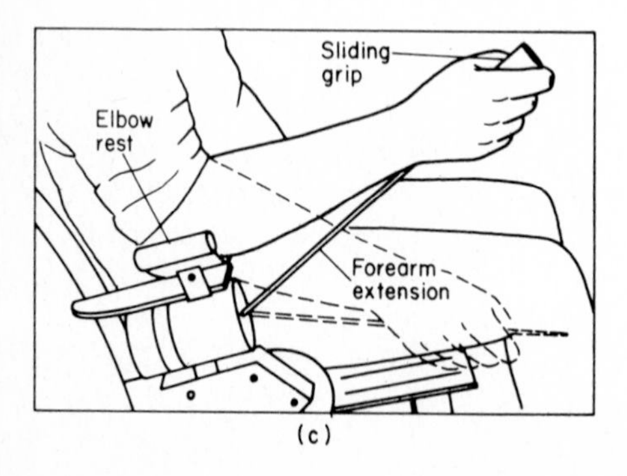

(c)

FIGURE 9-15
Spring-centered thumb, *a*; hand, *b*; and forearm, *c*, stick-type controls. (*Source: Hammerton & Tickner, 1966, fig. 1.*)

produced reliably superior performance to either the thumb or forearm controls. With no time delay and high gain, the results still favored the hand control, but the differences were smaller. With low gain and no lag, there were virtually no differences between the conditions. Thus, the differences became more apparent as the difficulty of the task increased (low to high gain, 0- to 2-s lag).

FOOT CONTROLS

Far and away, hand controls are more widely employed than are foot controls. One reason for this, as Kroemer (1971) points out, is that the general tenor of human factors handbooks implies that the feet are slower and less accurate than the hands. Kroemer hastens to add that this belief is neither supported nor discredited by experimental results. Besides this, foot controls often restrict the posture of the user. Having to maintain one foot (or both) on a control makes it more difficult to shift around in the seat or change position of the leg. Anyone who has taken a long drive on the open highway can attest to this. Operating foot controls from a standing position requires people to balance all their weight on the other foot. Despite all this, foot controls will continue to have a place in the control tasks of seated operators.

There are several important design parameters that affect performance with foot controls, such as whether they require a thrust with or without ankle action; the location of the fulcrum (if the pedal is hinged); the angle of the foot to the tibia bone of the leg; the load (the force required); and the placement of the control relative to the user. As with hand controls, our pupose is not to provide detailed design guidance on all aspects of control design, but rather to illustrate how specific features have a direct bearing on how adequately the control is used. Various criteria have been employed to evaluate alternative designs, such as reaction time, travel time, speed of operation, precision, force produced, and subjective preference. Chapter 12 will include a discussion of the placement of foot controls relative to the operator; here we will discuss design of foot controls and their placement relative to other controls.

Pedal Design Considerations

For illustrative purposes we will present certain of the results of an experiment by Ayoub and Trombley (1967) in which one of the factors they varied was the location of the fulcrum of the pedal. In one set of experimental conditions, the pedal, to be activated, had to be moved through an arc of 12°, no matter where the fulcrum was located (constant angle). The closer the fulcrum is to the heel, the further the ball of the foot must travel to achieve a 12° arc. In a second set of conditions, the pedal, to be activated, had to be moved a constant distance (0.75 in or 1.9 cm). In this case, the closer the fulcrum is to the heel, the smaller the arc will be to achieve the activation distance.

The mean travel times for various fulcrum locations are shown in Figure 9-16. It can be seen that for a constant angle of movement (of 12°), the optimum location of

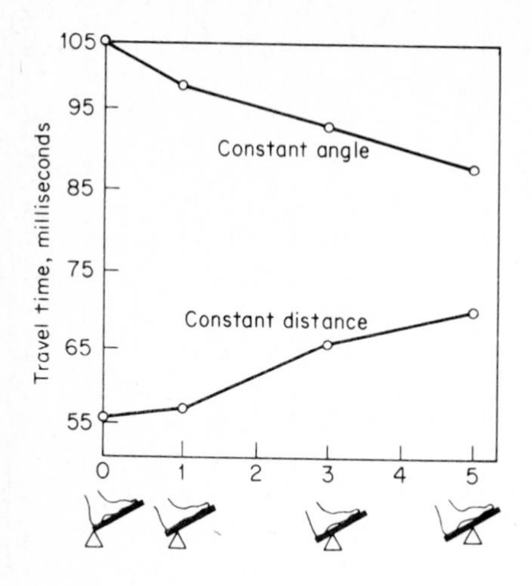

FIGURE 9-16
Mean travel time in pedal movement as related to location of the fulcrum for conditions of constant angle of movement (12°) and constant distance of movement (0.75 in or 1.9 cm). (*Source: Adapted from Ayoub & Trombley, 1967, fig. 13. Copyright 1967 by American Institute of Industrial Engineers, Inc., 25 Technology Park, Atlanta, Norcross, Ga. 30071, and reproduced with permission.*)

the fulcrum is forward of the ankle (about a third of the distance between the ankle and the ball of the foot), whereas for a constant distance of movement the optimum location of the fulcrum is at the heel. However, as Kroemer points out, the somewhat inconsistent results of various experiments with pedals (in part based on differences in experimental conditions) preclude any general statements as to what pedal allows the fastest activation or highest frequency in repetitive operation under a variety of specified conditions.

Foot Controls for Discrete Control Action Foot control mechanisms usually are used to control one or a couple of functions. There are some indications, however, that the feet can be used for more varied control functions. In one investigation, for example, Kroemer developed an arrangement of 12 foot positions (targets) for subjects, as illustrated in Figure 9-17. Using procedures that need not be described here, he was able to measure the speed and accuracy with which subjects could hit the various targets with their feet. By and large, he found that after a short learning period the subject could perform the task with considerable accuracy and with very short travel time (averaging about 0.1 s). Although forward movements were slightly faster than backward or lateral movements, these differences were not of any practical consequence. His results strongly suggest that it might be possible to assign control tasks to the feet which heretofore have been considered in the domain of the hands. Anyone

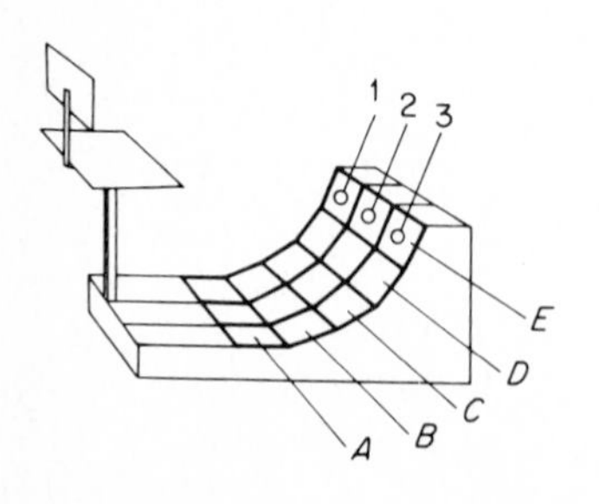

FIGURE 9-17
Arrangement of foot targets used in experiment by Kroemer in studying the speed and accuracy of discrete foot motions. (*Source: Kroemer, 1971, fig. 7.*)

who can play, or has watched someone play, the 13 pedals of an organ can attest to the precision and timing of foot movements that can be attained with practice.

Automobile Brake and Accelerator Pedals The most commonly used foot controls are the brake and accelerator controls of automobiles. Aside from their individual characteristics, an important factor in their use is that of their relative positions. In most automobiles the accelerator is lower than the brake pedal, thus requiring the lifting of the foot from the accelerator, its lateral movement, and then the depression of the brake. Several investigators, however, have found that movement time from accelerator to brake is shorter when the accelerator and brake are at the same level (Davies & Watts, 1970; Glass & Suggs, 1977; Snyder, 1976). Glass and Suggs, in fact, found that movement time was optimum when the brake was 1 to 2 in (2.5 to 5.0 cm) *below* the accelerator. In this configuration with the accelerator partially depressed, one could merely slide the foot to the brake without lifting it. The difference in time between the optimum placement and a typical truck configuration (with the brake 4 in or 10.2 cm above the accelerator) could reduce stopping distances by approximately 6 ft (1.8 m) when traveling at 55 mi/h (90 km/h). Even more important, Glass and Suggs report that when the brake was even with or above the accelerator, several subjects caught their foot on the edge of the brake pedal during upward movement from the accelerator. This resulted in an additional 0.3 s, or 24 ft (7.5 m), when traveling at 55 mi/h (90 km/h).

DATA ENTRY DEVICES

Our insatiable appetite for information has brought about a proliferation of machines and devices for storing, transmitting, and analyzing that information we hold so dear. These machines, such as the typewriter, telephone, computer, and calculator, require alphabetic or numerical information, or both, as input. Several types of devices have been used for the input of alphanumeric information, such as knobs, levers, thumb wheels, and push-button keyboards. Miner and Revesman (1962), for example, compared four different types of data entry devices for entering numerical data, and found the results shown in Table 9-4. The keyboard devices were superior (in terms of speed and accuracy) to the knobs or levers. The 10-key keyboard, such as used on hand calculators, was best. With very few exceptions, keyboard entry devices are generally

TABLE 9-4
COMPARISON OF DEVICES FOR ENTERING NUMERICAL DATA

Type of device	Average time to enter 10 digits, s	Percent of error
10-key keyboard, as on hand calculators	12	0.6
10-by-10 matrix keyboard	13	1.2
10-lever device, one for each digit	17	2.3
10 rotary knobs, one for each digit	18	2.3

Source: Miner and Revesman, 1962.

found to be superior to other data entry devices (e.g., thumb wheels, knobs, levers). For this reason we will concentrate our attention on keyboard design.

The speed and accuracy with which alphanumeric information can be entered by use of a keyboard, or any other device for that matter, is dependent on the quality of the data presented to the operator. As would be expected on the basis of common sense, data entry speed and accuracy are generally greater when the data presented to the operator are clear and legible. As discussed by Seibel (1972), speed and accuracy will also be greater if (1) the operator is familiar with the format of information to be entered; (2) upper- and lowercase characters are used for written text; and (3) long messages or strings of digits are divided into chunks, i.e., 123 456 rather than 123456.

Seibel also points out that there is no consistent difference in data entry rate between the usual alphanumeric and straight numeric keypunching. In addition, messages composed of random sequences of just 10 of the possible alphabetic characters are neither entered more rapidly nor more accurately than random sequences of all 26 characters.

Given this bit of introduction we will now turn to a brief discussion of some design features of keyboard entry devices.

Chord versus Sequential Data Entry

There are of course many types of data entry devices, and variations of each type. Most such devices are *sequential* in the sense that individual characters are to be entered in a specific sequence, there being a specific *key* or other device for every character. On the other hand, *chord* keyboards are those in which a single stimulus symbol (as a letter or number) requires the simultaneous activation of two or more keys.

Chord keyboards are not nearly as common as sequential keyboards, but are used with stenotype machines, pianos, and some mail-sorting machines. Comparisons between chord and sequential data entry must take into account that more information (in terms of bits) can be entered with a single chord stroke than with a single sequential entry key stroke. Thus, even if the stroke rates (number of strokes per minute) were less for chord than for sequential devices, the chord device probably would still result in a higher rate of information entry. In general, a skilled operator on a typewriter can out-perform a skilled operator on a stenograph chord keyboard in terms of strokes per minute. But in terms of information (bits) per minute, the chord keyboard wins hands down (no pun intended).

An actual comparison of the two types of keyboards was made by Bowen and Guinness (1965) in the context of a simulated mail-sorting task. In this study the encoding of mail by memory into various classes (each with a three-digit numerical code) was done by some subjects with a regular typewriter and by other subjects with a small (12-key) chord keyboard (requiring the use of one to four keys simultaneously) or a large (24-key) chord keyboard (requiring use of one to three keys). Some of the results are summarized in Table 9-5.

Both chord keyboards resulted in higher numbers of items sorted correctly than resulted from the typewriter. The superiority of the large over the small chord keyboard

TABLE 9-5
CHORD VERSUS SEQUENTIAL DATA ENTRY

Type of keyboard	Number sorted per minute	
	Correct	Incorrect
Sequential, typewriter	40.4	4.2
Chord, small	55.3	7.8
Chord, large	49.0	3.3

Source: Bowen and Guinness, 1965.

in reduced errors probably can be attributed to the fact that it did not require such difficult finger patterns. Bowen and Guinness suggest that the apparent advantage of chord over sequential encoding, in the context of their study, may be attributable to the fact that when using a learned memory code, one pairs it immediately with a *unit* response–not a response *spread out over time*, as called for in sequential keying.

Although chord keyboards seem to offer advantages over sequential keyboards, chord keyboards generally are more difficult to learn–in particular, it is more difficult to learn their coding system (i.e., the combinations of keys that represent individual items). Seibel (1972), however, recommends that chord keyboards be explored more fully for data entry purposes, especially data entry into digital computers.

Keyboard Arrangement

Since most keyboards are sequential, we will discuss briefly some aspects of alphabetic and numeric keyboard arrangements.

Alphabetic Keyboards The standard typewriter keyboard (called QWERTY because of the sequence of letters in the top line) appears to have been intentionally designed to slow the operator's key-stroke rate. The most commonly occurring letters (at least in English) are generally delegated to the weakest, slowest fingers and frequently occurring letter combinations are on the same hand. You might ask, "How could they have been so stupid?" The answer is, they weren't. The QWERTY keyboard was, in fact, intentionally designed to slow the rate of keying. Way back when, mechanical typewriters could not function as fast as people could stroke the keys and keys often jammed. QWERTY slowed the operators and thus reduced the frequency of jammed keys. By the time technology caught up with human keying capabilities, it was too late to change. Various alternative arrangements of the conventional typewriter keyboard have been proposed. The most notable was by Dvorak (1943) over a quarter of a century ago. Overall, the research suggests that the Dvorak keyboard might be better than the QWERTY keyboard, but the difference is not great enough to justify switching from one to another and retraining millions of people to type with the new arrangement. It seems safe, therefore, to conclude that the QWERTY keyboard is the de facto standard arrangement despite the fact that it may be less than totally efficient.

TABLE 9-6
COMMON NUMERIC KEYBOARDS

Adding machine	Telephone
7 8 9	1 2 3
4 5 6	4 5 6
1 2 3	7 8 9
0	0

Numeric Keyboards The two most common numerical data entry keyboards consist of three rows of three digits with the zero below, although there are two different arrangements of the numerals. One arrangement is used on many adding machines and the other is used on push-button telephones. These two arrangements are shown in Table 9-6. There is little question but that the highly practiced person will perform about equally well with either arrangement. The difference in arrangement becomes more important for people who only make occasional entries, and for people who must alternate between the different arrangements (and with the boom in pocket calculators this "switch-hitter" group may be quite large). At least three studies (using postal clerks, housewives, and air traffic controllers) have found the telephone arrangement, while only slightly faster (approximately 0.05 s per digit), somewhat more accurate than the adding machine arrangement (Conrad, 1967; Conrad & Hull, 1968; Paul, Sarlanis, & Buckley, 1965). Conrad and Hull, for example, report keying errors of 8.2 percent with the adding machine compared to 6.4 percent with the telephone arrangement. Unfortunately for us switch-hitters they found that these error rates were markedly better than those obtained from a group of subjects who alternated between the keyboards. This, of course, just reinforces the advantages of standardization in control display arrangements.

Although the telephone push-button arrangement results in more accurate performance than does the adding machine arrangement, how does it compare to the dial telephone? Pollard and Cooper (1978) found that even after almost two years of extensive experience with the push-button telephone, operators still made more errors than a control group of dial telephone operators (2.96 percent versus 1.60 percent). Data entry time, however, is significantly longer with a dial telephone.

Text-Editing Devices for Pointing

Technology is quickly replacing the time-consuming process of cut-and-paste, type-and-retype text editing. Soon computer-based text-editing, or word-processing, machines will be commonplace in almost every business office. (It can happen none too soon for your authors, who have already used a bathtubful of correction fluid and a mile of tape writing and editing this chapter.) This computer-based technology has introduced a new wrinkle in the data entry field. Now there is a need for a method of pointing by which operators can indicate to the computer their selection of some element (letter, word, sentence) on the computer display.

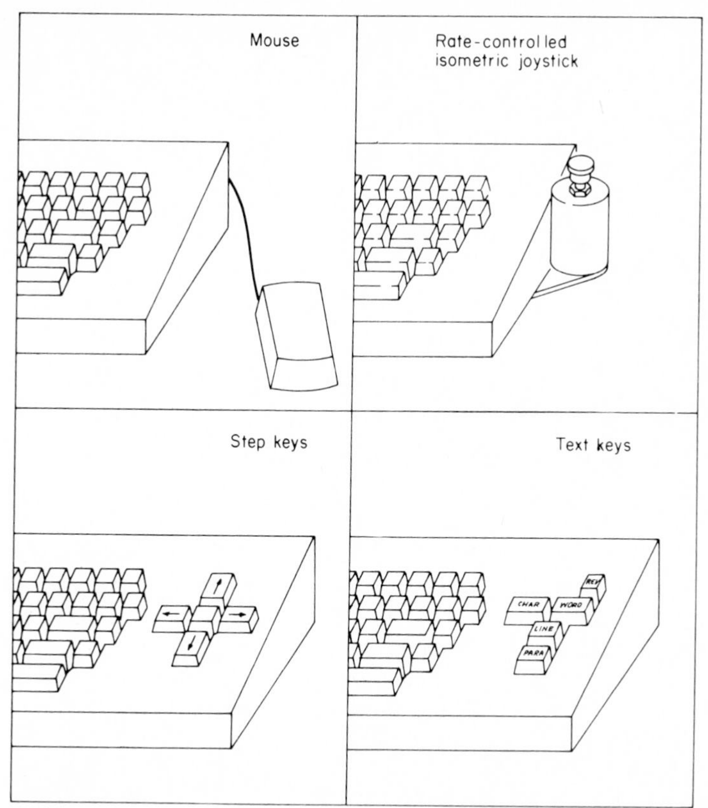

FIGURE 9-18
Pointing devices used by Card et al. for pointing to elements on computer displays. The mouse has two small wheels mounted at right angles on its undercarriage. (*Source: Card, English, & Burr, 1978, fig. 1.*)

Card, English, and Burr (1978) compared pointing performance using the mouse, isometric joystick, step keys, and text keys shown in Figure 9-18. The *mouse* is a device with two small wheels mounted at right angles to each other on its undercarriage. As the mouse is moved over the table, a cursor moves simultaneously on the computer screen. The results for movement time to a target and error rate are shown in Table 9-7. The mouse was found to be the fastest and also to have the lowest error rates.

TABLE 9-7
COMPARISON OF TEXT-EDITING DEVICES

Device	Time, s	Error, %
Mouse	1.66	5
Joystick	1.83	11
Step keys	2.51	13
Text keys	2.26	9

Source: Card, English, and Burr, 1978.

SPECIAL CONTROL DEVICES

The hands and feet are traditionally the parts of the body delegated the task of operating controls. Microcomputer technology, however, has opened new vistas of possibility for control devices and application. We cannot hope to cover fully all the new developments in this field; our purpose therefore is merely to introduce some of the devices that are being developed. Our emphasis will be on those devices which involve challenging human factors problems.

Teleoperators

The technological explosion of recent decades has made it possible to explore outer space and the depths of the ocean, thus exposing humans to new and hazardous environments. These and other developments require some type of "extension" of human control functions across distance or through some physical barrier, as in the control of lunar vehicles or of underwater activities. The control systems for performing such functions have been called *teleoperators* (Corliss & Johnson, 1968). Teleoperators are general-purpose man-machine systems that augment the physical skills of the operator. A person is always in the teleoperator control link on a real time basis. Robots are not considered teleoperators because they operate autonomously and sometimes (as in the science fiction movie *2001: A Space Odyssey*) counter to human interests.

Corliss and Johnson classify teleoperators into four categories: (1) manipulators; (2) prosthetic and orthotic devices; (3) amplifiers; and (4) walking machines. An example of a manipulator teleoperator is shown in Figure 9-19. Similar types of devices are commonly used today for handling dangerous materials (such as radioactive, explosive, or biologically toxic materials) or working in hostile environments such as involved in undersea work. An example of an amplifier teleoperator is *HandiMan,* a model of which is shown in Figure 9-20. This is an exoskeletal device which, with an accompanying power source, would make it possible to "amplify" a person's own physical movements so as to lift a weight of up to 1500 lb (680 kg).

Human Factors Considerations The engineering design of various types of teleoperators must of course take completely into account the nature of human physical responses and anthropometric characteristics in order that the device can simulate the human motions as they are executed by an individual. In this regard, however, particular consideration needs to be given to the nature of the feedback given to the operator. In many teleoperator applications, the operators are denied direct physical contact with the objects being manipulated by the teleoperator device; thus they lose their usual kinesthetic feedback. Feedback regarding the amount of force being exerted by the teleoperator, however, can be fed back to the operator through the use of servo systems or direct mechanical linkage. With the accelerated use of manipulator teleoperators, more and more applications deny the operator direct visual access to the effector end of the teleoperator. In such cases, television cameras are commonly used to extend the operator's vision. In this regard, Kama (1965) reports that direct viewing

FIGURE 9-19
Experimental remote-handling equipment used in various studies. (*Source: USAF, AFHRL.*)

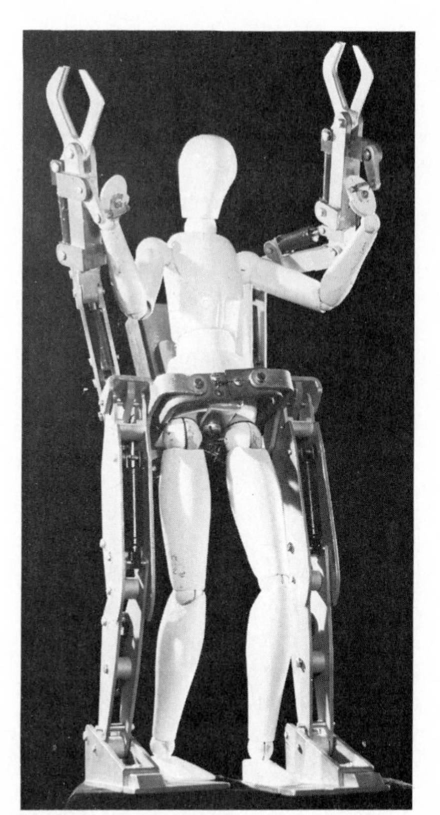

FIGURE 9-20
A model of *HandiMan,* a teleoperator that mimics the control responses of a human operator and that is capable of amplifying those responses by a ratio of 25:1 up to a force of 1500 lb (680 kg). (*Source: General Electric Company Research and Development Center, Schenectady, N.Y.*)

of the object being manipulated is not really necessary; indirect viewing (e.g., TV) usually will suffice.

The use of servos and television feedback introduces interesting human factors problems: a time delay may develop between the operator's control action and the feedback received about the device's response; television can distort distances and depth cues; and teleoperators can execute movement patterns which cannot be duplicated by people (e.g., joint extension or 360° rotation of the wrist).

Knowles (1962) suggests that the user of teleoperators needs to have some meaningful identification or integration with the equipment, or as Corliss and Johnson (1968) put it, the teleoperator controls should be "built along anthropomorphic lines." A manifestation of this principle comes from the study by Crawford (1964), in which he found that in the remote handling of disks to be placed in holes in a form board (by using a variation of the device shown in Figure 9-19), subjects did better with a joystick than with multiple-lever controls. Manipulations of the single joystick actually controlled six different control movements, simulating shoulder pivot, shoulder rotation, elbow pivot, wrist pivot, wrist rotation, and grip. This single joystick provides a higher degree of integration of the control with the operator than the independent use of six separate multiple levers. (This probably is, in a sense, another manifestation of the principle of compatibility.)

Prosthetic Devices The most widely employed control interface for prosthetic limbs is the electromyographic (EMG) signal (Sheridan & Mann, 1978). An example of such a device, shown in Figure 9-21, is the experimental model of the "Utah Arm" developed by Stephen Jacobsen, a professor of bioengineering at the University of Utah, and his associates. The device, described by Jerard and Jacobsen (1979), uses shoulder EMGs as control sites. These EMG signals serve as estimates of the vector of the natural limb torques, and by the use of mathematically derived equations trigger corresponding movements of the artificial limb. For an above-elbow amputee the device controls elbow flexion, humeral (i.e., shoulder) rotation, and wrist rotation. The device is based on certain theoretical formulations that differ from those used with typical prosthetic devices, but may in the long run offer the possibility of improved control.

Speech-Activated Control

The use of speech to converse with computers is a stock item in science fiction books and movies. The advantages are obvious: the operator is not tied to a console; the operator's hands are free to do other chores (like pour a cup of coffee); and, at least in science fiction, only minimal training of operators is required. Although such casual conversations with machines are not yet upon us, recent advances in speech recognition has spawned new and novel solutions to system problems.

Virtually all speech-recognition systems in active use today are of the "isolated word (or phrase)" type (Lane & Harris, 1980). Most perform recognition in one of two ways. The most common involves comparison of the frequency spectrum of the voice

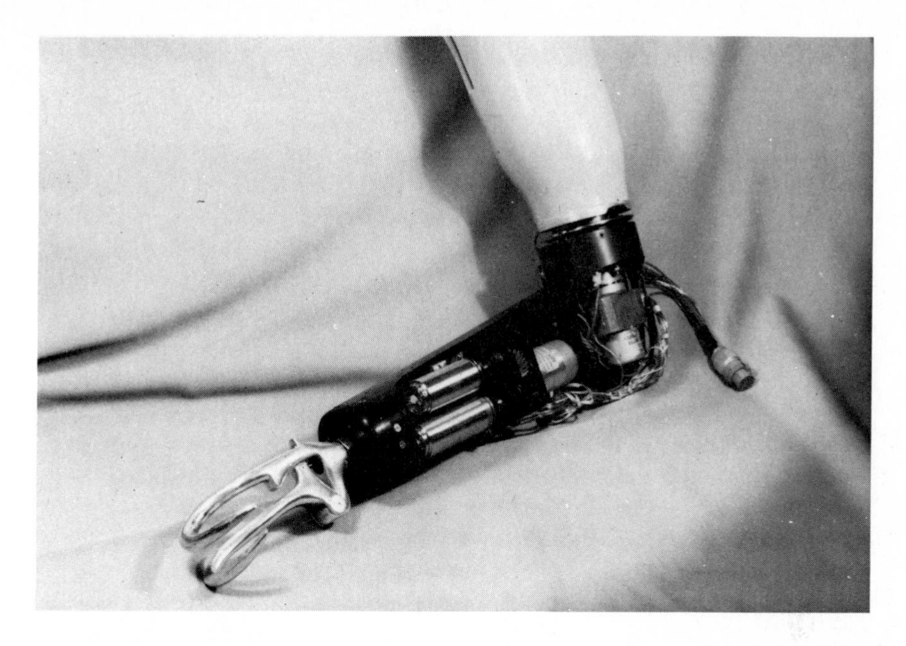

FIGURE 9-21
Illustration of the "Utah Arm" for above-elbow amputee, developed by S. C. Jacobsen, a professor of bioengineering at the University of Utah, and his associates. (*Photograph courtesy of Dr. Jacobsen.*)

signal (see Chapter 6) with the spectral characteristics of each vocabulary item stored in the computer until an acceptable match is found. The second approach focuses on isolating certain key features of an utterance, e.g., stops and fricatives (see Chapter 6), and comparing these to the stored vocabulary items. In terms of accuracy, both approaches are essentially equal (Lane & Harris).

These approaches have inherent limitations. First, the systems can handle only a limited predefined vocabulary. Accuracy is very high with 30 to 50 items and is acceptable with up to 100 items if they are sufficiently dissimilar (Lane and Harris). By placing constraints on what items can be said and in what order, very large vocabularies can be generated. For example, the first word or phrase in a voice command can be limited to one of, say, 30 to 50 preestablished items. When it is recognized, a second set of 30 to 50 items, composed of those items which are allowed to follow the first item, are now searched to determine the second item in the message. This process can be repeated, thus allowing for vocabulary of several hundred item strings (Coler, Plummer, Huff, & Hitchock, 1977).

A second limitation of isolated word recognition (IWR) systems is that the speaker must set off the items (words or phrases) with pauses. For example, the command "stop-come-back" must be spoken without any pause between the words to be recognized by these systems as one item. If the system expects this command as a single item, any pause between words will render it unrecognizable. Conversely, if the computer expects three items and *no* pause is made between "come" and "back", it will not be recognized. (Work is advancing on systems which do not require pauses for system recognition; these are called *continuous speech recognition systems.*) The need to control speech pauses makes it mandatory that the operators of IWR systems be trained to discipline their voice and speech.

In practice, not only must the operator be trained, but so must the system itself. As pointed out in Chapter 6, people have different speech patterns, inflections, frequency modulations, etc.; not everyone sounds the same. The speech recognition system must develop a complete vocabulary for each operator. This is usually done by having the operator repeat each item in the system's vocabulary 10 or so times to develop a good representation of that item for that speaker in the computer's memory.

People's voices change owing to fatigue and stress. The speech-recognition system, trained to recognize a rested calm voice, may not recognize the speaker after the voice becomes fatigued or in a stress situation. Imagine controlling a car with a speech recognition system; a child runs in front of your car and you yell "STOP!" Unless you yelled "stop" in the same way when you first trained the system, it is very likely that the car will not respond. And yelling louder will not help.

Other problems are inherent in speech-recognition systems as well; background noise may activate the system or render a valid command unrecognizable. Applied Psychological Services (1979), for example, reporting an evaluation of a speech-recognition system for sorting mail sacks in the U.S. Postal Service, found that background noise and talking resulted in quite a few inadvertent activations of the system. Overall, they found that the mean percentage of inputs rejected, misinterpreted, and not responded to was approximately 5 percent. The rate varied for individual items in the vocabulary (38-item vocabulary) with the high being 37 percent for "Chicago." As for input time, the speech recognition system was slower than the original hand keyboard used. The authors point out, however, that the operators could handle the mail sacks faster using the speech system because their hands were free. Considering bags of mail processed per unit of time, however, both systems were approximately equal.

New applications for speech-recognition systems are being found every week, air traffic control (Connolly, 1979), inspection, and mail sorting to name a few. Lane and Harris, as well as Skriver (1979) suggest that voice input becomes superior, in terms of time and errors, to manual input in those situations in which the operator is under high task demands, such as performing complex perceptual-motor coordination tasks or performing under speed stress.

Eye-Activated Control

Advances in equipment used to track eye movements can be used to convert eye movements into control inputs. The operator simply has to look at a target and the system being controlled would follow the position of the gaze. Poulton (1974) reports that tracking by eye is similar to tracking by hand in terms of reaction time, anticipation, and smoothness. Accuracy has been reported to be approximately ±10 minutes of visual arc at the center of the visual field (Chapanis & Kinkade, 1972; Coluccio & Mason, 1970). There are some disadvantages of eye control including these: the position of the operator's head must be fixed, or at least determined and used to establish the point of gaze; eye control is difficult in a vibrating or acceleratory environment because either one will produce eye movements which the operator is unable to prevent; and distractions which might cause operators to shift their gaze will degrade performance.

To date, most eye-activated control applications have been in the military field, for example in aiming guns. Other applications could be in control of teleoperators, instrument pointing, and visual search tasks.

Direct Brain Communication

The concept here is that the operator need only "think" a command and the system would respond. The control signals would be electroencephalographic (EEG) potentials (brain waves) picked up by electrodes attached to the scalp. The system would be similar to speech recognition systems in that the system would be trained to recognize a limited vocabulary of commands. Pinneo (1975), for example, reported on preliminary results of a 15-word vocabulary system in which 34 percent of 900 trials were correctly classified using EEG alone.

DISCUSSION

We have discussed various aspects of controls and their influence on operator performance. One might think that the selection or design of controls has "all been done"; however, numerous examples of inappropriately designed or selected controls can still be found today–for example, in nuclear power plants, heavy-equipment cabs, and numerous consumer products. A wide variety of controls and control mechanisms are available "off-the-shelf," but as we saw in this chapter the best design often depends on the particular demands of the task and the criteria used to judge performance. For those who view "controls" as somehow less glamorous than other areas of human factors, we hope the discussion of speech-recognition systems and direct brain communication illustrates that exciting things are still happening in this area. For example, touch-sensitive cathode-ray-tube (CRT) displays are opening up a new control area, "computer-generated controls." With such systems, push buttons, sliders, and keyboards can be shown on a display; the operator then makes a response by touching the control as displayed on the screen and other controls can then be displayed to the operator. In this way, an entire wall of controls can be condensed into one CRT display. This is fast becoming a challenging area for human factors research and application.

REFERENCES

Air Force System Command Design Handbook 1-3, Human Factors Engineering. 3d ed. U.S. Air Force, June 1980.

Applied Psychological Services Inc. Assessment of the use of voiced word recognition device in the sack sorting operation. *JSAS Catalog of Selected Documents in Psychology*, 1979, *9*(4). (Ms. No. 1973)

Ayoub, M. M., & Trombley, D. J. Experimental determination of an optimal foot pedal design. *Journal of Industrial Engineering*, 1967, *17*, 550–559.

Bowen, H. M., & Guinness, G. V. Preliminary experiments on keyboard design for semiautomatic mail sorting. *Journal of Applied Psychology*, 1965, *49*(3), 194–198.

Bradley, J. V. Tactual coding of cylindrical knobs. *Human Factors*, 1967, *9*(5), 483–496.

Bradley, J. V. Desirable dimensions for concentric controls. *Human Factors*, 1969, *11*(3), 213–226.

Card, S., English, W., & Burr, B. Evaluation of mouse, rate-controlled isometric joystick, step keys, and text keys for test selection on a CRT. *Ergonomics*, 1978, *21*, 601–613.

Carter, R. Knobology underwater. *Human Factors*, 1978, *20*, 641–647.

Chapanis, A., & Kinkade, R. Design of controls. In H. P. Van Cott & R. Kinkade (Eds.), *Human engineering guide to equipment design* (Rev. ed.). Washington, D. C.: U.S. Government Printing Office, 1972.

Coler, C., Plummer, R., Huff, E., & Hitchock, M. Automatic speech recognition research at NASA Ames Research Center. In M. Curran, R. Breaux, & E. Huff (Eds.), *Voice technology for interactive real-time command/control systems applications, proceedings of a symposium*. Moffett Field, Calif.: NASA Ames Research Center, 1977.

Coluccio, T., & Mason, K. Measurement of target position in a realtime display by use of the Oculometer eye direction monitor. Paper presented at the Human Factors Society Annual Meeting, Oct. 1970.

Connolly, D. *Voice data entry in air traffic control* (FAA-NA-79-20). Federal Aviation Administration, Atlantic City, N.J.: National Aviation Facilities Experimental Center, 1979.

Conrad, R. Performance with different push-button arrangements, *HET PTT BEDRIJF DEEL*, 1967, *15*, 110–113.

Conrad, R., & Hull, A. J. The preferred layout for numerical data-entry keysets. *Ergonomics*, 1968, *11*(2), 165–173.

Corliss, W., & Johnson, E. *Teleoperator controls* (NASA SP-5070). Washington, D. C.: NASA Office of Technology Utilization, 1968.

Crawford, B. M. Joy stick vs. multiple levers for remote manipulator control. *Human Factors*, 1964, *6*(1), 39–48.

Davies, B. T., & Watts, J. M., Jr. Further investigations of movement time between brake and accelerator pedals in automobiles. *Human Factors*, 1970, *12*(6), 559–561.

Davis, L. E. Human factors in design of manual machine controls. *Mechanical Engineering*, October 1949, *71*, 811–816.

Dvorak, A. There is a better typewriter keyboard. *National Business Education Quarterly*, 1943, *12*, 51–58.

Gamst, F. Human factors analysis of the diesel-electric locomotive cab. *Human Factors*, 1975, *17*, 149–156.

Glass, S., & Suggs, C. Optimization of vehicle-brake pedal foot travel time. *Applied Ergonomics*, 1977, *8*, 215–218.

Hammerton, M., & Tickner, A. An investigation into the comparative suitability of forearm, hand, and thumb controls in acquisition tasks. *Ergonomics,* 1966, *9,* 125-130.

Hartman, B. O. *The effect of joystick length on pursuit tracking* (Report 279). Fort Knox, Ky.: U.S. Army Medical Research Laboratory, Nov. 1956.

Howland, D., & Noble, M. The effect of physical constraints on a control on tracking performance. *Journal of Experimental Psychology,* 1955, *46*, 353-360.

Hunt, D. P. *The coding of aircraft controls* (Technical Report 53-221). U.S. Air Force, Wright Air Development Center, August 1953.

Jenkins, W. The discrimination and reproduction of motor adjustments with various types of aircraft controls. *American Journal of Psychology,* 1947, *60*, 397-406.

Jenkins, W. The tactual discrimination of shapes for coding aircraft-type controls. In P. Fitts (Ed.), *Psychological research on equipment design* (Research Report 19). Army Air Force, Aviation Psychology Program, 1947.

Jenkins, W., & Connor, M. B. Some design factors in making settings on a linear scale. *Journal of Applied Psychology,* 1949, *33*, 395-409.

Jenkins, W., & Karr, A. C. The use of a joy-stick in making settings on a simulated scope face. *Journal of Applied Psychology,* 1954, *38*, 457-461.

Jerard, R. B., & Jacobsen, S. C. Laboratory evaluation of a unified theory for simultaneous multiple axis artificial arm control (Paper 79-WA-810-8). Winter Annual Meeting of the American Society of Mechanical Engineers, New York: Dec. 2-7, 1979.

Kama, W. N. *Effect of augmented television depth cues on the terminal phase of remote driving* (Technical Report 65-6). U. S. Air Force, Aerospace Medical Research Laboratory, April 1965.

Katchmer, L. T. *Physical force problems: 1. Hand crank performance for various crank radii and torque load combinations* (Technical Memorandum 3-57). Aberdeen, Md.: Aberdeen Proving Ground, Human Engineering Laboratory, March 1957.

Knowles, W. B. *Human engineering in remote handling* (TDR 62-58) U.S. Air Force, AMRL, August 1962.

Kolesnik, P. E. *A comparison of operability and readability of four types of rotary selector switches* (T5-1187/3111). Autonetics Division, North American Aviation (now North American Rockwell), Inc., June 1965.

Kroemer, K. H. E. Foot operation of controls. *Ergonomics,* 1971, *14*(3), 333-361.

Lane, N., & Harris, S. Conversations with weapon systems: Crewstation applications of interactive voice technology. Warminster, Pa.: Naval Air Development Center, 1980.

McFarland, R. A. *Human factors in air transport design.* New York: McGraw-Hill, 1946.

Miner, F. J., & Revesman, S. L. Evaluation of input devices for a data setting task. *Journal of Applied Psychology,* 1962, *46*, 332-336.

Paul, L., Sarlanis, K., & Buckley, E. A human factors comparison of two data entry keyboards. Paper presented at Sixth Annual Symposium of the Professional Group on Human Factors in Electronics, IEEE, May 1965.

Pinneo, L. Direct brain computer communication feasible. *Human Factors Society Bulletin,* 1975, *18*(2), 1-2.

Pollard, D., & Cooper, M. An extended comparison of telephone keying and dialing performance. *Ergonomics,* 1978, *21*, 1027-1034.

Poulton, E. *Tracking skill and manual control.* New York: Academic Press, 1974.

Rockway, M. *Effects of variation in control deadspace and gain on tracking performance* (Technical Report 57-326). U.S. Air Force, Wright Air Development Center, September 1957.

Rockway, M., & Franks, P. *Effects of variations in control backlash and gain on tracking performance* (Technical Report 58-553). U.S. Air Force, Wright Air Development Center, January 1959.

Rogers, J. Discrete tracking performance with limited velocity resolution. *Human Factors,* 1970, *12*, 331–339.

Seibel, R. Data entry devices and procedures. In H. P. Van Cott and R. G. Kinkade (Eds.), *Human engineering guide to equipment design.* Washington, D. C.: U.S. Government Printing Office, 1972.

Seminara, J., Gonzalez, W., & Parsons, S. *Human factors review of nuclear power plant control room design* (EPRI NP-309). Palo Alto, Calif.: Electric Power Research Institute, 1977.

Sheridan, T. B., & Mann, R. W. Design of control devices for people with severe motor impairment. *Human Factors,* 1978, *20*(3), 321–328.

Skriver, C. *Vocal and manual response modes: Comparison using a time-sharing paradigm* (NADC-79-127-60). Warminster, Pa.: Naval Air Development Center, 1979.

Snyder, H. Braking movement time and accelerator-brake separation. *Human Factors,* 1976, *18*, 201–204.

Swartz, P., Norris, E. B., & Spragg, S. D. S. Performance on a following tracking task as a function of radius of control cranks. *Journal of Psychology,* 1954, *37*, 163–171.

CHAPTER 10

HAND TOOLS AND DEVICES

Some people still believe that the use of tools for pounding, digging, scraping, and cutting is what distinguishes humans from apes. Actually, there is considerable evidence (Washburn, 1960) that the use of such hand tools and other devices existed with prehuman primates almost a million years ago. Today hand tools are crafted for uncounted specific applications as well as for general purpose activities. Until recently, however, human factors had largely ignored the design of hand tools and other devices, concentrating on more sophisticated equipment. Perhaps the attitude was that a million years of evolutionary experience would produce tools and devices uniquely adapted for human use. Actually, a million years is no guarantee of proper design. Indeed, many hand tools and devices are not designed for efficient safe operations by humans–especially for repetitive operation.

Improperly designed tools and devices have several undesirable consequences. They result in accidents and injury. Ayoub, Purswell, and Hicks (1977), for example, reviewed injury data from several states in the United States and found that injuries resulting from hand tool use accounted for 5 to 10 percent of all compensable injuries.

Powered hand tools, however, accounted for only 21 to 29 percent of all hand-tool-related injuries. The most common tools implicated in injuries were knives, wrenches, and hammers. Such injury data, for the most part, represent single-incident traumatic events such as smashing a finger or cutting the palm of the hand. Other more insidious consequences of improper tool design are cumulative-effect traumas such as tenosynovitis, "trigger finger," ischemia, vibration-induced white finger, and even "tennis elbow." These conditions (which we will discuss later) usually do not show up on accident-injury reports but often lead to reduced work output, poorer-quality work, increased absenteeism, and single-incident traumatic injuries.

The proper design of hand tools and devices requires an understanding of technical, anatomical, kinesiological, anthropometric, physiological, and hygienic considerations. Tools cannot be designed in isolation. Often a redesign of the work space, as for example changing the height of the work surface or repositioning the work piece relative to the worker, can compensate for a less than optimally designed tool.

An important aspect of hand tool use that is often overlooked is training. It is usually assumed that workers know how to use hammers, pliers, saws, wrenches, etc. Actually, nothing could be further from the truth. As a rule, workers do not always know the proper way to use tools. A good training program in tool usage can often reduce the incidence of injuries and cumulative-effect traumas.

In this chapter we will concentrate on the human factors considerations in hand tool and device design. Our emphasis will be on factors that contribute to cumulative-effect trauma and both the quality and quantity of productivity. We will start with a discussion of the anatomy and functioning of the human hand. This will allow us to formulate some general human factors principles of hand tool and device design. To close the chapter, a few examples of applying human factors principles to specific tools and devices will be presented.

THE HUMAN HAND

The human hand is a complex structure composed of bones, arteries, nerves, ligaments, and tendons as shown in Figure 10-1. The fingers are flexed by muscles in the forearm. The muscles are connected to the fingers by way of tendons which pass through a channel in the wrist. This channel is formed by the bones of the back of the hand on one side, and the transverse carpal ligament on the other. The resulting channel is called the *carpal tunnel*. Through this tunnel passes a whole host of vulnerable anatomical structures including the radial artery and median nerve. Running over the outside of the transverse carpal ligament are the ulnar artery and ulnar nerve. This artery and nerve pass beside a small bone in the wrist called the *pisiform bone*. All this may seem a bit clinical, but a clear understanding of the relationships of these structures is important to appreciate the consequences of improper hand tool design.

The bones of the wrist connect to the two long bones of the forearm—the *ulna* and the *radius*. The radius connects to the thumb side of the wrist and the ulna connects to the little finger side of the wrist. The configuration of the wrist joint permits movements in only two planes, each one at an approximately 90° angle to the other. The first plane allows palmar flexion, or when performed in the opposite direction,

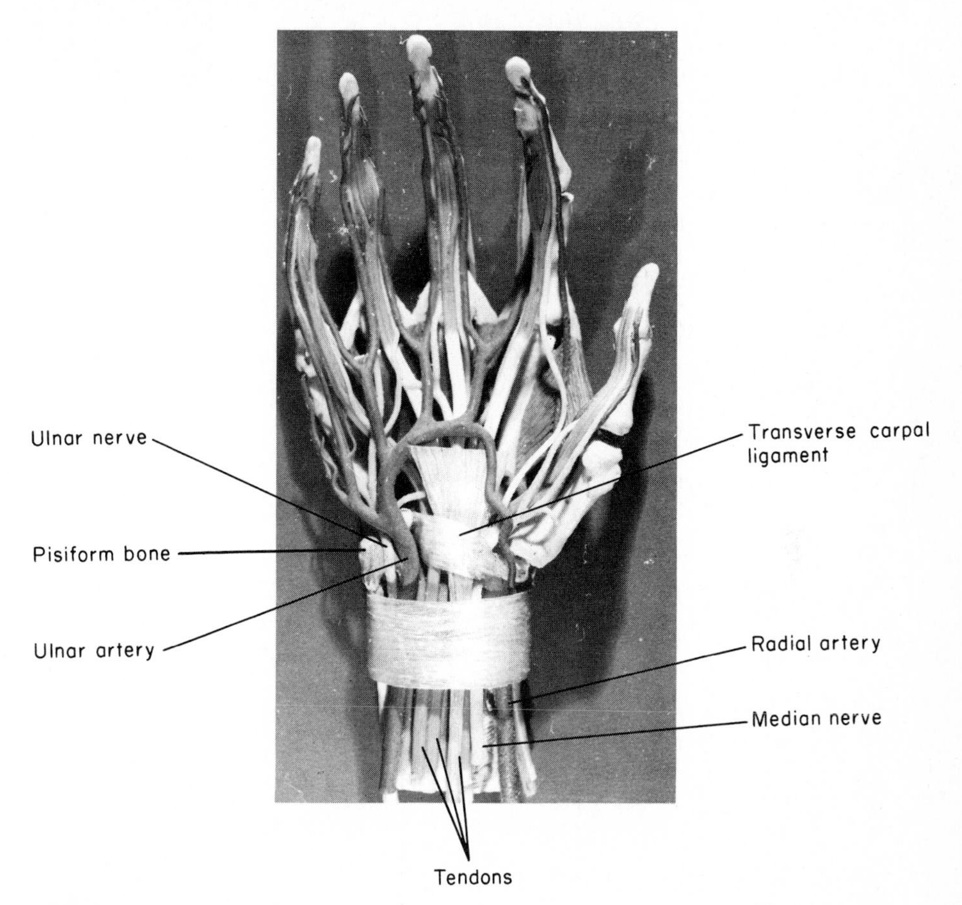

FIGURE 10-1
The anatomy of the hand as seen from the palm side. Shown are the blood vessels, tendons, and nerves underneath the ligaments of the wrist. (*Source: Tichauer, 1978, fig. 50.*)

dorsiflexion, as shown in Figure 10-2*a*. The second movement plane, shown in Figure 10-2*b*, consists of either ulnar deviation or radial deviation of the hand.

The ulna and radius of the forearm connect to the *humerus* of the upper arm as shown in Figure 10-3. The *bicep muscle* connects to the radius. When the arm is extended, the bicep muscles will pull the radius strongly against the humerus. This can cause friction and heat in the joint. The bicep is both a flexor of the forearm and an outward rotator of the wrist. This can be seen by bending your arm 90° at the elbow and rotating your wrist outward. Notice how the bicep muscle contracts and bulges. Thus any movement that requires a strong pull and simultaneous inward rotation of the hand should be avoided. As pointed out by Tichauer and Gage (1977), good practical design can be observed in the operation of a corkscrew. A wine steward pulls hard at the cork while rotating the right forearm outward.

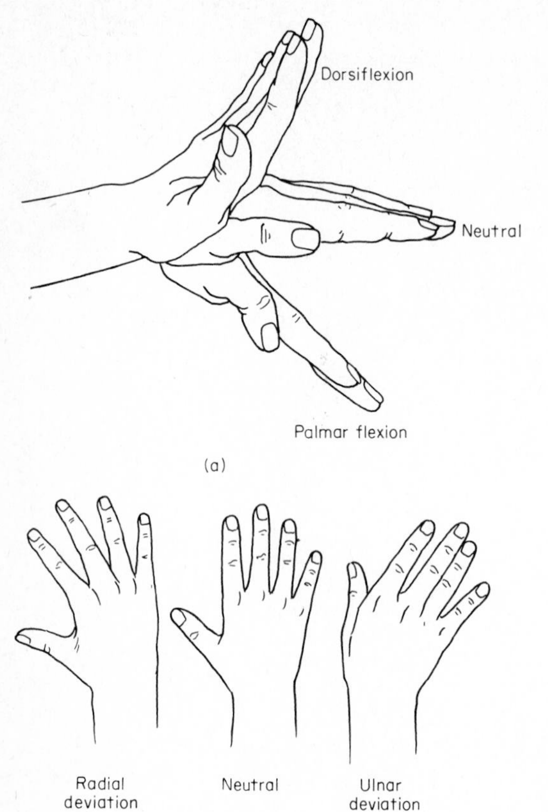

FIGURE 10-2
Movements of the wrist joint about two axes.

FIGURE 10-3
The elbow joint showing the connection of the bicep to the radius. (*Source: Tichauer, 1978, fig. 31.*)

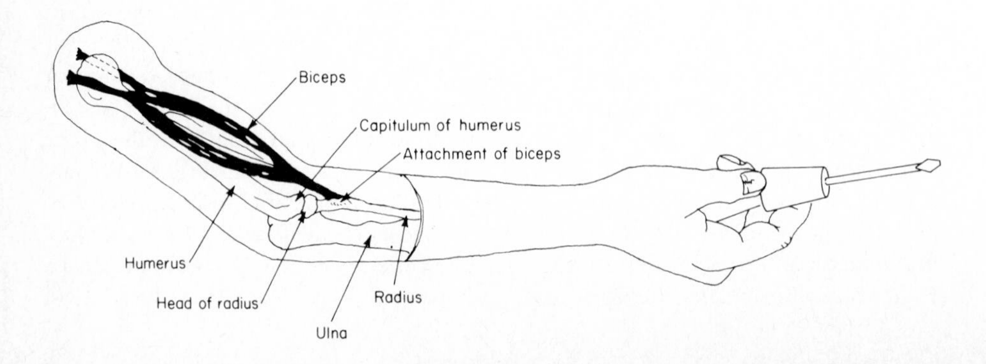

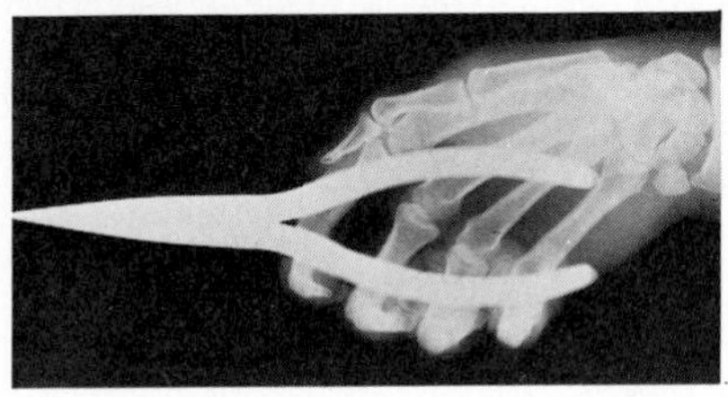

(*a*) Conventional design

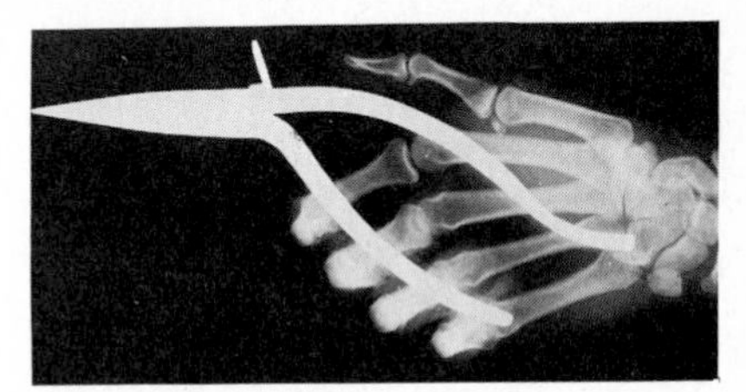

(*b*) Redesigned pliers

FIGURE 10-4
X-rays of hand using conventional pliers in a wiring operation, *a*, and in using a redesigned model, *b*. The redesigned model is more anatomically correct. (*Source: Damon, 1965; Tichauer, 1966; photographs courtesy of Western Electric Company, Kansas City.*)

PRINCIPLES OF HAND TOOL AND DEVICE DESIGN

It will not be possible to list or discuss all the many principles of hand tool design here. The major principles will be presented as they relate to the biomechanics of the human hand.

Maintain a Straight Wrist

The flexor tendons of the fingers pass through the carpal tunnel of the wrist. When the wrist is aligned with the forearm, all is well. However, if the wrist is bent, especially in palmar flexion or ulnar deviation (or both), problems occur. The tendons bend and bunch up in the carpal tunnel. Continued use will cause *tenosynovitis*, an inflammation of the tendon sheaths of the wrist. A common type of motion which can lead to tenosynovitis is "clothes wringing" (Tichauer, 1978), in which, for example, the wringing is done by a clockwise movement of the right fist and counterclockwise action of the left. This same type of motion is involved in inserting screws in holes, manipulating rotating controls such as found on steering handles of motorcycles, and looping wire while using pliers.

A key rule is to avoid ulnar deviation. The x-ray of a hand holding conventional pliers in Figure 10-4*a* shows a classic case of ulnar deviation. Redesigning the pliers, as shown in Figure 10-4*b*, so that the handles bend rather than the wrist, allows a more natural alignment of wrist and forearm. Tichauer (1976) reports on a comparison of two groups of 40 electronics assembly trainees, one using conventional straight pliers and the other using the redesigned bent pliers, during a 12-week training program. Figure 10-5 shows the incidence of tenosynovitis and "tennis elbow" in the two groups over the 12 weeks of training. There was a sharp increase in symptoms in the tenth and twelfth weeks for the group using the straight pliers while no such increase occurred in the group using the redesigned bent pliers.

The idea of bending the tool, not the wrist, has been applied to other things by John Bennett (Emanuel, Mills, & Bennett, 1980), who patented the idea of bent handles ($19° \pm 5°$) for all tools and sports equipment. Examples of this design are shown in Figure 10-6 for a push broom and for a hammer. Although Emanuel et al. (1980) do not report the results of any systematic evaluation of such handles, they report the experiences of several individuals and organizations that have used tools with such han-

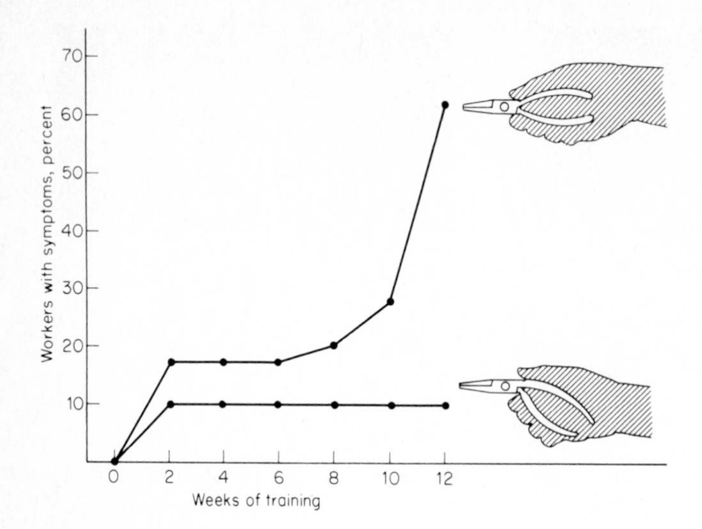

FIGURE 10-5
Comparison of two groups of trainees using different pliers. Shows percent of workers with tenosynovitis, epicondylitis (tennis elbow), and carpal tunnel irritation. (*Source: Adapted from Tichauer, 1976. Reprinted with permission from* Industrial Engineering *magazine, 1976, copyright American Institute of Industrial Engineers, Inc., 25 Technology Park/Atlanta, Norcross, GA 30092.*)

dles, all with positive results. The U.S. Forest Service, for example, has tested the concept on a set of 19 types of tools such as knives, axes, hoes, shovels, and shears, and preliminary results indicate some decrease in fatigue and in general a preference for the bent handle over the straight handle. The experiences of individuals with existing injuries or wrist problems suggest that the design might be particularly useful for certain handicapped persons. As an aside, a license to manufacture a softball bat with such a handle has been granted, and the bat is now legal for the game.

FIGURE 10-6
Examples of the *Bennett* handle that helps the user keep the wrist straight while using the tool. (*Source: Emanuel, Mills, & Bennett, 1980, fig. 1.*)

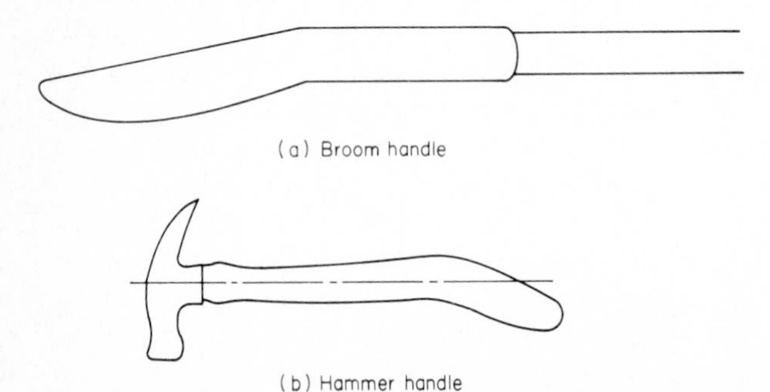

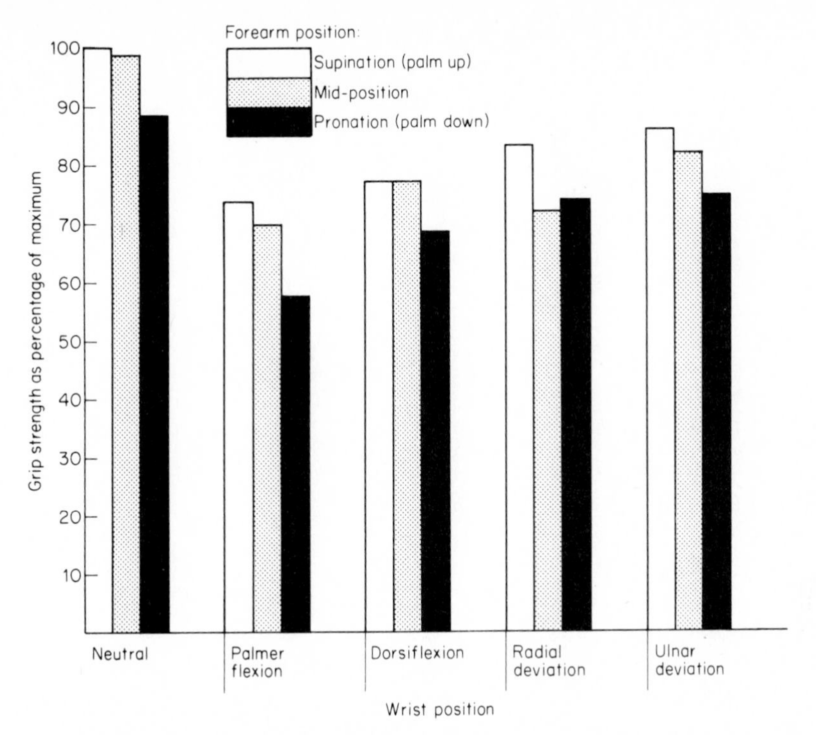

FIGURE 10-7
Grip strength as a function of wrist and forearm position. Grip strength is the average maximal grip sustained for three s, expressed as a percentage of neutral supinated grip strength. (*Source: based on data from Terrell & Purswell, 1976, table 1.*)

In addition to ulnar deviation, other types of wrist bending can also cause problems. Radial deviation, particularly if combined with pronation and dorsiflexion, increases pressure between the head of the radius and the capitulum of the humerus in the elbow (see Figure 10-3). This can lead to "tennis elbow" (*epicondylitis*). Many wire brushes have to be held in this way when being used overhead.

We have seen that ulnar deviation can cause tenosynovitis and that radial deviation can lead to tennis elbow. As if that were not enough, Terrell and Purswell (1976) report that grip strength is reduced if the wrist is bent in any direction. Figure 10-7 shows the grip strength as a percent of the maximum grip strength (achieved with the wrist in a neutral position with supinated forearm). Reductions in grip strength may increase the likelihood that the user will lose control of the tool or drop it, leading to an injury or poor-quality work. Attempts to maintain a strong grip will increase fatigue.

Figure 10-8 shows a man using a hacksaw in a manner that violates the straight-wrist principle. The man's right hand is in ulnar deviation while his left hand is in dorsiflexion with a touch of radial deviation. After a few hours of sawing like this, he will be in no shape to play the piano!

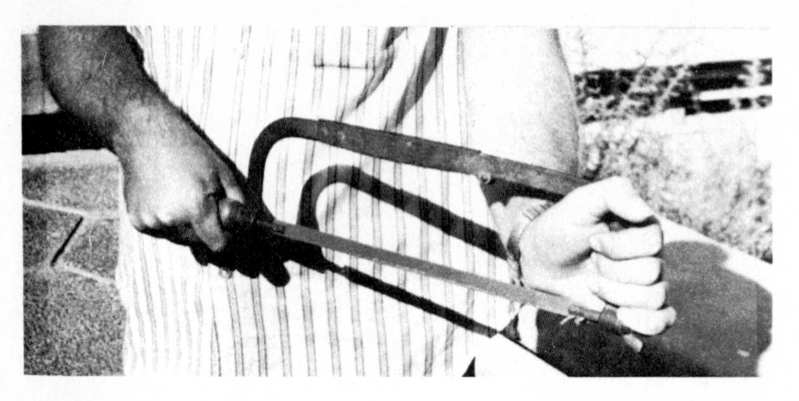

FIGURE 10-8
Man using a hacksaw and violating the straight wrist principle. The right hand is in ulnar deviation while the left hand is in dorsiflexion. (*Source: Greenberg & Chaffin, 1977, fig. 35.*)

Avoid Tissue Compression Stress

Often in the operation of a hand tool or device considerable force is applied with the hand as when squeezing pliers or scraping paint with a paint scraper. Such actions concentrate considerable compressive force on the palm of the hand. Particularly pressure-sensitive areas are those overlying critical blood vessels and nerves, specifically the ulnar and radial arteries. Figure 10-9*a* shows a hand holding a conventional paint scraper. The handle digs into the palm and obstructs blood flow through the ulnar artery. This obstruction of blood flow, or *ischemia*, leads to numbness and tingling of the fingers. Tichauer (1978) reports that affected workers will often take temporary breaks from work to relieve the symptoms. Thrombosis of the ulnar artery has also been reported (Tichauer, 1978).

If possible, handles should be designed to have large contact surfaces to distribute the force over a larger area, and to direct it to less-sensitive areas such as the tough tissue between thumb and index finger. Figure 10-9*b*, for example, shows an improved

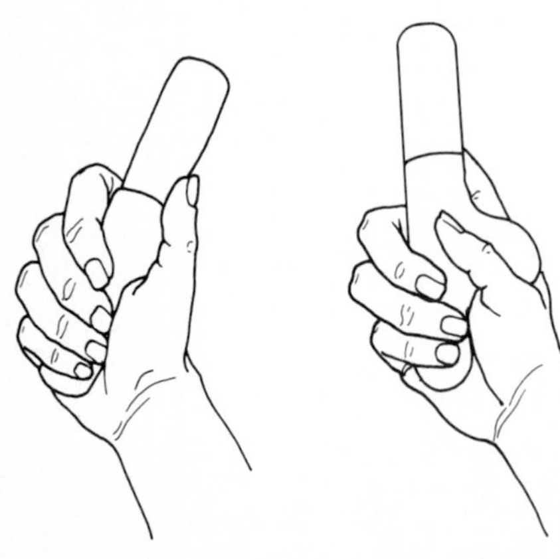

FIGURE 10-9
A conventional paint scraper that presses on the ulnar artery, and a modified handle which rests on the tough tissues between thumb and index finger, thus preventing pressure on the critical areas of the hand. (*Source: Tichauer, 1967.*)

paint scraper handle that rests on the tissue between thumb and finger, thus preventing pressure on the critical areas of the palm.

In a similar vein (no pun intended), the palm of the hand should never be used as a hammer. Not only will such action damage the arteries, nerves, and tendons of the hand, but the shock waves generated may travel to other body regions such as the elbow or shoulder.

Related to compression stress is the use of finger grooves on tool handles. As anyone who has ever watched a professional basketball game knows, hands come in a wide variety of sizes. A person with thick fingers using a tool with finger grooves often finds that the ridges of the grooves dig into the fingers. A small-handed person may put two fingers into one groove, thereby squeezing the fingers together. It is for this reason that Tichauer (1978) recommends not using deep finger grooves or recesses in tool handles if repetitive high finger forces are required.

Avoid Repetitive Finger Action

Occasionally, if the index finger is used excessively for operating triggers, a condition known as *trigger finger* develops. The afflicted person typically can flex but cannot extend the finger actively. It must be passively straightened, and when it is, an audible click may be heard. The condition seems to occur most frequently if the handle of the tool or device is so large that the distal phalanx (segment) of the finger has to be flexed while the middle phalanx must be kept straight (Tichauer, 1978).

As a rule, frequent use of the index finger should be avoided, and thumb-operated controls should be used. The thumb is the only finger that is flexed, abducted, and opposed by strong, short muscles located entirely within the palm of the hand. One must be careful, however, not to hyperextend the thumb such as shown in Figure 10-10*a*. This causes pain and inflammation. Preferable to thumb controls is the incorporation of a *finger-strip* control, as shown in Figure 10-10*b*, which allows several fingers to share the load and frees the thumb to grip and guide the tool.

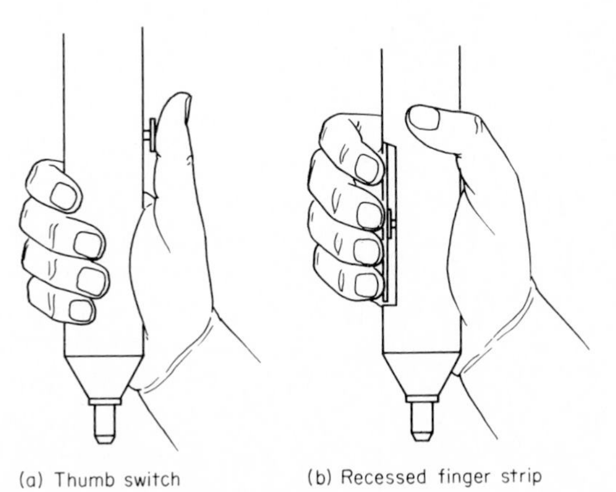

FIGURE 10-10
Thumb-operated and finger-strip-operated pneumatic tool. Thumb operation results in overextension of the thumb. Finger-strip control allows all the fingers to share the load and the thumb to grip and guide the tool.

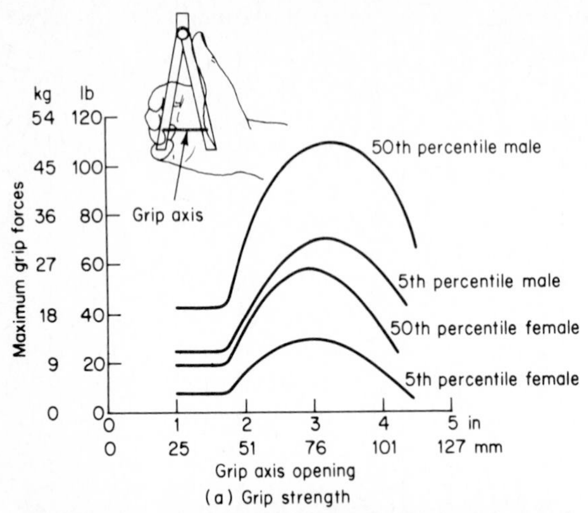

(a) Grip strength

(b) Pop-riveting gun

FIGURE 10-11
(*a*) Maximum grip strength for various handle openings. Subjects were 50 male and 50 female electronics manufacturing employees.
(*b*) A pop-riveting gun which to be operated must be grasped and squeezed in the fully open position. The outside edges of the handles are 6 in (150 mm) apart. (*Source: Greenberg & Chaffin, 1977, figs. 27 & 28.*)

The grip strength of the hand is related to the size of the object being gripped, as shown in Figure 10-11*a*. Maximum grip strength, for both males and females, occurs with a grip axis between 2.5 and 3.5 in (66 to 85 mm) (Greenberg & Chaffin, 1977). Figure 10-11*b*, however, shows a pop-riveting gun which must be gripped in the fully open position—a distance of 6 in (150 mm) between the outside edges—and squeezed. This is almost an impossible task for most women to perform.

Design for Safe Operation

Designing tools and devices for safe operation would include eliminating pinching hazards and sharp corners and edges. This can be done by putting guards over pinch points or stops to prevent handles from fully closing and pinching the palm of the hand. Sharp corners and edges can be rounded. Power tools such as saws and drills can be designed with brake devices so that when the trigger is released the blade or bit stops quickly. Proper placement of the power switch for quick operation can also reduce accidents with power tools. Each type of tool presents its own set of safety considerations. The designer must consider, in detail, how the tool will be used by the operator, and also how it is likely to be *misused* by the user.

Remember Women and Left-Handers

Women make up approximately 50 percent and left-handers make up approximately 8 to 10 percent of the world population (Barsley, 1970). Many hand tools and devices are not designed to accommodate these populations. Ducharme (1977) reports that in the Air Force the average hand length of women is almost 2 cm (0.8 in) shorter than that of the average male. Less than 1 percent of Air Force men have a hand that is as short as the average woman's hand. Further, grip strength of women is on the average only about two-thirds that of men (Konz, 1979). These differences obviously have implications for tool design.

Ducharme surveyed 1400 Air Force women working in the craft skills and asked them to rate the adequacy of the tools they used. Table 10-1 summarizes some of the major complaints and the tools that engendered them. Only those tools that were judged inadequate by at least 10 percent of the women in one or more craft fields are included. The percentages of women in each field who considered each tool inadequate is also shown in Table 10-1. From scanning the complaints it is obvious that almost all reported inadequacies can be traced to the fact that most women have smaller hands and less grip strength than most men. With women assuming a greater role in traditionally male-dominated occupations, the design of hand tools and devices must

TABLE 10-1
TOOLS RATED INADEQUATE BY AIR FORCE FEMALES WORKING IN THE CRAFT SKILLS AND SOME OF THE REASONS GIVEN FOR THEIR INADEQUACY

Tool	Rated inadequate because	Percentage rating tool as inadequate
Wire strippers	Hard to hold in hand; handles too far apart to squeeze; too heavy; clumsy; too hard to squeeze; fingers get pinched	12[a]; 18[b]; 19[c]; 15[d]; 18[e]
Crimping tool	Handles too far apart; too hard to squeeze; not able to manipulate	14[a]; 14[c]; 25[d]
Soldering iron	Too heavy; handle too large; clumsy, too bulky	17[a]; 15[c]
Soldering gun	Too heavy; can't reach trigger; hard to hold in hand	15[a]
Twist wire pliers	Too large to grip; handles too far apart	29[b]
Metal shears	Too large; need two hands to cut	22[f]
Rivet cutter	Too hard to squeeze; awkward	17[f]
Caulking gun	Hard trigger; awkward	11[g]

[a] Communications-electronic systems field.
[b] Missile-electronics maintenance field.
[c] Avionics systems field.
[d] Aircraft accessory maintenance field.
[e] Mechanical/electrical field.
[f] Metalworking field.
[g] Structural/pavements field.

Source: Ducharme, 1977.

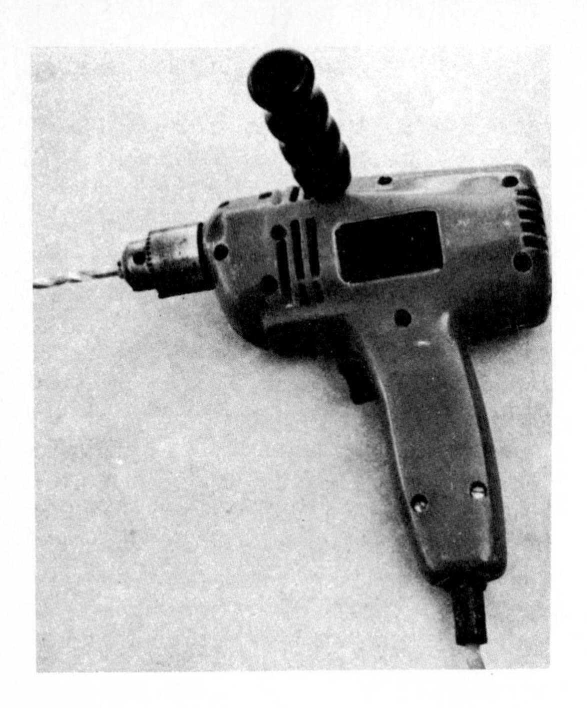

FIGURE 10-12
Drill with handle designed for right-handed operation. (*Source: Greenberg & Chaffin, 1977, fig. 25.*)

take into consideration the anthropometric and ergonomic differences that exist between men and women.

Another neglected group is left-handers. Actually, there are degrees of handedness from strong right-handed through ambidextrous to strong left-handed. Tools should be designed so that they can be used in the operator's preferred hand. Often, designs preclude the use of tools by left-handers. For example, the drill shown in Figure 10-12 requires left-handers to operate the drill with the right hand. If a threaded fastener were provided in the right side of the drill housing, the handle could be moved for left-handed use.

Laveson and Meyer (1976) point out that some serrated knives have the beveled (cutting) edge on the left side. The left-hander presses down and to the left; the beveled edge then fails to cut, since the pressure is on the wrong side. A serrated bread knife, when used by a left-hander, will often tear rather than cut the bread. The difficulty could be corrected by double-edge beveling.

VIBRATION

Typically in the human factors literature the topic of *vibration* refers to whole-body vibration such as heavy equipment operators experience as they drive over a bumpy road. We will discuss this type of vibration in Chapter 16. Our focus here, however, is with hand vibration induced by such power tools as chain saws, pneumatic drills, grinding tools, and chipping hammers.

Vibration-Induced White Finger

In the early 1900s reports began to appear that linked hand-held vibratory tools with "vascular spasm" in the hands (Taylor, 1974). Since then numerous other reports and investigations have been published on what has become known as *vibration-induced white finger* (VWF).[1]

VWF is a complex phenomenon whose exact pathology is not fully understood. The primary symptom is a reduction in blood flow to the fingers and hand. This is caused by the smooth muscles of the blood vessels in the hand and fingers constricting and thereby reducing the flow of blood. Workers afflicted with VWF will have *vascular attacks* wherein the hand or fingers *blanch* (turn white). The feeling is the same as when your hand or foot "falls asleep." The person complains of "pins and needles" (i.e., tingling), numbness, or pain. These attacks seem to be brought on by cold. They are especially prevalent among workers who use vibratory tools in cold environments. The attacks can last minutes or hours.

Workers afflicted with VWF appear to have reduced blood flow in the hands under normal conditions in addition to experiencing vascular attacks. Koradecka (1977), for example, examined pneumatic drill operators, manual grinder operators, and control groups working under similar microclimatic conditions but not exposed to vibration. He found that the exposed groups exhibited a 38 to 67 percent mean reduction in hand skin blood flow compared with control subjects.

One consequence of this reduced blood flow is a reduction in skin temperature. Figure 10-13 shows the mean finger skin temperature of vibration-exposed and control subjects. The vibration-exposed group starts out with a lower skin temperature but when the hand is heated (10 min in 113°F or 45°C water), both groups attain the same temperature. Cooling the hands, however, restores the initial difference which remains throughout recovery. This lower skin temperature is often associated with a decrease in sensitivity and fine finger dexterity, and a loss of grip.

Although most cases of VWF are not debilitating, advanced cases have been known to lead to gangrene of the fingertips (Walton, 1974). The prevalence of VWF, of course, varies with the type and amount of exposure and other individual differences. Not everyone exposed to hand-held vibratory tools develops VWF; some persons, however, develop the symptoms in only a few months (Leonida, 1977), while others do so only after years of exposure.

Vibration induced from hand-held tools has also been implicated in numerous other diseases (Williams, 1975), including neuritis and decalcification and cysts of the radial and ulnar bones. Together, all the various symptoms and diseases including VWF associated with hand-held vibratory tools are collectively called the *vibration syndrome* (Taylor, 1974).

[1] Several other names have also been used in the literature: *Raynaud's phenomenon of occupational origin, dead hand, dead finger, white finger, occupational vasomotor traumatic neurosis,* and *traumatic vasospastic disease* (TVD).

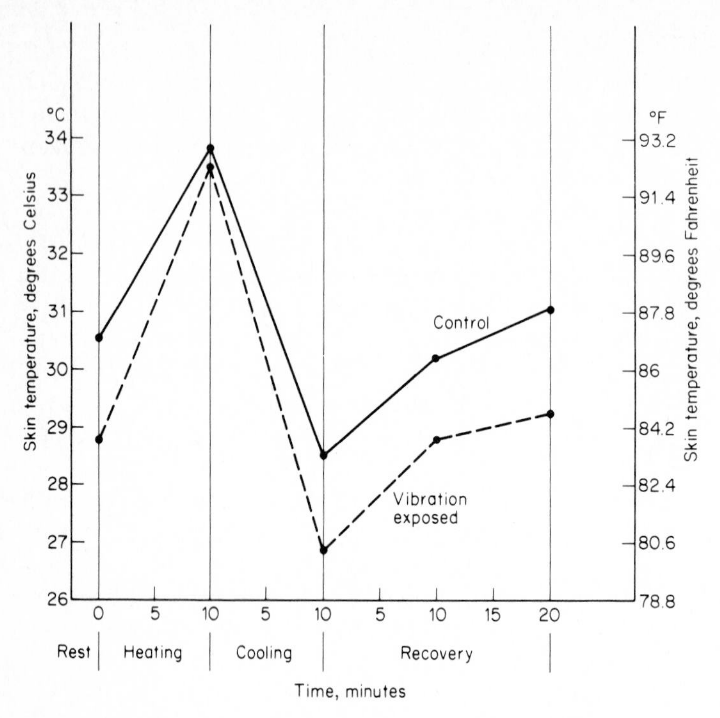

FIGURE 10-13
Skin temperature of the fourth finger of the left hand during thermal testing of a group of workers occupationally exposed to hand vibration and a control group not exposed. (*Source: Koradecks, 1977, fig. 13.*)

A Question of Standards

A number of proposals recommend safe exposure limits to hand-transmitted vibration. Figure 10-14 shows six such proposals for maximum vibration levels for continuous daily exposure. Clearly, the variation between them is so great that the use of any one of them should be done with caution. The basis of these proposals is almost entirely that of subjective assessment of what vibration levels are "noticeable," "unpleasant," etc.

The International Standards Organization (ISO) has developed a set of guidelines which are intended to take into account duration of exposure and interruptions in exposure. Hempstock and O'Connor (1977), however, state that there is only limited medical and epidemiological data supporting the ISO standards. Further, as accuracy of vibration measurement is still a problem, the proposal should not be taken too literally.

The British Standards Institute has taken a more simplified approach to setting hand-induced vibration standards. Only the total duration of exposure to vibration is considered important, and interruptions or irregularities in exposure are not quantified in any way. The two curves that form the basis of the recommended British standard

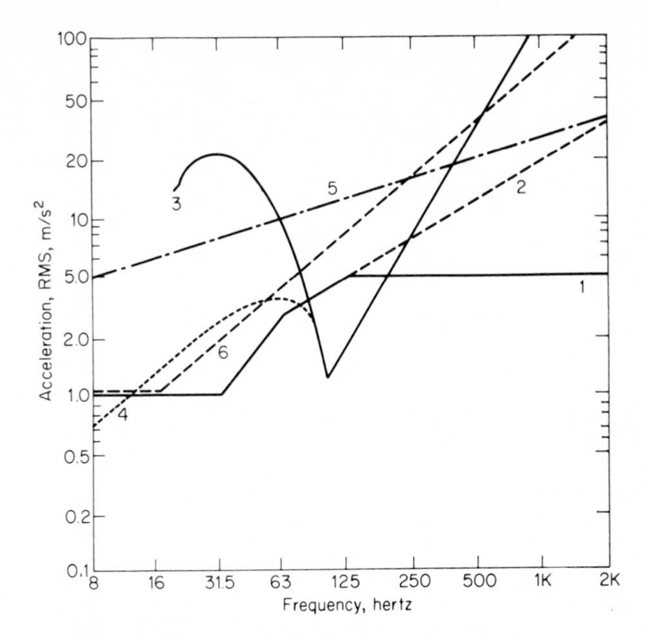

FIGURE 10-14
Maximum vibration levels for continuous daily exposure: (1) Czechoslovakian Hygiene Regulation No. 33; (2) Teisinger and Louda, Czechoslovakia 1966; (3) USSR Hygiene Regulation 1955; (4) USSR Gataninas Proposal after 1955; (5) USSR Hygiene Regulation 1966; (6) Draft British Standard 1975 400 min/d. (*Source: Hempstock & O'Connor, 1977, fig. 1.*)

are presented in Figure 10-15. The 400 min curve is the recommended limit for continuous daily exposure and the 150 min curve is the proposed limit for interrupted daily exposure for any vibration repeated regularly. The British standard is somewhat at odds with the ISO procedure especially if the exposure is interrupted.

It is not possible at the present time to resolve which, if either, of the two approaches is more accurate. Hempstock and O'Connor, however, suggest that the assessment of risk would be easier to evaluate if the energy input into the hand were measured rather than the vibration on the tool itself. This suggestion assumes that equal amounts of energy cause equal amounts of damage. Given the current state of affairs we should probably accept the standards for what they are—tentative recommendations offering provisional guidelines rather than definite damage risk criteria.

In 1971 the National Swedish Board of Occupational Safety and Health set maximum permissible vibration levels for chain saws sold in Sweden. The maximum permissible levels were decreased each year from a maximum 80 newtons (80 N) in 1971 to 50 N in 1973.[2] Actually manufacturers were able to reduce the levels below 30 N by improved design and use of vibration-damping rubber elements between the body of

[2] A *newton* is a unit of force.

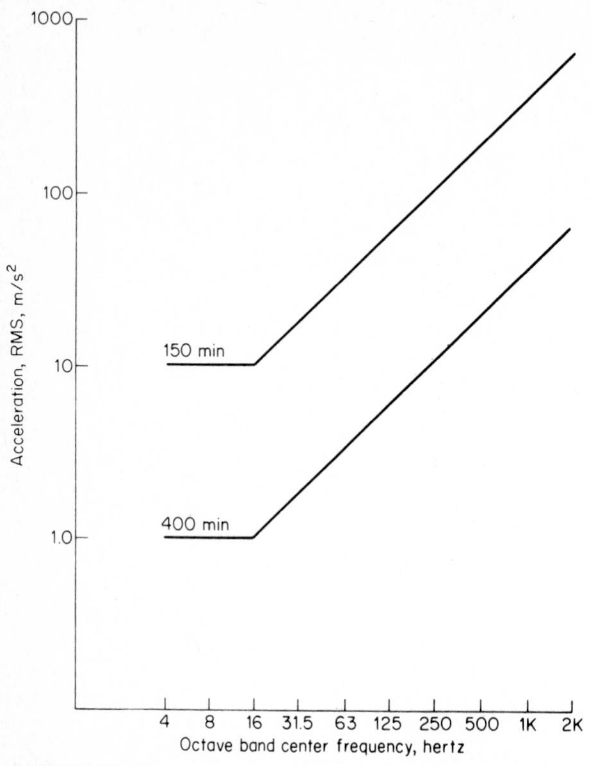

FIGURE 10-15
Recommended exposure limit curves for hand-transmitted vibration. (*Source: British Standards Institution Draft DD43.*)

the saw and the handles (Axelsson, 1977). The percentage of chain saw operators in northern Sweden with VWF decreased from 49 percent in 1967 to 28 percent in 1974. Of the 28 percent in 1974, 19 percent showed significant improvement in the severity of VWF (Axelsson). Although other factors, such as better heating of workers' cabins and warmer working clothes, probably contributed to this reduction, Axelsson and others (Taylor, Pearson, & Keighley, 1977) believe that the reduction in the vibration level of the saws was the principal cause for the reduced incidence of VWF.

GLOVES

A chapter on hand tools and devices would not be complete without at least a brief discussion of gloves. Gloves are often used in conjunction with hand tools for protection against abrasions, cuts, punctures, and temperature extremes. Gloves come in an amazing number of varieties. In general, however, they can be distinguished in terms of the material used for construction (e.g., cotton, leather, vinyl, neoprene, asbestos, and even metal); the cut (i.e., gunn cut and clute cut as shown in Figure 10-16); the design of the thumb (i.e., straight or wing as shown in Figure 10-16); and the type of wristband (e.g., knit, band top, gauntlet, or extended length).

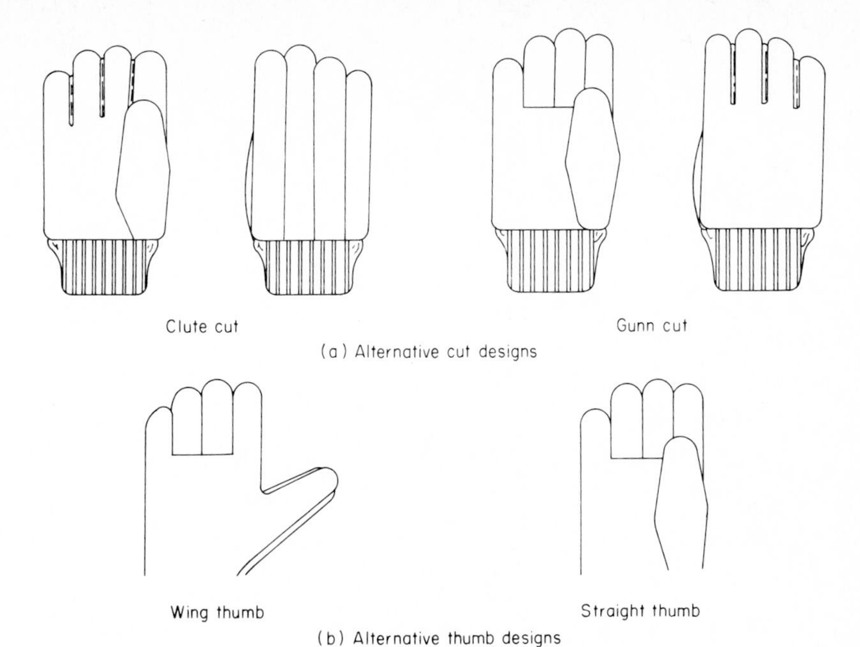

FIGURE 10-16
Common styles of work gloves distinguished by cut and thumb design.

Effect on Manual Performance

All the research on the effects of wearing gloves on manual performance have concentrated on comparing gloves of different materials. No studies exist which evaluate glove design independent of the material. We do not know, therefore, whether there is any difference in performance while wearing gunn or clute cut gloves.

Research on the effects of various glove materials on manual performance indicates that the nature of the task influences the relative superiority of the various materials. On some tasks there will be no differences in performance wearing different gloves, while on other tasks large differences will be found. In general, however, performance on tasks requiring fine motor control and tactile feedback will be adversely affected by wearing gloves, as compared to performance while bare-handed. An exception to this was noted in Chapter 5 wherein surface irregularities can be detected more accurately wearing thin cloth gloves than with bare fingers.

A study by Weidman (1970) is a good example of research in this area. Weidman compared four gloves [leather with canvas back, heavy-duty terry cloth, neoprene, and polyvinyl chloride (PVC)] and a bare-hand condition on performance of a maintenance-type task. The entire task was divided into five subtasks:

1 Using a key on a key ring to open a lock to get into a box
2 Replacing 16 television tubes by hand
3 Using a socket wrench and open-end wrench to remove 20 bolts
4 Using a screwdriver to remove four screws and replace two tubes
5 Using a trigger-type solder gun to remove two leads

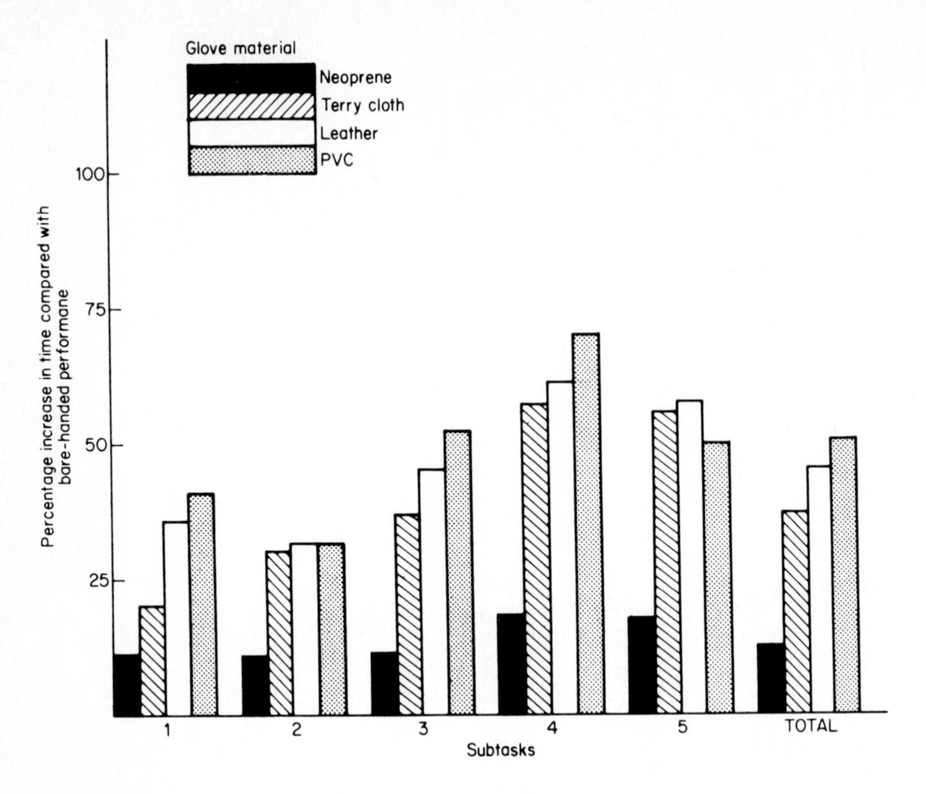

FIGURE 10-17
Performance (time to complete) on a maintenance-type task while wearing gloves constructed of five different materials. Performance is shown as percentage increase compared to bare-handed performance. (See text for a description of the subtasks.) (*Source: Weidman, 1970.*)

The results are shown in Figure 10-17 as percentage increase in task completion times compared to the bare-hand condition. For subtasks 1 and 5 there was no statistically significant decrement, or differential decrement, caused by any type of glove. This was caused by the overall short time required to complete these subtasks (less than 50 s). For subtasks 2, 3, and 4, and total time, the results showed that performance did not differ between neoprene and bare-handed. Leather, terry cloth, and PVC, however, were consistently inferior to either neoprene or bare hands, but did not differ from each other.

Other Considerations

Tichauer (1978) indicates that if a work glove is too thick between the fingers it may cause a worker to grasp a tool with insufficient force, resulting in the tool slipping out of the hand. In addition, he points out, irritants entrapped unintentionally in the glove may work into the skin, producing in some cases benign tumors. Chemicals might be soaked up in the glove material and cause dermatitides or other tissue damage.

These potential hazards speak to a need for careful evaluation of the work environment and the selection of suitable gloves. Knitted wristbands can reduce the intrusion of potential irritants. Neoprene- or PVC-coated gloves can reduce absorption of chemicals. The selection of proper glove material must also take into account the type of mechanical hazards (cuts, punctures, abrasions) and thermal hazards likely to be encountered on the job. Some materials are better suited to handle certain hazards than are others (Coletta, Arons, Ashley, & Drennan, 1976). Gloves must be given as careful thought as should be given the design or selection of the hand tools to be used.

EVALUATION OF ALTERNATIVE DESIGNS

In recent years there has been an increase in the application of human factors to hand tool and device design. In this section we would like to present a few examples. The emphasis will not be on the innovation per se, but rather on the methodology used to develop and evaluate the new designs. In Chapter 2 we discussed laboratory and real-life research as well as types of criteria that can be used to measure possible "effects" of new designs. In several of the examples to follow, combinations of laboratory and real-life research methods were employed to provide a comprehensive evaluation of alternative designs. We will try to illustrate the kinds of information sought and the insights provided by the various methodologies employed.

Toothbrush Design

Guilfoyle (1977) provides an informative description of the human factors development and evaluation of what ultimately became the *Reach* toothbrush (trademarked name) shown in Figure 10-18. Although maybe not the most sophisticated of tools, the toothbrush is probably one of the most widely used hand-held tools in the world. Since the first bone shaft and hog hair bristle toothbrush in 1780, the only major development had been the introduction of nylon filament bristles in the 1930s.

DuPont approached Applied Ergonomics, Inc., to design an improved toothbrush. The first step was to review relevant literature and conduct a series of consultations with dentists. From this preliminary investigation they discovered that no human factors research had been applied to toothbrush design and that the primary objective of a toothbrush was removal of plaque from the teeth. A secondary objective was gingival (gum) massage. Since there was little available literature on the public's dental care habits, a questionnaire was distributed to 300 adults. The information obtained helped the designers focus on basic dental care problems.

After analyzing the results of the questionnaire, the design team began collecting basic physical dimensions on existing toothbrushes and on the anthropometric characteristics of consumers (measurements of hands, teeth, and mouths). To obtain more information on how users handle a toothbrush, a series of time-motion film studies were made of people brushing their teeth. The design team studied how much time was devoted to brushing different mouth areas and the stroke directions used. The team also examined the way people manipulated the brush.

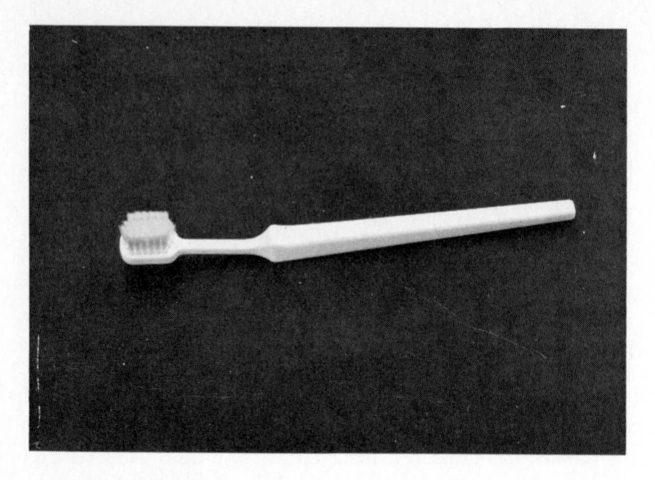

FIGURE 10-18
Reach toothbrush designed by Applied Ergonomics, Inc., and manufactured by Johnson and Johnson. Innovative features include a small bristle head and an angled (12°) and shaped handle with contoured thumb rest area.

The time-motion studies made it clear that a contoured grip was undesirable because it hindered hand movements. A round handle was briefly considered but was dropped because it would not resist rotational forces. Another reason the round handle was dropped illustrates the need to consider the entire system into which a tool or device will be placed. In this case, the diameter of the handle would have to be small enough to fit into the standard bathroom toothbrush holder. To do so would have made the handle too weak.

In addition to the time-motion studies, other laboratory studies were done relating brushing time and bristle diameter to plaque removal. Based on the information obtained from all the varied sources, two prototype toothbrushes were designed and a sufficient number manufactured for testing. Both prototypes had individual characteristics that required testing.

These two prototypes were then tested against two other commercially available toothbrushes for removing plaque. Both prototypes proved superior. Interviews with the test subjects indicated a preference for the bristle head of one prototype and the handle design of the other. The features were blended into the final product. Plaque removal tests with the final design demonstrated superior removal compared to other conventionally designed brushes. Some of the features incorporated into the Reach toothbrush include a small bilevel bristle head to concentrate brushing on a small area, an angled (12°) and shaped handle for easier manipulation, and a contoured thumb area which makes brushing easier. As an aside, your authors have both tried the Reach toothbrush and have found that it does take a little practice to learn to use it effectively.

This design and evaluation process illustrates the use of past literature, expert (dentist) opinions, questionnaire data, objective laboratory studies (time-motion and plaque-removal studies), and user opinion data to arrive at an improved tool design.

Multiple-Function Dental Syringe

Those of us who perhaps have not properly used our toothbrushes may have been on the receiving end of this tool. When a dentist fills a cavity the dental assistant often works alongside evacuating (with a suction vacuum) tooth debris being drilled by the dentist, as well as providing air, water, and an air-water spray. Currently two instruments are used, a three-way air, water, and spray syringe, and a separate suction device as shown in Figure 10-19*a*. Evans, Lucaccini, Hazell, and Lucas (1973) reasoned that reducing the number of instruments handled by the dental assistant by combining the suction and three-way syringe into a single four-way device would (1) free a second hand to assist the dentist, (2) reduce the number of instruments in the mouth, (3) increase visibility, and (4) reduce patient discomfort.

The four-way device is shown in Figure 10-19*b*. The air and water streams are each operated by depressing small spring-loaded levers with the thumb. Spray is achieved by activating both levers simultaneously. Suction is controlled by rotating thumb wheels located to the sides of the levers. The thumb wheels are joined by a common shaft, providing duplicate control functions to the operator, thus accommodating left-handed as well as right-handed operators—a plus. The tip has a 45° angle bend and rotates to provide access to difficult-to-reach areas of the mouth.

The evaluation of the new instrument consisted of mechanical performance tests, mock clinical tests (simulated operations), and actual field tests (Evans et al., 1973). We will briefly review each of these to illustrate the criteria used and the insights gained.

The mechanical performance tests were conducted to determine whether the device satisfied basic requirements for air, water, and suction pressure.

The mock (or simulated) clinical trials consisted of conducting restorative dental operations on a dental manikin instead of a live patient. The purpose was to be sure that the instrument would function satisfactorily and safely before proceeding to the live field tests. In addition, the reliability of the device under sustained use was evaluated.

Field tests were conducted to compare the three-way plus suction system with the new four-way device by actual dentists and dental assistants working on live patients. At each of five locations, four days of training in the use of the new device were given. On the fifth day one operation was done with the conventional system, and one operation was performed using the four-way device. Videotape and motion picture records were taken during each operation. The number of hand movements was counted; time to complete the cutting portion of the operation was recorded; energy costs for the dental assistant were calculated from samples of expired air; and finally operator preferences and suggestions were obtained. Here we see again the use of multiple criteria for evaluation, and again the inclusion of subjective user preference data. The results shown in Table 10-2 were found.

The interviews with dentists and dental assistants showed a highly favorable attitude toward the new device, but several suggestions for redesign were made, including adding a plastic cover to the tip end to protect against chipping tooth enamel; reducing the size and length of the handle to increase ease of handling (especially for female users); replacing the "on-off" action control levers with variable controls; changing the

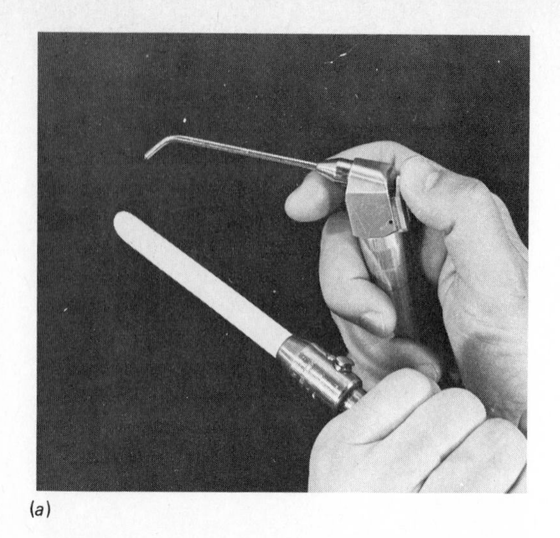

(*a*)

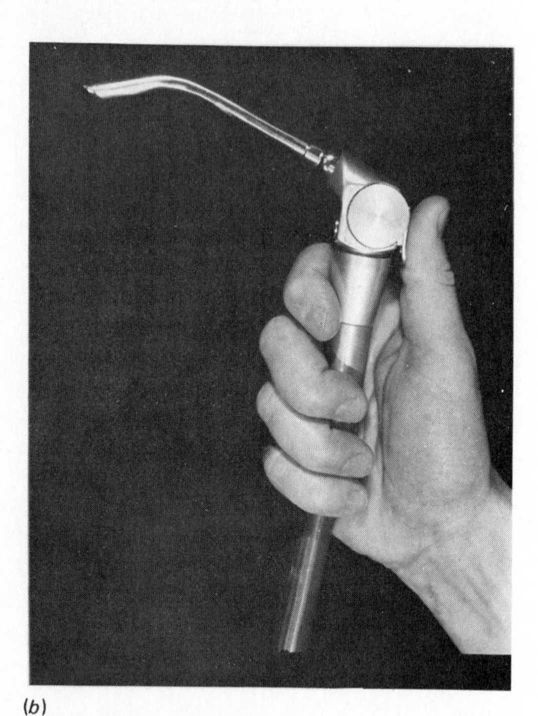

(*b*)

FIGURE 10-19
(*a*) The conventional three-way syringe (air, water, and spray) with separate suction used in oral dentistry. (*b*) The four-way (air, water, spray, and suction) device designed and evaluated by Evans et al. (*Source: Evans, Lucaccini, Hazell, & Lucas, 1973, figs. 1 & 3.*)

TABLE 10-2
COMPARISON OF TWO MULTIPLE-FUNCTION DENTAL SYRINGES

Criteria	Four-way	Three-way plus suction
Hand movements	42	65
Cutting time, s	84	91
Energy expenditure, kcal/(min)(kg)	0.023	0.023

Source: Evans, Lucaccini, Hazell, & Lucas, 1973.

direction of lever activation from downward to upward to conform to thumb motion; and separating and shape-coding the control levers to minimize inadvertent activation. These kinds of suggestions attest to the value of user inputs in system design and evaluation.

Writing Instruments

Kao, in a series of studies, evaluated various types of writing instruments (1976, 1979) and alternative designs (1977). In two studies, ball-point pens, fountain pens, and pencils were compared on a number of criteria. An attitudinal survey of users (Kao, 1976) showed that ball-point pens were most preferred, followed in order by pencils, and finally fountain pens. The survey addressed several subjective perceptions of the users including legibility, writing ease, and writing speed. In 1979, Kao followed up his subjective evaluation with laboratory tests. The dependent variables were objective measures of writing speed and writing pressure. He included felt-tip pens in the evaluation.

The results are important both for practical reasons (e.g., whether to invest $10 in a fountain pen or pick up a $1.98 ball-point pen) and as an illustration of trade-offs that we often discover when evaluating alternative designs using multiple criteria. The results are shown in Figure 10-20. The time measure represents the time required to write 10 lowercase *a*'s. The results indicate that for faster writing time the ball-point pen is the best instrument and the fountain pen the worst. On the other hand, for more comfortable and less fatiguing handwriting performance, the felt-tip is clearly the best and the ball-point is now the worst. So here we have a trade-off: speed versus fatigue. Given the relatively small differences in time between the instruments and large differences in pressure, Kao (1979) concludes that probably the felt-tip pen is the best "all-round" type of instrument for handwriting.

FIGURE 10-20
Comparison of writing speed and writing pressure with four writing instruments. Time represents the time required to write 10 lowercase *a*'s. (*Source: Kao, 1979, figs. 1 & 2.*)

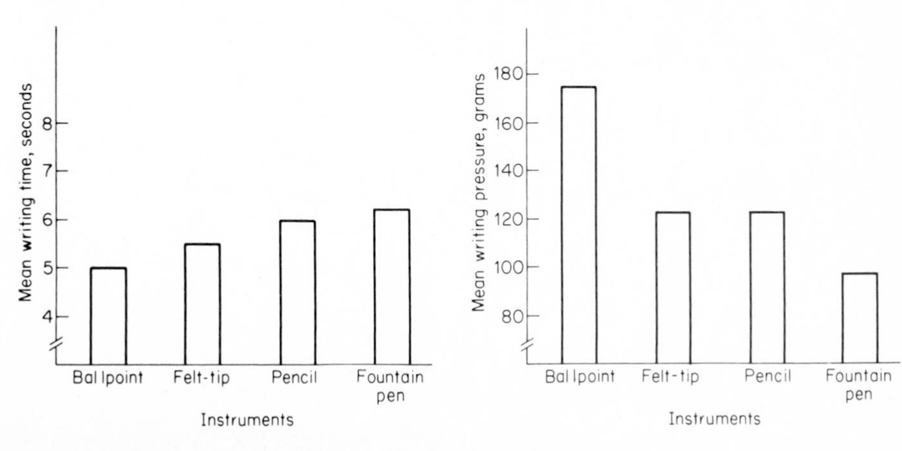

Some people might think that finding trade-offs among the designs being evaluated would make the evaluator's life more difficult–and it does. The consolation is knowing that a more complete understanding of the pros and cons of each design was achieved by using multiple criteria. If Kao had assessed only writing time, he might have erroneously concluded that the ball-point pen was the most efficient instrument.

An Improved Surgical Instrument

Some problems in the use of the conventional design of bayonet forceps as used by neurosurgeons led to the development and evaluation of a new design (Miller, Ransohoff, & Tichauer, 1971). The original and new designs are shown in Figure 10-21. The problems with the original design included a tendency for it to roll in the fingers while in use, and for fatigue to develop in the flexor muscles of the fingers and thumb. The new design was intended to overcome these disadvantages. The two designs were evaluated in use by four experienced surgeons, each instrument being tested during approximately 25 procedures. Measures were taken of stabilizing time, i.e., the time to grasp the instrument securely, and electromyogram recordings of the flexor muscles of the finger and of the opposer muscles of the thumb were taken. An example of electromyograms recorded while using the original and improved design is shown in Figure 10-21.

The results provide a good contrast with those reported in the previous example. In this case, the new design proved superior on all measures; stabilizing time was reduced by about 25 percent, and the work loads on the fingers and thumbs were reduced by about 38 and 42 percent, respectively.

Padded Gloves

Usually in designing a glove there is a trade-off between protection and manual dexterity–the more protection the less dexterity. Krohn et al. (1979), however, sought to provide added protection without decreasing dexterity. The application was somewhat unusual–low seam coal mines (under 48 in or 122 cm in vertical height). People work-

FIGURE 10-21
Examples of electromyograms of the finger flexor muscle during the "grasp" phase of using an original and an improved design of a bayonet forceps as used by neurosurgeons. (*Source: Adapted from Miller, Ransohoff, & Tichauer, 1971, figs. 1 & 6.*)

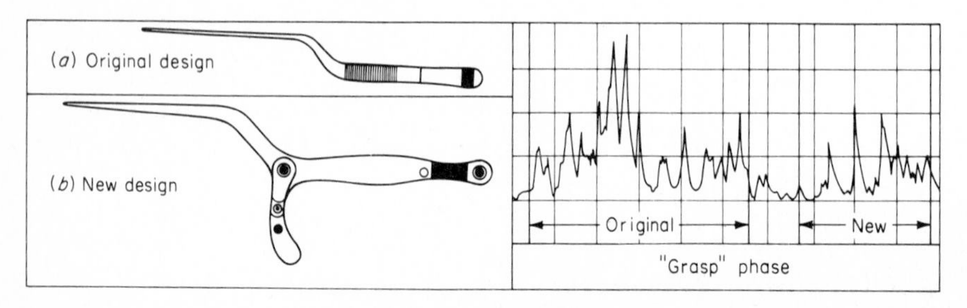

ing in such environments can crawl only on their hands and knees. Most wear common leather gloves which do not give much protection against sharp rubble on the floor. Some hold small blocks of wood in their hands for protection while crawling.

Krohn et al. (1979) altered a pair of common leather gloves by attaching an additional pad of leather to the palm of the glove. The padded gloves were compared with standard unpadded gloves. Once again we see the use of multiple measures in a variety of research settings. Laboratory tasks included a manual dexterity test (assembling nuts and bolts) and a measure of protection afforded by the gloves. For the protection test, subjects assumed a crawling posture and cylindrical rods were slowly raised under the glove until the subject indicated that it was "uncomfortable." The height of the rods above the surface served as the measure of protection.

In addition to laboratory tests, subjects wore the gloves and performed typical manual mining tasks (crawling, shoveling, timber setting, and pushing a weighted object) in a low seam mine simulator. The results indicated that adding the padding did increase the protection afforded (the rods could be raised 41 percent higher wearing the padded gloves); in addition, there were no significant losses of dexterity on the laboratory test or any of the simulator tasks.

DISCUSSION

In bygone centuries people used various tools and hand devices to accomplish certain objectives, such as making things or performing certain tasks, including clubbing one's attackers. Various types of hand tools and other hand-held devices still serve—and will continue to serve—many purposes. In the future we can look forward to new hand tools designed for specialized applications. To be of real value these new tools must be designed with full cognizance of the operational system (and its limitations and constraints) into which the device will be placed. In addition, the designer must consider the human factors, ergonomic, and biomechanical aspects of the user-tool interface.

The evaluation of new tool designs will require creativity to devise methodologies for reliable and valid measurement of relevant criteria. Combinations of various research strategies, including laboratory, simulation, and "real-life" field testing, will be required. Although hand tools have been around for a million years, there is still considerable work left in order to adapt them optimally for human use.

REFERENCES

Axelsson, S. Progress in solving the problem of hand-arm vibration for chain saw operators in Sweden, 1967 to date. In D. Wasserman & W. Taylor (Eds.), *Proceedings of the International Occupational Hand-Arm Vibration Conference* (No. 77-170). Cincinnati, Ohio: Dept. of Health, Education and Welfare, National Institute for Occupational Safety and Health, 1977.

Ayoub, M., Purswell, J., & Hicks, J. Data collection for hand tool injury: An approach. In V. Pezoldt (Ed.), *Rare event/accident research methodology.* Washington, D. C.: National Bureau of Standards, 1977.

Barsley, M. *Left-handed man in a right-handed world.* London: Pitman, 1970.

British Standards Institution. *Guide for the evaluation of human exposure to hand-arm system vibration.* British Standards Institution Draft for Development DD43, 1975.

Coletta, G., Arons, I., Ashley, L., & Drennan, A. *The development of criteria for fire-fighter's gloves* (Vol. I: Glove requirements) (No. 77-134-A). Cincinnati, Ohio: Department of Health, Education and Welfare, National Institute for Occupational Safety and Health, 1976.

Damon, F. The use of biomechanics in manufacturing operations. *The Western Electric Engineer,* 1965, *9*(4).

Ducharme, R. Women workers rate "male" tools inadequate. *Human Factors Society Bulletin,* 1977, *20*(4), 1–2.

Emanuel, J., Mills, S., & Bennett, J. In search of a better handle. *Proceedings of the symposium: Human factors and industrial design in consumer products,* Medford, Mass.: Tufts University, 1980.

Evans, T., Lucaccini, L., Hazell, J., & Lucas, R. Evaluation of dental hand instruments. *Human Factors,* 1973, *15,* 401–406.

Greenberg, L., & Chaffin, D. *Workers and their tools.* Midland, Mich.: Pendell Publishing Co., 1977.

Guilfoyle, J. Look what design has done for the toothbrush. *Industrial Design,* 1977, *24,* 34–38.

Hempstock, T., & O'Connor, D. Evaluation of human exposure to hand-transmitted vibration. In D. Wasserman & W. Taylor (Eds.), *Proceedings of the International Occupational Hand-Arm Vibration Conference* (No. 77-170). Cincinnati, Ohio: Department of Health, Education, and Welfare, National Institute for Occupational Safety and Health, 1977.

International Standards Organization. *Draft proposal for guide for the measurement and evaluation of human exposure to vibration transmitted to the hand* (ISO/TC108/SC4). September 1975.

Kao, H. An analysis of user preference toward handwriting instruments. *Perceptual and Motor Skills,* 1976, *43,* 522.

Kao, H. Ergonomics in penpoint design. *Acta Psychologica Taiwanica,* 1977, *18,* 49–52.

Kao, H. Differential effects of writing instruments on handwriting performance. *Acta Psychologica Taiwanica,* 1979, *21,* 9–13.

Konz, S. *Work design.* Columbus, Ohio: Grid Inc., 1979.

Koradecka, D. Peripheral blood circulation under the influence of occupational exposure to hand-transmitted vibration. In D. Wasserman and W. Taylor (Eds.), *Proceedings of the International Occupational Hand-Arm Vibration Conference* (No. 77-170). Cincinnati, Ohio: U.S. Department of Health, Education, and Welfare, National Institute for Occupational Safety and Health, 1977.

Krohn, G., Sanders, M., Volkmer, K., Wick, D., Miller, H., Beith, B., & Blake, T. *Experiments on personal equipment for low seam coal miners: II. Dexterity, protection and performance with padded gloves.* Pittsburgh, Pa.: Bureau of Mines, 1979.

Laveson, J., & Meyer, R. Left out "lefties" in design. *Proceedings of the 20th Annual Meeting of the Human Factors Society.* Santa Monica, Calif.: Human Factors Society, 1976.

Leonida, D. Ecological elements of vibration (chipper's) syndrome. In D. Wasserman and W. Taylor (Eds.), *Proceedings of the International Occupational Hand-Arm Vibration Conference* (No. 77-170). Cincinnati, Ohio: U.S. Department of Health, Education, and Welfare, National Institute for Occupational Safety and Health, 1977.

Miller, M., Ransohoff, J., & Tichauer, E. Ergonomic evaluation of a redesigned surgical instrument. *Applied Ergonomics,* 1971, *2*, 194–197.

Taylor, W. The vibration syndrome: Introduction. In W. Taylor (Ed.), *The vibration syndrome.* New York: Academic Press, 1974.

Taylor, W., Pearson, J., & Keighley, G. A longitudinal study of Raynaud's phenomenon in chain saw operators. In D. Wasserman & W. Taylor (Eds.), *Proceedings of the International Occupational Hand-Arm Vibration Conference* (No. 77-170). Cincinnati, Ohio: U.S. Department of Health, Education, and Welfare, National Institute of Occupational Safety and Health, 1977.

Terrell, R., & Purswell, J. The influence of forearm and wrist orientation on static grip strength as a design criterion for hand tools. *Proceedings of the 20th Annual Meeting of the Human Factors Society.* Santa Monica, Calif.: Human Factors Society, 1976.

Tichauer, E. Some aspects of stress on forearm and hand in industry. *Journal of Occupational Medicine,* 1966, *8*(2), 63–71.

Tichauer, E. Ergonomics: The state of the art. *American Industrial Hygiene Association Journal,* 1967, *28*, 105–116.

Tichauer, E. Biomechanics sustains occupational safety and health. *Industrial Engineering,* Feb. 1976.

Tichauer, E. *The biomechanical basis of ergonomics,* New York: Wiley, 1978.

Tichauer, E., & Gage, H. Ergonomic principles basic to hand tool design. *American Industrial Hygiene Association Journal,* 1977, *38*, 622–634.

Walton, K. The pathology of Raynaud's phenomenon of occupational origin. In W. Taylor (Ed.), *The vibration syndrome.* New York: Academic Press, 1974.

Washburn, S. Tools and human evolution. *Scientific American,* 1960, *203*, 3–15.

Wasserman, D., & Taylor, W. *Proceeding of the International Occupational Hand-Arm Vibration Conference* (No. 77-170). Cincinnati, Ohio: U.S. Department of Health, Education, and Welfare, National Institute for Occupational Safety and Health, 1977.

Weidman, B. *Effect of safety gloves on simulated work tasks* (AD 738981). Springfield, Va.: National Technical Information Service, 1970.

Williams, N. Biological effects of segmental vibration. *Journal of Occupational Medicine,* 1975, *17*, 37–39.

PART FOUR

WORK SPACE AND ARRANGEMENT

CHAPTER **11**

APPLIED ANTHROPOMETRY AND WORK SPACE

Every day of our lives we use some physical facilities that have (or should have!) some relationship to our basic physical features and dimensions—facilities such as chairs, seats, tables, desks, workplaces, and clothing. As we know from universal experience, the comfort, physical welfare, and performance of people can be influenced for better or worse by the extent to which such facilities "fit" people.

ANTHROPOMETRY[1]

What Kroemer (1978) refers to as *engineering anthropometry* deals with the application of scientific physical measurement methods to human subjects for the development of engineering design standards. It includes static and functional (dynamic) measurements of dimensions and physical characteristics of the body as they occupy space, move, and apply energy to physical objects, as a function of age, sex, occupation, ethnic origin, and other demographic variables. (*Biomechanics*, as discussed in Chapter 7, deals more with the physics of structure and behavior of the body.) Although we cannot reproduce here the voluminous anthropometric data that are available, we will at least illustrate such data. As indicated above, such data fall into two general classes, namely *static* and *functional* (*dynamic*) dimensions.

Static Body Dimensions

Static (*structural*) body dimensions are taken with the body of the subjects in fixed (static), standardized positions. Many different body features can be measured. The

[1] The interested reader is referred to compilations of anthropometric data such as Damon, Stoudt, and McFarland (1966); NASA Anthropometric Source Book (1978); U.S. Public Health Service (1965); and Ayoub and Halcomb (1976).

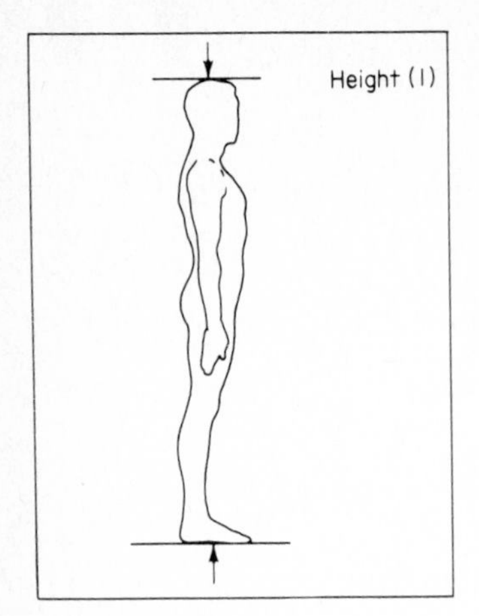

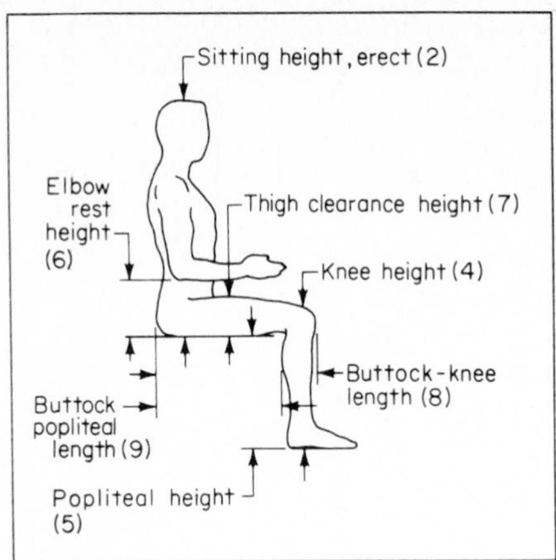

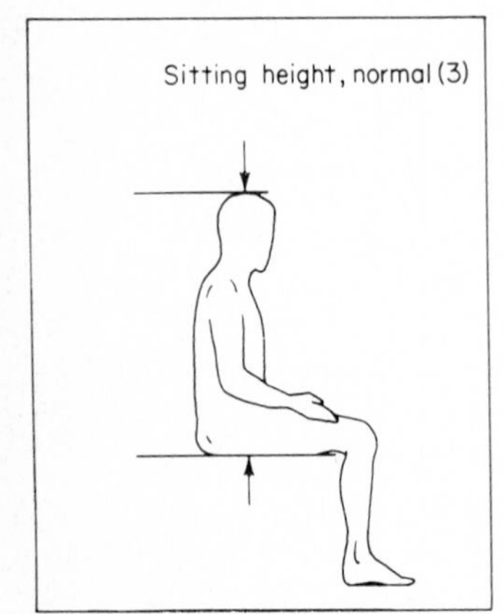

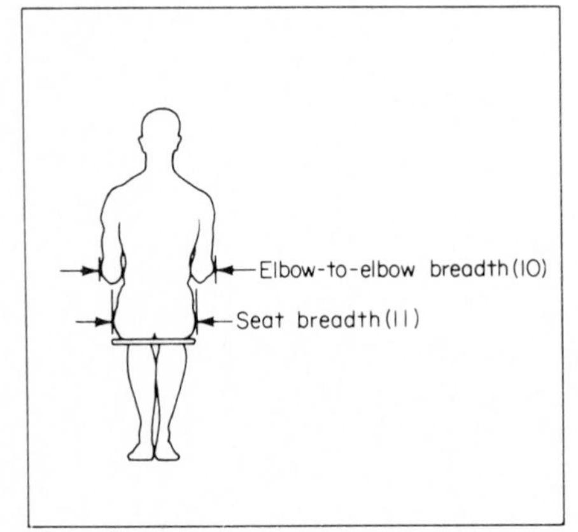

FIGURE 11-1
Diagrams of static body features measured in National Health Survey of anthropometric measurements of 6672 adults. See Table 11-1 for selected data based on the survey.

NASA *Anthropometric Source Book (Vol. II)* (1978), for example, illustrates 973 such measurements, and presents data on certain of these measurements from 91 worldwide surveys. Certain measurements would of course have very specific applications, such as in designing helmets, earphones, or pince-nez glasses. However, measurements of certain body features probably have rather general utility, and summary data on some of these features will be presented for illustrative purposes. These data come from a survey by the United States Public Health Service (1965) of a representative sample of 6672 adult males and females. The specific body features measured are shown in Figure 11-1; for each of these (plus weight), data on the 5th, 50th, and 95th percentiles are given in Table 11-1. It should be pointed out that these values cover ages from 18 to 79 and that most of the measurements given did vary with age, particularly weight and height. (The average weight of both males and females rises from the early twen-

TABLE 11-1
SELECTED STRUCTURAL BODY DIMENSIONS AND WEIGHTS OF ADULTS*

	Dimensions, in						Dimensions, cm†					
	Male, percentile			Female, percentile			Male percentile			Female, percentile		
Body feature	5th	50th	95th	5th	50th	95th	5th	50th	95th	5th	50th	95th
1 Height	63.6	68.3	72.8	59.0	62.9	67.1	162	173	185	150	160	170
2 Sitting height, erect	33.2	35.7	38.0	30.9	33.4	35.7	84	91	97	79	85	91
3 Sitting height, normal	31.6	34.1	36.6	29.6	32.3	34.7	80	87	93	75	82	88
4 Knee height	19.3	21.4	23.4	17.9	19.6	21.5	49	54	59	46	50	55
5 Popliteal height	15.5	17.3	19.3	14.0	15.7	17.5	39	44	49	36	40	45
6 Elbow-rest height	7.4	9.5	11.6	7.1	9.2	11.0	19	24	30	18	23	28
7 Thigh-clearance height	4.3	5.7	6.9	4.1	5.4	6.9	11	15	18	10	14	18
8 Buttock-knee length	21.3	23.3	25.2	20.4	22.4	24.6	54	59	64	52	57	63
9 Buttock-popliteal length	17.3	19.5	21.6	17.0	18.9	21.0	44	50	55	43	48	53
10 Elbow-to-elbow breadth	13.7	16.5	19.9	12.3	15.1	19.3	35	42	51	31	38	49
11 Seat breadth	12.2	14.0	15.9	12.3	14.3	17.1	31	36	40	31	36	43
12 Weight*	120	166	217	104	137	199	58	75	98	47	62	90

*Weight given in pounds (first six columns) and kilograms (last six columns).
†Centimeter values rounded to whole numbers.
Source: U.S. Public Health Service, 1965.

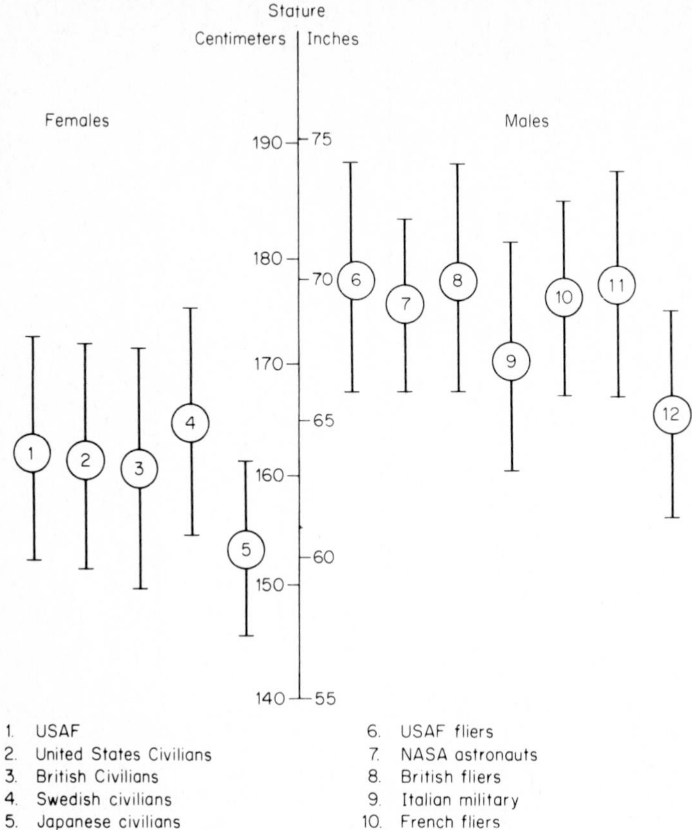

FIGURE 11-2
Range of variability (5th to 95th percentile) in stature of selected populations. [*Source: Adapted from NASA Anthropometric source book (Vol. I), fig. 11, p. II-40.*]

ties, and then later declines, peaking for males at about age 50 and for females at about 60. The average height for both sexes tapers off systematically from the early or mid-twenties.)

In the use of anthropometric data one needs to be aware of the fact that such data for different groups of people vary considerably. Some indications of such differences are represented in Figure 11-2. This shows, for each of various samples of people (including those of different nationalities), the range of variability from the 5th to the 95th percentiles of the sample.

Functional Body Dimensions

What are referred to as *functional* or *dynamic* body dimensions are taken under conditions in which the body is involved in some physical movement. Although static body

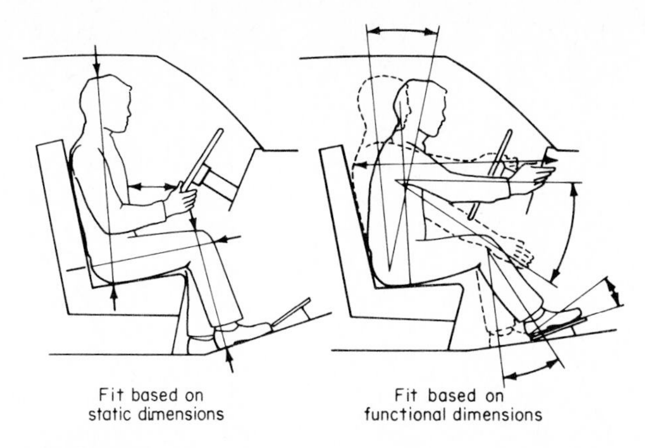

FIGURE 11-3
Illustration of the difference in the application of static versus functional body dimensions in the context of vehicular cab design. The use of static dimensions tends to focus on clearances of body dimensions with the surroundings, whereas the use of functional measurements tends to focus on the functions of the operations involved. (*Source: Adapted from Damon, Stoudt, & McFarland, 1966.*)

dimensions are useful for certain design purposes, functional dimensions are probably more widely useful for most design problems. In most circumstances in life people are not inert (not even when sleeping). Rather, in most work and nonwork situations people are functional—whether operating a steering wheel, assembling a mouse trap, or reaching across the table for the salt. Figure 11-3 illustrates the difference in the application of structural versus functional body dimensions to the design of a vehicle cab.

The central postulate of the emphasis on use of functional dimensions relates to the fact that in performing physical functions, the individual body members normally operate not independently but rather in concert. The practical limit of arm reach, for example, is not the sole consequence of arm length; it is also affected in part by shoulder movement, partial trunk rotation, possible bending of the back, and the function that is to be performed by the hand. Some examples of functional anthropometry are given later in this chapter.

THE MATTER OF POSTURE

Anthropometric data have implications for the design of work spaces, seats, equipment, etc., but before we get into such matters we will discuss the matter of *posture*, since the design of such items can affect the posture of people in their work and other situations. Posture can influence both the comfort and the physical conditions of people.

Posture and Comfort

Corlett and Bishop (1976) express the opinion that industrial organizations have not paid too much attention to the comfort of workers, but believe that providing for the comfort of workers would be beneficial, even in terms of production performance. They introduced the term *industrial comfort* as a concept with a threshold level below which a worker would not be distracted from the work, and developed a procedure for obtaining the judgments of people regarding their level of comfort, both in an overall sense and regarding sensations of pain associated with specific areas of the body. In one use of this procedure they derived *discomfort scores* for spot welders before and after certain changes in their work layout, and found that after the changes were made the discomfort scores were considerably lower.

To introduce a bit of semantic confusion, Hockenberry (1979) makes the point (when referring to seating) that the notion of *comfort* is not a very meaningful concept, and that it is more realistic to consider comfort as the absence of discomfort, just as cold is the absence of heat. Semantic niceties aside, it probably is most reasonable to be concerned with reducing discomfort rather than with creating *comfortable* facilities.

Posture and Physical Condition

The most important possible physical consequence of improper posture is with respect to spinal problems. Grandjean et al. (1973) estimate that 50 percent of adults suffer backaches during at least one period of their lives, and state that the main reason for frequent backaches is a pathological degeneration of the discs, which lie between the bony vertebrae and act as an elastic cushion between the vertebrae, thus giving the spinal column its flexibility. Improper postures used in such activities as stooping, lifting, and carrying loads tend to wear out the discs. Nachemson and Elfström (1970) measured the intradiscal pressure in different postures, and found that in either standing or sitting the pressures increased with increasing degrees of bending the back. The manner in which the pressure occurs is illustrated later in this chapter in Figure 11-13 (page 334) and Figure 11-18 (page 339). Nachemson and Elfström also reported a very sharp increase in such pressure when a person lifted a 20-kg weight with straight knees and a bent back, as opposed to the recommended practice of lifting with bent knees and a straight back.

Thus, one of the objectives in the use of anthropometric data would be to design the things people use so as to enhance the possibility of maintaining proper postures.

THE USE OF ANTHROPOMETRIC DATA

As indicated above, anthropometric data can have a wide range of applications in the design of physical equipment and facilities. In this regard, although static anthropometric data have certain uses, it is becoming increasingly evident that functional anthropometric data probably have greater potential use. Whatever type of data would be most relevant for a particular design problem, in most circumstances it is important to use data that are based on samples of subjects that are similar to the population who will ultimately use the item in question. However, Kroemer (1978) points out that comprehensive anthropometric data are still missing for population groups such as females, children, the elderly, and the handicapped. Most such data are based on young males in military service. In some instances the "population" of interest is expected to consist of "people at large," implying that the design must accommodate a broad spectrum of people.

Principles in the Application of Anthropometric Data

In the application of anthropometric data there are certain principles that may be relevant, each one being appropriate to certain types of design problems.

Design for Extreme Individuals In the design of certain aspects of physical facilities there is some *limiting* factor that argues for a design that specifically would accommodate individuals at one extreme or the other of some anthropometric characteristic, on the grounds that such a design *also* would accommodate virtually the entire population. A *minimum* dimension, or other aspect, of a facility would usually be based on an *upper* percentile value of the relevant anthropometric feature of the sample used, such as the 95th percentile. Perhaps most typically a minimum dimension would be used to establish clearances, such as for doors, escape hatches, and passageways. If the physical facility in question accommodates large individuals (say, the 95th percentile), it also would accommodate all those smaller in size. The minimum weight carried by supporting devices (a trapeze, rope ladder, or other support) is another example. On the other hand, *maximum* dimensions of some facility would be predicted on *lower* percentiles (say, the 5th) of the distribution of people on the relevant anthropometric feature. The distance of control devices from an operator is an example; if those with short functional arm reach can reach a control, persons with longer arm reach generally could also do so. In setting such maximums and minimums it is frequently the practice to use the 95th and 5th percentile values, if the accommodation of 100 percent would incur trade-off costs out of proportion to the additional benefits to be derived. To take an absurd case, we do not build 8½-ft doorways for the rare eight-footers, or dining-room chairs for the rare 400-lb guest. There are circumstances, however, in which designs that accommodate all people can be achieved without appreciable trade-off costs.

Design for Adjustable Range Certain features of equipment or facilities preferably should be adjustable in order to accommodate people of varying sizes. The forward-backward adjustments of automobile seats and the vertical adjustments of typists' chairs are examples. In the design of adjustable items such as these, it is fairly common practice to do so for the range of cases from the 5th to the 95th percentiles. The example given in Figure 11-4 illustrates seat adjustability requirements (i.e., the range of adjustment that should be provided for in a seat) to accommodate different segments of the population according to their sitting height. This illustrates the point that the amount of seat adjustment required to accommodate the extreme cases (such as below the 5th percentile and above the 95th) is disproportionate to the additional numbers of individuals who would be accommodated.

In this connection, as in other contexts, trade-off considerations may be in order. The military practice of rejecting persons who are extremely short or tall is dictated, in

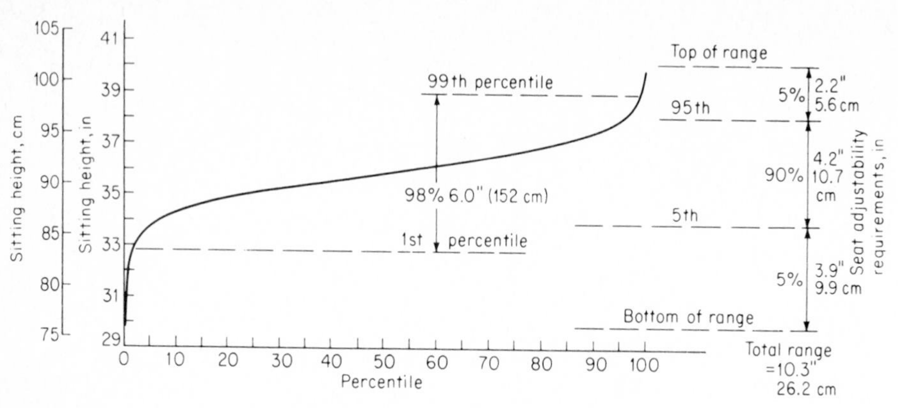

FIGURE 11-4
Illustration of the relationship between percentiles of cases and anthropometric measurements. This figure shows, specifically, the cumulative percentiles (along the base line) of people whose sitting height is at or below the values on the left vertical scale. The right vertical scale shows the corresponding seat adjustability requirements for various percentile groups. The differences in seat adjustability requirements are disproportionate at the extremes to the additional percentiles that would be accommodated. (*Source: Hertzberg, 1960.*)

part, by the fact that the requirement for smaller or larger items of clothing, shoes, etc., would impose an additional administrative load in supplying such items; this is trading off the possible utility of such people in the service for a certain degree of administrative simplicity.

Design for the Average While we frequently hear of the *average* or *typical* person, this is, in one sense, an illusive, will-o'-the-wisp concept. In the domain of human anthropometry there are few, if any, people who would really qualify as average—average in each and every respect. In connection with this, Hertzberg (1960) indicates that in a survey of over 4000 Air Force men there were *none* who fell within the (approximately) 30 percent central (average) range on all 10 of a series of measurements. Since the concept of the average person is then something of a myth, there is some rationale for the common proposition that physical equipment should not be designed for this mythical individual. Recognizing this, however, we would like to make a case here for the use of *average* values in the design of *certain* types of equipment or facilities, specifically those for which, for legitimate reasons, it is not appropriate to pitch the design at an extreme value (minimum or maximum) or feasible to provide for an adjustable range. As an example, the check-out counter of a supermarket built for the average customer probably would discommode customers less in general than one built either for the circus midget or for Goliath. This is not to say that it would be optimum for all people, but that, collectively, it would cause less inconvenience and difficulty than one which might be lower or higher.

Discussion The discussion of the above principles generally refers to the application of anthropometric data for single dimensions and to what percentage of individuals would be accommodated with specified specifications in terms of such dimensions.

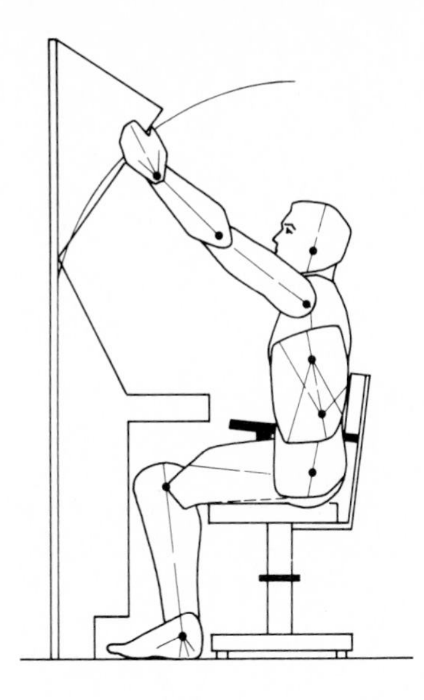

FIGURE 11-5
An articulated anthropometric scale model, such as used in the development of design of work spaces. (*Source: Meyer, 1979, fig. 5.*)

(The term *accommodated percentage* is sometimes used in this context.) The accommodated percentage can be determined in a straightforward manner for individual dimensions, but the problem becomes more complex when several dimensions need to be considered in combination. The complications arise from the interrelationships between and among the dimensions, some of which have low correlations with each other. Thus, an individual who falls within an accommodated percentage on one dimension might fall outside the accommodated percentages on another. Bittner (1974), for example, derived estimates of the percentages of naval aviators who would be excluded in the development of specifications of cockpit dimensions if both the 5th and 95th percentiles were used for each of 13 human dimensions, and found that such specifications would exclude 52 percent (as contrasted with 10 percent using the 5th and 95th percentiles in combination for a single dimension such as sitting height). The exclusion of such a high percentage would of course argue for a return to the drawing board, to redesign the work space to accommodate a larger percentage.

In the application of anthropometric data it is sometimes the practice to use physical models, such as the articulated model illustrated by Meyer (1979) and shown in Figure 11-5. Such models usually represent specific percentiles of the population.

WORK-SPACE DIMENSIONS

Human work space can consist of many different physical situations, many of which are specifically designed for workers. Certain aspects of the design of such spaces are discussed and illustrated below. (There are some work-space problems that designers cannot do much about, such as the plumber working under an already existing stopped-up sink.)

Work Space for Personnel When Seated

There are millions of people whose work activities are carried out while seated in a fixed location. The space within which such an individual works is sometimes referred to as the *work-space envelope*. This envelope should, of course, be designed on a situational basis, considering the particular activities to be performed and the types of people who are to use the space. To illustrate the types of data that would be relevant in designing specific work-space envelopes, however, the results of a couple of anthropometric studies will be shown.

Functional Arm Reach Envelope In a study by Roth, Ayoub, and Halcomb (1977) a device called the *Ayoub Reach Anthropometer* was used to measure the functional arm reach of subjects at various lateral angles from a dead-ahead seated position (from −45° left to +120° right) and at levels ranging from −60° to +90° from a seat reference point (SRP). The specific nature of the device need not be described, but it provided for measuring 114 *grip-center* reach points. The arm reaches of the subjects were measured under both restrained and unrestrained conditions. In the restrained conditions the shoulders of the subjects were held back against the seat back, whereas in the unrestrained condition the subjects could move their shoulders. The subjects were selected to be reasonably representative of adults in terms of height and weight.

Some of the results are shown in Figure 11-6, in particular the envelope for the 5th percentiles of males and females for each condition. Each line represents the 5th percentile value at specified levels above a seat reference point. (Data are given for only certain levels.) On the basis of the data for either sex, one could envision a three-dimensional space the outer surface of which would include the values represented by the data in the figure. Such a three-dimensional space—a work-space envelope—would then represent the limits of convenient arm reach for the 5th percentile of the population, but such limits would of course accommodate 95 percent of the population.

Effects of Manual Task on Work-Space Envelope The nature of the manual task to be performed influences the boundaries of the work-space envelope. For example, if an individual simply has to activate push buttons or toggle switches a "fingertip" measurement is appropriate, as contrasted with the requirement to use knobs to grasp levers, for which a "thumb tip" measurement is used (according to data summarized by Bullock, 1974, such measurements are about 5 or 6 cm or a couple of inches shorter than fingertip measurements). In turn, a hand-grasp or griplike action (such as the "grip-center" measurement used by Roth et al., 1977) limits the reach further (by 5 cm or more according to Bullock).

Even different hand-grasp actions influence the space envelope, as demonstrated years ago by the classic study by Dempster (1955), who had male subjects use eight different hand grasps in an anthropometric study. (These involved grasping a handle-like device with the hand in one of eight fixed orientations, namely, supine, prone, inverted, and at five spatial angles.) Photographic traces of contours of the hand were made as the hand moved over a series of frontal planes spaced at 6-in intervals. From these a *kinetosphere* was developed for each grasp, showing the mean contours of the tracings as photographed from each of three angles, top (transverse), front (coronal),

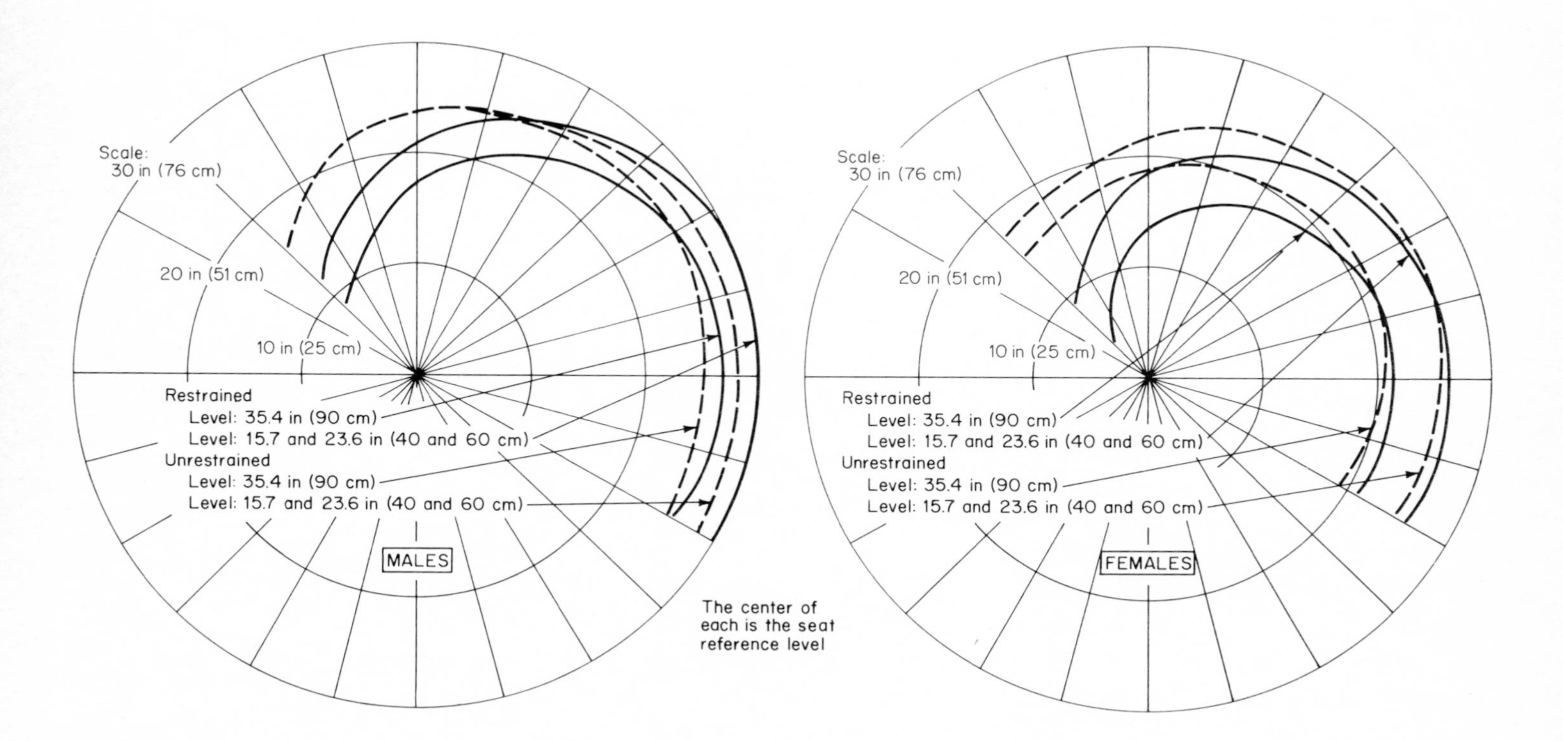

FIGURE 11-6
Functional arm reach for the 5th percentiles of males and females at specified levels above a seat reference point. The three-dimensional space represented by these data can be considered as forming a work-space envelope that would accommodate 95 percent of the population. (*Source: Roth, Ayoub, & Halcomb, 1977, for description of procedures used in obtaining data. Appreciation is expressed to Dr. M. M. Ayoub and his associates for special tabulations used in the preparation of this figure.*)

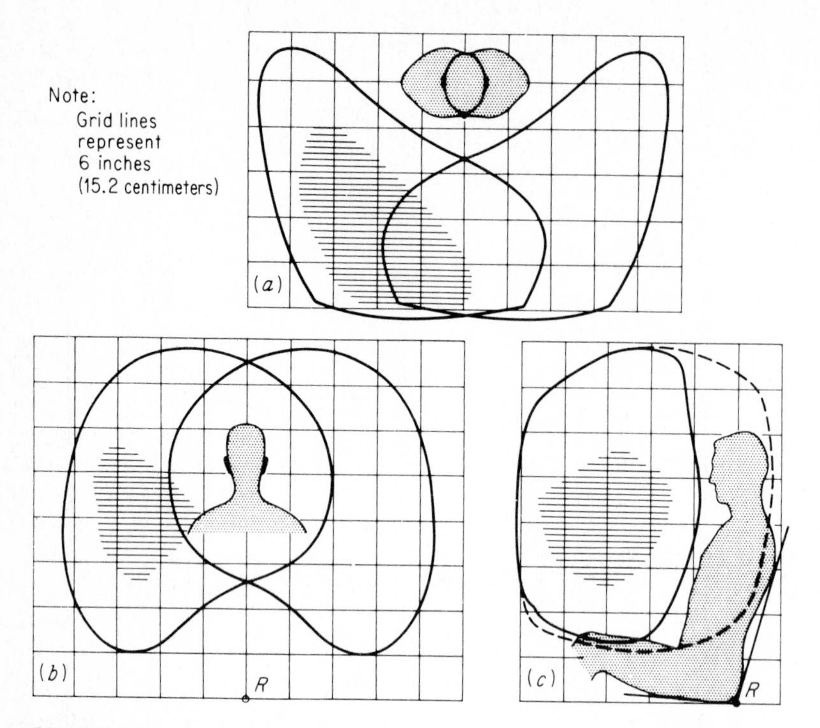

FIGURE 11-7
Strophosphere resulting from superimposition of kinetospheres of range of hand movements with a number of hand-grasp positions in three-dimensional space. The shaded areas depict the region common to all hand motions (prone, supine, inverted, and several different angles of grasps), probably the optimum region, collectively, of the different types of hand manipulations. (*Source: Adapted from Dempster, 1955.*)

and side (sagittal). Although the kinetospheres for the several types of grip will not be illustrated, they were substantially different. These were, however, combined to form *strophospheres* as shown in Figure 11-7. The shaded areas define the region that is common to the hand motions made with the various hand grips and therefore indicate those three-dimensional spaces within the work-space envelope within which the various types of hand grips in question could most adequately be executed by people.

Effects of Apparel on Work-Space Envelope The apparel worn by people can restrict the movements of people and the distances they can reach, and therefore can influence the size of the work-space envelope. In a survey of truck and bus drivers, for example, Sanders (1977) found that winter jackets restricted reach by approximately 2 in.

Discussion of Work-Space Envelopes In the case of individuals who typically remain at fixed work stations, *work-space envelopes* are the three-dimensional spaces around them within which their physical activities are carried out. The dimensions of such envelopes clearly should be based on the anthropometric and biomechanical characteristics of the people in question. We can see that various factors can influence the convenient work space of people, factors such as variability in the sizes of people, sex, the type of manual action to be performed, and the matter of restraints. In practice, of course, restraints usually are used only in vehicles. Although restraints are seldom used in stationary work spaces, as a matter of good design practice, and for safety, convenience, and comfort, controls should be placed well within the space envelope for the 95th percentile reach. Still another factor that can influence the envelope is the angle of the backrest. Although an inclined backrest makes controls behind the seat reference point more accessible, those forward and to the side of the midline are less accessible.

Brief mention should be made of the work space of special groups of people, in particular those with relevant handicaps. Rozier (1977), for example, reports that with a below-elbow prosthesis for below-elbow amputees there was an average decrease in the work space of 45 percent, and with an above-elbow prosthesis for above-elbow amputees there was an average decrease of 83 percent. Thus, even with a prosthesis, amputees have very restricted work-space envelopes.

Most desks and tables for writing, reading, and other office activities have flat (horizontal) surfaces, although certain ones (such as drafting tables) have slanted surfaces. There is evidence from a study by Eastman and Kamon (1976), however, that (where feasible) a slanted surface can provide far better posture and less discomfort than a flat surface. Their subjects performed writing and reading tasks on flat and slanted surfaces (12° and 24° slant). They sat on an adjustable bar stool with their feet on an adjustable footrest. They obtained electromyograms (EMGs); ratings of pain, "feeling fine" and fatigue; and films of postural changes. Collectively these criteria indicated that the use of slanted desk surfaces resulted in more exact posture, less back movement, and reduced fatigue and discomfort. Certain results are summarized in Table 11-2.

Minimum Requirements for Restricted Spaces

People sometimes find themselves working in, or moving through, some restricted and sometimes awkward spaces (such as an astronaut crawling through an escape hatch). For certain types of restricted spaces dynamic anthropometric data have been derived that provide minimum values. Some such data are given in Figure 11-8 for illustration.

TABLE 11-2
TRUNK MOVEMENTS AND RATINGS OF DISCOMFORT OF SUBJECTS WRITING AND READING AT FLAT AND SLANTED DESK SURFACES

	Surface slant		
Criterion	**0°**	**12°**	**24°**
Ratings of back discomfort	1.6	1.3	1.0
Trunk movements (horizontal)	13.2	12.5	9.7

Source: Eastman and Kamon, 1976.

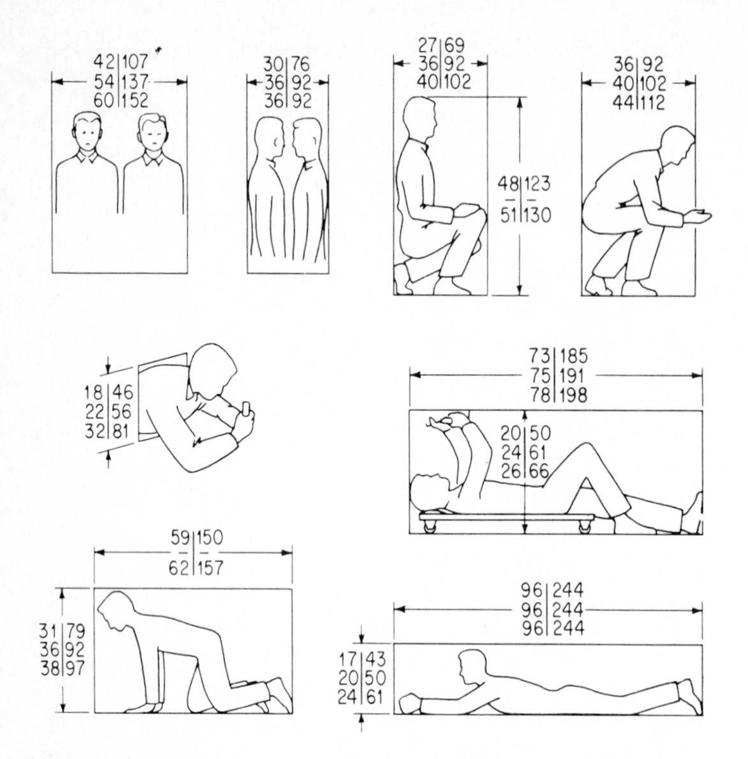

FIGURE 11-8
Clearances for certain work spaces that individuals may be required to work in or pass through. *Note:* The three dimensions given (inches at left, centimeters at right) are (from top to bottom in each case) minimum, best (with normal clothing), and with heavy clothing (such as arctic). (*Source: Adapted from Rigby, Cooper, & Spickard, 1961.*)

Note that the dimensions given include those applicable to individuals with heavy clothing. In most cases such clothing adds 4 to 6 in (10 to 15 cm), and in the case of a vertical escape hatch it adds 10 in (25 cm) to the requirements.

WORK SURFACES

Within the three-dimensional envelope of a work space, more specific considerations of work-area design relate to specific features of horizontal, vertical, and sloping work surfaces. These features of the work situation also preferably should be determined on the basis of anthropometric considerations of the people who are to use the facilities in question.

Horizontal Work Surface

Many types of manual activities are carried out on horizontal surfaces such as workbenches, desks, tables, and kitchen counters. For such work surfaces, the *normal* and *maximum* areas have been proposed by Barnes (1963), based on the measurements of 30 men. These two areas are shown in Figure 11-9 and have been described as follows:

1 *Normal area:* This is the area that can be conveniently reached with a sweep of the forearm, the upper arm hanging in a natural position at the side.

2 *Maximum area:* This is the area that could be reached by extending the arm from the shoulder.

Related investigations by Squires (1956), however, have served as the basis for proposing a somewhat different work-surface contour that takes into account the dynamic interaction of the movement of the forearm as the elbow also is moving. The area that is so circumscribed is superimposed over the area proposed by Barnes in Figure 11-9. The fact that the normal work area proposed by Barnes has gained wide acceptance probably indicates that it is reasonably adequate, although the shallower area proposed by Squires is more compatible with dynamic anthropometric realities. In addition, Tichauer (1978) points out that a posture that requires substantial forward extension of the forearm may impose greater stress on the elbow joint and increased likelihood of physical impairment than one in which the elbow is in more of a bent position when performing work. Because of this he argues for minimizing the distance between the chair and the work space, thus permitting the angle at the elbow to be reduced.

FIGURE 11-9
Dimensions (in inches and centimeters) of normal and maximum working areas in horizontal plane proposed by Barnes, with normal work area proposed by Squires superimposed to show differences. (*Source: Barnes, 1963; Squires, 1956.*)

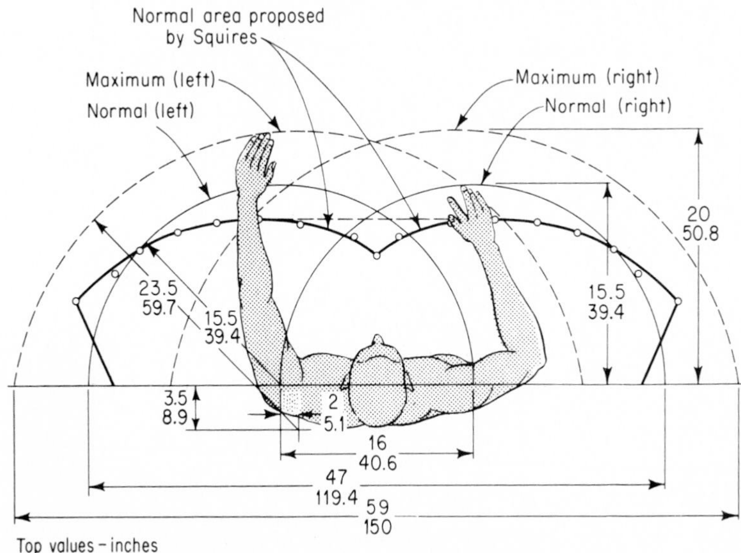

Work-Surface Height: Seated

The wide range of tasks performed by seated personnel at tables, desks, workbenches, etc., plus the range of individual differences, obviously precludes the establishment of any single, universal height that would be appropriate for such surfaces. However, considering body structure and biomechanics, at least one can set forth a guiding principle that should be applied, namely, that the work surface (or, really, the location of the devices or objects to be used continuously) should be at such a level that the arms can hang in a reasonably natural, relaxed position from the shoulder. Generally this would mean that the forearm would be approximately horizontal or sloping down slightly when performing most simple manual tasks. When the work surface requires the upper arm to be held at an angle with the torso (rather than being in a relaxed vertical position), fatigue and discomfort may occur, along with reduced work efficiency (Tichauer, 1978).

Such a principle would suggest that work surfaces generally should be lower than is reflected by common practice. With respect to desk heights, Bex (1971), on the basis of a European survey, reports that the most common heights have in fact been reduced from about 30 in (76 cm) in 1958 to about 28½ in (72 cm) in 1970. But on the basis of his own and other anthropometric data he argues for a further reduction of fixed desk heights to about 27 in (68.5 cm). However, he urges that, when feasible, adjustable desk heights be provided, the range to be from about 23 to 30 in (58 to 76 cm).

The case for adjustable work-surface heights is supported on the basis of three factors, as follows: individual differences in physical dimensions (especially seated elbow height); individual differences in preference; and differences in the tasks being carried out. In this regard, for example, Ward and Kirk (1970) carried out a survey of the preferences of British homemakers with respect to working-surface heights when performing three different types of tasks. The percents preferring work surfaces at certain levels (relative to the elbow) when performing these tasks are given in Table 11-3.

The mean preferred work-surface heights were: A–23.7 in (60.2 cm); B–25.3 in (64.3 cm); and C–24.2 in (14.7 cm), but it is clear that there were individual differences in preferences as to relative level for each type of task, presumably because of the nature of the muscular involvement of the tasks.

TABLE 11-3
PERCENT OF SUBJECTS EXPRESSING PREFERRED LEVELS IN PERFORMING THREE KITCHEN TASKS

	Level relative to elbow		
Type of task	Lower	Even	Above
A. Working above surface (peeling vegetables, slicing bread, etc.)	54	14	32
B. Working on surface (spreading butter, chopping ingredients, etc.)	16	11	73
C. Exerting pressure (ironing, rolling pastry, etc.)	41	9	50

Source: Ward and Kirk, 1970.

TABLE 11-4
RECOMMENDATIONS BY AYOUB FOR SEATED WORK-SURFACE HEIGHTS FOR FOUR TYPES OF TASKS

Type of task (for seated person)	Male		Female	
	in	cm	in	cm
A. Fine work (e.g., fine assembly)	39.0-41.5	99-105	35.0-37.5	89-95
B. Precision work (e.g., mechanical assembly)	35.0-37.0	89-94	32.5-34.5	82-87
C. Writing, or light assembly	29.0-31.0	74-78	27.5-29.5	70-75
D. Coarse or medium work	27.0-28.5	69-72	26.0-27.5	66-70

Source: Ayoub, 1973.

The implications of the nature of the task with respect to work-surface height are emphasized further by Ayoub (1973), who offers the guidelines shown in Table 11-4 for four types of tasks, these being based on average anthropometric dimensions.

Ayoub's recommendations for fine and precision work actually are in contradiction to the general recommendations that work surfaces be below elbow height. Specifically he recommends that the levels for fine and precision work be 6 and 2 in above elbow height, respectively. In such instances there should be provision for the arms to rest on the work surface, and within close visual range.

The work-surface height that would be most appropriate for individuals, or for people generally, however, is closely tied in with seat height (to be discussed later), the thickness of the work surface, and the thickness of the thigh, as shown in Figure 11-10. Dannhaus, Bittner, and Ayoub (1976) point out that the clearance between

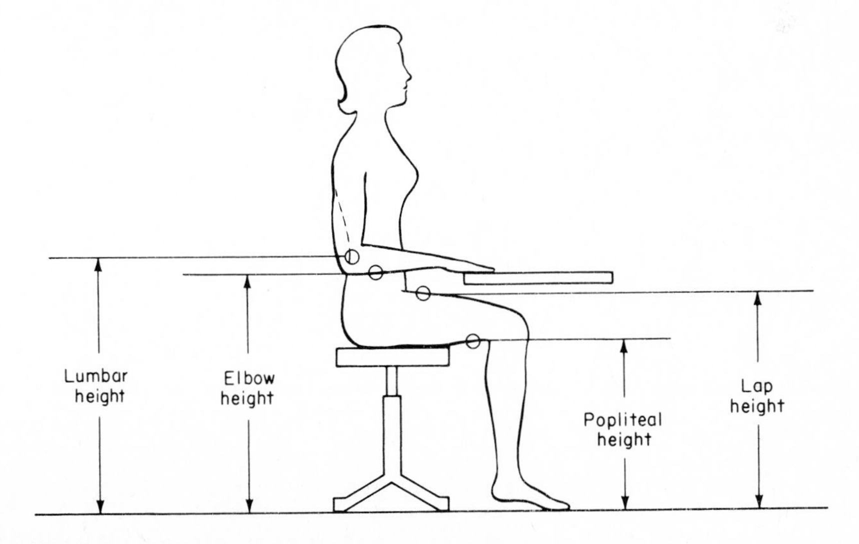

FIGURE 11-10
Illustration of the relationships of certain body dimensions and of work surface and seat height. If feasible, the work-surface height should be adjustable for the individual's dimensions and preferences. (*Source: Bex, 1971, fig. 4.*)

the seat and the underside of the work surface should accommodate the thighs of the largest users when the seat height is at the middle of its adjustability if it is adjustable. The combinations of variables involved make it almost impossible to design a fixed work surface and seat arrangement that would be fully suitable for people of all sizes. Thus, where feasible, some adjustable features should be provided, such as seat height, foot position (as by the use of a footrest), or work-surface height.

Work-Surface Height: Standing

Some experimental evidence relating to work-surface height for persons working in a standing posture comes from a study by Ellis (1951). Using a manipulation test of turning over wooden disks, he varied the work-surface height in relation to the distance from floor to fingertip of each subject. From his results he concluded that for standing, a work surface normally should be a bit below elbow height. Barnes proposes 2 to 4 in or 5 to 10 cm below elbow for at least light assembly or similar manipulatory tasks. In terms of height above the floor, these would represent a range of average values for males from about 42 or 41 in (107 or 104 cm) to 36 in (81 cm), and for females from about 38 in (97 cm) to 33 in (84 cm).

Although this principle is quite well supported, there are data that suggest some variations in the most appropriate heights for various types of tasks. Ward, for example, used four methods for assessing work-surface heights of women performing various kitchen tasks, the heights being 30, 33, 36, and 39 in (76, 84, 91, and 99 cm). The methods used were electromyography, anthropometry, "center of weight" determination, and expressed preferences. On the basis of data obtained from women of three size groups (small, medium, and large) she proposed ranges of heights for six different tasks; see Table 11-5.

These ranges—and corresponding ranges for performing other tasks in other circumstances—are of course very much a function of individual differences, which are always with us.

Further indication of the fact that the nature of the activity influences the desirable standing work-surface height is reflected by the guidelines proposed by Ayoub for three types of tasks, as shown in Table 11-6 (based on average dimensions).

Although many work-surface heights do not lend themselves to on-the-spot adjustments in height, there sometimes are ways and means of matching facilities to people, as in selecting or building facilities for individuals (as counter tops, workbenches, etc.),

TABLE 11-5
RANGES OF STANDING WORK-SURFACE HEIGHTS PROPOSED BY WARD AND KIRK FOR SIX KITCHEN TASKS

	At sink		At work top		At stove	
	Wash up	Peel potatoes	Iron	Slice potatoes	Frying	Boiling
in	36–42	36–42	33–39	36–39	33–39	33–39
cm	90–105	90–105	85–100	90–100	85–100	85–100

Source: Ward & Kirk, 1970.

TABLE 11-6
RECOMMENDATIONS BY AYOUB FOR STANDING WORK-SURFACE HEIGHTS FOR THREE TYPES OF TASKS

Type of task (for standing person)	Male		Female	
	in	cm	in	cm
A. Precision work, elbows supported	43.0-47.0	109-119	40.5-44.5	103-113
B. Light assembly work	39.0-43.0	99-109	34.5-38.5	87-98
C. Heavy work	33.5-39.5	85-101	31.0-37.0	78-94

Source: Ayoub, 1973.

placing blocks under legs of benches or tables, having mechanically adjustable legs, or having low platforms (a few inches high) for people to stand on.

Work Surfaces for Standing or Sitting

It is usually desirable to provide the opportunity to perform a job in either a standing or a sitting posture, or to permit workers to alternate their postures. In such instances the work-surface height should permit a relaxed position of the upper arm, and the chair and footrest should permit such a posture, as illustrated in Figure 11-11.

THE SCIENCE OF SEATING

Whether at work, at home, at horse races, on buses, or elsewhere, the members of the human race spend a major fraction of their lives sitting down. As we know from experience, the chairs and seats we use cover the gamut of comfort; they can also vary in their influence on the performance of people who use them when carrying out some types of work activities.

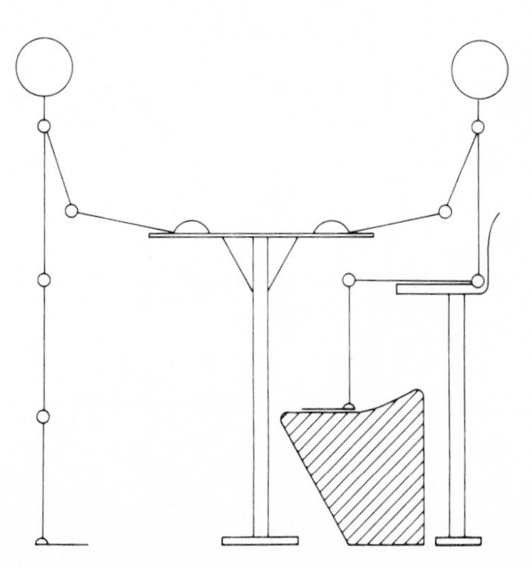

FIGURE 11-11
Illustration of a workplace that would permit the worker to stand or sit, and still maintain an appropriate relationship with the work surface. (*Source: Tichauer, 1973, pp. 138-139.*)

Principles of Seat Design

The relative comfort and functional utility of chairs and seats are, of course, the consequence of their physical design in relationship to the physical structure and biomechanics of the human body. The uses of chairs and seats (from TV lounge chairs to stadium bleachers) obviously require different designs, and the range of individual differences complicates the design problem. Granting that compromises sometimes are necessary in the design of seating facilities, there nonetheless are certain general guidelines that may aid in the selection of designs that are sufficiently optimum for the purposes in mind.

Weight Distribution Various seating studies have led to the conclusion that the seat should be so designed that the weight of the body is distributed primarily throughout the buttock region surrounding the ischial tuberosities. (The *ischial tuberosities* are the bone protuberances of the buttock area, commonly known as *sitting bones*.) The weight should be distributed rather evenly throughout the buttocks area, but minimized under the thighs. Such distribution can be accomplished by proper contouring of the seat pan in combination with other features of the seat such as seat height, seat angle, and seat back. Figure 11-12 shows what is considered to be a desir-

FIGURE 11-12
Representation of what is considered to be a desirable distribution of weight on the buttocks, showing equal-pressure contours from the ischial tuberosities to the periphery. The upper value for each contour is g/cm² and the lower value is lb/in². (*Source: Adapted from Rebiffé, 1969, fig. 7, p. 256.*)

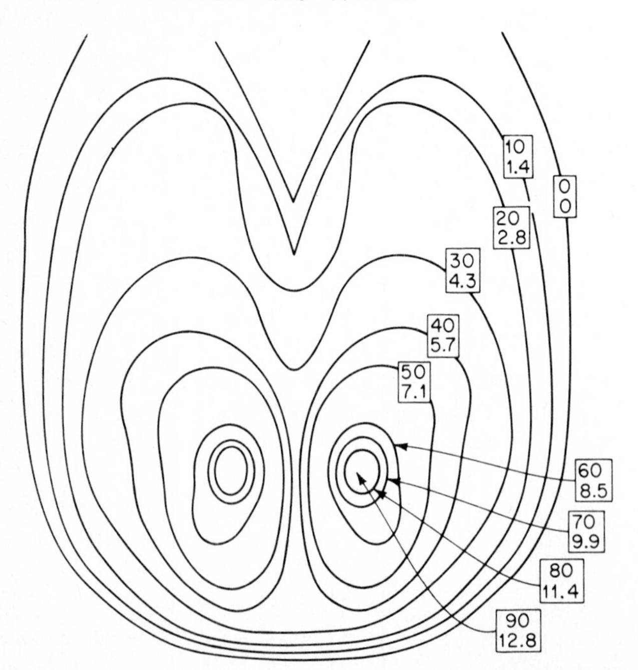

able distribution of weight for a person when driving a vehicle, each of the lines representing an equal pressure contour, these ranging from the pressure on the ischial tuberosities of 90 grams per square centimeter (90 g/cm^2) (12.8 lb/in^2) to the outer contours of 10 g/cm^2 (1.4 lb/in^2).

Seat Height The height of a seat should be such as to avoid excessive pressure on the thigh. In general, this means that the front edge of the seat should be a bit lower than the distance from the floor to the thigh when seated (i.e., *popliteal height*). In this regard Tichauer (1978) recommends that the front edge be at least 2 in (about 5 cm) below the *popliteal crease* (the crease at the back of the hollow of the knee).

Referring back to Table 11-1, we see that the 5th percentile of popliteal heights for males and females are 15.5 and 14.0 in (39 and 36 cm), respectively. Although seat heights a bit below these values would suit the 5th percentile individuals, such seats (if fixed) would complicate the mechanics of sitting for the taller members of the human clan by a chain reaction starting with the knee angle that can cause such an individual to sit with his lumbar back area in a convex rather than concave posture. By taking into account the fact that heels typically add an inch or more to the 5th percentile values (and more for females), it has become fairly common practice to use seat heights of around 17 in (43 cm). This jibes quite well with the recommendation by Grandjean, et al. (1973) of 43 cm (16.8 in) for multipurpose chairs which have a slanting seat. Where feasible, of course, adjustable seat heights (perhaps from 15 to 19 in or 38 to 48 cm) should be provided to accommodate persons of various heights. When adjustable seats are not practical, footrests for the shorter members of the clan can be used to minimize the pressure on the underside of the thigh.

Seat Depth and Width The preferable depth and width of seats would depend in part on the type of seat (whether a multipurpose chair, typing chair, lounge chair, etc.). In general terms, however, the depth should be set to be suitable for small persons (to provide clearance for the calf of the leg and to minimize thigh pressure) and the width should be set to be suitable for large persons. On the basis of comfort ratings for chairs of various designs, Grandjean et al. (1973) recommend that for multipurpose chairs depths not exceed 43 cm (16.8 in) and that the width of the seat surface be not less than 40 cm (15.7 in), although such a seat width (or perhaps one a bit wider, say 17 in) would do the trick for individual seats; if people are to be lined up in a row, or seats are to be adjacent to each other, elbow-to-elbow breadth values need to be taken into account, with even 95th-percentile values around 19 and 20 in producing a moderate sardine effect. (And for bundled-up football observers the sardine effect is amplified further.) In any event these are approximate minimum values for chairs with arms on them (for well-fed friends, you should have even wider lounge chairs).

Seat and Back Support Seats should provide for correct curvature of the *lumbrosacral* section of the spine (the *lumbar*, or lower back area) in order to keep the spinal column in a state of balance. This is illustrated in Figure 11-13. Figure 11-13*a* shows a condition of kyphosis which would occur without a backrest or with an inadequate backrest; this forward-leaning posture would produce excessive pressure between the

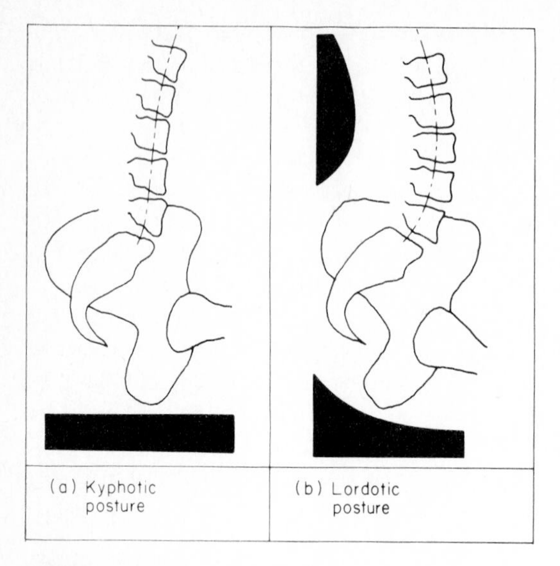

FIGURE 11-13
Illustration of the effect of seat design on the posture of the spine. *(a)* A kyphotic posture which would occur without a seat back or with an inadequate back; this "leaning-forward" posture accentuates the pressure between the vertebrae. *(b)* A lordotic posture which would result from adequate lumbar support and which would minimize pressure between the vertebrae. *(Source; Adapted from Hockenberry, 1979, fig. 1.)*

vertebrae. Figure 11-13*b*, which illustrates the lordosis type of posture with adequate lumbrosacral back support, represents the more desirable posture.

Such posture is made possible with two seat features. One concerns the angle of the seat and back. The seat should be at a moderate angle, and the back should have an angle of from about 95° to 105° or more with the seat. These angles, however, should depend in part on the use of the seat. For example, in the case of an office seat, as illustrated in Figure 11-14, the recommended seat angle is about 3° and the back angle (the angle between the back and the seat) is 100°. For resting and reading, however, Grandjean, Boni, and Krestzschmer (1969) found that most people prefer greater angles (as reported in a later discussion of chairs for resting and reading). The second feature for a satisfactory lordotic posture is adequate support for the lumbar area.

Dannhaus et al. (1976), in their very thorough summary of data relating to seat design, make the point that the height and width of the backrest should depend on the user's necessity of moving the arms and shoulders. Without specifying the details, they state that if mobility is required a small backrest should be used, but if mobility is not required a larger backrest can be used. In either case it should be designed with a convex lumbar support.

Armrests are generally desirable unless the tasks require free mobility of the trunk, shoulders, and arms. When possible they should be adjustable, but when fixed their optimum height from the seat ranges from about 7 to 9 in (18 to 23 cm).

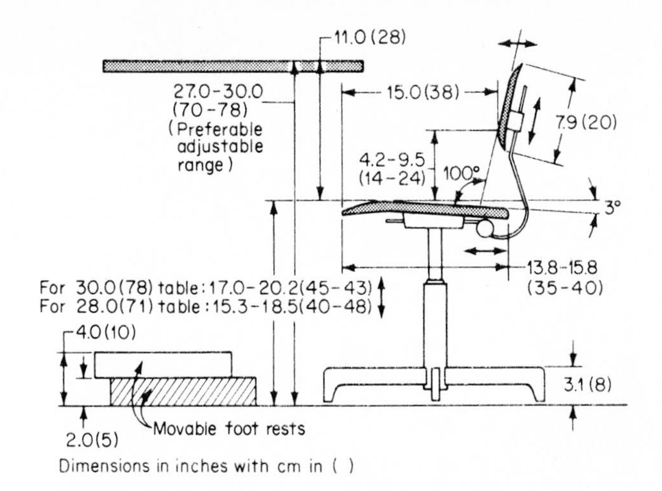

FIGURE 11-14
Recommended dimensions (in inches) of adjustable features of office seat. Note ranges of adjustability of seat height in relation to two table heights, 30 and 28 in (78 and 71 cm). To maintain approximately 11 in (28 cm) between seat height and work surface, the seat should have a range of adjustability that would depend on the work-surface height, as indicated. With short persons, footrests may be required if work-surface height is high. (Data in inches converted from centimeters.) (*Source: Adapted from Burandt & Grandjean, 1963.*)

Seat Designs for Varied Purposes

Since the specific features of seats need to be determined for the particular use, we will illustrate a few selected examples.

Office Chairs For people doing office work at desks Burandt and Grandjean (1963) have proposed the design features shown in Figure 11-14, these features being based on a substantial amount of data relative to the comfort of people using such chairs. A chair for more general use is illustrated in Figure 11-15. Hockenberry (1979) reports that the design of this chair was based on anthropometric data developed by Ridder back in 1959, which he considers to represent the most accurate, scientific measurements of the buttock and back contours of the adult American male and female population.

Multipurpose Chairs In the study by Grandjean et al. mentioned earlier (1973), 25 men and 25 women were asked to rate the comfort of 11 parts of the body when testing 12 different designs of multipurpose chairs. In addition, each subject compared every chair with every other one and rated the overall comfort by the paired-comparison method. The contours of the two most preferred chairs are shown in Figure 11-16, along with design recommendations made on the basis of the analysis of the results of all the data. The recommendations include foam rubber of 2 to 4 cm (about 0.75 to 1.5 in) on the entire seat.

FIGURE 11-15
Illustration of a general-purpose office chair designed on the basis of relevant anthropometric measurements of adult American males and females. (*Source: Hockenberry, 1979. Photograph courtesy of Steelcase Inc., Grand Rapids, Mich.*)

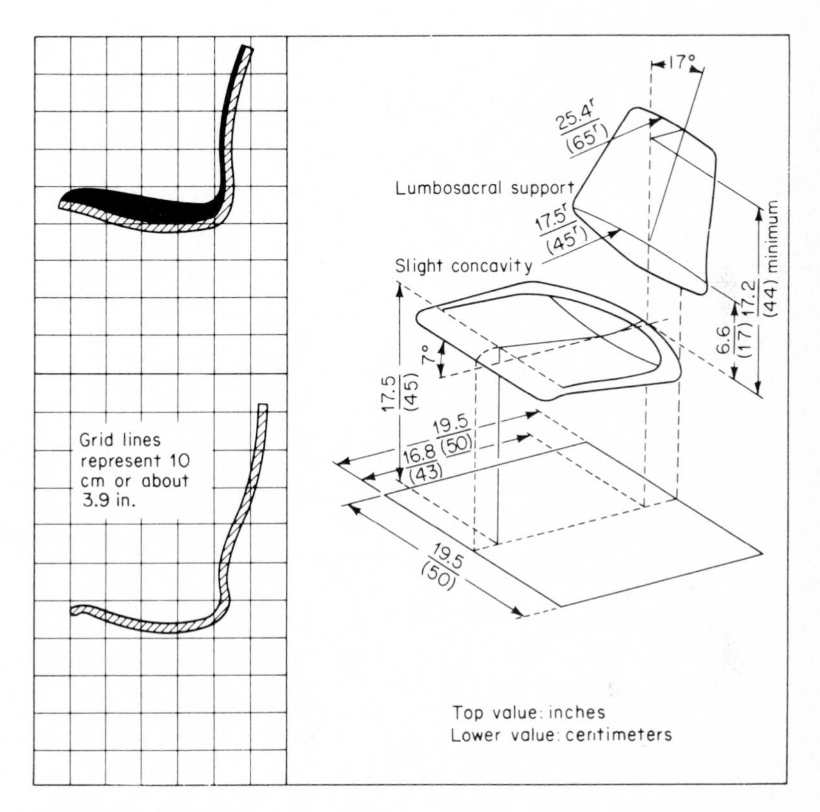

FIGURE 11-16
Contours of the 2 multipurpose chairs (of 12) judged to be most comfortable by 50 subjects, and the design features recommended for multipurpose chairs based on the study. (*Source: Grandjean, Hünting, Wotzka, & Shärer, 1973, figs. 2, 6, 13.*)

TABLE 11-7
RANGES OF DIMENSIONS OF SEATS FOR READING AND RESTING PREFERRED BY SUBJECTS

	Reading	Resting
Seat inclination,°	23-24	25-26
Backrest inclination, °	101-104	105-108
Seat height, cm	39-40	37-38
Seat height, in	15.3-15.7	14.6-15.0

Source: Grandjean, Boni, & Krestzschmer, 1969.

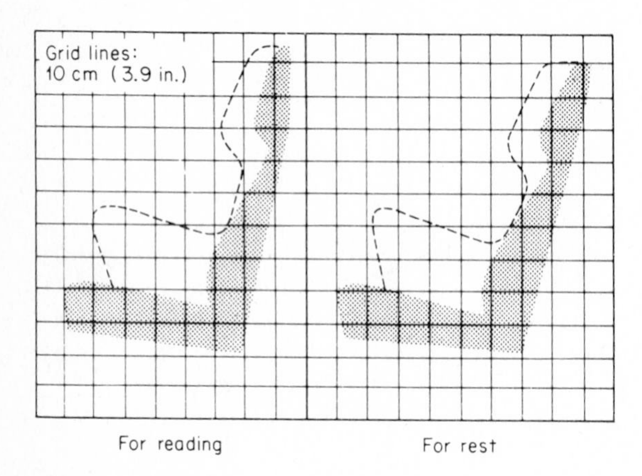

FIGURE 11-17 Profiles of seats proposed for reading and resting. The dotted lines correspond to the armrests and a possible outer contour. The shaded area shows the surface of the seat, including upholstering to be 6 cm (2.5 in) thick. *(Source: Grandjean, Boni, & Krestzschmer, 1969, fig. 2, p. 310.)*

Chairs for Resting and Reading The desirable features of chairs for relaxation and reading are of course different from those of chairs used for more active use. A study was carried out by Grandjean et al. (1969) in which they used a *seating machine* for eliciting judgments of subjects about the comfort of various seat designs. The *seating machine* consisted of features that could be adjusted to virtually any profile. Without summarizing all of the results, they found that the angles and dimensions shown in Table 11-7 were preferred by more subjects than others for the two purposes of reading and resting. Profiles for two such chairs are shown in Figure 11-17. Note in these the distinct angles of the backs and seats to provide full back support with particular support for the lower (lumbar) sections of the spine.

Automobile Driver Seats The desirability of adequate back support—for whatever activity is involved—is further illustrated in the case of automobile driver seats, as illustrated by the sketches in Figure 11-18. With unsatisfactory support (as shown in Figure 11-18*b*), the angles between the vertebrae (shown in the inset) can generate discomfort and conceivably cause spinal complications. The angles of the various body

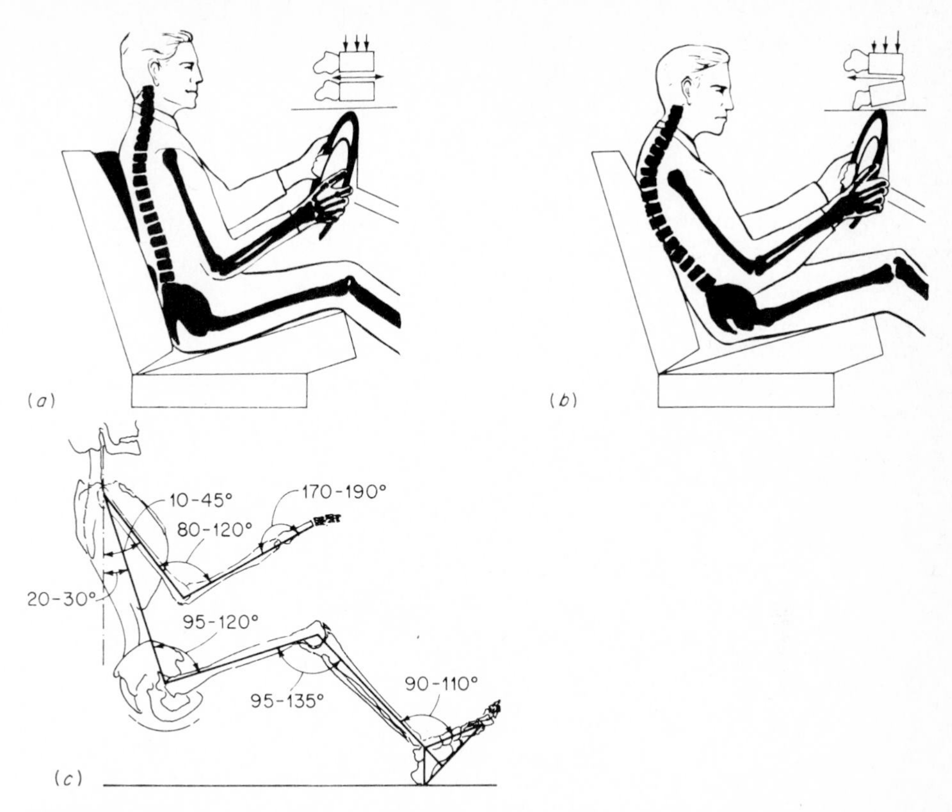

FIGURE 11-18
Illustrations of anthropometric considerations in designing seats for automobile drivers. Parts *a* and *b* illustrate desirable and undesirable postures as related to the spine. (Note in the inset of *b* the uneven pressures between the vertebrae caused by the angle between them.) Part *c* illustrates the angles of body members that would provide a reasonably satisfactory posture while driving. (*Source:* a *and* b, *Rizzi, 1969, fig. 5, p. 230;* c, *Rebiffé, 1969, fig. 1, p. 247.*)

joints shown in Figure 11-18*c* are those which are proposed by Rebiffé (1969) as providing the basic driving posture which he considers to be desirable in terms of anthropometric considerations.

DISCUSSION

It must be granted that tables of anthropometric data do not make for very enthralling bedtime reading (although they might help those who have insomnia). But when we consider the application of such data to the design of the physical facilities and objects we use, we can see that such intrinsically uninteresting data play a live, active role in the real, dynamic world in which we live and work.

In the application of anthropometric data to specific design problems there can be no nicely honed set of procedures to follow, because of the variations in the circumstances in question and in the types of individuals for whom the facilities would be designed. As a general approach, however, the following suggestions are offered (based in part on those of Hertzberg, 1960):

1 Determine the body dimensions important in the design (e.g., sitting height as a basic factor in seat-to-roof dimensions in automobiles).

2 Define the population to use the equipment or facilities. This establishes the dimensional range that needs to be considered (e.g., children, women, United States civilians, different age groups, world populations, different races).

3 Determine what "principle" should be applied (e.g., design for extreme individuals, for an adjustable range, or for the "average").

4 When relevant, select the percentage of the population to be accommodated (e.g., 90 percent, 95 percent) or whatever is relevant to the problem.

5 Locate anthropometric tables appropriate for the population, and extract relevant values.

6 If special clothing is to be worn, add appropriate allowances (some of which are available in the anthropometric literature).

REFERENCES

Ayoub, M. M. Work place design and posture. *Human Factors,* 1973, *15*(3), 265–268.

Barnes, R. M. *Motion and time study* (5th ed.). New York: Wiley, 1963.

Bex, F. H. A. Desk heights. *Applied Ergonomics,* 1971, *2*(3), 138–140.

Bittner, A. C., Jr. Reduction in user population as the result of imposed anthropometric limits: Monte Carlo estimation. (TP-74-6.) Point Mugu, Calif.: Naval Missile Center, 1974.

Bullock, M. I. The determination of functional arm reach boundaries for operation of manual controls. *Ergonomics,* 1974, *17*(3), 375–388.

Burandt, V., & Grandjean, E. Sitting habits of office employees. *Ergonomics,* 1963, *6*(2), 217–228.

Corlett, E. N., & Bishop, R. P. A technique for assessing postural discomfort. *Ergonomics,* 1976, *19*(2), 175–182.

Damon, A., Stoudt, H. W., & McFarland, R. A. *The human body in equipment design.* Cambridge, Mass.: Harvard, 1966.

Dannhaus, D., Bittner, A., Jr., & Ayoub, M. M. Seating, console, and workplace design. In M. M. Ayoub and C. A. Halcomb (Eds.), *Improved seat console design: Final report.* Lubbock, Tex.: Institute for Biotechnology, Texas Tech University, Oct. 1976.

Dempster, W. T. *Space requirements of the seated operator* (Technical Report 55-159). U.S. Air Force, WADC, July 1955.

Eastman, M. C., & Kamon, E. Posture and subjective evaluation at flat and slanted desks. *Human Factors,* 1976, *18*(1), 15–26.

Ellis, D. S. Speed of manipulative performance as a function of work-surface height. *Journal of Applied Psychology,* 1951, *35*, 289–296.

Grandjean, E., Boni, A., & Krestzschmer, H. The development of a rest chair profile for healthy and notalgic people. *Ergonomics,* 1969, *12*(2), 307–315.

Grandjean, E., Hünting, W., Wotzka, G., & Shärer, R. An ergonomic investigation of multipurpose chairs. *Human Factors,* 1973, *15*(3), 247–255.

Hertzberg, H. T. E. Dynamic anthropometry of working positions. *Human Factors,* August 1960, *2*(3), 147–155.

Hockenberry, J. Comfort = "the absence of discomfort." *C P News* (Newsletter of the Consumer Products Technical Group), Human Factors Society, 1979, *4*(2).

Kroemer, K. H. E. Functional anthropometry. *Proceedings of the Human Factors Society,* October 16–19, 1978, pp. 680–683.

Meyer, R. P. Articulated anthropometric models. *C P News* (Newsletter of the Consumer Products Technical Group), Human Factors Society, 1979, *4*(2).

Nachemson, A., & Elfström, G. Intravital dynamic pressure measurements in lumbar discs. *Scandinavian Journal of Rehabilitation Medicine,* Suppl. 1, 1970.

National Aeronautics and Space Administration (NASA). *Anthropometric source book (Vol. I) Anthropometry for designers; (Vol. II) A handbook of anthropometric data; (Vol. III) Annotated bibliography* (NASA Reference Publication 1024). 1978.

Rebiffé, P. R. Le siège du conducteur: son adaptation aux exigences fonctionnelles et anthropométriques. *Ergonomics,* 1969, *12*(2), 246–261.

Ridder, C. A. *Basic design measurements for sitting.* Fayetteville, Ark.: University of Arkansas, (Bulletin 616, Department of Home Economics, Agricultural Experiment Station) 1959.

Rigby, L. V., Cooper, J. I., & Spickard, W. A. Guide to integrated system design for maintainability (Technical Report 61-424). U.S. Air Force, ASD, Oct. 1961.

Rizzi, V. M. Entwicklung eines verschiebbaren rückenprofils für auto-und ruthesitze. *Ergonomics,* 1969, *12*(2), 226–233.

Roth, J. T., Ayoub, M. M., & Halcomb, C. G. Seating, console and workplace design: Seated operator reach profiles. *Proceedings of the Human Factors Society, 21st annual meeting,* Oct. 17–20, 1977, pp. 83–87.

Rozier, C. K. Three-dimensional work space for the amputee. *Human Factors,* 1977, *19*(6), 525–533.

Sanders, M. S. Anthropometric survey of truck and bus drivers: Anthropometry, control reach and control force. Westlake Village, Calif.: Canyon Research Group, Inc., 1977.

Squires, P. C. *The shape of the normal work area* (Report 275). New London, Conn.: Navy Department, Bureau of Medicine and Surgery, Medical Research Laboratory, July 23, 1956.

Tichauer, E. R. In *The industrial environment: Its evaluation and control.* Washington, D. C.: Department of Health, Education and Welfare, National Institute for Occupational Safety and Health, 1973.

Tichauer, E. R. *The biomechanical basis of ergonomics.* New York: Wiley, 1978.

United States Public Health Service. *Weight, height, and selected body dimensions of adults: United States, 1960–1962* (USPHS Publication 1000, ser. 11, no. 8). Data from National Health Survey, June 1965.

Ward, J. S., & Kirk, N. S. The relation between some anthropometric dimensions and preferred working surface heights in the kitchen. *Ergonomics,* 1970, *13*(6), 783–797.

CHAPTER **12**

PHYSICAL SPACE AND ARRANGEMENT

A large proportion of the lifetime of most people is spent within physical space environments that are man-made, ranging from "local" situations in which people find themselves (such as at workplaces, in a kitchen, or in an automobile), through intermediate types of situations (such as office buildings, homes, and theaters), to general environments (such as communities). Our common experience points up the effects that the designs of such space and facilities can have on people, including their performance, their comfort, and even their physical well-being. Since we cannot deal intensively with the many human factors aspects of the arrangement of the space and facilities people use, we will touch on only certain aspects, with particular attention to work stations.

CONSIDERATIONS IN LOCATION OF COMPONENTS

In the case of many systems and facilities there are various "components" that need to be located within the system or facility. (We will use the term *component* in this discussion to refer to virtually any relevant feature, such as displays, controls, materials, machines, work areas, and rooms.) It is reasonable to hypothesize that any given component has a generally "optimum" location for serving its purpose. This optimum would be predicated on the human sensory, anthropometric, and biomechanical characteristics that are concerned (reading a visual display, activating a foot push button, etc.), or on the performance of some operational activity (such as reaching for parts, preparing food in a restaurant, or storing material in a warehouse). Preferably, of course, components should be placed in their optimum locations, but since this frequently is not possible, priorities sometimes must be established. These priorities, however, do not descend from heaven like manna but must be otherwise determined, usually on the basis of some factors such as those mentioned below.

Guiding Principles of Arrangement

Before we touch on a few methods that are used in trying to figure out what should go where, however, let us set down a few general guides (in addition to the idea of *optimum location*) that may be helpful. Depending on the circumstance, these guidelines can deal with either or both of two separate but interrelated phases, as follows: that concerned with the general location of components (such as specific components within a fixed work space or larger components that might be located in a more general work area such as an office), and that concerned with the specific arrangement of components.

Importance Principle This principle deals with *operational importance*, that is, the degree to which the performance of the activity with the component is vital to the achievement of the objectives of the system or some other consideration. The determination of importance is largely a matter of judgment, but a warning light in an automobile to indicate engine malfunction or low oil supply should be directly in front of the driver.

Frequency-of-Use Principle As implied by the name, this concept applies to the frequency with which some component is used, such as having the activation control of a punch press in a convenient location since it is used very frequently.

Functional Principle The *functional principle* of arrangement provides for the grouping of components according to their function, such as the grouping together of displays, or controls, that are functionally related in the operation of the system. Thus, temperature indicators and temperature controls might well be grouped together, and electric power distribution instruments and controls usually should be in the same general location.

Sequence-of-Use Principle In the use of certain items, there are sequences or patterns of relationship that frequently occur in the operation of the equipment. In applying this principle, the items would be so arranged as to take advantage of such patterns; thus, items used in sequence would be in close physical relationship with each other. An example is given later in Figure 12-11 (page 355).

Discussion In putting together, like pieces of a jigsaw puzzle, the various components of a system, it is manifest that no single guideline can, or should, be applied consistently, across all situations. But, in a very general way, and in addition to the optimum premise, the notions of *importance* and *frequency* probably are particularly applicable to the more basic phase of *locating* components in a *general area* in the work space; in turn, the sequence-of-use and functional principles tend to apply more to the *arrangement* of components *within* a general area.

The application of these various principles of arrangement of components generally has had to be predicated on rational, judgmental considerations, since there has been little empirical evidence available regarding the evaluation of these principles. There are, however, some data from at least one study that cast a bit of light on this matter. The study in question, by Fowler, Williams, Fowler, and Young (1968), consisted of

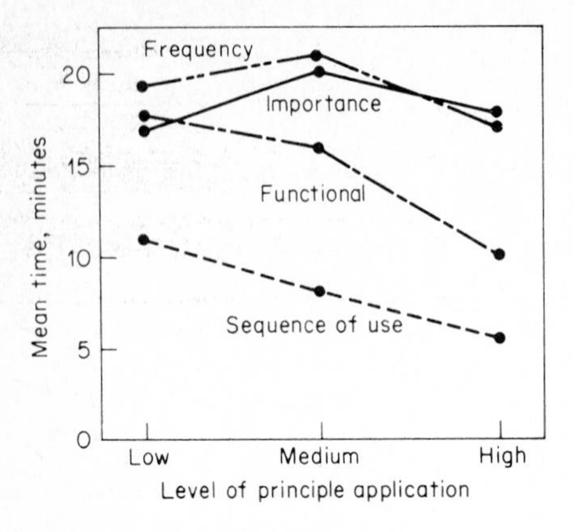

FIGURE 12-1
Time required to carry out a standard simulated task in the use of controls and displays arranged on the basis of four principles. In the case of each principle, three control panels were used, these varying in terms of the rated "level" or degree to which the principle has been applied in the panel design. (*Source: Adapted from Fowler, Williams, Fowler, & Young, 1968.*)

the evaluation of various control panel layouts in which the controls and displays had been arranged following each of the four principles described above. The panels included 126 standard military controls and displays. The arrangement of these following the four principles need not be described, but it should be added that, for each principle, three control panels were developed, varying in terms of three "levels" of application of the principle (based on a scoring scheme), these being high, medium, and low. The 200 male college student subjects used the various arrangements in a simulated task. Their performance was measured in terms of time and errors, the results for the time criterion being shown in Figure 12-1. This figure (and corresponding data regarding errors) showed a clear superiority for the sequence-of-use principle. Even the arrangements that were based on the "low" and "medium" application levels of this principle were better than, or equal to, the "high" levels for the other principles.

Although the sequence-of-use principle came out on top in this study, this principle obviously could be applied only in circumstances (as in this study) in which the operational requirements actually do involve the use of the components in question in rather consistent sequences. Where this is the case, this principle certainly should be followed.

GENERAL LOCATION OF COMPONENTS

As indicated above, it is reasonable to assume that any given component in a system or facility would have some reasonably optimum location, as predicated on whatever sensory, anthropometric, biomechanical, or other considerations are relevant. Although the optimum locations of some specific components probably would depend upon situational factors, some generalizations can be made about certain classes of components. Certain of these will be discussed below.

Visual Displays

The normal line of sight is usually considered to be about 15° below the horizon. Visual sensitivity accompanied by moderate eye and head movements permits fairly convenient visual scanning of an area around the normal line of sight. The area for most

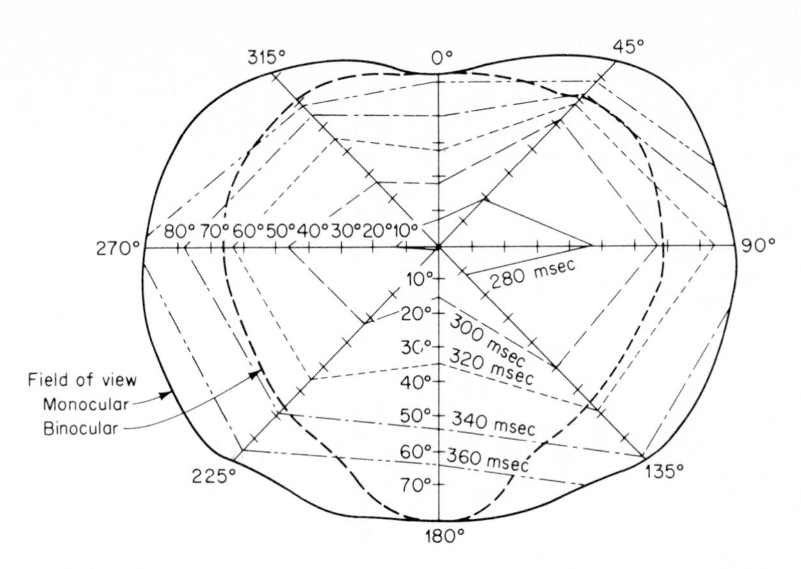

FIGURE 12-2
Isoresponse times within the visual field. Each line depicts an area within which the mean response time to lights is about the same. (*Source: Adapted from Haines & Gilliland, 1973, fig. 2. Copyright 1973 by the American Psychological Association and reproduced by permission.*)

convenient visual regard (and therefore generally preferred for visual displays) has generally been considered to be defined by a circle roughly 10 to 15° in radius around the normal line of sight.

However, there are indications that the area of most effective visual regard is not a circle around the line of sight but rather is more oval. Such an indication is found in the research of Haines and Gilliland (1973), who measured subjects' times of response to a small light that flashed on in various locations of the visual field. Figure 12-2 shows the mean response times to the lights in the various regions of the visual field. Each boundary on that figure indicates the region within which mean response time can be expected to be about the same (the *isoresponse* time region). The generally concentric lines tend to form ovals—but slightly lopsided ovals, being flatter above the line of sight than below.

The subjects in this study detected the lights even toward the outer fringes of this area, but the primary implications of this (and some other) research are that critical visual displays should be placed within a reasonably moderate oval around the normal line of sight.

Hand Controls

The optimum location of hand-control devices is, of course, a function of the type of control, the mode of operation, and the appropriate criterion of performance (accuracy, speed, force, etc.). Certain preceding chapters have dealt with some tangents to this matter, such as the discussion of the work-space envelope in Chapter 11.

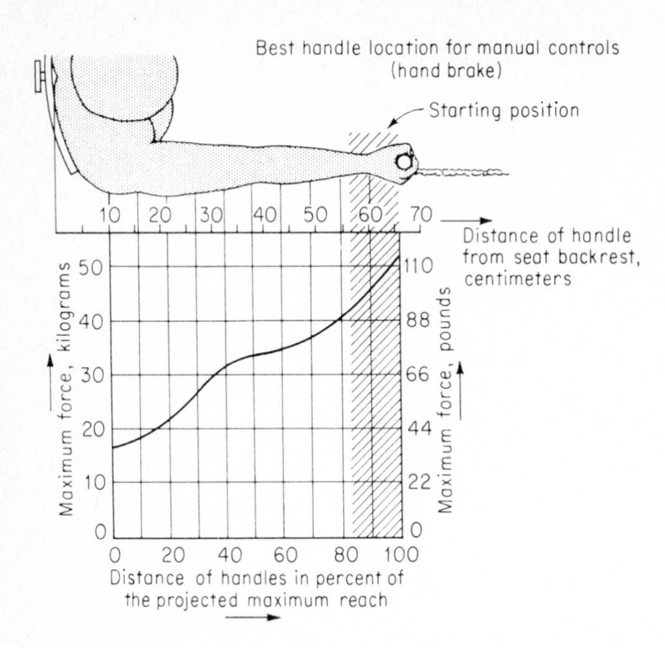

FIGURE 12-3
Relationship between maximum pulling force (such as on a hand brake) and location of control handle. (*Source: Dupuis, Preuschen, & Schulte, 1955.*)

Controls That Require Force Many controls are easily activated, and so a major consideration in their location is essentially one of ease of reach. Controls that require at least a moderate force to apply (such as certain control levers and hand brakes in some tractors), however, bring in another factor, that of force that can be exerted in a given direction with, say, the hand in a given position. Investigations by Dupuis (1957) and by Dupuis, Preuschen, and Schulte (1955) dealt with this question, specifically the pulling force that can be exerted, when seated, when the hand is at various distances from the body (actually, from a seat reference point). Figure 12-3, which illustrates the results, shows the serious reduction in effective force as the arm is flexed when pulling toward the body. The maximum force that can be exerted by pulling is about 57 to 66 cm forward from the seat reference point, and this span, of course, defines the optimum location of a lever control (such as a hand brake) if the pulling force is to be reasonably high.

According to Kroemer (1970) the best location for cranks and levers (especially those to be operated continuously to and fro) is in front of the sitting or standing operator so that the handle travels at about waist height in the sagittal plane (i.e., the vertical plane from front to back) passing through the shoulder.

Controls on Panels Many controls are positioned on panels or in areas forward of the person who is to use them. Because of the anthropometric and biomechanical characteristics of people, controls in certain locations can be operated more effectively

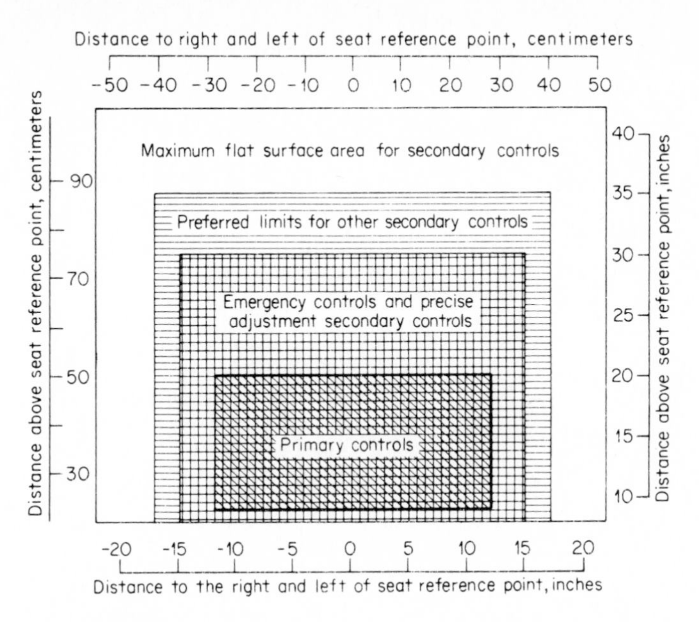

FIGURE 12-4
Preferred vertical surface areas and limits for different classes of manual controls. (*Source: Adapted from Human Factors Engineering, 1980.*)

than those in other locations. Figure 12-4 shows one set of preferred areas for four classes of control devices as based on relative priorities (Human Factors Engineering, 1980). Although this proposed arrangement is based on data for military personnel, it would of course have more general applicability.

When there are many controls and displays to arrange in a console or panel, the use of angled side panels may place more of the controls within convenient access. The advantage of this was demonstrated empirically by Siegel and Brown (1958) in a study in which subjects, using a 48-in front panel with side panels at 35, 45, 55, and 65°, followed a sequence of verbal instructions to use the controls on the panels. A number of criteria were obtained, including objective criteria of average number of seat movements, average seat displacement, average body movements (number and extent), average number of arm extensions (part and full), subjective criteria based on the subjects' responses of degree of ease or difficulty, judgments that the panels should be wider apart or closer together, and preference ranking for the four angles.

Only some of the data will be presented, but they characterize the results generally. Figure 12-5 shows data for four of the criteria for the four angles. The criterion scales have been converted here to fairly arbitrary values, and only the "desirable" and "undesirable" directions are indicated. It can be seen, however, that all four criteria were best for the 65°-angle side panels. The consistency across all criteria (these and the others) was quite evident.

Although Figures 12-4 and 12-5 provide general guidelines for locating controls, including consideration of relative priorities, it should be added that the operational requirements of certain specific types of controls sometimes impose impossible

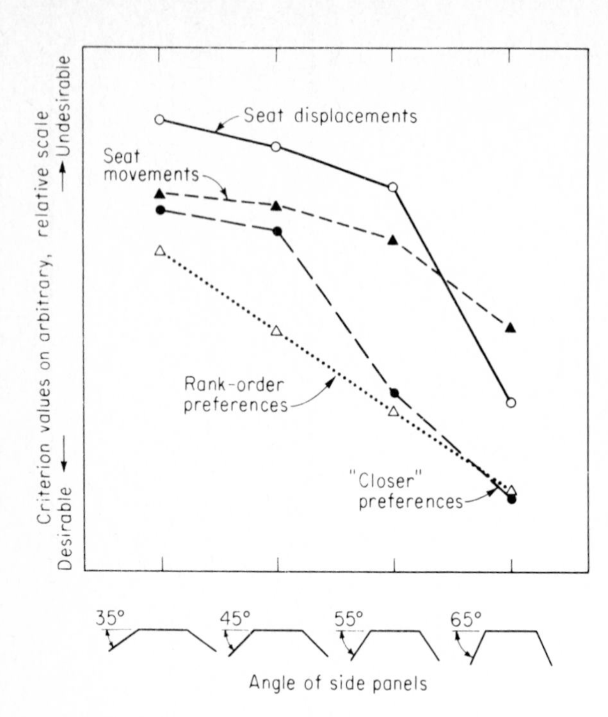

FIGURE 12-5
Representation of criteria from study relating to angles of side panels of console. The criterion values are all converted to an arbitrary scale for comparative purposes, but they all indicate the desirability of the 65° angle panels over the others. (*Source: Adapted from Siegel & Brown, 1958, courtesy of Applied Psychological Services, Wayne, Pa.*)

constraints on their location. This was illustrated, for example, by the results of a study by Sharp and Hornseth (1965) in which seated subjects operated each of 3 types of controls (knobs, toggle switches, and push buttons) at each of 12 locations in each of 3 consoles (far, middle, and close). The controls of the close console were positioned for convenience of reach of individuals of small build (about the 5th percentile of males), and only data for this console will be given here.

Figure 12-6*a* shows the time in seconds to activate the three types of controls when located at various angles from the center position, that part indicating that activation time was minimum for all three controls at about 25° from center. Figure 12-6*b* in turn shows the areas of the shortest performance times for each type of control (e.g., the area within which times were *within 5 percent* of *minimum* times for the control in question); the smaller area for the toggle switch suggests that the selection of a location for such devices may be more critical than for the other devices if time is of the essence. However, as suggested by the investigators, a well-designed system preferably should not impose response-time requirements on operators in which differences of 0.1, 0.2, or 0.3 s are crucial.

Foot Controls

Since only the most loose-jointed among us can put their feet behind their head, the location of foot controls generally needs to be in the fairly conventional areas, such as those depicted in Figure 12-7. These areas, differentiated as optimal and maximal, for

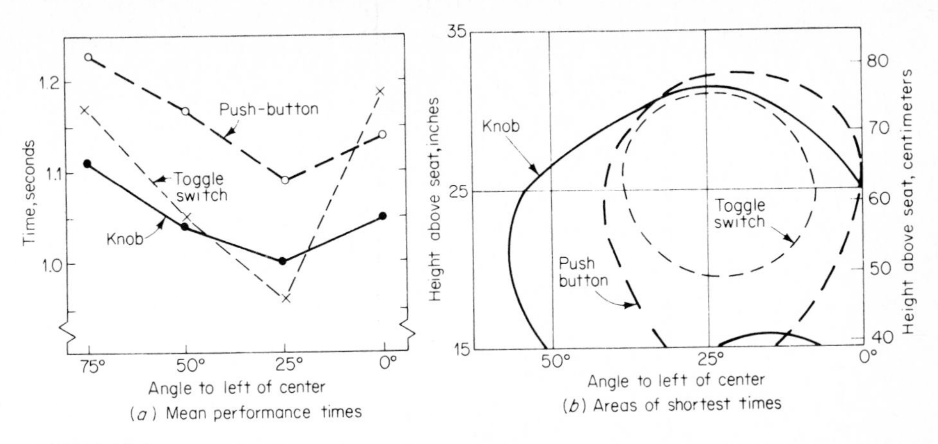

FIGURE 12-6
Data on time to activate push button and toggle switch and time to reach knob, when in various positions. Data are for *left* hand, and for close console. Part *a* gives mean times for controls positioned 25 in above seat reference level. Part *b* gives contours of the area for which times were *shortest* for the particular control (e.g., times within 5 percent of the minimum time for the control). (*Source: Adapted from Sharp & Hornseth, 1965.*)

FIGURE 12-7
Optimal and maximal vertical and forward pedal space for seated operators. (*Source: Adapted from Human Factors Engineering, 1980.*)

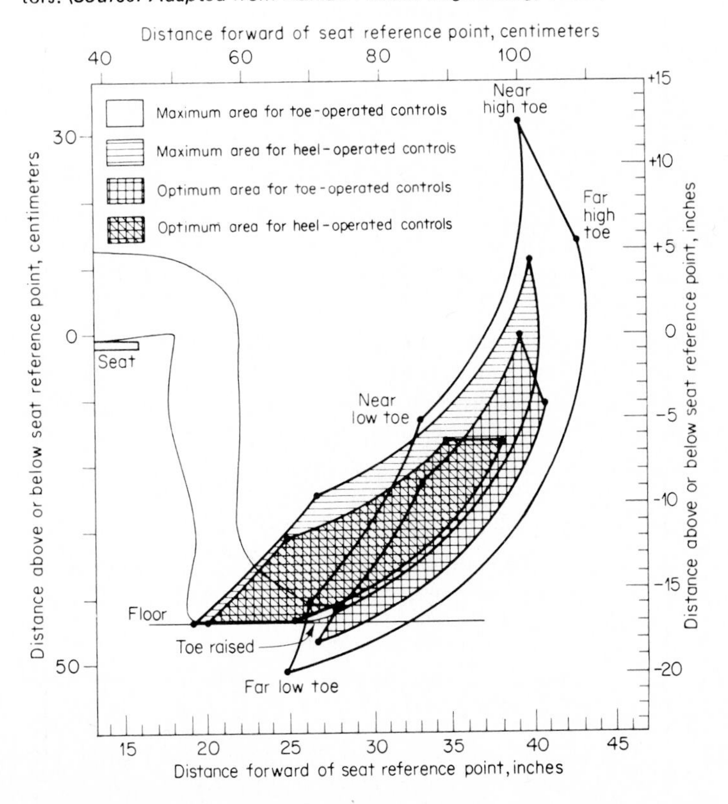

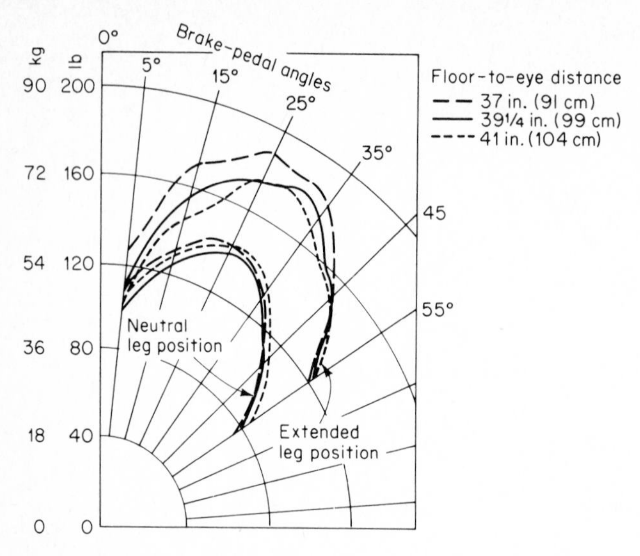

FIGURE 12-8
Mean maximum brake-pedal forces exerted by 100 Air Force pilots in two leg positions and for various brake-pedal angles and three "floor-to-eye" distances. (The seat level was adjusted for three conditions so the eye level was at the three specified distances above the floor.) (*Source: Adapted from Hertzberg & Burke, 1971, fig. 6.*)

toe-operated and heel-operated controls, have been delineated on the basis of dynamic anthropometric data. The maximum areas indicated require a fair amount of thigh or leg movement or both, and preferably should be avoided as locations for frequent or continual pedal use. Incidentally, Figure 12-7 is predicated on the use of a horizontal seat pan; with an angular seat pan (and more angled backrest) the pedal locations need to be manipulated accordingly (although such adjustments have been published, they will not be given here).

The areas given in Figure 12-7 generally apply to foot controls that do not require substantial force. For applying considerable force, a pedal preferably should be fairly well forward. This point is illustrated by Figure 12-8, which shows the mean maximum brake-pedal forces for 100 Air Force pilots when the leg was in a "normal" position and in an extended position (with the leg essentially forward from the seat). The figure shows that the maximum brake-pedal forces were clearly greater for the extended position for three seat levels that were used in the experiment. (The seat level was adjusted so the "floor-to-eye" distances were standardized.) As an aside, Figure 12-8 also shows that the forces for both leg positions were greatest when the foot angle on the pedal was between about 15 and 35° from the vertical.

SPECIFIC ARRANGEMENT OF COMPONENTS

As indicated in the above discussion of the general arrangement of components, there are certain general areas that are most suitable (or even required) for locating various

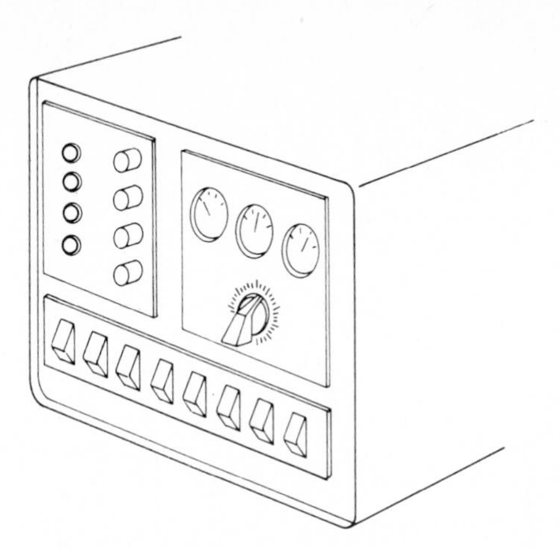

FIGURE 12-9
Example of controls and displays grouped by function, in which the different groups are clearly indicated.

types of components, such as visual displays, hand controls, and foot controls. Within the constraints imposed by these considerations the next process is that of arranging components within the areas that are appropriate for them. In this process the arrangement of groups of components can be based on the principles of sequence or function. Where there are common sequences, or at least frequent relationships, in the use of displays, controls, or other components, the layout usually should be such as to facilitate the sequential process—as in hand movements, eye movements, etc. Where there are no fixed or common sequences, the components should be grouped on the basis of function. In such instances the various groups should be clearly indicated by borders, color, shading, or otherwise. An example of such groups is shown in Figure 12-9. When relevant, of course, compatibility principles should be followed in arranging components.

In many circumstances, especially when there are very few components to be placed, there is no particular problem in arranging them in a satisfactory manner. When there are a number of components, or when a satisfactory arrangement is not very obvious, some systematic procedure, including the collection and analysis of relevant data, may be in order. When a modification of an existing system is being developed, data relating to an existing model may be appropriate. Such data can be obtained by various methods, such as the use of film; observation; the use of eye-movement recordings (with eye cameras or other related devices); and interviews with experienced personnel (including questions to elicit their opinions, such as about frequency or importance of various activities, or about the desirable arrangement of components).

On the other hand, in the case of new systems or facilities (without current counterparts), information about the activities to be performed (such as frequency, sequence, and other activity parameters) needs to be inferred from whatever tentative drawings, plans, or concepts are available.

Summarizing and Using Activity Data

In some circumstances the needs for at least certain data on which to base design decisions can be partially fulfilled by simply summarizing whatever relevant activity data

TABLE 12-1
ILLUSTRATION OF METHOD OF COMBINING RATINGS OF IMPORTANCE AND FREQUENCY FOR USE IN ARRANGING DISPLAYS

	Display				
	A	*B*	*C*	*D*	*E*
1. Average rating, importance	3.0	1.7	1.6	2.5	1.1
2. Average rating, frequency	3.0	1.2	1.7	1.0	2.7
3. Sum of 1 and 2	6.0	2.9	3.3	3.5	3.8
4. Average of 1 and 2	3.0	1.45	1.65	1.75	1.9

are obtained (such as frequency of use of a brake pedal) or by deriving average values (such as the average time spent by nurses walking to and from patients' rooms). In other cases, however, it may be desirable to derive some index of priorities to provide guidance in locating components. In such instances the designer has to decide what the appropriate basis for the priorities (rated importance, frequency of use, etc.) should be. When two or more such factors seem relevant, however, it is possible to combine them into a single index, either by adding the ratings on, say, importance and frequency, or by multiplying them. Table 12-1 shows, for each of five displays (*A, B, C, D,* and *E*), the average ratings on these two factors (a rating of 3 being high and a rating of 1 being low); in addition, the sum and the average of the two ratings are given for each display.

Although the average values shown in Table 12-1 are based on equal weighting of importance and frequency, in some circumstances differential weighting might well be in order.

Links as Indexes of Interrelationships

The operational relationships between people and between people and physical components usually can be expressed in terms of *link* values. Link values can be developed for a wide range of such relationships, although they fall generally into three classes, namely, communication links, control links, and movement links. Communication and control links can be considered as functional. Movement links generally reflect sequential movements from one component to another. Some versions of the three types follow:

I Communication links
- **A** Visual (person to person or equipment to person)
- **B** Auditory, voice (person to person, person to equipment, or equipment to person)
- **C** Auditory, nonvoice (equipment to person)
- **D** Touch (person to person or person to equipment)

II Control links
  E Control (person to equipment)
III Movement links (movements from one location to another)
  F Eye movements
  G Manual movements, foot movements, or both
  H Body movements

Link indexes can be used as aids in connection with the general location of components or with their relative arrangements. In some circumstances they can be used as the bases for assignment of priorities.

Derivation of Link Values In the case of existing systems or facilities, certain link values can be derived from empirical observation or films, especially link values of frequency (such as frequency of control actions or of movements from one component to another). In the case of systems or facilities being developed, frequency usually would have to be estimated. Link values of importance almost of necessity need to be based on judgment. When link values are derived for both importance and frequency, it is usually the practice to compute a composite link value by multiplying or adding the importance and frequency values of the individual links, in much the same manner as illustrated above in the case of priority indexes.

Link values in operational procedures sometimes can be derived by a graphic approach in which the sequential steps in an operation are recorded. Subsequently the functional links and the sequential links (showing relations in operation between all pairs of components) can be tallied in the manner presented by Haygood, Teel, and Greening (1964) and illustrated in Table 12-2. This table actually presents data for only a few of the "panels" that were considered as components, but these will at least illustrate the nature of the results. Incidentally, the purpose of this investigation was to compare a computerized approach to the development of link values with a more conventional graphic approach. Both methods produced substantially the same results in link values, but the computerized approach was more economical of time and cost.

TABLE 12-2
LINK ANALYSIS OF CERTAIN PANELS OF FLIGHT SYSTEM CHECK-OUT CONSOLE

	Functional links		Sequential links among panels				
	Visual	Control	2	3	4	5	6
1. Master selector	175	884	348	97	44	7	1
2. Programmer	31	637		32	11	10	3
3. Recorder	51	81			0	1	2
4. Meter panel	63	0				8	0
5. Power supply	5	49					0
6. Oscilloscope	12	12					

Source: Haygood, Teel, & Greening, 1964, tables 1 and 2.

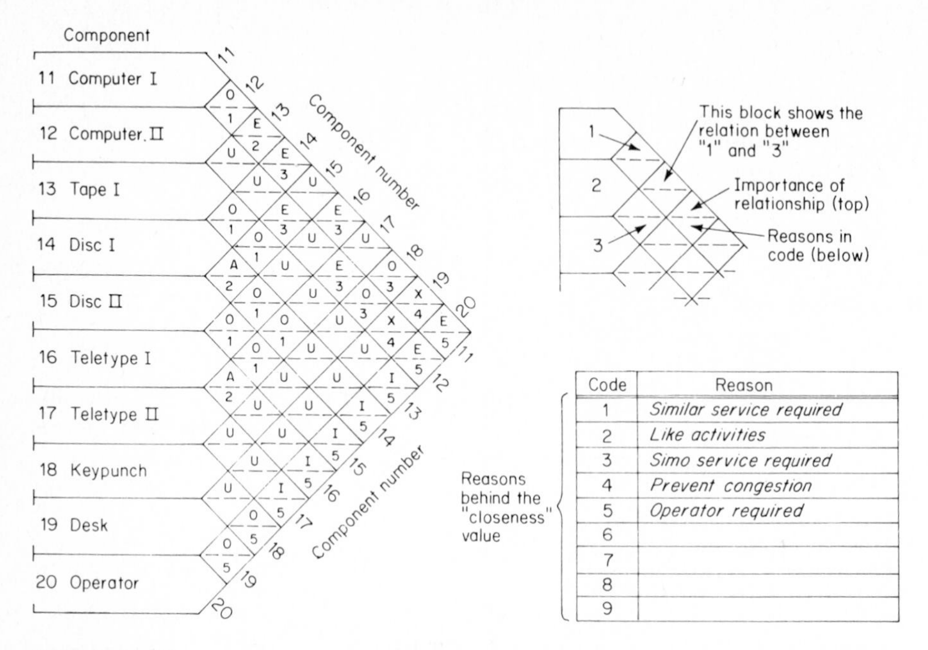

FIGURE 12-10
Illustration of a chart of the link values of the relationships between pairs of components in a hypothetical minicomputer laboratory. See text for discussion of link values and the codes. (*Source: Cullinane, 1977, fig. 1.*)

Another example of link analysis is presented by Cullinane (1977), as illustrated in Figure 12-10, this representing a hypothetical minicomputer laboratory. In this procedure the relationships between all pairs of components are rated using the following rating scale regarding the relationships:

A – It is ABSOLUTELY ESSENTIAL for the activities to be located close together.
E – It is ESSENTIAL for two activities to be located close together.
I – It is IMPORTANT that the two activities be located close together.
O – Ordinary closeness is ACCEPTABLE for the two activities being considered.
U – A link does not exist, and it is UNIMPORTANT whether two activities are placed together.
X – It is UNDESIRABLE for two activities to be placed together.

The chart in Figure 12-10 shows the link values in the upper triangular part of the square formed by the down-sloping line of one component and the up-sloping line of another (these being *A, E, I, O, U,* or *X*), and in the lower triangular part of the square a code explaining the reasons for the relationship. (In the example in Figure 12-10, the codes used for this purpose are in the inset in the lower right-hand corner.)

Use of Activity Data in Arranging Components

In the use of activity data for developing a reasonably optimal arrangement of components, the typical approach is through some form of physical stimulation. However, in some circumstances, more systematic, quantitative methods can be used.

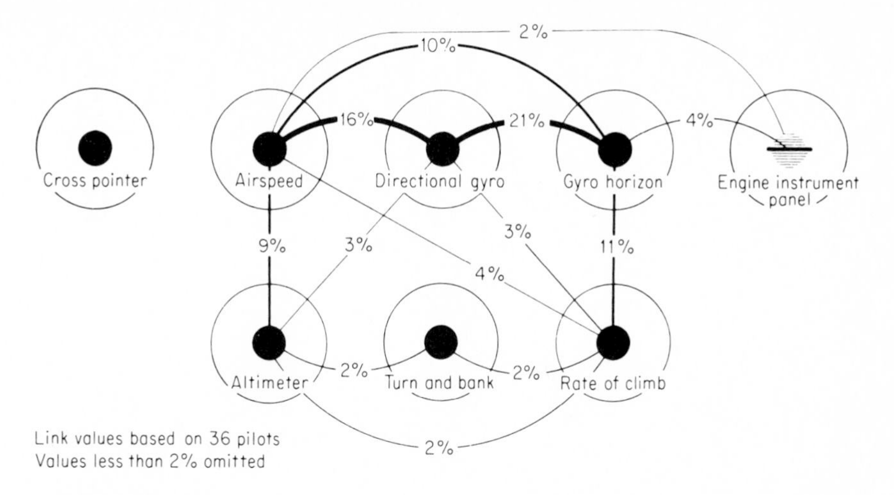

FIGURE 12-11
Eye-movement link values between aircraft instruments during climbing maneuver with constant heading. (*Source: From U.S. Air Force, AMRL, Behavioral Sciences Laboratory.*)

Arrangement by Physical Simulation In the use of a physical simulation approach, the various components are juggled around on paper (in graphic form) or in the form of models or mock-ups, until an arrangement is achieved that is judged to be reasonably optimum for whatever considerations are relevant (for example, the optimum location of components, their priorities, or their functional link values or sequential link values.)

An example of an arrangement developed by essentially graphic methods comes from a study of the layout of aircraft instruments (Jones, Milton, & Fitts, 1949). The basic data for the study came from the use of an eye camera that records the reflection of the light from the cornea of the eye. Recordings were made of the eye movements of 36 pilots during various maneuvers. Figure 12-11 shows the sequential link values between the various instruments (expressed as percentages of eye shifts between instruments) for one particular maneuver. Data such as shown in these figures were used to develop a standard instrument arrangement (basically the one shown) which has been accepted by the U.S. Air Force, the U.S. Navy, American commercial airlines, the Royal Canadian Air Force, and the Royal Air Force (U.K.).

Quantitative Solutions to Arrangement Problems Especially in simple problems of arranging components, it would be gilding the lily to apply sophisticated quantitative methods. But with complex systems that have many components, some quantitative attack may well be justified. One such method is that of *linear programming*. This is a method that results in the optimizing of some criterion or dependent variable by manipulation of various independent variables. The optimum in some cases would be the minimum, and in other cases the maximum—whichever is the desired value in terms of the criterion.

An example of this technique draws upon data from the following two sources: (1) data on frequencies with which a pilot made task responses, with eight controls, in

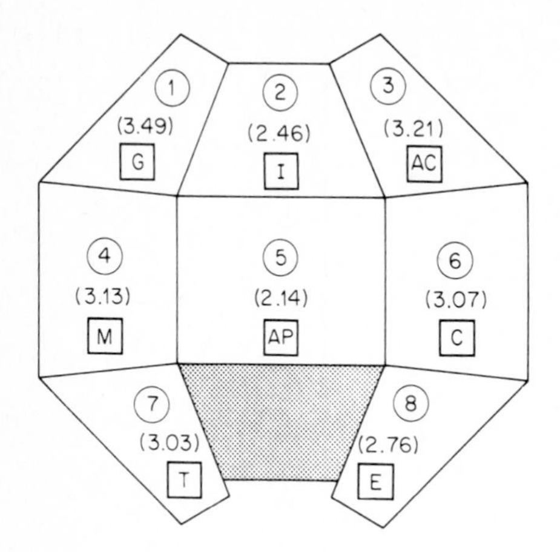

FIGURE 12-12
Perspective drawing of eight areas used in application of linear programming to the arrangement of eight aircraft controls. (*Source: Huebner & Ryack, 1961.*) Mean accuracy scores of blind-positioning movements are given in parentheses for the eight target areas. (*Source: Based on data from Fitts, 1947.*) The letters in boxes are symbols for the eight controls, their locations shown here representing the optimum linear programming solution.

flying simulated cargo missions in a C-131 aircraft, over 139 1-min periods (Deininger, 1958); and (2) data on the accuracy of manual blind-positioning responses in various areas, based on the study by Fitts (1947); for this particular purpose accuracy data for 8 of the 20 target areas were used. The accuracy scores are average errors in inches in reaching to targets in the eight areas. Linear programming, when used in the analysis of the data by Huebner and Ryack (1961), involved the derivation of a *utility cost* rating for each of eight controls in each of the eight areas shown in Figure 12-12. Each one was computed by multiplying the *frequency* of responses involving each control (from Deininger) by the *accuracy* of responses in each area (from Fitts). For example, the values for the *C* control (cross-pointer set) for the eight areas range from 42.8 to 69.8, each such value being the product of the *frequency* value for the control (in this case 20) and the mean *accuracy* scores for the areas as shown in Figure 12-12. For any possible arrangement of controls one can derive a *total cost*, this being the sum of the utility costs of the several controls in their locations specified by that arrangement. By linear programming it was possible to identify the particular arrangement for which the total cost was minimum. The resulting optimum arrangement is that shown in Figure 12-12, the individual controls simply being identified by a letter code. Note that each control is not in its *own* optimum location, but that *collectively* the derived arrangement is optimum in terms of the criterion of total cost. (Control *C*, mentioned above, was thus placed in space 6, even though its utility cost in that location was less.)

Another example of a computerized approach for designing a facility is described by Bonney and Williams (1977), this dealing with the arrangement of control devices and other items at the "pulpit" of a bar and rod mill. Figure 12-13*a* shows the existing layout, the items being located on a wall panel, a left desk, a right desk, and the floor. Figure 12-13*b* shows the arrangement based on a computer program that was intended to optimize the collective locations of the items. Without going into details, the computerized arrangement was estimated to reduce limb movements from 35 m to 31 m, and provided for 98 percent of the operational time in postures and positions with a specified "comfort rating," as compared with 50 percent of the time for the existing arrangement.

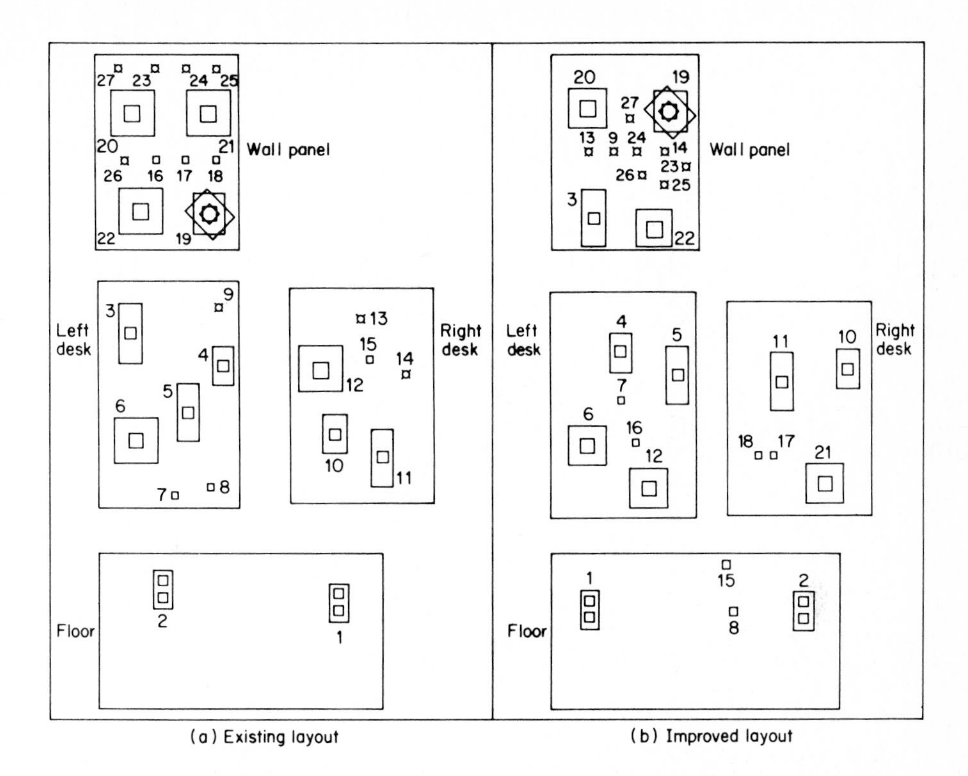

Key

1. pedal 1	6. lever 4	11. lever 6	16. button 4	21. lever 9	26. lamp 7
2. pedal 2	7. button 1	12. lever 7	17. button 5	22. lever 10	27. lamp 8
3. lever 1	8. button 2	13. lamp 2	18. button 6	23. lamp 4	
4. lever 2	9. lamp 1	14. lamp 3	19. knob 1	24. lamp 5	
5. lever 3	10. lever 5	15. button 3	20. lever 8	25. lamp 6	

FIGURE 12-13

(*a*) Existing arrangement of controls and other items of the "pulpit" of a bar and rod mill in Great Britain. (*b*) An improved arrangement developed with a computer program to optimize the location of the items. (*Source: Bonney & Williams, 1977, figs. 3 and 5.*)

Although quantitative solutions for design assignment problems may well be warranted in connection with some complex design problems, Francis and White (1974, chap. 6) point out that in many circumstances such procedures are not warranted, and that simpler approaches can produce substantially similar results. In connection with the design problem represented in Figure 12-12, for example, an approach such as the following might be used:

Put the control with highest frequency in the area with the lowest error.

Put the control with the second highest frequency in the area with the second lowest error.

Put the control with the third highest frequency in the area with the third lowest error.

Etc., to the eighth control.

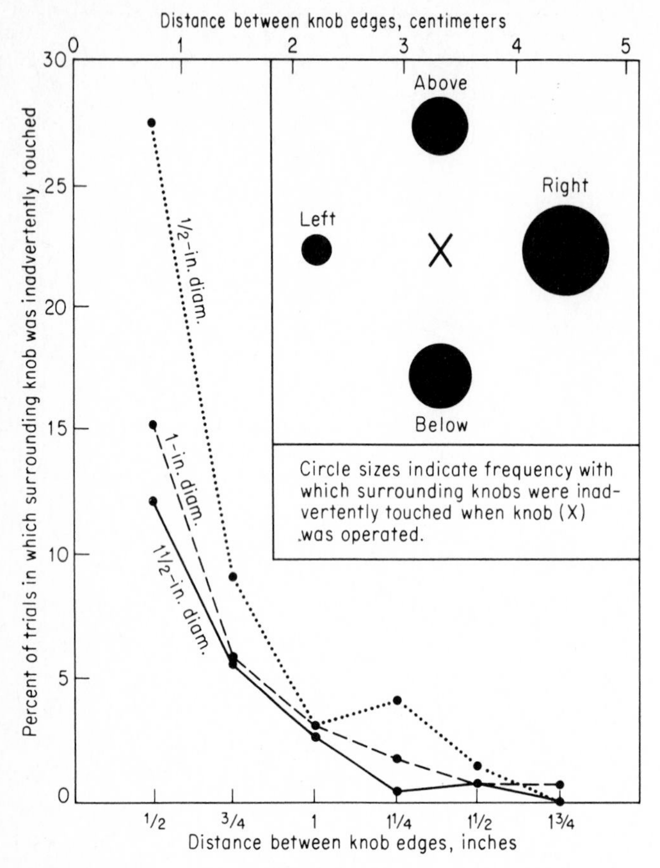

FIGURE 12-14
Frequency of inadvertent touching errors for knobs of various diameters as a function of the distance between knob edges. Areas of circles in inset indicate relatively the frequency with which the four surrounding knobs were inadvertently touched when knob *x* was being operated. (*Source: Bradley, 1969, fig. 2.*)

SPACING OF CONTROL DEVICES

Although we have talked about minimizing the distances between components, such as the sequential links between controls, there are obvious lower-bound constraints that need to be respected, such as the physical space required in the operation of individual controls to avoid touching other controls. Whatever lower-bound constraints there might be would be predicated on the combination of anthropometric factors (such as of the fingers and hands) and on the precision of normal psychomotor movements made in the use of control devices.

An illustration of the effects of such factors is given in Figure 12-14, this showing inadvertent "touching errors" in the use of knobs of various diameters as a function of the distances between their edges. In this instance the figure shows that errors dropped sharply with increasing distances between knobs up to about 1 in (2.5 cm), while be-

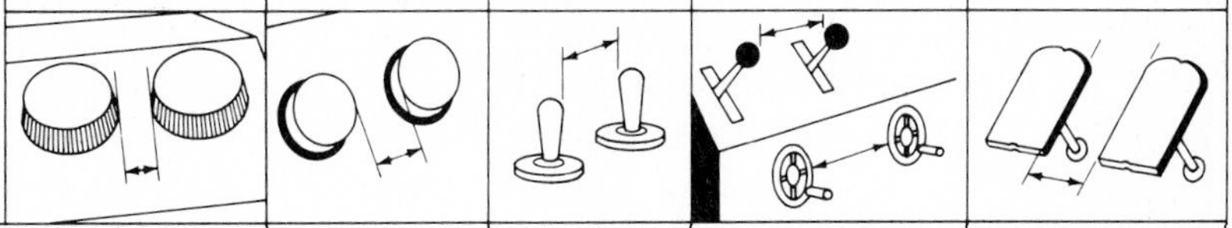

Number of body members and type of use		Knobs	Push buttons	Toggle switches	Cranks, levers	Pedals
1, randomly	in	2(1)	2(1/2)	2(3/4)	4(2)	6(4)
	cm	5(2.5)	5(1.3)	5(1.8)	10(5)	15(10)
1, sequentially	in		1(1/4)	1(1/2)		4(2)
	cm		2.5(.6)	2.5(1.3)		10(5)
2, simultaneously	in	5(3)			5(3)	
	cm	12.7(7.6)			12.7(7.6)	
2, randomly, sequentially	in		1/2(1/2)	3/4(5/8)		
	cm		1.3(1.3)	1.8(1.6)		

FIGURE 12-15
Recommended separation (in inches and centimeters) between adjacent controls. Preferred separations are given for certain types of use with corresponding minimum separation in parentheses. (*Source: Adapted from Chapanis, 1972.*)

yond that distance performance improved at a much slower rate. When separate comparisons were made between knob centers (rather than edges), however, performance was more nearly error-free for ½-in-diameter knobs than for the larger knobs. This suggests that when panel space is at a high premium, the smaller-diameter knobs are to be preferred. By referring again to Figure 12-14, it can be seen that touching errors were greatest for the knob to the right of the one to be operated, and were minimal for the one on the left.

On the basis of various studies of control devices Chapanis (1972, chap. 8) set forth certain recommended distances (preferred and minimal) between pairs of similar devices, as given in Figure 12-15.

GENERAL GUIDELINES IN DESIGNING INDIVIDUAL WORKPLACES

In designing workplaces some compromises are almost inevitable because of competing priorities. In this regard, however, appropriate link values can aid in the trade-off process. Some general guidelines for designing workplaces that involve displays and controls are given below (Van Cott & Kinkade, 1972, chap. 9):

- *First priority:* Primary visual tasks
- *Second priority:* Primary controls that interact with primary visual tasks
- *Third priority:* Control-display relationships (put controls near associated displays, compatible movement relationships, etc.)
- *Fourth priority:* Arrangement of elements to be used in sequence
- *Fifth priority:* Convenient location of elements that are used frequently
- *Sixth priority:* Consistency with other layouts within the system or in other systems

EXAMPLES OF INDIVIDUAL WORKPLACES

Although it is not feasible to illustrate the physical arrangement of very many types of workplaces, a few examples will be given below—in particular, certain ones for which some design guidelines have been developed or that represent special design problems.

Consoles

Consoles of one sort or another are used for operators of various types of systems, the consoles usually including both displays and controls. Figure 12-16 illustrates the design features of consoles that are recommended on the basis of anthropometric characteristics of people along with consideration of their visual and psychomotor skills.

FIGURE 12-16
Recommended design features of consoles for seated operators. (These features are designed to be suitable for persons ranging from the 5th to the 95th percentiles of an Air Force population.) (*Source: Adapted from Van Cott & Kinkade, 1972, fig. 9-6.*)

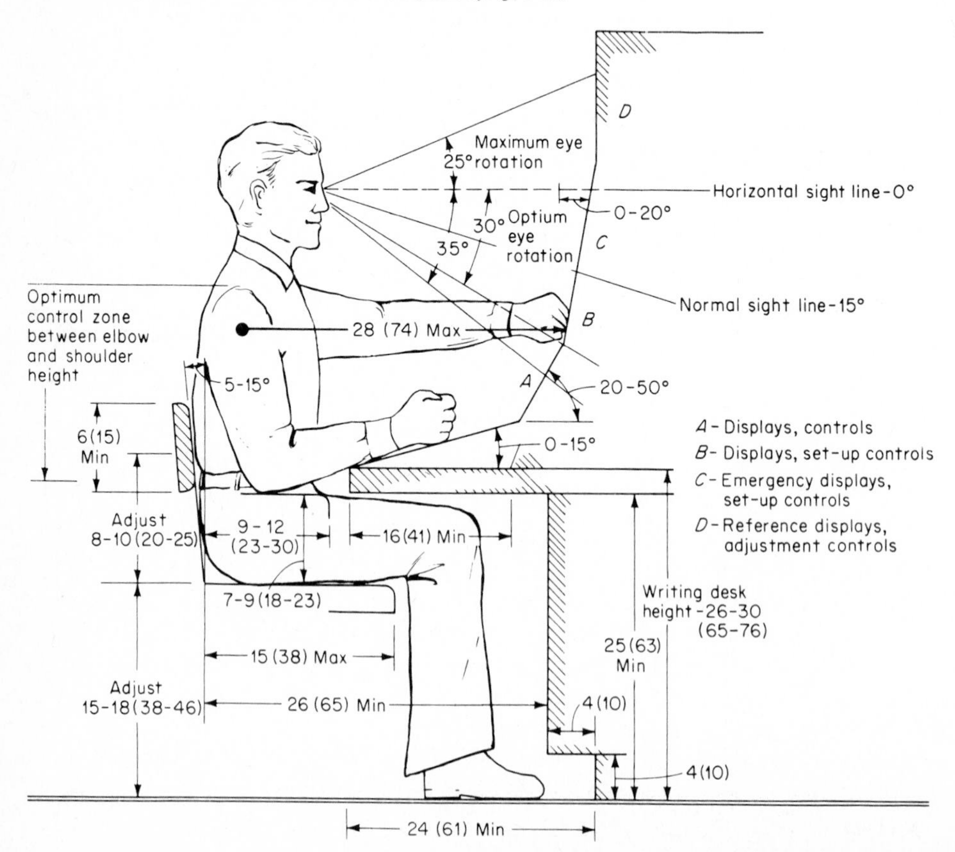

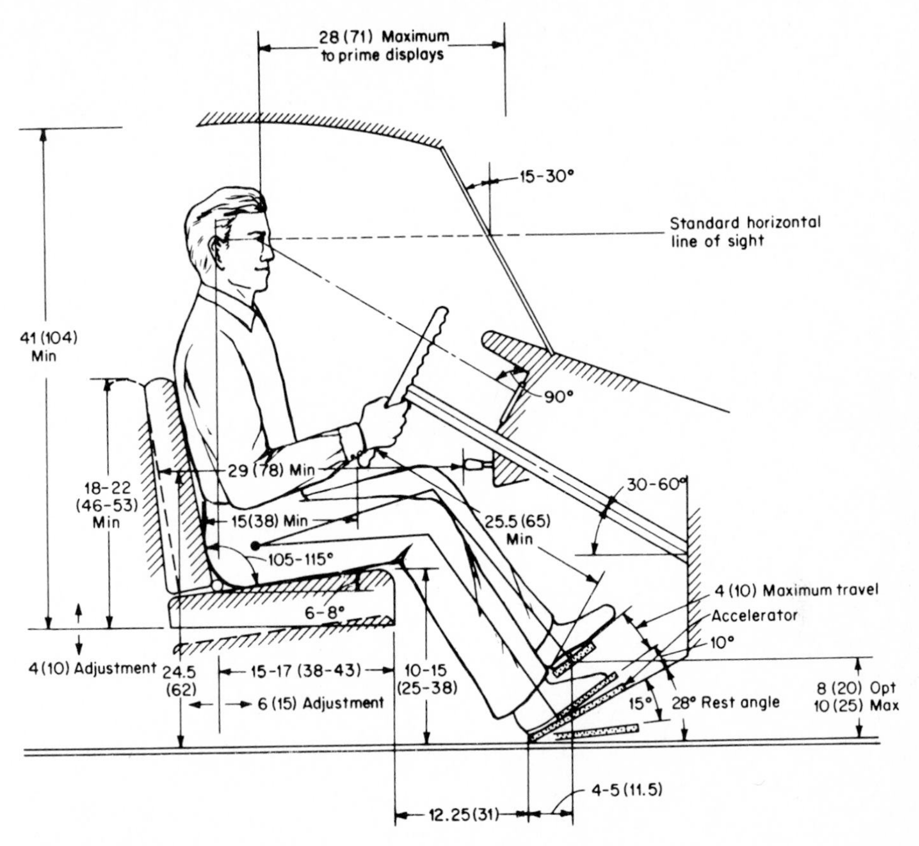

FIGURE 12-17
Recommended design features of vehicle cabs. (These features are designed to be suitable for persons between the 5th and 95th percentiles of operators.) (*Source: Adapted from Van Cott & Kinkade, 1972, fig. 9-11.*)

Vehicle Cabs

A similar set of recommended design features of vehicle cabs is shown in Figure 12-17, this also representing something of a general composite. One particular feature should be given special attention, namely, the visual field that can be viewed from the driver's position. (In Figure 12-17, with a vertically adjustable seat, the eye position of most drivers would be within 2 in above or below the position shown in the figure.) Other important features are the seat height, depth, and back angle; leg and knee clearance; and hand and foot reach requirements for control action.

Crane Controls and Cabs

In the iron and steel industries, as well as certain others, traveling overhead cranes are used to move materials from one place to another. The operational requirements of this moving control would suggest the need for unobstructed vision, easily operated

(*a*) Original crane cab

(*b*) Original crane controls

(*c*) Improved crane cab

(*d*) Improved crane controls

FIGURE 12-18
Illustration of original and improved cab and controls for overhead traveling crane used in steel mill. (*Source: Laner, 1961; Sell, 1958. Courtesy of the British Iron and Steel Research Association.*)

controls that respond with acceptable fidelity to the operator's responses, and a relatively comfortable posture. The British Iron and Steel Research Association has interested itself in the design and arrangement of equipment for use in iron and steel operations, including this one. On the basis of some research, plus the application of sound human factors principles, a modification was made of an original crane cab along with its associated controls. The before-and-after designs are given in Figure 12-18. The improved visibility, better arrangement of the controls, and posture of the operator can be seen from these photographs.

THE INTEGRATION OF CONTROL SYSTEMS

In the case of arrangement of work areas for control operations (the control of some equipment or system) the primary items to be "arranged" consist of controls and displays. It is in this connection that the appropriate integration of displays and controls needs to take place. In many circumstances the control operations are relatively simple and the arrangement of the displays and controls is therefore fairly straightforward. In the case of some systems, however, the control operations are extremely complex and the consequences of possible human error may be very serious or even catastrophic. The control processes in some chemical, petroleum, metal processing, and utility plants, for example, are very intricate and can be potentially hazardous.

The accident at the Three Mile Island nuclear power plant in Pennsylvania in 1979 gives some impression of the potential disasters that could result from accidents in such facilities. The nature of the operations involved in such a system are such that its "control" could not be reduced to the simplicity of adjusting the thermostat in the living room. However, good human factors design of the control system of such a facility could at least make the control manageable. In connection with the control room of that facility there is extensive evidence that it was poorly designed from the human factors point of view. The chairman of the presidential commission investigating the accident is quoted as saying, "It looks to me like this is very bad human engineering." Subsequent investigations confirmed this opinion. No single picture can illustrate all the specific human factors deficiencies of that control system, but Figure 12-19, which shows the control room, may at least give some impression of the problems of controlling such a facility and of the importance of appropriate design of control systems from the human factors point of view.

FIGURE 12-19
The control room of the Three Mile Island nuclear power plant. See text for discussion. (*Photograph courtesy of Dr. Thomas B. Malone, Essex Corporation.*)

DISCUSSION

The variety of physical situations in which people find themselves is of course almost infinite, and we have here been able to touch on only a few aspects of the design of such situations. The basic point that is intended is that the design of such situations can have an effect upon any of a number of relevant criteria, such as the performance of intended functions (such as operation of a processing operation), safety, and comfort. Thus, in the design process the designers need to utilize whatever anthropometric and performance data are available (plus good judgment) in designing the system or facility for suitable human use.

REFERENCES

Bonney, M. C., & Williams, R. W. CAPABLE: A computer program to layout controls and panels. *Ergonomics,* 1977, *20*(3), 297–316.

Bradley, J. V. Optimum knob crowding. *Human Factors,* 1969, *11*(3), 227–238.

Chapanis, A. Design of Controls. In H. P. Van Cott & R. G. Kinkade (Eds.), *Human engineering guide to equipment design* (Rev. ed.). Washington, D. C.: U.S. Government Printing Office, 1972.

Cullinane, T. P. Minimizing cost and effort in performing a link analysis. *Human Factors,* 1977, *19*(2), 151–156.

Deininger, R. L. *Process sampling, workplace arrangements, and operator activity levels.* Unpublished report, Engineering Psychology Branch, U.S. Air Force, WADD, 1958.

Dupuis, H. *Farm tractor operation and human stresses.* Paper presented at the meeting of the American Society of Agricultural Engineers, Chicago: Dec. 15–18, 1957.

Dupuis, H., Preuschen, R., & Schulte, B. *Zweckmäbige gestaltung des schlepperführerstandes.* Dortmund, Germany: Max Planck Institutes für Arbeitphysiologie, 1955.

Fitts, P. M. A study of location discrimination ability. In P. M. Fitts (Ed.), *Psychological research on equipment design* (Research Report 19). Army Air Force, Aviation Psychology Program, 1947.

Fowler, R. L., Williams, W. E., Fowler, M. G., & Young, D. D. An investigation of the relationship between operator performance and operator panel layout for continuous tasks (Technical Report 68-170). U.S. Air Force AMRL, December 1968. (AD-692 126).

Francis, R. L., & White, J. A. *Facility layout and location: An analytical approach.* Englewood Cliffs, N. J.: Prentice-Hall, 1974.

Haines, R. F., & Gilliland, K. Response time in the full visual field. *Journal of Applied Psychology,* 1973, *58*(3), 289–295.

Haygood, R. C., Teel, K. S., & Greening, C. P. Link analysis by computer. *Human Factors,* 1964, *6*(1), 63–78.

Hertzberg, H. T. E., & Burke, F. E. Foot forces exerted at various aircraft brake pedal angles. *Human Factors,* 1971, *13*(5), 445–456.

Huebner, W. J., Jr., & Ryack, B. L. Linear programming and work place arrangement: Solution of assignment problems by the product technique (Technical Report 61-143). U.S. Air Force, Air Research and Development Command, WADD, March 1961.

Human factors engineering (AFSC DH 1-3, U.S. Air Force Systems Command Design Handbook 1-3, 3d ed., rev. 1). Wright Patterson Air Force Base, Ohio: Aeronautical Systems Division, June 25, 1980.

Jones, R. E., Milton, J. L., & Fitts, P. M. *Eye fixations of aircraft pilots: IV. Frequency, duration, and sequence of fixations during routine instrument flight* (U.S. Air Force Technical Report 5975), 1949.

Kroemer, K. H. E. Human strength: Terminology, measurement, and interpretation of data. *Human Factors,* 1970, *12*(3), 297–313.

Laner, S. *Ergonomics in the steel industry.* (Report 19/61, List 120) The British Iron and Steel Research Association, Nov.-Dec. 1961.

Sell, R. G. The ergonomic aspects of the design of cranes. *Journal of the Iron and Steel Institute,* 1958, *190,* 171–177.

Sharp, E., & Hornseth, J. P. *The effects of control location upon performance time for knob, toggle switch, and push button* (Technical Report 65-41). AMRL, October 1965.

Siegel, A. I., & Brown, F. R. An experimental study of control console design. *Ergonomics,* 1958, *1,* 251–257.

Van Cott, H. P., & Kinkade, R. G. Design of individual workplaces. In H. P. Van Cott & R. G. Kinkade (Eds.), *Human engineering guide to equipment design* (rev. ed.). Washington, D. C.: U.S. Government Printing Office, 1972.

PART FIVE

ENVIRONMENT

CHAPTER **13**

ILLUMINATION

In many aspects of life we depend upon the sun as our source of illumination, as in driving in the daylight, playing golf, and picking tomatoes. When human activities are carried on indoors or at night, however, it is usually necessary to provide some form of artificial illumination. The design of such illumination systems, as we shall see, have an impact on the performance and comfort of those using the environment, as well as the affective responses of the people to the environment. Illuminating engineering is both an art and a science. The scientific aspects include the measurement of various lighting parameters and the design of energy efficient lighting systems. The artistic side comes into play when combining light sources to create, for example, a particular mood in a restaurant, highlight a display in a store, or complement a particular color scheme. It is not our intention here to make anyone an illuminating engineer, but rather our aim is to familiarize you with the basic concepts in the field and to illustrate the importance of proper illumination from a human factors point of view. We will leave the artistic aspect to the artists and concentrate on the science aspects of illumination.

THE NATURE OF LIGHT

Light, according to the Illuminating Engineering Society (IES), is "radiant energy that is capable of exciting the retina (of the eye) and producing a visual sensation" (IES Nomenclature Committee).

The entire radiant energy (electromagnetic) spectrum consists of waves of radiant energy that vary from about 1/1 billion of a millionth m to about 100 million m in length. This tremendous range includes cosmic rays; gamma rays; x-rays; ultraviolet rays; the visible spectrum; infrared rays; radar; FM, TV, and radio broadcast waves; and power transmission—as illustrated in Figure 13-1.

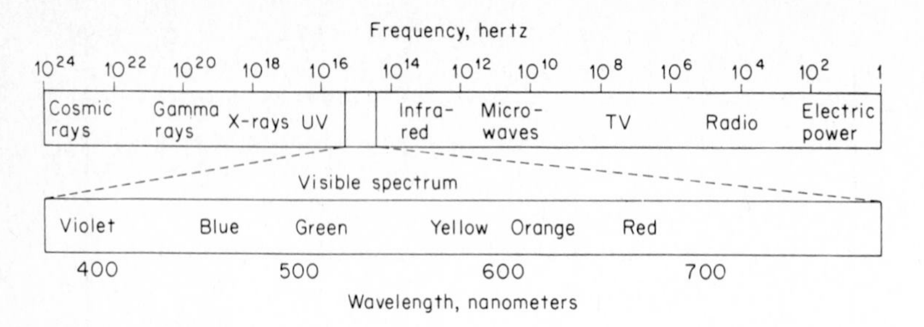

FIGURE 13-1
The radiant energy (electromagnetic) spectrum, showing the visible spectrum. (*Source: Adapted from* Light and color, *1968, p. 5.*)

The visible spectrum ranges from about 380 to 780 nanometers (nm). The *nanometer* (formerly referred to as a *millimicron*) is a unit of wavelength equal to 10^{-9} (one-billionth) m. Light can be thought of as the aspect of radiant energy that is visible; it is then basically psychophysical in nature rather than purely physical or purely psychological.

Scientists today use two theories to explain the nature of light. The *electromagnetic wave theory* holds that light energy is transmitted in the form of electromagnetic waves. The movement of light through space seems to be explained best by this theory. The *quantum theory*, on the other hand, conceives of light energy as being emitted in discrete "bundles," ejected in straight lines. The effect of light on matter (e.g., a barrier layer cell in a light meter) is best explained by this theory.

Color

Variations in wavelength within the visible spectrum give rise to the perception of color, the violets being around 400 nm, blending into the blues (around 450 nm), the greens (around 500 nm), the yellow-oranges (around 600 nm), and the reds (around 700 nm and above).

Just as the ear is not equally sensitive to all frequencies of sound, the eye is not equally sensitive to all wavelengths of light. Unlike the ear, however, the eye actually has two sensitivity curves depending on the overall level of illumination. At high levels of illumination, the rods and cones both function (*photopic vision*) and the eye is most sensitive to light wavelengths around 550 nm (green). As illumination levels decrease, however, the cones cease to function and the rods take over the entire job of seeing (*scotopic vision*) and the eye becomes most sensitive to wavelengths around 500 nm (blue-green). This shift in sensitivity from photopic to scotopic vision is called the *Purkinje effect*. One practical application of this effect is that to increase the probability of detection at night, targets can be made blue-green in color.

Light comes to us from two sources: *incandescent bodies* ("hot" sources, such as the sun, luminaires, or a flame) and *luminescent bodies* ("cold" sources, i.e., the objects we see in our environment, which reflect light to us). A hot light source that includes all wavelengths in about equal proportions is called *white light*. Most light sources, as luminaires, have spectra that include most wavelengths but that tend to have more energy in certain areas of the spectrum than in others. These differences make the lights appear yellowish, greenish, bluish, etc. Illuminating engineers often speak of the *color temperature* of a light source. The *color temperature* is the temperature at which a blackbody radiator must be operated to have the same (or in the case of fluorescent lamps, similar) color appearance as that of the light source. The noon sun, i.e., daylight, has a color temperature of 5500 degrees kelvin (K).

As light from a hot source falls upon an object, some specific combination of wavelengths is absorbed by the object. The light that is so reflected is the effect of the interaction of the spectral characteristics of the light source with the spectral absorption characteristics of the object. If a colored object is viewed under white light, it will be seen in its *natural* color. If it is viewed under a light that has a concentration of energy in a limited segment of the spectrum, the reflected light may alter the apparent color of the object, such as a blue car appearing green when viewed under yellow sodium lights in a parking lot.

The light that is reflected from an object—and that produces our sensations of the object's color—can be described in terms of three characteristics: the *dominant* wavelength; *saturation* (i.e., the predominance of a narrow range of wavelengths, as contrasted with an admixture of various wavelengths); and *luminance*. These physical characteristics of the light, in turn, influence our perceptions of color in terms of three corresponding attributes, respectively: *hue, saturation* (the attribute that determines the degree of difference of a color from a gray with the same lightness); and *lightness* (the attribute associated with the relative amount of the incident light). These three attributes are depicted in the color cone shown in Figure 13-2. In the color cone, hue is indicated by position around the circumference. Saturation (sometimes called *purity* or *chroma*) is shown in the color cone as the radius. A *saturated* color consists of a

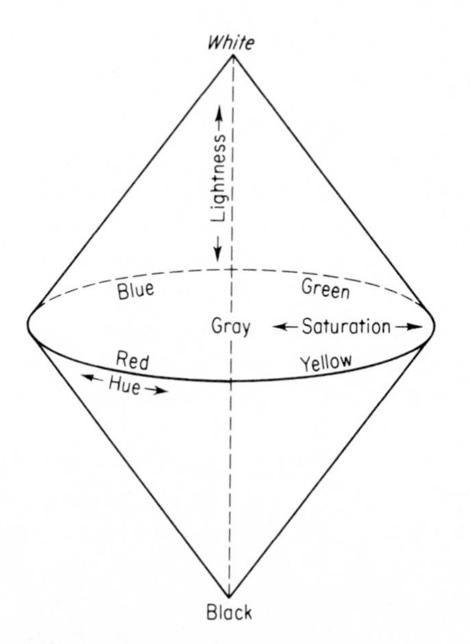

FIGURE 13-2
The color cone. Hue is shown on the circumference, lightness (from light to dark) on the vertical, and saturation on the radius from circumference to the center.

single hue and would be positioned on the circumference of the cone. Colors toward the center are mixtures of various hues, and while they may have a dominant hue, they do not appear to be pure. Lightness (sometimes referred to as *value* or *brightness*) is shown on the vertical dimension, the center of which ranges from white through varying levels of gray, to black. Any color of a given hue and saturation can be varied in its lightness. Although there is a general relationship between the luminance of light and the subjective response of lightness, all colors that reflect equal total amounts of light energy are not necessarily perceived as equal in lightness. This is due to the fact that the eye is differentially sensitive to various wavelengths as we discussed before.

Color Systems It is very difficult to describe a particular color when one considers the variations in hue, saturation, and lightness that are possible. Although difficult, it is often necessary to communicate such information without actually reproducing the color. For example, a design specification may have to describe the colors for, say, visual codes or labels. To assist in this communication process various color systems have been developed to serve as standards for describing colors. Two types of color systems are used.

First are those which consist of color plates or color chips for use as standards in characterizing colorants. The Munsell color system (Munsell Color Co., 1929), the Ostwald color system (Container Corp. of America, 1942), and the Maerz and Paul *Dictionary of Color* (1950) are of this type. Some of these (for example, the Munsell and Ostwald systems) correspond substantially to the color cone and provide standard nomenclature to identify selected colors of certain hues, saturations, and lightness levels. The Ostwald system, for example, identifies 680 color samples (28 color chips for each of 24 hues, plus 8 representing lightness levels on the white-to-black continuum).

The CIE color system, developed by the Commission Internationale de l'Eclairage (Optical Society of America), provides for designating colors in terms of their relative percentages of each of the three primary colors of light, namely, *red* (X), *green* (Y), and *blue* (Z). All possible colors can be designated with the CIE system on a chromaticity diagram, whether they are emitted, transmitted, or reflected.

The Measurement of Light

There are many concepts and terms that relate to the measurement of light (*photometry*). We will define a few of them here and show their interrelationships. One source of confusion to the photometric novice is the multitude of different measurement units used in the field. Part of this problem stems from the existence of two systems of measurement, the "United States" system and the International System of Units (or SI units). We bow to the rest of the world and the scientific community and encourage the use of SI units. The United States units will be mentioned, however, and conversions given to aid those of us who were weaned on footcandles and foot-lamberts.

Luminous intensity or *candlepower* is the intensity of a point source of light. The unit of measure of luminous intensity is the *candela* (cd).[1] The rate at which light is emitted from a source is called *luminous flux*. The unit of measure of luminous flux is

[1] The *candela* is defined as the luminous intensity of $^1/_{600,000}$ m^2 of projected area of a blackbody radiator operating at the temperature of the solidification of platinum (2047 K).

the *lumen* (lm). *Luminous flux* is a somewhat esoteric concept which is similar to other flow rates such as gallons per minute (gal/min). Time is implied in the unit of luminous flux. Thus, one would say that light is emitted from a 100-W incandescent lamp at a rate of 1740 lm. The rate of flow of light from a 1-cd source is 12.57 lm.

We now have a source of some luminous intensity emitting luminous flux in all directions. Imagine the source as being placed inside, at the center, of a sphere. The amount of light striking any point on the inside surface of the sphere is called *illumination* or *illuminance.* It is measured in terms of luminous flux per unit area, as for example, lumens per square foot (lm/ft^2) or lumens per square meter (lm/m^2). To obscure matters a bit, special names have been given to units of illuminance. One lumen per square foot is called a *footcandle* (fc), a United States unit, whereas one lumen per square meter is called a *lux* (lx), an SI unit. One footcandle equals 10.76 lx; however, an accepted practice for some purposes is to consider 1 fc to equal 10 lx and to forget the fraction (*IES Lighting Handbook*, 1981).

Figure 13-3 illustrates the relationship between candelas, footcandles, and lux. Our 1-cd source emits 12.57 lm and the total surface area of the sphere equals 12.57 times the radius squared (i.e., $4\pi r^2$). At a radius of 1 m, therefore, our 1-cd source is distributing 12.57 lm evenly over the 12.57 m^2 of surface area. Thus, the amount of light on any point is 1 lm/m^2, or 1 lx. The same logic holds at a radius of 1 ft for the footcandle.

FIGURE 13-3
Illustration of the distribution of light from a light source following the inverse-square law. (*Source:* Light measurement and control, *1965, p. 5.*)

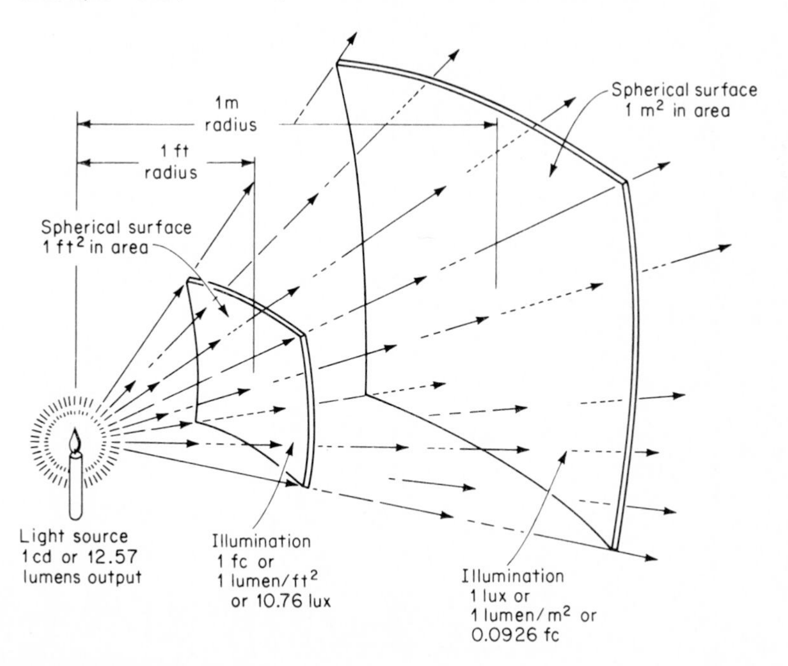

The amount of illumination from a point source follows the *inverse-square law:*

$$\text{Illuminance (lx)} = \frac{\text{candlepower (cd)}}{D^2}$$

where D is the distance from the source in meters. At 2 m a 1-cd source would produce ¼ lx and at 3 m it would produce ⅑ lx.

We now have light striking a surface (*illuminance*), but what happens to light that strikes a surface? Some of it is absorbed, and some of it is reflected. The light that is reflected from the surface of objects is what allows us to "see" objects—their configuration and color. The amount of light per unit area leaving a surface is called *luminance*. The light leaving the surface may be reflected by the surface or emitted by the surface, as would occur with a fluorescent light panel. The amount of light can be measured in terms of luminous flux (lumens) or luminous intensity (candelas). When the amount of light is measured in lumens and area is in square feet, the United States unit of luminance is the *footlambert* (fL). When the amount of light is measured in candelas and the area is in square meters, the SI unit of luminance is *candela per square meter* (cd/m^2). One footlambert is equal to 3.43 cd/m^2. Another unit of luminance found in some older literature is the *millilambert* (mL) which is equal to 0.929 fL, and for practical purposes can be considered as an equivalent measure.

The ratio of the amount of light reflected by a surface (*luminance*) to the amount of light striking the surface (*illuminance*) is called reflectance. In SI units:

$$\text{Reflectance} = \frac{\text{luminance (cd/m}^2\text{)}}{\text{illuminance (lx)}}$$

and is expressed as a unitless proportion. From the formula we can see that if a perfect reflecting surface (i.e., reflectance = 1.0) were illuminated by 1 lx, the surface luminance would be 1 cd/m^2. In like manner, if the surface were illuminated by 1 fc, its luminance would be 1 fL.

By way of summary, Table 13-1 lists some photometric terms and the corresponding United States and SI unit of measure associated with each.

TABLE 13-1
PHOTOMETRIC TERMS AND THE CORRESPONDING UNITED STATES AND SI UNITS OF MEASURE ASSOCIATED WITH EACH

Photometric term	United States units	SI units
Luminous intensity (or candlepower)	Candela, cd	Candela, cd
Luminous flux	Lumen, lm	Lumen, lm
Illuminance	Footcandle, fc	Lux, lx
Luminance	Footlambert, fL Millilambert, mL	Candela per square meter, cd/m^2

LAMPS AND LUMINAIRES

The term *lamp* is a generic term for an artificial source of light. A *luminaire*, however, is a complete lighting unit consisting of a lamp or lamps together with the parts designed to distribute the light, to position and protect the lamps, and to connect the lamps to the power supply (IES Nomenclature Committee, 1979). A trip to a lighting store will quickly convince you that lamps come in hundreds of configurations and sizes, and that luminaire designs are limited only by the taste of the designer, ranging from simple globes to elaborate crystal chandeliers.

In the following two sections of this chapter, we will introduce some of the generic types of lamps and luminaires and discuss some of the implications of choosing one or another of the generic types described.

Lamps

There are two main classes of lamps: *incandescent filament lamps* in which light is produced by electrical heating of a filament or by combustion of gases within a thin mesh mantle; and *gas discharge lamps* in which light is produced by the passage of an electric current through a gas.

Gas discharge lamps are of three types: high-intensity discharge lamps (HID) including mercury, metal halide, and high-pressure sodium lamps; low-pressure sodium lamps; and fluorescent lamps. In the case of fluorescent lamps, the radiation created by the electric current passing through the gas is invisible to the eye (it is ultraviolet radiation). This radiation, however, is used to excite a phosphor coating on the inner surface of the bulb, which in turn produces visible radiation. Different phosphors can create different colors of illumination.

The various types of lamps differ markedly in terms of their characteristic spectral distribution, or what we plain folk call *color*. Figure 13-4 shows representative distributions for several types of lamps. These differences can affect task performance and the subjective impressions of the people in the illumination environment. Delaney, Hughes, McNelis, Sarver, and Soules (1978), for example, tested subjects under various types of fluorescent illumination (cool white, deluxe cool white, warm white, deluxe warm white, etc.). They found that subjects made fewer color discrimination errors under deluxe lamps than under standard prime color lamps. In addition, subjects preferred the cool white color temperatures (5000 K) in terms of clarity and brightness.

Lion (1964) found that subjects performed three manipulative tasks under fluorescent lighting faster, and with less detrimental effect on accuracy, than under incandescent tungsten lighting. This, however, may be due to differences in the distribution of light rather than color. Fluorescent lamps tend to produce more diffuse lighting with less glare than do incandescent lamps.

In today's world, energy conservation has become a prime goal in system design and this is especially true in design of lighting systems. The efficiency of a light source, called *lamp efficacy*, is measured in terms of the amount of light produced [lumens

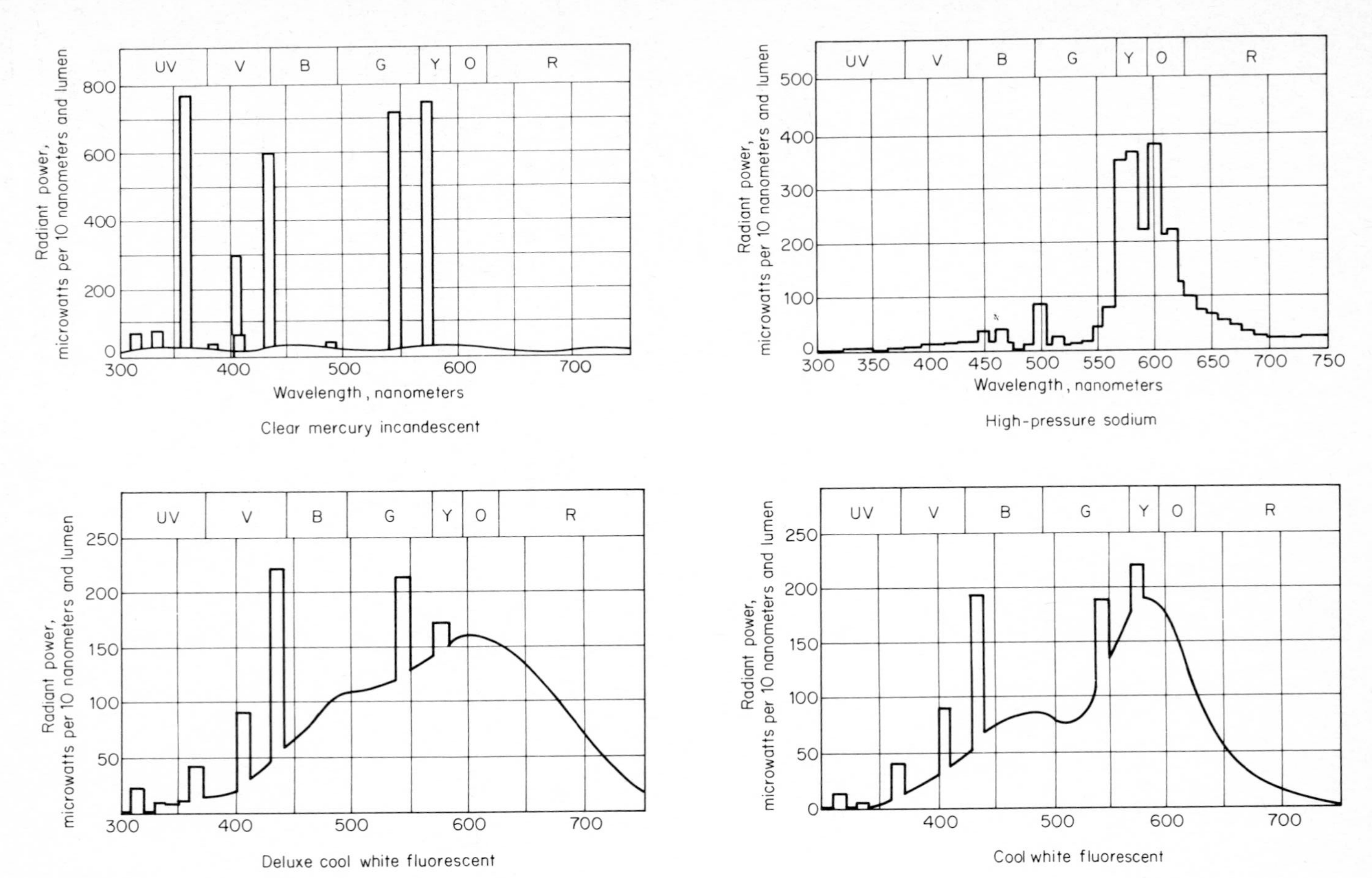

FIGURE 13-4
Representative sprectral distributions (chromaticity diagrams) for clear mercury, high pressure sodium, and two fluorescent lamps. (*Source:* IES Lighting Fundamentals Course, *1976, figs. 4-5, 4-13, 4-23.*)

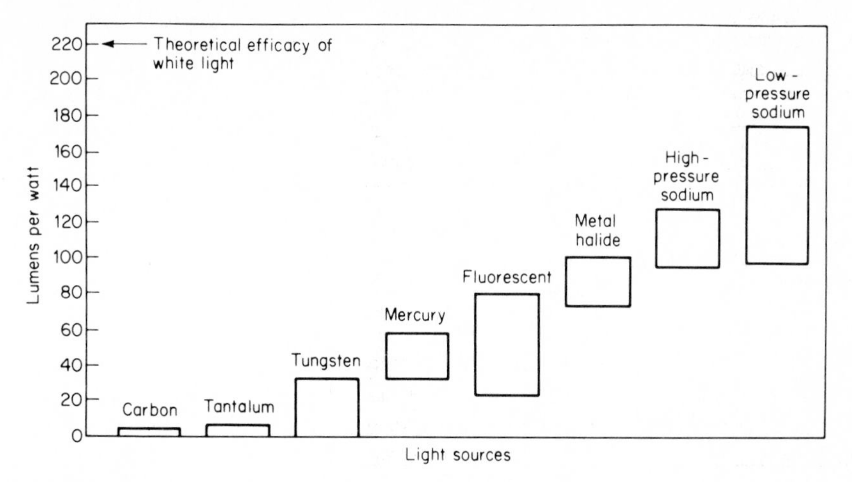

FIGURE 13-5
Lamp efficacy ranges for common light sources. (*Source:* IES Lighting Fundamentals Course, *1976, fig. 3-4.*)

(lm)] per unit of power consumed [watts (W)]. Figure 13-5 shows some typical lamp efficacy values for various types of light sources.

Switching from incandescent to fluorescent lights can result in real dollar savings. For example, consider the comparison of two light sources (incandescent and fluorescent) shown in Table 13-2. The fluorescent tube represents a 41 percent energy savings, lasts 20 times longer, and gives 30 percent more light to boot! And who said there was no such thing as a free lunch?

Incidentally, don't be fooled into thinking all incandescent bulbs of the same wattage give off the same amount of light–they don't! Bulb manufacturers are now required to put the tested lumen output on their packages. For example, a long-life 100-W bulb may give off as little as 1470 lm compared to perhaps 1740 lm for a standard 100-W bulb. Proper choice of light source often can result in substantial energy and cost savings–a worthy goal for any human factors design effort.

TABLE 13-2
COMPARISON OF TWO LIGHT SOURCES

	Incandescent	**Deluxe fluorescent**
Watts	75	30 (44 total input)
Rated life	750 h	15,000 h
Amount of light	1180 lm	1530 lm

Source: IES Lighting Fundamentals Course, 1976.

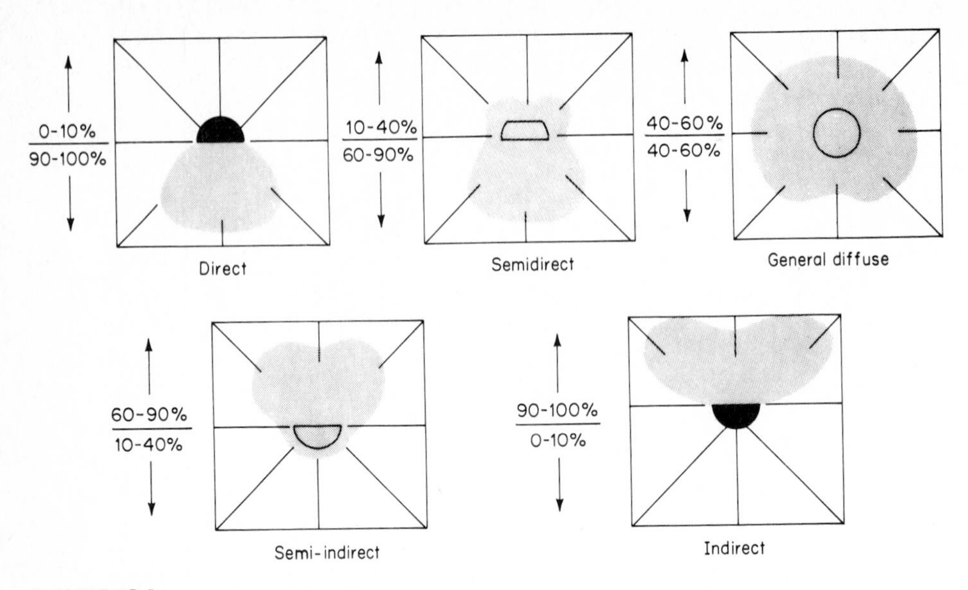

FIGURE 13-6
Types of luminaires based on the proportion of lumens emitted above and below the horizontal. (*Source:* IES Lighting Fundamentals Course, *1976, fig. 5-6.*)

Luminaires

Luminaires are classified into five categories based on the proportion of light (lumens) emitted above and below the horizontal, as shown in Figure 13-6. In selecting a particular type of luminaire for use, consideration must be given to the pattern of light distribution, glare, task illumination, shadowing, and energy efficiency. Various devices can be incorporated into a luminaire to control the distribution of light, including lenses, diffusers, shielding, and reflectors. Choice of an efficient luminaire is a complex decision and should be done by a qualified, experienced person after an analysis of the lighting needs and physical environment has been made.

THE CONCEPT OF VISIBILITY

Visibility refers to how well something can be seen by the human eye. Visibility, therefore, involves human judgment. There is no device that can measure visibility directly; a human must always be involved in its determination. One key factor influencing visibility of a target is how well it stands out from its background, that is, its *contrast*. We defined contrast in Chapter 4. Although contrast is related to visibility, it is not the same as visibility. DiLaura (1978*a*) provides a simple illustration of this. Imagine a target with a contrast equal to 0.5 placed on a stage and lit up with a flashlight held in the balcony. It would be hardly visible. Now illuminate this same 0.5 contrast target with a stage light 10,000 times the intensity of the flashlight—some contrast, but hardly the same level of visibility. From this we can see that both contrast and luminance are important for visibility. Another factor is the size of the target; an elephant on the stage would be more visible than a mouse.

Visibility, we said, was how well something can be seen, but seeing can be defined in different ways. For example, *seeing the elephant* might mean "detecting its presence on stage" or it could mean "recognizing the elephant as an Indian elephant." These different definitions, or *information criteria*, obviously then have a bearing on the visibility level of the target. One last variable we should mention is exposure time to the target. The more time you have to look at the elephant the easier it becomes to recognize it as an Indian elephant (they have smaller ears).

Incidentally, the visibility of a target should not be a function of the observer; visibility should be a measure of the task itself. We would like a measure of task visibility that does not change each time a different person measures it.

Measuring Task Visibility

To determine the level of visibility for a particular target we must compare it against some standard level of visibility, much as we do when we determine length by comparing it to a standard meter stick. In the case of visibility this standard must be specified in terms of the target, the illumination environment, and the information requirements of the task. The choice of these conditions is somewhat arbitrary, but because of the extensive research carried out by Blackwell (Blackwell, 1959, 1961, 1964, 1967; Blackwell & Blackwell 1968, 1971) the standard target has become a uniformly luminous disc which subtends a visual angle of 4 minutes (approximately 1.1 mm at a distance of 1 m) presented in pulses of ⅕ s on a uniformly luminous screen. The information requirement is to detect the presence of the disc. Incidentally, the pulses are used to simulate the manner in which we usually acquire visual information (DiLaura, 1978b).

Blackwell sought to determine the *threshold of visibility* for the standard target, that is, the point at which subjects could detect the target 50 percent of the time when it was actually present. To do this, he had observers view a uniform screen having a given level of luminance. The standard 4-minute luminous disc was then presented at the center of the screen in ⅕-s pulses. The observer adjusted the physical contrast of the disc (i.e., increased or decreased the luminance of the disc) until it was judged to be just barely visible. An interesting phenomenon occurred when the screen luminance was changed. Blackwell found that the observers had to readjust the disc's contrast so that it was again at threshold. If screen luminance was decreased, the contrast of the disc had to be increased to bring it back to threshold. The higher the screen, or background, luminance, the lower was the contrast required to bring the disc to threshold.

Blackwell performed this experiment on approximately 100 young observers and determined the average relationship between background luminance and threshold disc contrast. This he called the *Visibility Reference Function* or *Visibility Level 1* (VL 1). This is shown in Figure 13-7.

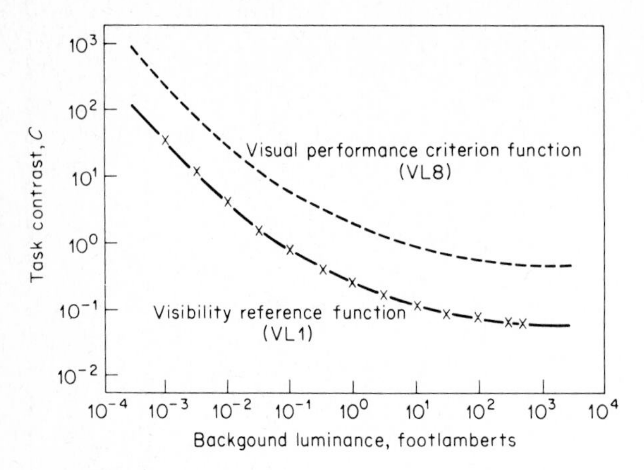

FIGURE 13-7
Two visibility level (VL) curves, each representing combinations of contrast and luminance required for equal visibility. VL 1 (Visibility Reference Function) is a basic curve relating to the luminance needed to achieve threshold visibility of a static task of discriminating a 4-minutes-of-arc luminous disc exposed for 1/5 s. In turn, VL 8 (Visual Performance Criterion Function) shows the level adjusted for three factors: dynamic versus static target; not knowing versus knowing where and when the target will appear; and requiring a 99 versus 50 percent probability of detection. (*Source: IES Lighting Handbook, 1972, fig. 3-24, as based on work of Blackwell.*)

Blackwell performed this basic experiment under a variety of conditions, for example using a dynamic (moving) target, telling and not telling subjects where and when the target might appear in the visual field, and requiring 99 percent probability of detection rather than just 50 percent. In each case the shape of the curve was the same but displaced upward by a constant multiple of the Visibility Reference Function. The multipliers are:

Dynamic presentation	2.78
Not knowing where or when	1.50
99 percent detection	1.90

Blackwell reasoned that with all three conditions present the curve would be approximately eight times Visibility Level 1 ($2.78 \times 1.50 \times 1.90 = 7.92$). This he termed VL 8 and it is also shown in Figure 13-7. Incidentally, no one has ever empirically determined that under these three conditions the resulting curve would indeed be eight times VL 1 (DiLaura, 1978c).

We now have a scale of visibility—visibility level (VL). Given the contrast of a 4-min luminous disc and the luminance of its background, one can determine its VL, i.e., the degree to which the target is above threshold, expressed as a multiple of threshold visibility—VL 2, VL 3.4, etc. This is just fine if all we ever did in life was detect the presence of little luminous dots. The problem, of course, is to somehow relate practical real-life visual tasks to our reference target.

This bit of magic is done using a *contrast-reducing visibility meter*, such as the Blackwell Visual Task Evaluator (VTE). The observer views the standard 4-min disc under a given background luminance through the VTE. The observer then reduces the contrast of the disc until it is at the threshold of detection. The amount of contrast reduction required to do this is recorded by the meter. Now the same observer views the practical target and, using the same meter, reduces its contrast to bring it to threshold based on the actual information requirement of the task. The meter keeps the background luminances of the practical target and the standard target the same. This procedure ultimately yields a measure of *equivalent contrast* ($\tilde{C}$). *Equivalent contrast* can be interpreted as the contrast of the 4-min luminous disc which makes it as difficult to detect as it is to "see" the practical task (keeping in mind the information criterion of the practical task). Whenever possible, equivalent contrast should be carried out under reference lighting conditions established by a photometric sphere. In this situation a practical target is placed at the center of a uniformly luminous sphere so that it is illuminated equally from all directions.

Follow the logic; given a practical task we can determine its equivalent contrast and measure the actual background luminance. Using Figure 13-7 we can locate the point corresponding to $\tilde{C}$ and background luminance, and then determine how many times greater than VL 1 is $\tilde{C}$ at the same background luminance. This gives us the VL for our practical task—under reference lighting conditions. The real lighting environment, however, is likely to illuminate the practical target so that it is more or less visually difficult than it was under the reference lighting conditions. With a little more magic, which we need not try to explain here, the VL of the task under its actual lighting environment can be determined. This is called *effective VL* (VL_{eff}) and is a general metric for expressing a practical target's suprathreshold visibility in a real lighting environment (DiLaura, 1978b).

This system of $\tilde{C}$ and VL forms the basis for establishing recommended levels of illumination as proposed by the Commission Internationale de l'Eclairage (CIE). The Illuminating Engineering Society of North America also used this system for recommending illumination levels (*IES Lighting Handbook*, 1972), but has since adopted a simpler approach which we will describe below (*IES Lighting Handbook*, 1981).

AMOUNT OF ILLUMINATION

There have been numerous studies relating the amount of illumination to task performance, and of course we cannot hope to review all of them. Most studies, however, find negative accelerating monotonic relationships which show increasing levels of il-

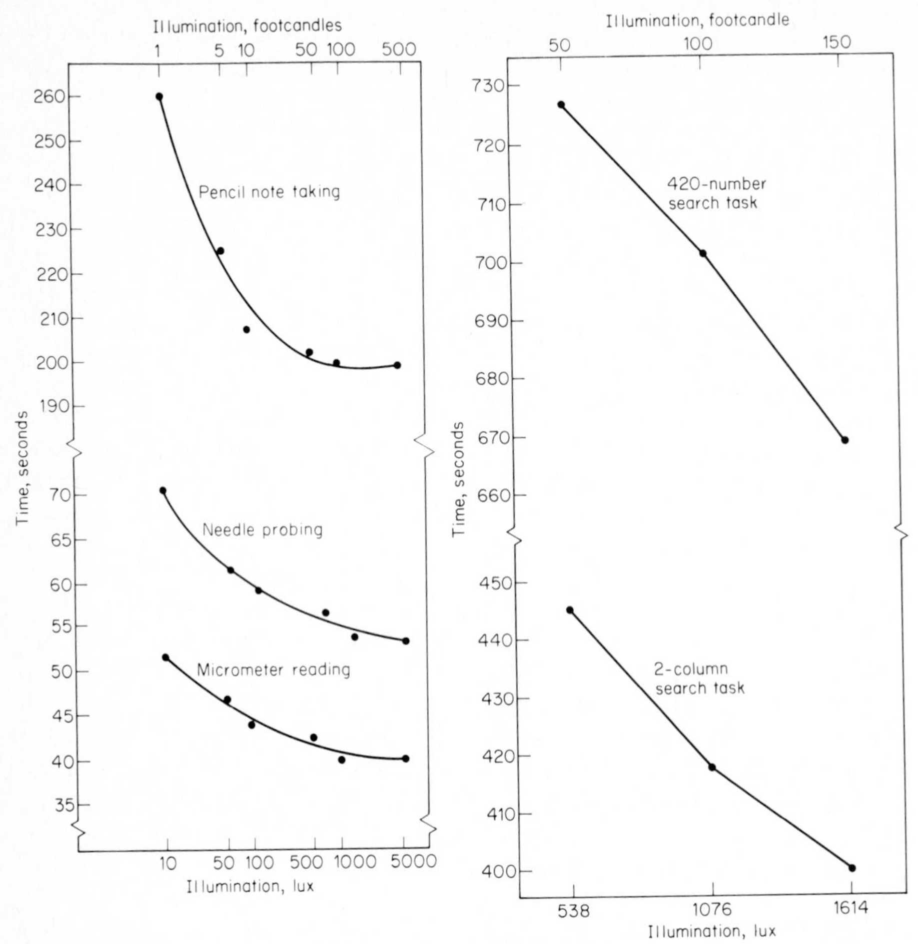

FIGURE 13-8
Relationship between amount of illumination and task completion time for three representative industrial tasks. (*Source: Bennett, Chitlangia, & Pangrekar, 1977, figs. 1, 3, and 4.*)

FIGURE 13-9
Relationship between amount of illumination and task completion time for two search tasks representative of many clerical tasks. (*Source: Adapted from Hughes & McNelis, 1978, table 1.*)

lumination resulting in smaller and smaller improvements in performance. The exact shape of the relationship depends in great part on the specific task being performed. Figure 13-8, for example, illustrates relationships between illumination level and time to complete several representative industrial tasks (Bennett, Chitlangia, & Pangrekar, 1977). Figure 13-9 shows relationships for two clerical search tasks, one in which a three-digit number had to be located on a work sheet containing 420 handwritten numbers, and one in which letter-digit combinations in one column had to be found in a second column (Hughes & McNelis, 1978). The unusual aspect of Figure 13-9 is that the curves do not seem to be leveling off over the levels of illumination used. We will comment on this in a moment.

There is some controversy over just how important amount of illumination is for task performance. Results from several sources (including Bennett et al., 1977; Hughes & McNelis) indicate that the age of the observer has a major impact on the results, with age taking its toll. Further, it appears that the amount of illumination may be more important for older persons. Ross (1978), after reviewing numerous studies, concludes that above 34 to 68 cd/m^2 (10 to 20 fL) of background luminance, other variables, notably age and print quality, are more important determiners of performance than amount of illumination. Ross further concludes that increasing illumination above 500 lx (50 fc) results in little additional improvement in task performance. This seems to be the case in Figure 13-8, but not in Figure 13-9.

One aspect of performance which cannot be ignored is motivation. The results in Figure 13-9, showing continued improvement beyond 1500 lx, may be due to providing an illumination level more preferred by the subjects, rather than actually reducing the visual difficulty of the task. Hughes and McNelis collected subjective evaluations of the three light levels from their subjects and indeed found some evidence that the higher illumination levels were more satisfying. Unfortunately, it is impossible to separate the effects of illumination level on motivation from the direct effects on task difficulty.

It is not always wise to provide high levels of illumination. Besides being wasteful of energy, too much illumination may create unwanted effects. Glare is one such effect which we will discuss in the next section. Zahn and Haines (1971) offer another example. They varied the luminance level of a centrally located diffuse screen upon which a search task was presented. The luminance varied from 29.1 cd/m^2 (8.5 fL) to 23,325 cd/m^2 (6800 fL). They measured reaction time to peripheral lights and found that reaction time increased with increased central task luminance, the increase being greatest for the more peripheral locations. Thus, if peripheral detection is important, too much central task luminance may actually degrade performance.

Another example of the unpleasant effects of high levels of illumination comes from a study by Sanders, Gustanski, and Lawton (1974). "High" and "low" levels of illumination were created in the hallways of a university and the noise level generated by the students waiting outside the classrooms was measured. The mean noise level under the high-illumination condition (20 large overhead fluorescent light panels lit) was 61.1 dBC, while under low illumination (two-thirds of the panels were turned off) the noise level generated was only 50.3 dBC. Here then, it appears, may be a silver lining in the energy crisis cloud–peace and quiet.

How Much Is Enough?

The problem of determining the level of illumination that should be provided for various visual tasks has occupied the attention of illuminating engineers, psychologists, and others for many years. One method, based on the work of Blackwell that we discussed previously, was used up to 1981 by the Illuminating Engineering Society to set recommended levels (*IES Lighting Handbook*, 1972). It involved determining the equivalent contrast of the task to be performed, and then, using the VL 8 curve in Figure 13-7, determining the appropriate background luminance. Knowing the reflectance of the background, it is then an easy matter to determine the level of illumination required to produce the desired background luminance.

All of that is history now, for the Illuminating Engineering Society has adopted a simpler approach which uses ranges of illuminance accompanied by weighting factors to select the specific recommended illuminance level (RQQ, 1980; *IES Lighting Handbook*, 1981).

The first step in the procedure is to identify the type of activity that will be performed in the area for which illumination recommendations are sought. The *IES Lighting Handbook* (1981) contains extensive tables listing various types of tasks, each referencing one of the illumination categories shown in Table 13-3. Although we cannot reproduce the lists of tasks here, the column headed *Type of Activity* in Table 13-3 contains enough information to give the reader a sense of what is involved in each category. Note that categories *A* through *C* do not involve visual tasks.

Table 13-3 lists, for each category, a range (low-middle-high) of illuminances. To decide which of the three values to choose, one must compute a correction factor using Table 13-4. For each characteristic in Table 13-4 (workers' ages, speed or accuracy, and reflectance of task background) a weight is assigned (−1, 0, +1). These weights are then added algebraically (i.e., +1 − 1 = 0) to obtain the total weighting factor (TWF). However, since categories *A* through *C* do not involve visual tasks, the *speed or accuracy* correction is not used for these categories and the average reflectance of the room, walls, and floor are used rather than task background reflectance. The following rules are then applied to select the low, middle, or high value in the category.

For categories *A* through *C*:

1. TWF = −1 or −2	Use low value
2. TWF = 0	Use middle value
3. TWF = +1 or +2	Use high value

For categories *D* through *I*:

1. TWF = −2 or −3	Use low value
2. TWF = −1, 0, or +1	Use middle value
3. TWF = +2 or +3	Use high value

This system can also be used if the equivalent contrast value of the task is known. A table is provided, but not reproduced here, that indicates the appropriate illumination category for any value of equivalent contrast. The calculation and use of the correction factors remains the same.

Where different activities are being performed in adjacent areas, the illumination levels should not vary from each other by more than a ratio of 5:1 (RQQ, 1980). Therefore, if an office requires 1000 lx, a corridor outside the office should have an illumination level of at least 200 lx.

Although high levels of illumination might be warranted in some circumstances, there are definite indications that for certain tasks high levels of illumination can weaken information cues (Logan & Berger, 1961) by suppressing the "visual gradients" of the pattern density of the objects being viewed. An illustration of this effect is given

TABLE 13-3
RECOMMENDED ILLUMINATION LEVELS FOR USE IN INTERIOR LIGHTING DESIGN

Category	Range of illuminances, lx (fc)	Type of activity
A	20-30-50* (2-3-5)*	Public areas with dark surroundings
B	50-75-100* (5-7.5-10)*	Simple orientation for short temporary visits
C	100-150-200* (10-15-20)*	Working spaces where visual tasks are only occasionally performed
D	200-300-500† (20-30-50)†	Performance of visual tasks of high contrast or large size: e.g., reading printed material, typed originals, handwriting in ink and good xerography; rough bench and machine work; ordinary inspection; rough assembly
E	500-750-1000† (50-75-100)†	Performance of visual tasks of medium contrast or small size: e.g., reading medium-pencil handwriting, poorly printed or reproduced material; medium bench and machine work; difficult inspection; medium assembly
F	1000-1500-2000† (100-150-200)†	Performance of visual tasks of low contrast or very small size: e.g., reading handwriting in hard pencil on poor quality paper and very poorly reproduced material; highly difficult inspection
G	2000-3000-5000‡ (200-300-500)‡	Performance of visual tasks of low contrast and very small size over a prolonged period: e.g., fine assembly; very difficult inspection; fine bench and machine work
H	5000-7500-10,000‡ (500-750-1000)‡	Performance of very prolonged and exacting visual tasks: e.g., the most difficult inspection; extra fine bench and machine work; extra fine assembly
I	10,000-15,000-20,000‡ (1000-1500-2000)‡	Performance of very special visual tasks of extremely low contrast and small size: e.g., surgical procedures

*General lighting throughout room.
†Illuminance on task.
‡Illuminance on task, obtained by a combination of general and local (supplementary) lighting.
Source: RQQ, 1980, table 1.

TABLE 13-4
WEIGHTING FACTORS TO BE CONSIDERED IN SELECTING THE SPECIFIC ILLUMINATION LEVELS WITHIN EACH CATEGORY OF TABLE 13-3

Task and worker characteristics	Weight		
	−1	0	+1
Age	Under 40	40–55	Over 55
Speed or accuracy	Not important	Important	Critical
Reflectance of task background	Greater than 70 percent	30 to 70 percent	Less than 30 percent

Source: RQQ, 1980, table 2.

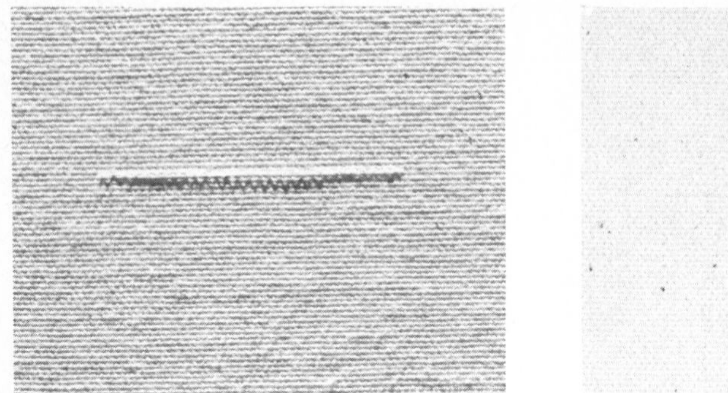

FIGURE 13-10
Illustration of the effects of general versus surface-grazing illumination on the visibility of a loose thread on cloth. (*Source: Faulkner & Murphy, 1973, figs. 3 and 4.*)

in Figure 13-10, which shows the effects of general versus "surface-grazing" illumination. Thus, in some circumstances lower levels of illumination (albeit of the appropriate type) would be better than high levels. The illumination recommendations, for at least some tasks, may have to be tempered by experience.

DISTRIBUTION OF LIGHT

There are bits and pieces of evidence that visual performance is generally (but not always) better if there is a reasonable level of general illumination at the work area.

Luminance Ratio

The *luminance ratio* is the ratio of the luminance of a given area (as the work area) to the surrounding area. The ratios recommended by the IES for various areas relative to the visual task, for both office and industrial situations, are given in Table 13-5. In this, the ratio for the task and its adjacent surroundings is given as 3:1 (except if the surroundings are very light in color).

TABLE 13-5
RECOMMENDED LUMINANCE RATIOS FOR OFFICES AND INDUSTRIAL SITUATIONS

	Recommended maximum luminance ratio	
Areas	**Office**	**Industrial**
Task and adjacent surroundings	3:1	
Task and adjacent darker surroundings		3:1
Task and adjacent lighter surroundings		1:3
Task and more remote darker surfaces	5:1	10:1
Task and more remote lighter surfaces	1:5	1:10
Luminaires (or windows, etc.) and surfaces adjacent to them		20:1
Anywhere within normal field of view		40:1

Source: IES Lighting Handbook, 1972, fig. 11-2, p. 11-3, and fig. 14-2, p. 14-3.

TABLE 13-6
RELATIONSHIP BETWEEN REFLECTION AND UTILIZATION COEFFICIENT

Reflection of surface, %				Utilization
Ceiling	Walls	Floor	Furniture	coefficient, %
65	40	12	28	29
85	72	85	50	57

Source: IES Lighting Handbook, fig. 5-17, p. 5–17.

Reflectance

The distribution of light within a room is not only a function of the amount of light and the location of the luminaires, but is also influenced by the reflectance of the walls, ceilings, and other room surfaces. Tied in with reflection is the concept of the *utilization coefficient*, which is the percentage of light that is reflected, collectively, by the surfaces in the room or area. The influence of reflection on the utilization coefficient is illustrated by the example shown in Table 13-6.

In order to contribute to the effective distribution and utilization of light in a room, it is generally desirable to use rather light walls, ceilings, and other surfaces. However, areas of high reflectance in the visual field can become sources of reflected glare. For this and other reasons (including practical considerations), the reflectances of surfaces in a room (such as an office) generally increase from the floor to the ceiling. Figure 13-11 illustrates the IES recommendations on this score, indicating for each type of surface the range of acceptable reflectance levels. Although that figure applies specifically to offices, essentially the same reflectance values can be applied to other work situations, as in industry.

FIGURE 13-11
Reflectances recommended for room and furniture surfaces in an office. (*Source: American national standard practice for office lighting, 1973, fig. 11.*)

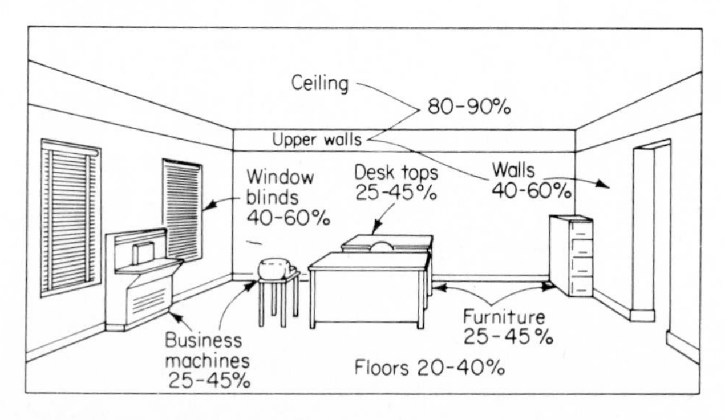

GLARE

Glare is produced by brightness within the field of vision that is sufficiently greater than the luminance to which the eyes are adapted to cause annoyance, discomfort, or loss in visual performance and visibility. *Direct* glare is caused by light sources in the field of view, and *reflected* or *specular* glare is caused by reflections of high brightness from polished or glossy surfaces that are reflected toward an individual. An example would be the reflections of overhead lights in a computer terminal CRT screen (Hultgren & Knave, 1974).

Variations in the level of glare have been categorized in terms of three types, as follows: (1) *discomfort glare* that produces discomfort, but does not necessarily interfere with visual performance or visibility; (2) *disability glare* that reduces visual performance and visibility and often is accompanied by discomfort; and finally (3) *blinding glare* that is so intense that for an appreciable length of time after it has been removed no object can be seen (IES Nomenclature Committee, 1979).

Discomfort Glare

Visual discomfort from glare is, unfortunately, a common experience, and a major concern in the design of luminaires and lighting installations should be to minimize such discomfort. Toward this end, there has been considerable research relating to glare and its effects on the subjective sensations of visual comfort and discomfort. As a direct result of such research, the IES has adopted a standard procedure for computing *discomfort glare ratings* (DGR) for luminaires and for tentatively planned interior lighting situations (*IES Lighting Handbook*, 1981).

DGR is a numerical assessment of the capacity of a number of sources of luminance, such as luminaires, in a given visual environment for producing discomfort. The calculation of DGR ratings for specific lighting layouts takes into account most of the situational factors that affect visual comfort, as follows: (1) room size and shape; (2) room surface reflectances; (3) illumination level; (4) luminaire type, size, and light distribution; (5) number and location of luminaires; (6) luminance of entire field of view; (7) observer location and line of sight; and (8) equipment and furniture. In addition, the procedures can take into account a ninth variable, if desired, namely, differences in individual glare sensitivity.

Let us digress for a moment to discuss further this notion of individual differences in glare sensitivity. Bennett (1977a, 1977b) used a psychophysical concept called *borderline between comfort and discomfort* (BCD) as a measure of glare sensitivity. Subjects adjusted the luminance of a potential glare source to an intensity such that it was not annoying or uncomfortable, but if it were any brighter it would be uncomfortable. Thus, the higher the BCD, the less the sensitivity. In one study, Bennett (1977b) varied background luminance, size of the glare source, and angle in the visual field and found the following correlations between the variables and BCD:

Background luminance	0.26
Size of glare source	−0.41
Angle in visual field	0.12

Translated, this indicates that the higher the background luminance, the smaller the size of the glare source, and the higher the angle in the visual field, the less glare is produced. Bennett makes the point, however, that these three factors together only account for 28 percent of the variance in BCD judgments. Differences between observers accounted for as much as 55 percent of the variance.

In another study, Bennett (1977a) related various demographic factors to BCD judgments. Not surprisingly, he found that glare sensitivity increased with age. Figure 13-12 shows the relationship between age and BCD found by Bennett (remember, as BCD goes down, glare sensitivity is increasing). Incidentally, he also found that blue-eyed people are more sensitive to glare than brown-eyed people. People with brown eyes had a median BCD of 1300 fL while blue-eyed people had a median of 1100 fL.

Let us now get back to DGR. The scheme for estimating DGR for any specific lighting layout is complex and will not be described here (see *IES Lighting Handbook*, 1981). The derived DGR value for any given circumstance, however, can be converted into a *visual comfort probability* (VCP). The VCP is the percentage of occupants who would rate the lighting system as comfortable if they were looking horizontally from a seated position in the least advantageous location in the room–considered to be at the center of the rear of the room.

The IES considers a VCP of 70 to be satisfactory from a discomfort glare standpoint. Since 70 percent of the people would be expected to rate the environment comfortable in the worst position, it is apparent that most occupants would be in comfortable visual locations.

FIGURE 13-12
Sensitivity to discomfort glare as a function of age. As BCD decreases, sensitivity is increasing. Data are based on a 1° glare source size and a background luminance of 1.6 fL. (*Source: Bennett, 1977a, fig. 1.*)

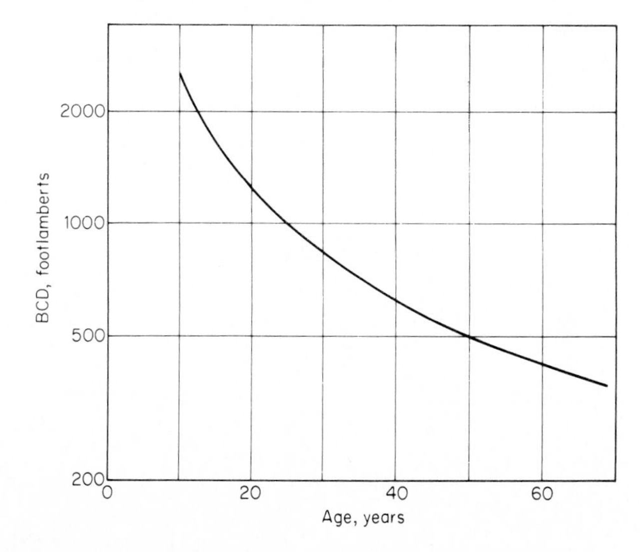

Although most of the interest in direct glare has been with respect to glare from luminaires, Hopkinson (1972) has carried out some research relating to glare from daylight (through windows) and presents a modified formula for computing a glare index for the daylighting of interiors. Windows, of course, can provide a useful source of illumination, so a basic problem is that of designing windows in order to provide daylight illumination (along with other desirable features, such as ventilation and possibly a view) with minimum glare.

Disability Glare from Direct Sources

The effects of direct glare on the visual performance of people (i.e., *disability glare*) are illustrated by the results of a study in which the subjects viewed test targets with a glare source of a 100-W inside-frosted tungsten-filament lamp in various positions in the field of vision (Luckiesh & Moss, 1927–1932). The test targets consisted of parallel bars of different sizes and contrasts with their background. The glare source was varied in position in relation to the direct line of vision, these positions being at 5, 10, 20, and 40° with the direct line of vision, as indicated in Figure 13-13. The effect of the glare on visual performance is shown as a percentage of the visual effectiveness that would be possible without the glare source. It will be seen that with the glare source at a 40°-angle, the visual effectiveness is 58 percent, this being reduced to 16 percent at an angle of 5°.

FIGURE 13-13
Effects of direct glare on visual effectiveness. The effects of glare become worse as the glare source gets closer to the line of sight. (*Source: Luckiesh & Moss, 1927–1932.*)

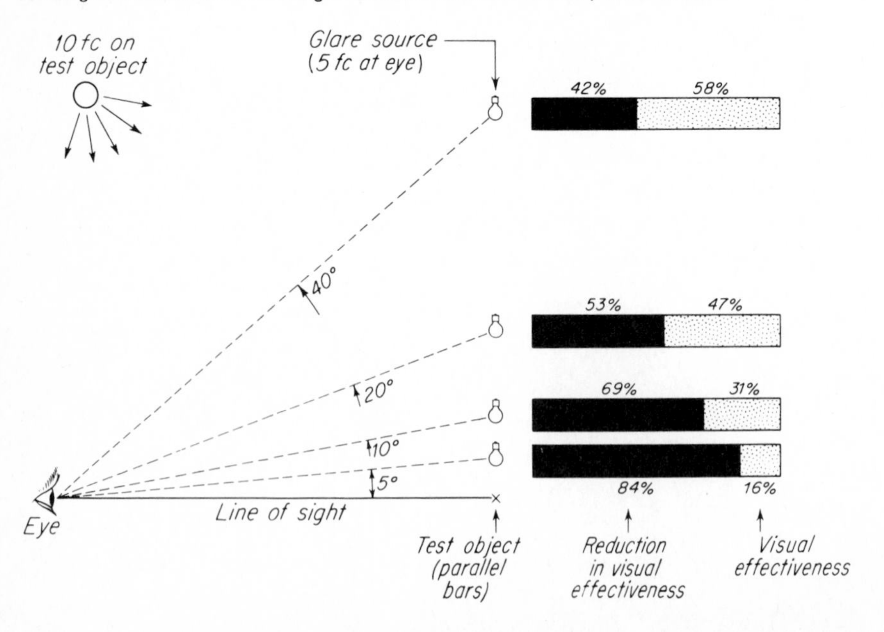

The Reduction of Glare

- *To reduce direct glare from luminaires* (1) select luminaires with low DGR; (2) reduce the luminance of the light sources (for example, by using several low-intensity luminaires instead of a few very bright ones); (3) position luminaires as far from line of sight as feasible; (4) increase luminance of area around any glare source, so the luminance (brightness) ratio is less; and (5) use light shields, hoods, and visors where glare source cannot be reduced.
- *To reduce direct glare from windows* (1) use windows that are set some distance above the floor; (2) construct an outdoor overhang above the window; (3) construct vertical fins by the window extending into the room (to restrict direct line of sight to the windows); (4) use light surrounds (to minimize contrast with light from window); and (5) use shades, blinds, or louvers.
- *To reduce reflected glare* (1) keep luminance level of luminaires as low as feasible; (2) provide good level of general illumination (such as with many small light sources and use of indirect lights); (3) use diffuse light, indirect light, baffles, window shades, etc.; (4) position light source or work area so reflected light will not be directed toward the eyes; and (5) use surfaces that diffuse light, such as flat paint, nonglossy paper, and crinkled finish on office machines; avoid bright metal, glass, glossy paper, etc.

DISCUSSION

We have come a long way in this chapter and have introduced many concepts which were probably unfamiliar to most readers, yet we have just barely scratched the surface of the field of illuminating engineering. The design of a lighting system to supply the proper amount of illumination, with the proper spectral composition, without creating glare, and to do it all in an energy-efficient manner, is quite an achievement. Obviously, our goal in this chapter was not to develop this talent, but rather to acquaint you with the major concepts in order to make you a more intelligent consumer of the expertise offered by illuminating engineers.

We will leave you with one last concept and an example. The concept is *phototropism*, the tendency of the eyes to turn toward a light. Store owners and department store window designers take advantage of this human tendency when they direct bright lights toward a particular part of the store (perhaps where the high profit items are) or at a particular item in the window. This is one example of the artistic aspect of illuminating engineering we talked about at the beginning of this chapter.

Phototropism, however, can have negative effects for task performance and safety if it draws the eyes away from the area of most important visual attention. Figure 13-14 (on page 392), for example, illustrates this phenomenon. By changing the position of the luminaire, the eye is drawn to the center of the machine, Figure 13-14*b*, rather than the glare spot at the end, Figure 13-14*a* (Hopkinson & Longmore, 1959).

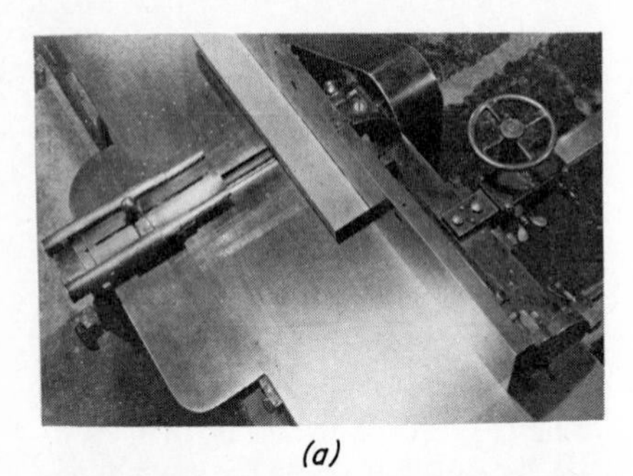

(a)

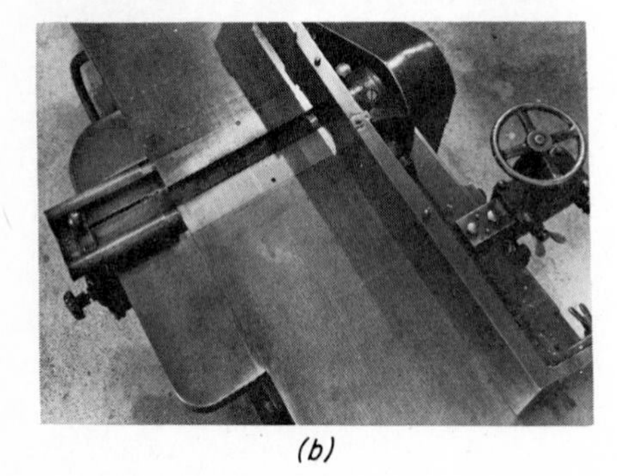

(b)

FIGURE 13-14
Effects of change of location of luminaires on wood planer. Original position *(a)* caused reflection from polished surface, which distracted attention from cutters. Relighting *(b)* gives lower brightness reflection, which makes cutter the center of attraction. (*Source: Hopkinson & Longmore, 1959, by permission of the Controller of H. M. Stationery Office.*)

REFERENCES

American national standard practice for office lighting. *Journal Illuminating Engineering Society*, 1973, *3*(1), 3–27.

Bennett, C. The demographic variables of discomfort glare. *Lighting Design and Application*, January 1977, pp. 22–24. (a)

Bennett, C. Discomfort glare: Concentrated sources parametric study of angulary small sources. *Journal Illuminating Engineering Society*, 1977, *7*(1), 2–14. (b)

Bennett, C., Chitlangia, A., & Pangrekar, A. Illumination levels and performance of practical visual tasks. *Proceedings of the Human Factors Society 21st Annual Meeting*, Santa Monica, Calif.: Human Factors Society, 1977.

Blackwell, H. R. Development and use of a quantitative method for specification of interior illumination levels on the basis of performance data. *Illuminating Engineering*, 1959, *54*, 317–353.

Blackwell, H. R. Development of visual task evaluators for use in specifying recommended illumination levels. *Illuminating Engineering*, 1961, *56*, 543–544.

Blackwell, H. R. Further validation studies of visual task evaluation. *Illuminating Engineering,* 1964, *59*(9), 627–641.

Blackwell, H. The evaluation of interior lighting on the basis of visual criteria. *Applied Optics,* 1967, *6*(9), 1443–1467.

Blackwell, H. R., & Blackwell, O. M. The effect of illumination quantity upon the performance of different visual tasks. *Illuminating Engineering,* 1968, *63*(3), 143–152.

Blackwell, O. M., & Blackwell, H. R. Visual performance data for 156 normal observers of various ages. *Journal Illuminating Engineering Society,* October 1971, *1*(1), 3–13.

Container Corporation of America. *Color harmony manual.* Chicago: Container Corporation of America, 1942.

Delaney, W., Hughes, P., McNelis, J., Sarver, J., & Soules, T. An examination of visual clarity with high color rendering of fluorescent light sources. *Journal Illuminating Engineering Society,* 1978, *2*(1), 74–84.

DiLaura, D. Visibility, human performance, and lighting design—equivalent contrast as a prelude to visibility level. *Lighting Design and Application,* February 1978, p. 8. (a)

DiLaura, D. Visibility, human performance, and lighting design—visibility level as a supra-threshold metric. *Lighting Design and Application,* April 1978, pp. 50–51. (b)

DiLaura, D. Visibility, human performance, and lighting design—the choice and use of a visibility level. *Lighting Design and Application,* June 1978, pp. 58–59. (c)

Faulkner, T. W., & Murphy, T. J. Lighting for difficult visual tasks. *Human Factors,* 1973, *15*(2), 149–162.

Hopkinson, R. G. Glare from daylighting in buildings. *Applied Ergonomics,* 1972, *3*(4), 206–215.

Hopkinson, R. G., & Longmore, J. Attention and distraction in the lighting of workplaces. *Ergonomics,* 1959, *2*, 321–334.

Hughes, P., & McNelis, J. Lighting, productivity, and work environment. *Lighting Design and Application,* May 1978, pp. 37–42.

Hultgren, G., & Knave, B. Discomfort glare and disturbances from light reflections in an office landscape with CRT display terminals. *Applied Ergonomics,* 1974, *5*(1), 2–8.

IES Lighting Fundamentals Course. New York: Illuminating Engineering Society, 1976.

IES lighting handbook (5th ed.). Illuminating Engineering Society, New York, 1972.

IES lighting handbook (Application Vol.). New York: Illuminating Engineering Society, 1981.

IES Nomenclature Committee. Proposed American national standard nomenclature and definitions for illuminating engineering (proposed revision of Z7.1R1973). *Journal Illuminating Engineering Society,* 1979, *9*(1), 2–46.

Light and color (TP-119). Nela Park, Cleveland, Ohio: Large Lamp Department, General Electric Company, August 1968.

Light measurement and control (TP-118). Nela Park, Cleveland, Ohio: Large Lamp Department, General Electric Company, March 1965.

Lion, J. S. The performance of manipulative and inspection tasks under tungsten and fluorescent lighting. *Ergonomics,* 1964, *7*(1), 51–61.

Logan, H. L., & Berger, E. Measurement of visual information cues. *Illuminating Engineering,* 1961, *56,* 393–403.

Luckiesh, M., & Moss, F. K. The new science of seeing. In *Interpreting the science of seeing into lighting practice* (Vol. 1). Cleveland: General Electric Co., 1927–1932.
Maerz, A., & Paul, M. R. *Dictionary of color* (2d ed.). New York: McGraw-Hill, 1950.
Munsell Color Co. *Munsell book of colors.* Baltimore: Munsell Color Co., 1929.
Optical Society of America, Committee on Colorimetry. *The science of color.* New York: Thomas Y. Crowell, 1953.
Ross, D. Task lighting—yet another view. *Lighting Design and Application*, May 1978, pp. 37–42.
RQQ. Selection of illuminance values for interior lighting design (RQQ Report No. 6). *Journal Illuminating Engineering Society*, 1980, *9*(3), 188–190.
Sanders, M., Gustanski, J., & Lawton, M. Effect of ambient illumination on noise level of groups. *Journal of Applied Psychology*, 1974, *59*(4), 527–528.
Zahn, J., & Haines, R. The influence of central search task luminance upon peripheral visual detection time. *Psychonomic Science*, 1971, *24*(6), 271–273.

CHAPTER **14**

ATMOSPHERIC CONDITIONS

Since the current model of the human organism is the result of evolutionary processes over millions of years, it has developed substantial adaptability to the environmental variables within the world in which we live, including its atmosphere. There are, however, limits to the human range of adaptability. In addition, science and technology are busy developing new kinds of environments for the human being, involving space capsules, cold-storage warehouses, and underwater caissons, and causing changes in our natural environment as the unintentional by-product of civilization—such as smog and air pollution. Since we cannot cover all aspects of the ambient environment, we will cover those that are of most general concern.

THE HEAT EXCHANGE PROCESS

The human body is continually making adjustments to maintain thermal equilibrium, that is, a balance between the net heat produced and the net heat loss to, or gain from, the environment. The primary factors that influence thermal regulation are

M – *metabolism:* the oxidation of food elements in the body

C – *convection:* the heat gain or loss by the mixing of air close to the body

E – *evaporation:* the heat loss by evaporation of body fluids, especially perspiration and exhaled breath

R – *radiation:* heat loss to, or gain from, surrounding environmental sources (the sun, walls, etc.) by direct radiation

Heat can also be exchanged by *conduction*, the direct contact of the body with some object, but conduction is a very negligible factor and usually is not taken into account in studying heat exchange.

The factors that influence the heat exchange process can be expressed in terms of the following formula, in which S (storage) refers to the gain or loss of body temperature.

$$\pm S \text{ (storage)} = M \text{ (metabolism)} \pm C \text{ (convection)} - E \text{ (evaporation)} \pm R \text{ (radiation)}$$

If the body is in a state of thermal balance, S (storage) is zero; otherwise S can be positive (representing an increase in body temperature) or negative (representing a decrease).

Factors That Affect Heat Exchange

The discussion above has given hints of the environmental variables that affect the heat-exchange process, but it may be useful here to pin these down specifically. These factors are air temperature; air humidity; air movement; and radiant temperature (temperature of walls, ceilings, and other surfaces in the area).

Heat Exchange under Various Conditions The relative importance of convection, radiation, and evaporation in maintaining thermal equilibrium depends very much upon the four factors listed above, these interactions being illustrated in Figure 14-1. This figure shows, for each of five conditions of air and wall temperature, the percentage of heat loss by evaporation, radiation, and convection. In particular, conditions *d*

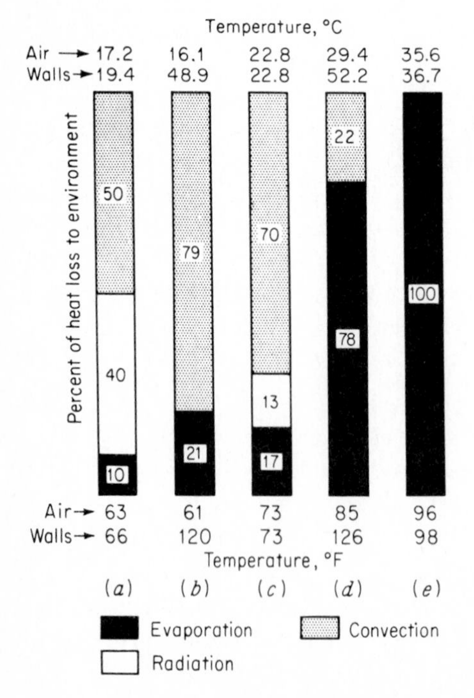

FIGURE 14-1
Percentage of heat loss to environment by evaporation, radiation, and convection under different conditions of air and wall temperature. (*Source: Adapted from Winslow & Herrington, 1949.*)

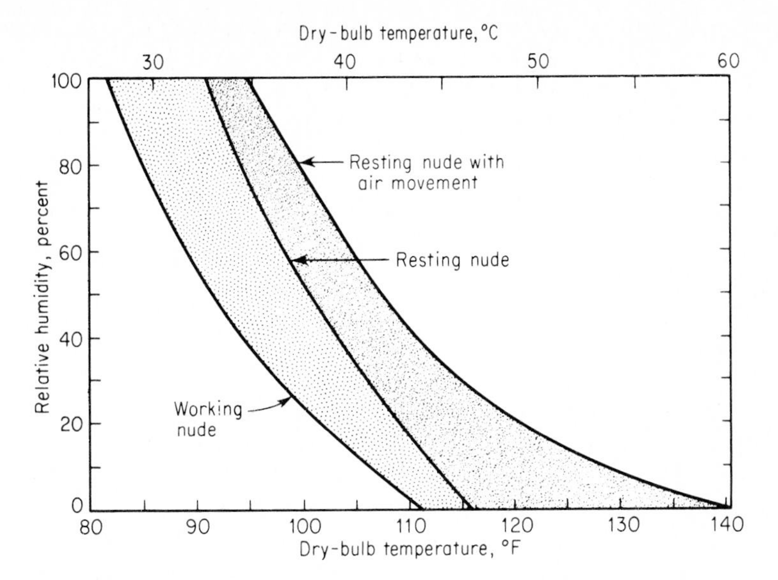

FIGURE 14-2
Approximate upper limits of tolerance for heat loss by evaporation. For any of the three conditions, temperature and humidity combinations to the right of the curve prevent evaporation. (*Source: Derived from data by Winslow et al., 1942.*)

and *e* illustrate the fact that, with high air and wall temperatures, convection and radiation cannot dissipate much body heat, and the burden of heat dissipation is thrown on the evaporative process. But, as we all know, evaporative heat loss is limited by the humidity; there is indeed truth to the old statement that "it isn't the heat–it's the humidity." To see the limiting effect of humidity on evaporation, let us look at Figure 14-2. This shows the upper limits of tolerance in relation to temperature and relative humidity for working and for resting nude subjects. For any one of the three curves in Figure 14-2, combinations of temperature and humidity to the right of the curve represent conditions which, if accentuated enough and long enough, could result in some physiological consequences ranging up to heatstroke or death. A comparison of the two curves at the right shows that air movement usually helps to make conditions more tolerable by exposing the surface of the body to more air.

Composite Indexes of Environmental Factors

Since there are different environmental factors that affect the heat-exchange process, it would be desirable to have some index of the stress imposed by combinations of environmental factors, and of the strain induced in the individual by exposure to the environment. The strain caused by environmental variables (including work) is measured with such physiological variables as body temperature, heart rate, and amount of sweat. A few of the composite indexes of environmental factors are discussed below.

Effective Temperature There are two indexes of effective temperature, both of which were developed under the sponsorship of the American Society of Heating, Refrigerating, and Air-Conditioning Engineers (ASHRAE, 1977). The original effective temperature (ET) scale was developed many years ago as an empirical sensory index, combining into a single value the effect of temperature and humidity on thermal sensations, with an adjustment for the effects of air movement. It was developed through a series of studies in which subjects compared the relative thermal sensations of various combinations of air conditions in adjoining rooms by passing back and forth from one room to the other. One objective was to identify different combinations of temperature and humidity that gave the same subjective sensation. Any given ET–such as 70°–was operationally characterized as the thermal sensation produced by a dry-bulb temperature of that same temperature (e.g., 70°) in combination with relative humidity (rh) of 100 percent. However, other combinations of dry-bulb temperature and rh that produced the same sensation under the experimental condition used have that same ET (e.g., 70°). In general terms, the "other" combinations for any given ET were characterized in terms of *higher* dry-bulb temperatures and *lower* rh values.

The original effective temperature (ET) scale has been demonstrated to overemphasize the effects of humidity in cool and neutral conditions and to underemphasize its effects in warm conditions, and it does not account fully for air movement in hot-humid conditions. Therefore, the ASHRAE has sponsored the development of a new effective temperature (ET*) scale. The nature of the ET* scale is very intricate and cannot be described in detail here. However, it is based in part on consideration of the physiology of heat regulation as it applies to comfort, temperature sensation, and health, especially as heat regulation depends upon evaporative heat loss. Evaporative heat loss consists of three parts. That heat which is lost from the lungs through respiration, and the heat lost through vaporized water diffusing through the skin layer, are both called *insensible* heat loss. The heat of vaporized sweat necessary for regulation of body temperature is called *sensible* heat loss. The body generally is in a state of thermal neutrality with respect to *regulatory* (i.e., sensible) heat loss when the dry-bulb temperature is about 77 degrees Fahrenheit (77°F) (25°C) and the rh is 50 percent. Holding rh constant (at 50 percent), higher or lower temperatures would alter the evaporative process, thus affecting the levels of skin "wettedness" at various temperatures. The ET* scale is based essentially on the resulting levels of wettedness. However, any given level of skin "wettedness" can be produced by different combinations of dry-bulb temperature and rh, and those combinations that produce the same level of wettedness then have the same ET* value.

Figure 14-3 includes the new ET* values. For the moment consider the dry-bulb temperatures (at the bottom of the figure) and the relative humidity lines that angle upward from left to right. Follow any given dry-bulb temperature line, such as 70°F, up to the 50-percent rh line. That intersection represents the ET* that corresponds with the dry-bulb temperature, such as 70. At that intersection the ET* line (a broken line) angles upward to the left and downward to the right. All the combinations of dry-bulb and humidity values on the ET* line represent combinations that are considered to be equal in terms of the bases used in the development of the scale as discussed above. (For our present purposes the other features of the figure can be dis-

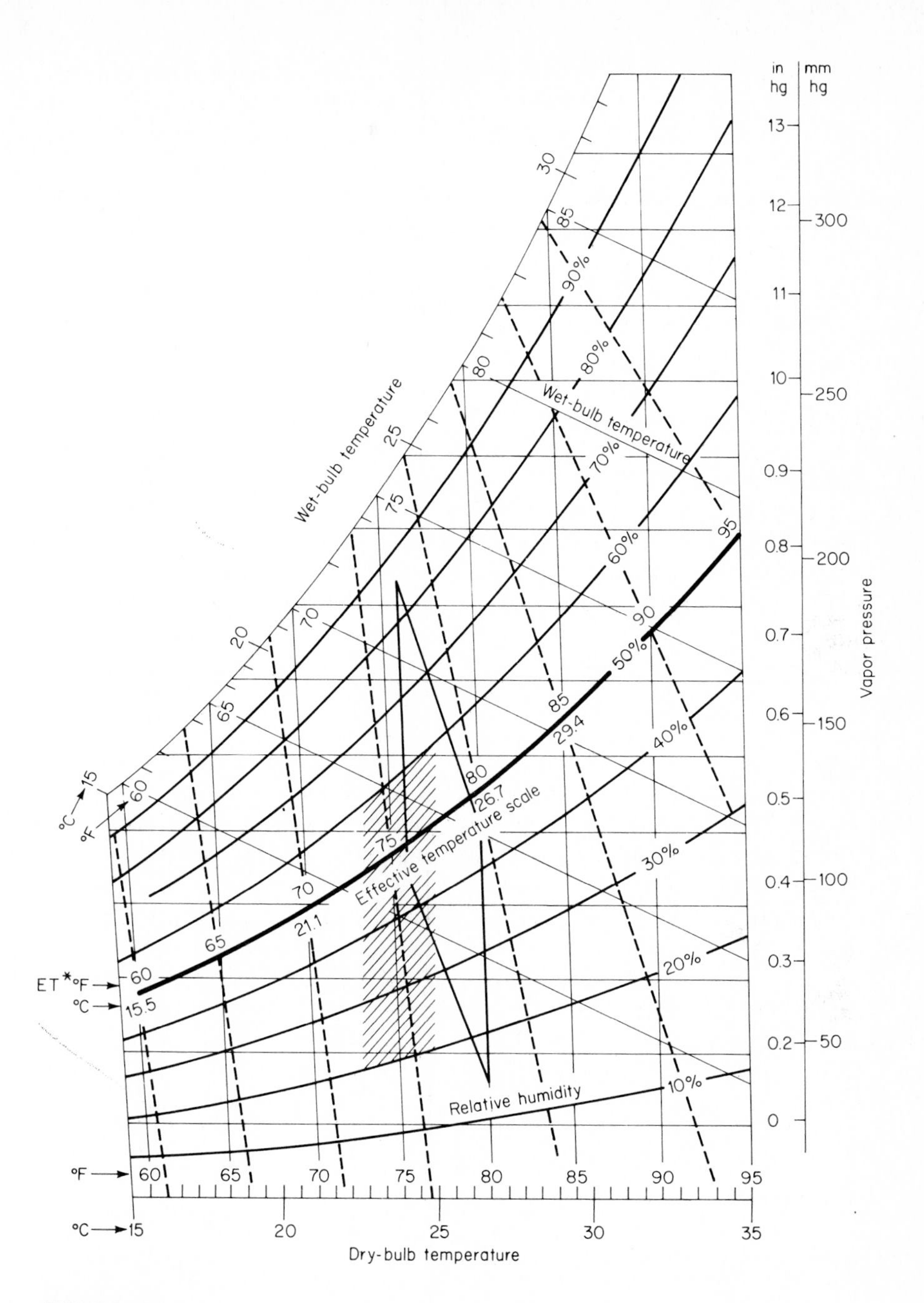

FIGURE 14-3
New effective temperature (ET*) scale. The broken lines represent the ET* values. See text for discussion. Vapor pressure is an index of humidity. Wet-bulb temperature is a measure of air temperature under conditions of 100 percent rh, obtained with an aspirated wet-wick temperature or sling thermometer. (*Source: Adapted from ASHRAE, 1977, fig. 16, p. 8.21.*)

regarded.) The ET* scale as illustrated applies for lightly clothed, sedentary individuals in spaces with low air movement and areas in which the effects of radiation would be limited (specifically, areas in which the mean radiant temperature equals air temperature).

It should be noted that the original ET scale had the loci of the ET lines at the intersection of corresponding dry-bulb temperatures in combination with 100-percent rh. Since the loci of the new ET* lines are at the 50-rh lines, the new ET* values would tend to be numerically higher than the older ET values for corresponding conditions. (Most of the later references to effective temperature are to the old ET scale rather than to the new ET* scale.)

Heat-Stress Index This index, originally developed by Belding and Hatch (1955, pp. 129-136), has been modified by Hatch (1963) and Hertig and Belding (1963). The index expresses the heat load in terms of the amount of perspiration that must be evaporated to maintain heat balance; this is referred to as E_{req} (the required evaporation heat loss). In turn, it is possible to determine the maximum heat that can be lost through evaporation E_{max}, from assumptions of body size, weight, and temperature and by taking into account water-vapor pressure of the environment and air velocity. While the details of these derivations will not be repeated, the *difference* between these two values [expressed in British thermal units per hour (Btu/h)] indicates the load that must be reduced or dissipated otherwise. The *otherwise* can take various forms, such as further reduction of convection or radiation sources, reduction of task demands by reducing physical requirements or by rest pauses, and by proper clothing.

To illustrate this general approach, Hertig and Belding present a hypothetical example of a task, as shown in Table 14-1. In this particular case, heat by radiation is the major source of the heat load and would be the primary aspect of the situation to do something about.

In connection with the heat-stress index, McKarnes and Brief (1966) have worked up a set of nomographs based on the modifications reported by Hertig and Belding, and have also elaborated on the theme by setting up a formula for estimation of allowable exposure time (AET) and minimum recovery time (MRT).

TABLE 14-1
ILLUSTRATION OF METHOD OF DERIVING HEAT-STRESS INDEX FOR A HYPOTHETICAL TASK

Source of heat load	Btu/h
Metabolism (based on type of activity)	800
Radiation	2800
Convection	60
E_{req}	3660
E_{max} (computed by formula)	2530
Difference	1130

Source: Hertig & Belding, 1963.

Operative Temperature This index takes into account the combined effects of radiation and convection, but not humidity and air flow. (An equation for deriving this index is included in the *ASHRAE Handbook*, 1977.)

Oxford Index The *Oxford index*, or WD index, is a simple weighting of wet-bulb and dry-bulb temperatures, as follows (Leithead & Lind, 1964, p. 82):

$$\text{WD} = 0.85w \text{ (wet-bulb temperature)} + 0.15d \text{ (dry-bulb temperature)}$$

Wet-bulb temperature is a measure of air temperature under conditions of 100 percent rh, obtained with an aspirated wet-wick or sling thermometer.

The Oxford index has been found to be a reasonably satisfactory index to equate climates with similar tolerance limits.

Wet-Bulb Globe Temperature (WBGT) This index consolidates into a single value the effects of four environmental factors, namely, dry-bulb temperature, relative humidity (or vapor pressure), mean radiant temperature, and air velocity. The formula for deriving WBGT as given by Azer and Hsu (1977) is

$$\text{WBGT} = 0.7T_{\text{nw}} + 0.3T_{\text{g}}$$

where T_{nw} = natural wet-bulb temperature obtained with wetted sensor exposed to natural air movement

T_{g} = temperature at center of 6-in (15-cm) diameter hollow copper sphere painted on outside with black matte finish (globe thermometer)

One feature of this measure that makes it attractive is the fact that the air velocity need not be measured directly, since its value is reflected in the measurement of the natural wet-bulb temperature.

The WBGT index is coming into more common use because it does take into account the combination of variables mentioned above, and because it is used in the proposed Occupational Safety and Health Administration (OSHA, 1974) standards for thermal comfort indoors.

CLO UNIT: A MEASURE OF INSULATION

Since there will be later reference to the clo unit, it is described here. The basic heat exchange process is of course influenced by the insulating effects of the clothing worn, varying from a bikini bathing suit to a parka as worn in the arctic. Such insulation is measured by the clo unit. The *clo unit* is a measure of the thermal insulation necessary to maintain in comfort a sitting, resting subject in a normally ventilated room at 70°F temperature and 50 percent relative humidity. Since the typical individual in the nude is comfortable at about 86°F, one clo unit would be required to produce an equal sensation at about 70°F; a clo unit has very roughly the amount of insulation required to compensate for a drop of about 16°F, and is approximately equivalent to the insulation of the clothing normally worn by men. To lend support to the old adage that there is nothing new under the sun, the Chinese have for years described the weather in terms of the number of suits required to keep warm, such as a *one-suit* day (reasonably comfortable), a *two-suit* day (a bit chilly), up to a limit of a *twelve-suit* day (which would be really bitter weather).

BODY CHANGES DURING THERMAL ADJUSTMENT

When the body changes from one thermal environment to another, certain physical adjustments are made by the body, especially the following:

- *Changes from warm environment to cold one:* (1) The skin becomes cool; (2) the blood is routed away from the skin and more to the central part of the body, where it is warmed before flowing back to the skin area; (3) rectal temperature rises slightly but then falls; and (4) shivering and "goose flesh" may occur. The body may stabilize with large areas of skin receiving little blood.
- *Changes from cool environment to warm one:* (1) More blood is routed to the surface of the body, resulting in higher skin temperature; (2) rectal temperature falls but with continued exposure rises; and (3) sweating may begin. The body may stabilize with increased sweating and increased blood flow to the surface of the body.

If the change from one situation to another is so extreme as to cause the body temperature to increase appreciably (producing a condition of *hyperthermia*) or to decrease appreciably (producing *hypothermia*) there may be serious physiological consequences, and in extreme situations death. Extended heat spells, for example, usually are accompanied by reports of death from heatstroke, especially of aged and infirm individuals.

Acclimatization to Heat and Cold

Acclimatization consists of a series of physiological adjustments that occur when an individual is habitually exposed to extreme thermal conditions, hot or cold, as the case may be. An illustration of the changes that occur during acclimatization to heat is given in Figure 14-4. The men involved in the study worked each of 9 days for 100

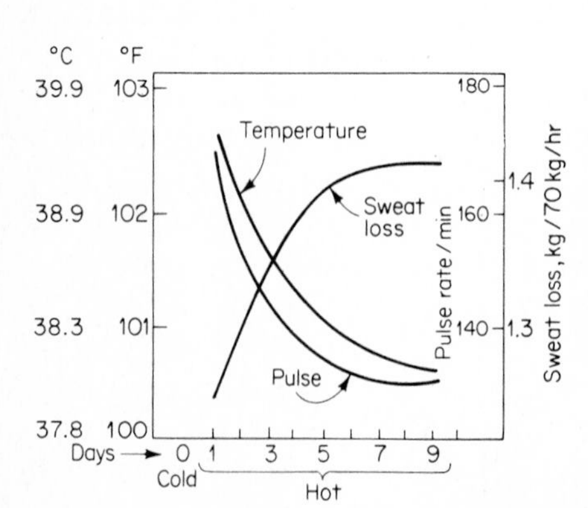

FIGURE 14-4
Changes in rectal temperature, pulse rate, and sweat loss of a group of men during 9 days of exposure to a hot climate (dry-bulb and wet-bulb temperatures of 120 and 80° F), during which they worked for 100 min at an energy expenditure of 300 kcal/h. Comparative temperature and pulse values are given for a preceding control day (day 0) during which the men worked in a cool climate. (*Source: Adapted from Lind & Bass, 1963.*)

min at an energy expenditure of 300 kcal/h in a hot climate (Lind & Bass, 1963). The figure shows changes in rectal temperature, pulse rate, and sweat loss. The changes in these physiological measurements are very distinct. Shapiro, Pandolf, and Goldman (1980) confirm the changes in physiological indices during acclimatization, in particular heart rate, rectal temperature, and skin temperature. Although there is some question about the time required to become fully acclimatized, it is evident that much acclimatization to heat occurs within 4 to 7 days, and reasonably complete acclimatization occurs usually in 12 to 14 successive days of heat exposure (Leithead & Lind, p. 21). Quite a bit of acclimatization to cold occurs within a week of exposure, but full acclimatization may take months or even years. Even complete acclimatization, however, does not fully protect an individual from extreme heat or cold, although acclimatized individuals can tolerate extremes better than those who are not acclimatized.

With regard to sex considerations, Shapiro, Pandolf, Avellini, Pimental, and Goldman (1980) and Shapiro, Pandolf, and Goldman (1980) conclude that females tend to tolerate hot-wet climates better than males and that males tend to tolerate hot-dry conditions better than females. Sex differences in certain thermoregulatory processes presumably account partially for these differences. Part of the differences can also be attributed to the differences in body size between males and females, especially in the fact that females have a higher ratio of body surface to their weight than males.

Although evidence regarding the *decay* (i.e., loss) of acclimatization is rather inconclusive, there are indications that within 2 or 3 weeks after working in a hot environment individuals lose most of their acclimatization. However, Pandolf, Burse, and Goldman (1977) concluded from their study that physical fitness can minimize the decay of acclimatization and limit the time required for reacclimatization.

THERMAL SENSATIONS

In the above discussion of Figure 14-3 reference was made to the ranges of effective temperature (ET*) that generally are associated with various sensations regarding the environment. In a more detailed study of thermal comfort, Rohles (1974) used a sample of 1600 college student subjects in an effort to develop a *modal comfort envelope* (MCE). The subjects were divided into 160 groups of 10 each, with 5 men and 5 women in each group. Those in each group were exposed to a different combination of 1 of 20 dry-bulb temperatures and 1 of 8 relative humidities (160 tests) for 3 h. The radiant temperature was equal to air temperature, and the air movement was less than 50 ft/min (15.2 m/min). The subjects were asked to rate their thermal sensation using the following scale: cold (1), cool (2), slightly cool (3), comfortable (4), slightly warm (5), warm (6), and hot (7). From these ratings the MCE shown in Figure 14-3 was developed. This envelope falls between the 75° and 80° ET* lines, forming a vertical diamond shape. That figure also includes a shaded area that represents the conditions given in the ASHRAE Standard for Thermal Comfort 55-66 as shown in the *ASHRAE Handbook* (1977). Although these two areas do overlap somewhat, the MCE is generally higher than the ASHRAE standard. Rohles suggests that this difference arises from differences in the experimental conditions used in their development. The

MCE study was carried out with subjects who wore clothing of 0.6 clo units, whereas those used in the ASHRAE standard study wore clothing of 0.8 to 1.0 clo units. In addition, the MCE applies to "sedentary" activity whereas the ASHRAE Standard 55-66 applies to office work, which is above the sedentary level. Rohles makes the point that the ASHRAE standard needs to be revised. However, since the two areas do overlap he recommends that that area be considered as the "design conditions for comfort," which is specified as an ET* of 76°F. Rohles points out that the MCE applies only to situations in which dry-bulb temperature and humidity are varied, and not to conditions of increased air movement or higher activity levels, or when different clothing is used.

Some indication of the influence of level of activity and of clothing on sensations of comfort is shown in Figure 14-5. This shows "comfort lines" for each of three levels of activity (sedentary, medium, and high) for persons with light and medium clothing. The differences in these lines (they are actually ET* lines) reflect the very definite effects on comfort of work activity and clothing.

Although much has been found out about the sensations of thermal comfort under various conditions, there are still many gaps in the available knowledge, especially with regard to the interactions among some of the variables that influence such sensations.

FIGURE 14-5
Comfort lines for persons engaged in three levels of work activity with light clothing (0.5 clo) and medium clothing (1.0 clo). These data are for a low level of air velocity. (*Source: Based on data from Fanger, 1970; reprinted, with adaptations, with permission from ASHRAE, 1977, chap. 8, fig. 18.*)

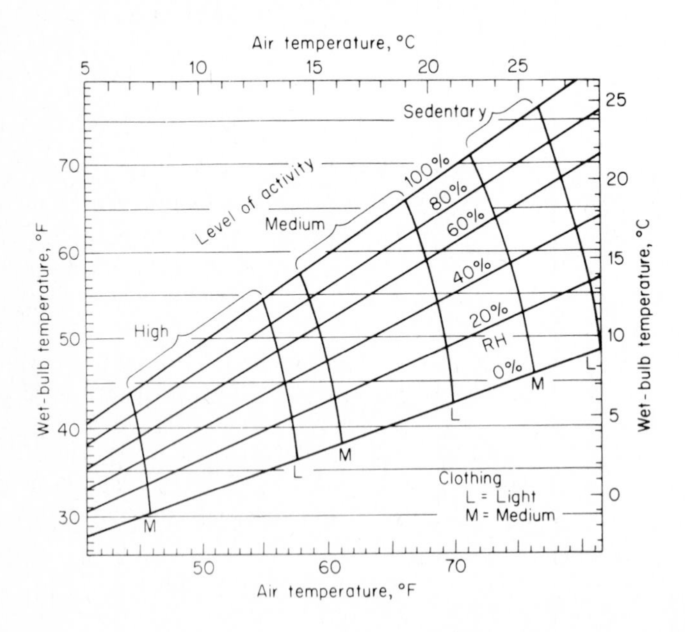

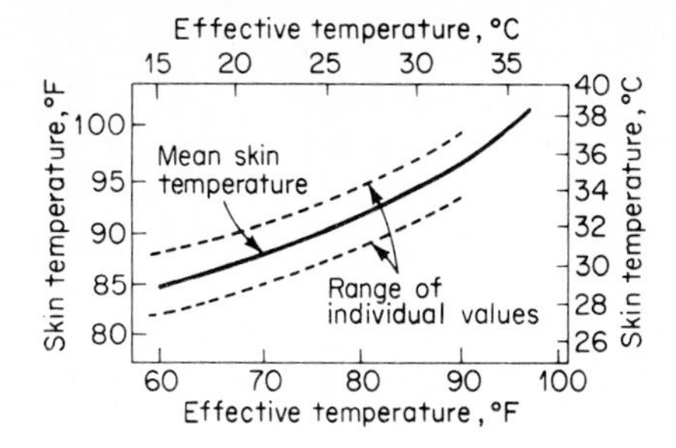

FIGURE 14-6
Relationship between effective temperature (ET) and skin temperature as consolidated from various studies. (*Source: Adapted from Leithead & Lind, 1964, fig. 6.*)

HEAT STRESS

Having discussed the heat-exchange process and certain indexes of environmental conditions, let us now turn to the effects of heat stress.

Physiological Effects

One of the most direct effects of heat stress is on the temperature of the body. However, there are several different measures of temperature, including oral temperature, rectal temperature, and skin temperature. The relationship among these is relatively limited except under conditions of high temperature and high humidity. Rectal temperature is generally considered to be a representative temperature of the core. Heat stress usually is accompanied by increases in rectal temperatures.

Recognizing that skin and rectal temperature cannot be used interchangeably, let us first see the relationship between ET and skin temperature as summarized from different studies by Leithead and Lind (1964) as shown in Figure 14-6. The upswing of the curves probably reflects the consequences of the flow of blood toward the surface of the skin to increase the dissipation of surplus heat with increasing effective temperatures. As a follow-up, Figure 14-7 represents the relation between ET and rectal temperature for one individual engaged in three levels of physical activity and for the average of three individuals at a single level of activity. In each instance the long, flat sections of the curves represent what Lind (1963a) refers to as the *prescriptive zone* for whatever activity is involved. As a prelude to a later discussion of the effects of work load, we can see the differential upswings for the three levels of work activity represented, the upswing starting at successively lower points with increasing work load. The middle, light line (that represents a single individual working at a level of 300 kcal) parallels the heavy line (that represents the average for three individuals), a fact which indicates considerable stability of the patterns of the curves.

As shown in Figure 14-7, the level of work activity contributes to the strain. This is illustrated further in Figure 14-8, which shows the metabolic cost of carrying various loads on the level and on a slope.

Levels of work in combination with environmental conditions that bring about the rise in core temperature (as shown in Figure 14-7) also induce other corresponding

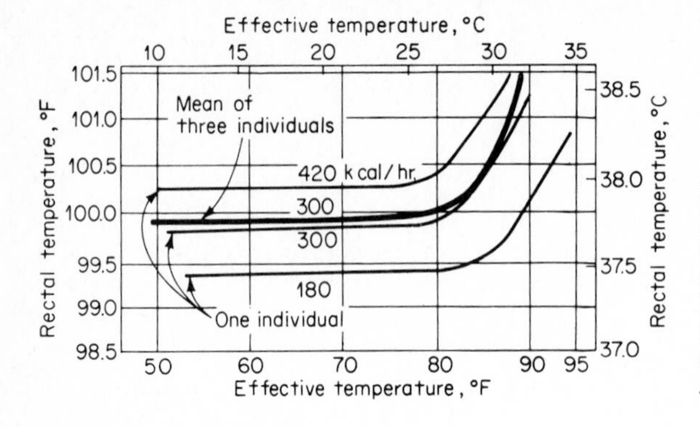

FIGURE 14-7
Rectal temperature as related to effective temperature (ET) for one individual working at three levels of work activity and for three individuals working at the same level (300 kcal/h). (*Source: Adapted from Lind, 1963a.*)

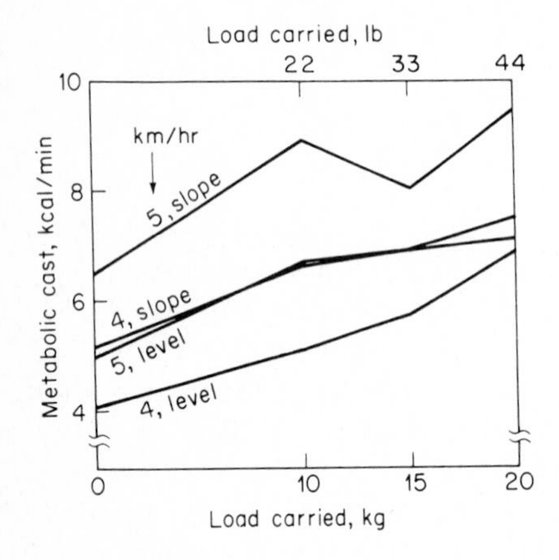

FIGURE 14-8
Metabolic cost of carrying various loads on the level and on a 4 percent slope at 4 and 5 km/h. (*Source: Adapted from Kamon & Belding, 1971, table 3.*)

physiological changes, which, if continued, can cause hypothermia, a condition that renders normal heat loss more difficult. Dehydration, such as from sweat, is another possible consequence of heat stress. In the case of 51 operating engineers working during mild summer weather in California's Central Valley, the average weight loss from water depletion per day was about 5 pounds 2 ounces (5 lb 2 oz). Such a deficit, of course, needs to be replaced during the evening and night.

Working Conditions in Blast Furnaces

Although most studies of heat stress have been carried out in laboratories, a major project was carried out in blast furnaces in Luxembourg and Belgium over a 4-year period to minimize the effects of heat. Without going into details, at one stage certain modifications were made in a couple of blast furnaces to reduce heat stress. Figure 14-9 shows the comparison of the heat stress of these two as related to the heat stress of one of the blast furnaces before any changes were made. The changes in design of the two clearly reflect appreciable reduction in heat stress.

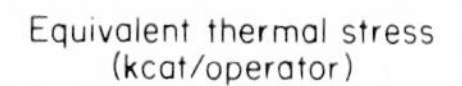

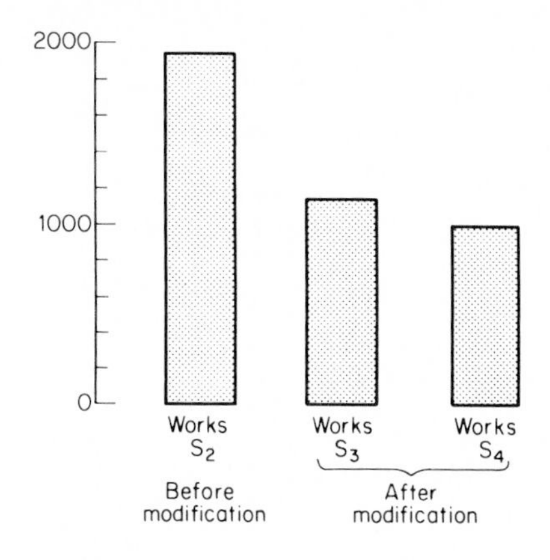

FIGURE 14-9
Heat stress of two blast furnaces that were partially redesigned (S_3 and S_4) as compared with that of a blast furnace before any modification. (*Source: Adapted from Vogt, Foehr, Kuntzinger, Seywert, Libert, Candas, & Van Peteghem, 1977, fig. 10, p. 179.*)

Heat Stress and Performance

The relationship between heat stress and performance is admittedly complicated, the effects being related in part to the type of work activity. A few relevant studies are summarized to give some impression of the effects of heat stress on performance.

Physical Work The effects of heat stress on performance of physical work are rather well documented. In this regard Leithead and Lind (1964, chap. 12) summarize the results of several industrial surveys that nail down the point that heat stress takes its toll on performance in physical work activities. Further evidence is reflected in data from Lind (1963b) presented in Figure 14-10 (tolerance times of men working at three levels of energy expenditure in relation to variations in the Oxford index).

Tracking Tasks An example of the effects of heat stress on tracking tasks is reported by Bell (1978) who had 72 male and 72 female subjects perform a tracking task along with a subsidiary test. The subsidiary task consisted of listening to a tape of two-digit numbers. (The subject was to press a telegraph key once if the second number

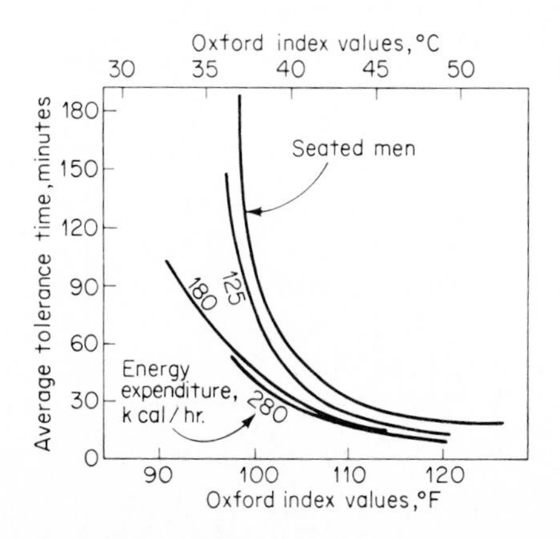

FIGURE 14-10
Average tolerance times of men seated, and of men working at three levels of energy expenditure, in relation to Oxford index values. Safe tolerance values should be taken to be no more than about 75 percent of the times shown. (*Source: Adapted from Lind, 1963b, as based on data from other studies.*)

TABLE 14-2
ERRORS IN A SUBSIDIARY TASK PERFORMED UNDER VARIOUS TEMPERATURES AND NOISE LEVELS

	Temperature		
Noise level, dB	**72°F (22°C)**	**84°F (29°C)**	**95°F (35°C)**
55	27	45	53
95	53	55	78

Source: Bell, 1978.

was lower than the first, and twice if it was higher). The tasks were performed under combinations of three temperatures and two noise levels. Although performance on the tracking task itself was not significantly different for the three temperatures or for the two noise conditions, performance on the subsidiary task did differ significantly for both environmental conditions, as shown in Table 14-2. The results indicate that exposure to high levels of temperature, noise, or both had detrimental effects on performance on the subsidiary task, presumably because of overload. As an aside, it should be pointed out that the effects of the combination of temperature and noise were considered to be additive, and not "interactive."

Typing In 1923 the New York State Commission on Ventilation reported the results of a very comprehensive study of the effects of heat and ventilation on people. In certain specific experiments males and females performed various tasks such as mental multiplication, arithmetic addition, cancellation tasks, and typing under two temperature conditions, 68°F (20°C) and 75°F (24°C), the subjects performing such tasks for as long as 2 weeks. The experimental conditions were complicated and so are not described here. The raw data laid around for over 50 years until Wyon (1974) exhumed it and reanalyzed it using more recently developed statistical procedures. The typing task was an "optional" task performed under "relaxed" conditions that Wyon claimed approximated most closely those during ordinary office work. The higher temperature condition generally resulted in decreases in typing performance up to 40 or 50 percent, the differences being so large that Wyon claimed that they cannot be ignored in a practical situation. However, it is still a bit surprising that the differences were found, since 75°F (24°C) is not a very high temperature.

Mental Activities The effects of heat stress on performance of mental activities are intertwined with the environmental conditions (such as ET) and duration of the work. A generalized pattern of the temperature-duration function related to mental performance has been developed by Wing (1965) on the basis of a very thorough analysis of the results of 15 previous studies. In this analysis he identified, from the several studies, the lowest temperature at which a statistically significant performance decrement occurred. These points, plotted at their respective exposure durations, were then

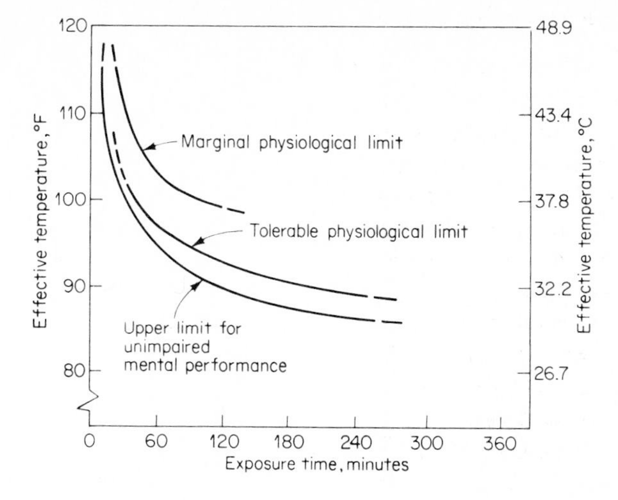

FIGURE 14-11
Tentative upper limit of effective temperature (ET) for unimpaired mental performance as related to exposure time; data are based on an analysis of 15 studies. Comparative curves of tolerable and marginal physiological limits are also given. (*Source: Adapted from Wing, 1965, fig. 9. See Wing's report for original sources.*)

used in drawing the curve shown in Figure 14-11. In addition, that figure shows, for comparative purposes, curves of tolerable and marginal physiological limits based on earlier studies by other investigators. Wing suggests that the thresholds for at least some mental tasks might be between the lower curve and the tolerable physiological-limit curve in the case of fully acclimatized or highly practiced individuals.

Performance on Industrial Jobs Although most of the studies of heat stress on performance have been carried out in laboratories, there have been a few studies in actual job situations, which, in general, support the indications of laboratory studies that heat stress does take its toll. An example is one reported by Tichauer (1962) dealing with the performance on the job of "picking" in cotton weaving. Figure 14-12 shows an index of the number of "picks" per day under various temperature conditions, the results showing the highest production levels within the range of 75 to 80°F (24 to 27°C) wet-bulb temperatures.

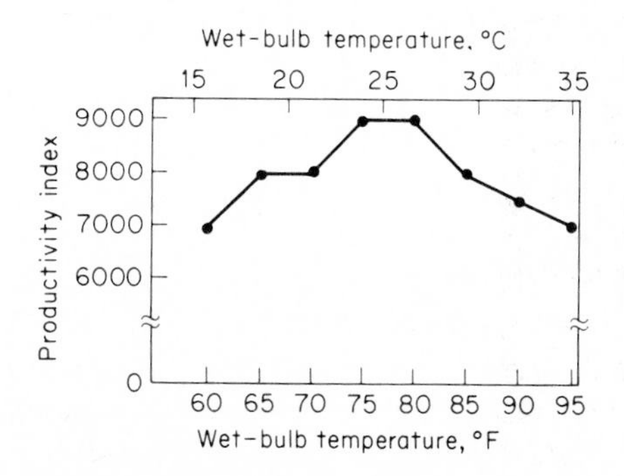

FIGURE 14-12
Relationship between temperature and performance of "pickers" in cotton weaving. Cotton fiber can be processed best in a warm, humid atmosphere, which of course is bad for human efficiency. In this instance, a wet-bulb temperature of 80° F (27° C) represents a satisfactory compromise. (*Source: Adapted from Tichauer, 1962.*)

TABLE 14-3
RECOMMENDED MAXIMUM WET-BULB GLOBE TEMPERATURES (WBGT) FOR VARIOUS WORK LOADS AND AIR VELOCITIES

	Air velocity	
Work load	Low: Below 300 ft/min (1.5 m/s)	High: 300 ft/min (1.5 m/s) or above
Light (200 kcal/h or below)	86° F (30.0° C)	90° F (32.2° C)
Moderate (201–300 kcal/h)	82° F (27.8° C)	87° F (30.6° C)
Heavy (above 300 kcal/h)	79° F (26.1° C)	84° F (28.9° C)

Source: Occupational Safety and Health Administration, 1974.

Recommended Limits of Heat Exposure

The illustrations cited above give some impression of the effects of heat stress on people. As a step in the direction of establishing some limits on heat exposure that would minimize the possibility of undesirable consequences, the Occupational Safety and Health Administration (OSHA, 1974) has recommended the WBGT values shown in Table 14-3 as maximums for work involving different work loads. In interpreting these recommendations keep in mind that they are based on WBGT values, not conventional dry-bulb temperatures. It should be added that these values generally apply to continuous work. Humans usually can tolerate higher values for short periods of time.

COLD

While civilization is generally reducing the requirement for many people to work in cold environments, there still are some circumstances where people must work, and live, in cold environments. These situations include outdoor work in winter, arctic locations (especially in military and exploration activities), cold-storage warehouses, and food lockers. As in the case of heat exposure, there are a number of interlaced factors that affect the tolerance, comfort, and performance of people in cold environments; these include the level of activity, degree of acclimatization, duration, and insulation.

Indexes Related to Effects of Cold

Wind Chill One of the indexes related to cold is a wind-chill index (Siple & Passel, 1945). It provides a means for making a quantitative comparison of combinations of temperature and wind speed. The quantitative value corresponds to a calorie scale (actually kilocalories per square meter per hour) but is converted into a sensation scale ranging from hot (about 80), to pleasant (about 200), cool (400), cold (800), bitterly cold (1200), and even colder values (including that at which exposed flesh freezes in 1 min or less). Although the wind-chill index itself is not given here, Table 14-4 shows the cooling effects of combinations of certain temperatures and wind speeds derived from the scale, these being expressed as *equivalent temperatures*. For example, an air temperature of 10°F (−12°C) with a 20-mi/h wind produces the same cooling effect as a temperature of −25°F (−33°C) under calm conditions.

TABLE 14-4
COOLING EFFECTS OF TEMPERATURE AND WIND SPEED

Wind speed		Equivalent temperature											
		Air temperature, °F						Air temperature, °C					
mi/h	km/h	40	20	10	0	−10	−30	4	−7	−12	−18	−23	−29
calm		40	20	10	0	−10	−20	4	−7	−12	−18	−23	−29
5	9	37	16	6	−5	−15	−26	3	−9	−14	−20	−26	−32
10	16	28	4	−9	−21	−33	−46	−2	−16	−23	−29	−36	−43
20	32	18	−10	−25	−39	−53	−67	−8	−23	−33	−37	−47	−55
30	49	13	−18	−33	−48	−63	−79	−11	−28	−36	−43	−52	−62
40	64	10	−21	−37	−53	−69	−85	−12	−29	−38	−47	−56	−65

Source: Siple & Passel, 1945

TABLE 14-5
SUBJECTIVE SENSATIONS ASSOCIATED WITH SKIN TEMPERATURES

Sensation	Mean skin temperature	Hand-skin temperature
Comfortable	92°F (33.3°C)	
Uncomfortably cold	88°F (31°C)	68°F (20°C)
Shivering cold	86°F (30°C)	
Extremely cold	84°F (29°C)	59°F (15°C)
Painful		41°F (5°C)

Source: ASHRAE, 1977, pp. 8–15.

Physiological Effects of Cold

With inadequate protection, exposure to the cold brings about a reduction of both core and shell temperatures. Continued exposure, of course, can bring about frostbite and other effects, and ultimately death. A fall in deep body (rectal) temperature below 95°F (35°C) threatens loss of control of body temperature regulation, while 82°F (28°C) is considered critical for survival (ASHRAE, 1977, pp. 8–15). Exposure to cold also reduces mean skin temperature and hand-skin temperature.

Subjective Sensations in Cold Conditions

Subjective sensations as related to skin temperature are reported in Table 14.5.

Performance in Cold

The primary interest in the effects of cold on performance relates to manual tasks of one sort or another. In this regard it is quite evident that performance on manual tasks is very directly related to hand-skin temperature (HST). However, Lockhart, Kiess, & Clegg (1975) provide some evidence to indicate that the lower boundary of HST varies somewhat with the task. In their experiment subjects performed six tasks, as follows: block packing (putting small blocks into a box); block stringing (stringing small blocks with holes using needle and thread); Craik screw (putting screws into holes in a metal bar); knot tying; Purdue Pegboard (assembling washer, collar, and washer on a metal peg); and screw tightening. With half the subjects the desired finger temperature was attained within 5 min of cold exposure (fast cooling), and with the other half within 50 min (slow cooling). Results with two of the tasks are shown in Figure 14-13.

Figure 14-13 shows marked deterioration in performance with reduced HST, and also shows that performance tended to be somewhat better for those who had experienced fast (rather than slow) hand cooling. Data from the other tasks led to the implication that for certain tasks performance impairment may start at about 65°F (18.3°C), and for others at 55°F (12.8°C). In some instances in which the fingers have been cooled rapidly impairment may not appear until HST is 48°F (8.9°C).

From studies such as that illustrated in Figure 14-13, it is clearly evident that when people are to perform manual tasks in the cold, provision should be made for maintaining hand and skin temperature at reasonable levels such as by permitting the individuals to warm their hands indoors or in some other manner. In this regard, for example, Lockhart and Kiess (1971) experimented with the use of infrared heaters for per-

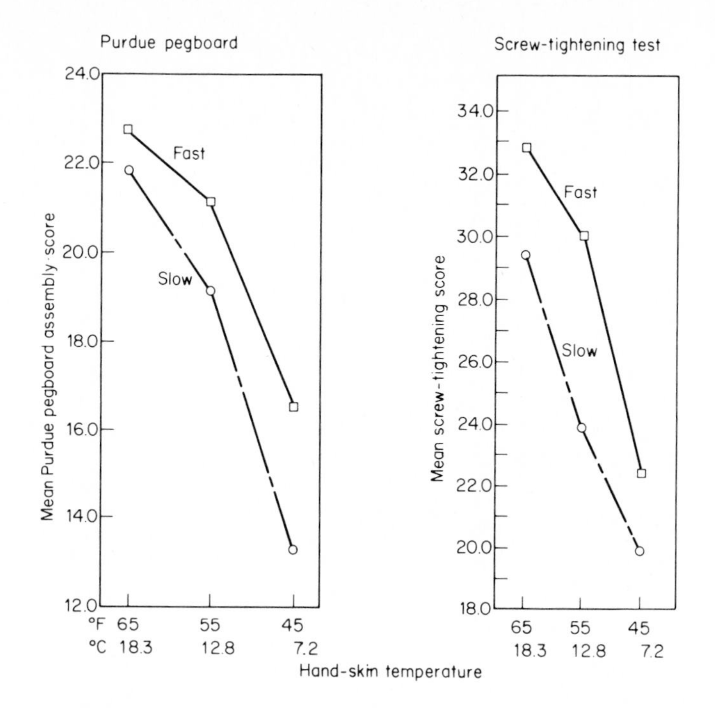

FIGURE 14-13
Deterioration in performance on two tasks as associated with reduction in hand-skin temperature (HST) and with fast cooling of the hands (within 5 min) or slow cooling (within 50 min). (*Source: Adapted from Lockhart, Kiess, & Clegg, 1975, fig. 2, p. 111. Copyright 1975 by the American Psychological Association. Reprinted by permission.*)

sons performing various tasks. One group used these heaters under 0°F (–18°C) conditions, another had no heaters under the same conditions, and a third control group performed the task indoors. The results for the Purdue Pegboard test are shown in Figure 14-14, and indicate that the auxiliary heating with the heaters was very effective in maintaining task performance.

In outdoor activities, wind chill is a factor to be contended with. Among its effects are increased numbness and increased reaction time. Although acclimatization can increase the tolerance to wind chill, it does not eliminate its effects on manual performance. But cold seems not to affect mental performance or visual performance.

COMFORT-HEALTH INDEX

Realizing the complexities of the relationships among the variables that influence the effects of atmospheric conditions, a comfort-health index (CHI) has been developed that summarizes information on temperature sensation, comfort, health, and physiological aspects for people engaged in sedentary tasks (ASHRAE, 1977, p. 8.30). This is given in Figure 14-15. It shows the sensory, physiological, and health responses typically associated with prolonged exposures at the ET* or CHI values, expressed as a dry-bulb temperature at 50 percent rh.

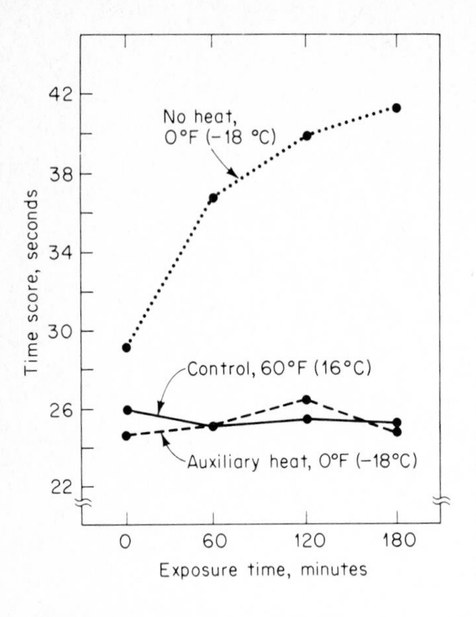

FIGURE 14-14
Time to perform the assembly part of the Purdue Pegboard Test under air conditions of 0°F (−18°C) with and without auxiliary warming of the hands, and for a control condition. (*Source: Adapted from Lockhart & Kiess, 1971, fig. 1.*)

FIGURE 14-15
Comfort-health index (CHI) expressed in terms of ET* and related sensory, physiological, and health responses in humans for prolonged exposures. (*Source: ASHRAE, 1977, chap. 8, fig. 23, p. 8.30; reprinted with permission.*)

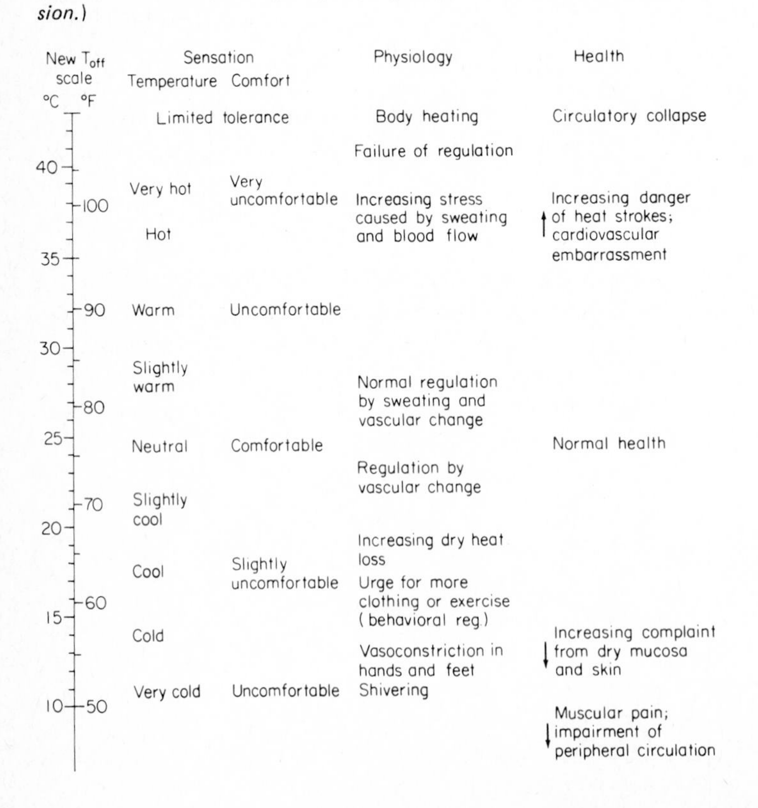

THE MANAGEMENT OF TEMPERATURE PROBLEMS

In the case of indoor situations, atmospheric control can be effected through heating, air conditioning, circulation of air, humidity control, insulation, and shielding against radiation, as well as by other techniques. Such atmospheric control is of course in the domain of the engineering professions and it is not appropriate or feasible to include here any discussion of the engineering aspects of such control. There are, however, two relevant personnel-related matters discussed below: clothing and rest periods.

The Thermal Effects of Clothing

We all know from personal experiences that in the case of warm and hot environments one should wear light, loose clothing. And those who live in hot climates have "discovered" that light-colored clothing (such as white) is cooler than dark clothing, this being because light-colored clothing reflects more heat.

Within relatively moderate temperature ranges the amount of clothing can influence sensations of comfort. Wyon, Fanger, Olesen, and Pedersen (1975), for example, had young male and female subjects perform various mental tests and tasks for 2½ h in which they wore either light clothing (0.6 clo) or heavy clothing (1.1 clo). The subjects indicated their thermal comfort on a dial apparatus, and the temperature condition of the chamber was adjusted up and down until the subjects indicated their own individual "comfort" zones on the dial. The mean preferred temperatures determined in this way are given in Table 14-6 for both clothing conditions. The differences in preferred temperatures for the two clothing conditions are quite evident.

In the case of cold temperature the use of warm clothing can extend the tolerance level of people, as illustrated in Figure 14-16. This shows the tolerable combination of exposure time and temperature for each of four insulation levels; the difference in tolerance between 1 and 4 clo units is upward of 60°F. This demonstrates the trade-off effects in terms of maintaining heat balance of exposure time and insulation.

In connection with the use of apparel in cold conditions, however, McIntyre and Griffiths (1975) report that adding additional garments does not compensate fully for the discomfort of cool conditions. Although their subjects did report increased feelings of warmth with long-sleeved woolen sweaters, the sweaters did not fully alleviate the feelings of discomfort. They attributed this to the fact that the subjects still reported that their feet felt cold. Thus, for adequate protection in the cold, attention should be given to providing footwear that is as warm as possible.

TABLE 14-6
PREFERRED TEMPERATURES FOR PERSONS WEARING LIGHT AND HEAVY CLOTHING

		Mean preferred temperature	
Clothing	**clo**	**Males**	**Females**
Light	0.6	74°F (23.4°C)	73°F (22.9°C)
Heavy	1.1	65°F (18.6°C)	66°F (18.9°C)

Source: Wyon et al., 1975.

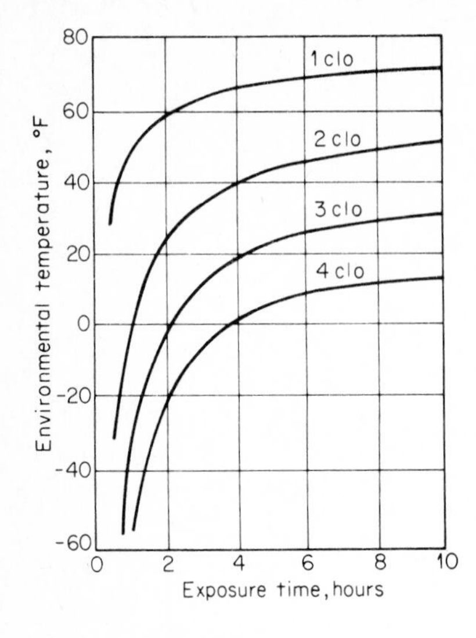

FIGURE 14-16
Exposure time and temperature that are tolerable for different levels of insulation (clo units). (*Source: Adapted from Burton & Edholm, 1955, as presented in Webb, 1964, p. 125.*)

Conductive Cooling

Some types of work are of necessity performed in hot environments, as in certain areas of steel mills and in deep mines. In this regard there have been some interesting developments in cooling the body by the use of air, water, and ice. The use of water and ice offers greater possible efficiency than air. Body cooling by water is carried out through a network of polyvinylchloride (PVC) tubing in jackets or other garments. Ice used in cooling usually is held in pockets also made of PVC, usually in the form of jackets. Dry ice has been used in at least experimental situations.

In discussing such techniques Shvartz (1975) points out that cooling is more effective with some parts of the body. Cooling of the head, for example, alleviates about 50 percent of the heat strain, while cooling of the upper torso, upper arms, and thighs reduces heat strain by more than 50 percent. The use of jackets filled with ice results in responses that are similar to those found in a temperate environment. Shvartz reports the heart rates shown in Table 14-7 for various cooling conditions.

One of the obvious problems associated with such techniques is the possible interference of the cooling apparatus with the work activities of people. In this regard Shvartz offers certain suggestions about the areas of the body that might be cooled by such methods for various types of work. These suggestions are given in Figure 14-17.

TABLE 14-7
HEART RATES AS ASSOCIATED WITH CONDUCTIVE COOLING WITH ICE

Condition	Heart rate
No cooling	158
Hood cooling (of the head)	133
Cooling of the head, upper arms, torso, and thighs	105

Source: Shvartz, 1975.

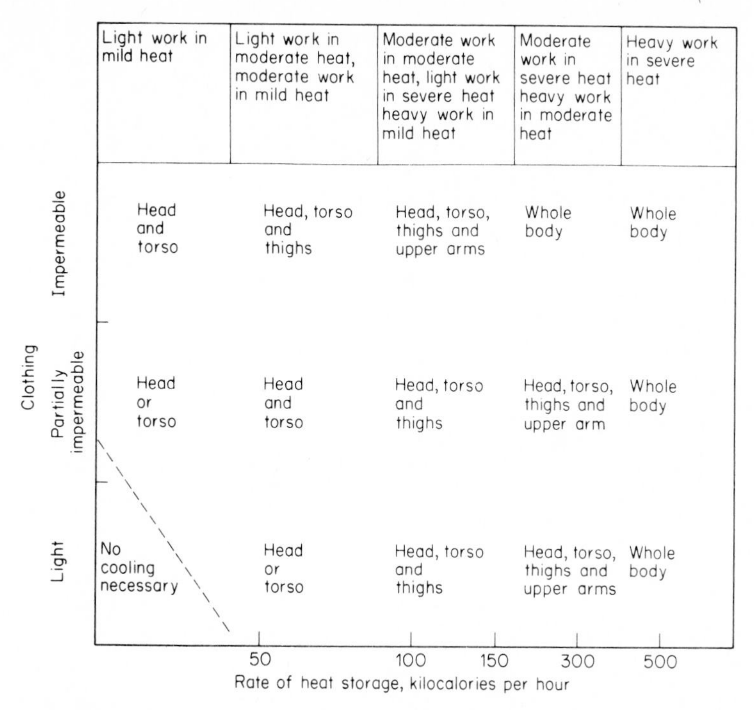

FIGURE 14-17
Areas of the body for which conductive cooling by water or ice is proposed for standing or walking operator in relation to heat stress, work load, and clothing. *Work load* refers to the following values: light = 150 to 280 kcal/h; moderate = 290 to 490 kcal/h; heavy = 500 to 650 kcal/h. (*Source: Shvartz, 1975, fig. 2, p. 422.*)

Although the apparatus involved in the use of such techniques can be complicated to use and may restrict work activity, there are indeed some circumstances in which such procedures might well be justified.

Rest Periods

Persons working in hot environments need to be provided with adequate rest periods in order to reduce the total heat stress to which they are subjected. The frequency and duration of the rest periods should depend on the environmental conditions and the nature of the work. In this regard, although there are no comprehensive guidelines available, some indications can be gleaned from Figure 14-5, which shows comfort lines for persons engaged in three levels of work activity. If these comfort lines are exceeded it is probable that rest periods should be provided.

When rest periods are provided the rest preferably should be in somewhat neutral situations to permit more complete recovery. Some support in this comes from Krajewski, Kamon, and Avellini (1979), who had subjects perform on a treadmill under

TABLE 14-8
SWEAT RATES DURING REST FOLLOWING TREADMILL EXERCISE UNDER TWO ATMOSPHERIC CONDITIONS

Treadmill condition	Resting condition	
	Same as treadmill	Neutral
Hot-dry	417	285
Warm-humid	428	325

Note: These data are for males, but the pattern was the same for females.
Source: Krajewski, Kamon, & Avellini, 1979.

hot-dry conditions and warm-hot conditions and then rest either in the same condition or in a neutral condition. As one indication of the advantage of the neutral condition they reported the sweat rate of the subjects under various combinations of conditions as shown in Table 14-8 above. Resting in the "treadmill" condition resulted in distinctly higher sweat rates (as an indication of strain) than resting in a neutral condition did.

Discussion

Where heat or cold environmental conditions could be undesirable for people who are to work or live in them, the optimum solution—when possible—is to modify the environments to make them more suitable for people. When this is not possible, certain other actions in the management of personnel may be desirable, such as selection of personnel who can tolerate the condition (sometimes by tryout for 4 or 5 days); permitting people to become gradually acclimatized; modifying the work activities (as by reducing energy requirements); rotating personnel from one job situation to another; providing adequate rest periods; having people wear appropriate clothing; in the case of hot environments, considering conductive cooling; and also in the case of hot environments, being sure that people drink enough water (or other liquids) to minimize dehydration.

AIR PRESSURE AND ALTITUDE

In the mundane lives of most mortals, the atmospheric variables that we complain about most are temperature and humidity. In less normal environments, however, other variables may play first fiddle. For people in high altitudes (e.g., mountainous areas and aircraft) and for people below sea level (e.g., diving and underwater construction work) air pressure and associated problems can be of paramount importance to well-being and to human performance.

The Atmosphere around Us

The atmosphere of the earth consists primarily of oxygen (21 percent) and nitrogen (78 percent), but also includes a bit of carbon dioxide (0.03 percent), plus other odds and ends. The density of the atmosphere, however, is reduced at higher altitudes; thus, at an altitude of, say, 20,000 ft, there is less air in a given volume than at sea level. Further, because of the weight of the atmosphere, the pressure decreases with altitude. Air pressure is frequently measured in pounds per square inch (lb/in^2) or in millimeters of mercury (mmHg). At sea level the air pressure is 14.71 lb/in^2 (760 mmHg). Figure 14-18 shows the pressures at other altitudes. A reduction of pressure by some ratio would mean that the volume occupied by a given amount of air would be increased inversely (such as doubling the volume if the pressure is reduced by half).

FIGURE 14-18
General effects of hypoxia at various altitude levels and equivalent pressure levels. (*Source: Adapted from Roth, 1968, vol. 3, sec. 10; USAF, 1954, 1960; Tufts University, 1952, part 7, chap. 2, sec. 14, tables 4-1 to 4-4; Balke, 1962.*)

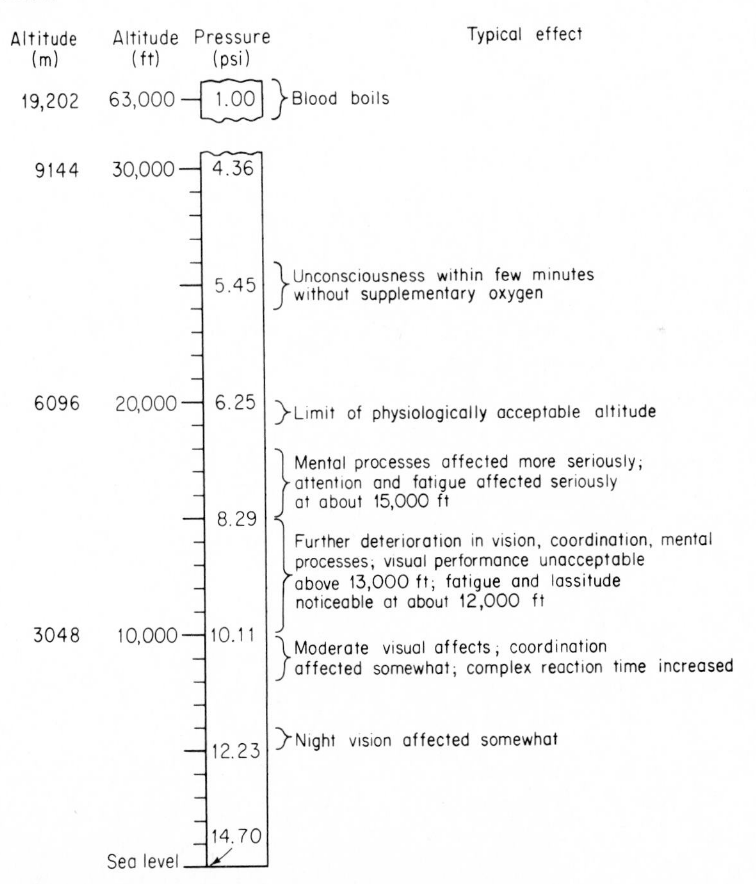

Air Pressure and Oxygen Supply

A primary function of the respiratory system is transporting oxygen from the lungs to the body tissue and picking up carbon dioxide on the return trip and carting it back to the lungs, where it is exhaled. Under normal circumstances (including near-sea-level pressure) the blood (actually the red blood cells) carries oxygen up to about 95 percent of the red blood cells' capacity. As air pressure is reduced, however, the amount of oxygen that the blood will absorb is reduced. For example, at approximately 10 lb/in^2 (equivalent to about a 10,000-ft altitude) the blood will hold about 90 percent of its potential capacity; at about 7.3 lb/in^2 (18,000 ft) the percentage drops to about 70. At 1.0 lb/in^2 (63,000 ft) the pressure is so low that the blood actually boils, like water in a teakettle.

Hypoxia

If the oxygen supply is reduced, a condition of *hypoxia* (also called *anoxia*) can occur, the effects varying with the degree of reduction. Some indication of these is given in Figure 14-19 as related to the amounts of oxygen that would be available at various altitudes (or their equivalents). Generally speaking, the effects below about 8000 ft are fairly nominal, but above that level, or at least above about 10,000 ft, the effects become progressively more serious, as shown in the figure. In connection with the effects of hypoxia, however, two qualifications are in order: (1) There are marked individual differences and (2) acclimatization does increase tolerance somewhat.

FIGURE 14-19
Physiological relationships between the percentage of oxygen in the atmosphere and pressure (and equivalent altitude). The white band under the *sea-level equivalent* curve shows the conditions in which human performance normally is unimpaired. (*Source: Adapted from Roth, 1968, fig. 1-2, based on data compiled by U. C. Luft and originally drawn by E. H. Green of the Garrett Corporation.*)

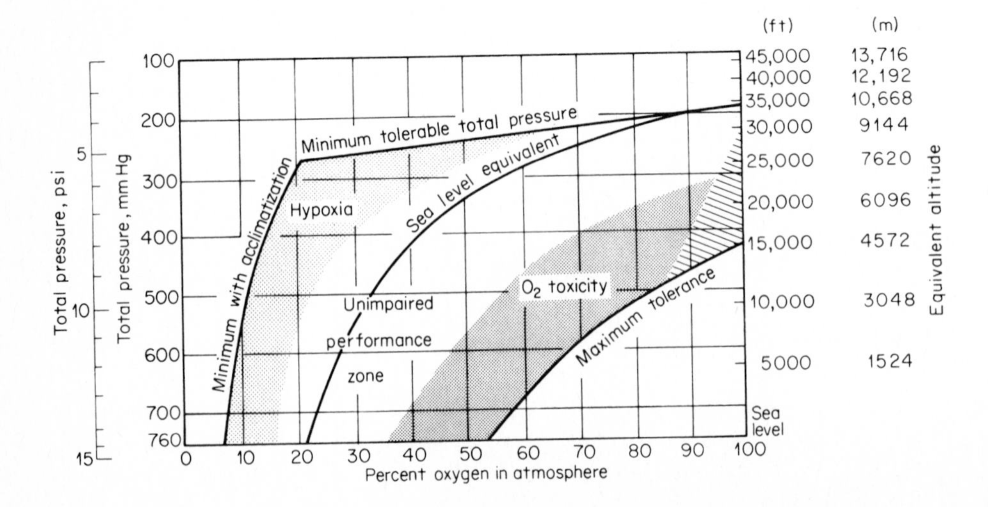

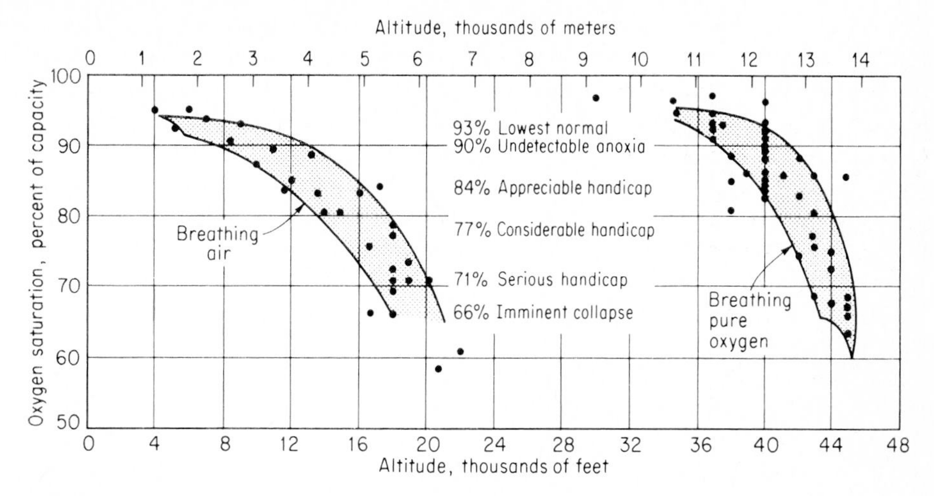

FIGURE 14-20
Relationship between altitude and oxygen saturation (percent of capacity) for subjects breathing air and those breathing pure oxygen. (*Source: USAF, 1954.*)

Use of Oxygen At altitudes where the hypoxia effects would normally be of some consequence, the use of oxygen masks can stave off the onset of hypoxia or minimize its degree. This is illustrated, for example, in Figure 14-20, which shows the relationship between altitude and percentage of oxygen capacity for subjects breathing air versus pure oxygen. This also shows the approximate degree of handicap for various percentage values. It can be seen, for example, that while breathing pure air reduces the percentage of capacity to about 70 at around 18,000 ft, the use of oxygen delays this amount of reduction up to about 42,000 ft.

Pressurization The ideal scheme for avoiding hypoxia at high altitudes is the use of a pressurized cabin that maintains the atmospheric conditions of some lower altitude. This is done in high-altitude civilian planes and in some military planes. The Air Force generally requires the use of oxygen equipment for aircraft that will operate at altitudes of 10,000 ft or above, or at 8000 ft on flights of 4 h or longer (*System Safety Design Handbook*, 1978), and at higher altitudes prescribes minimum differentials in pressure for different types of planes and flights (USAF, 1960). Pressure suits usually are prescribed for high-altitude aircraft, in part as protection against decompression.

Decompression

It has been indicated that the volume of gas expands or contracts in proportion to the pressure applied to it (this is *Boyle's law*). The atmosphere within the human body is not immune from this law. Thus, as an individual changes from one pressure to another the atmosphere in body tissues and cavities expands or contracts. In this connection, as the external air pressure is increased, that within the body follows suit fairly closely, and (aside from some discomfort here and there) there are no serious consequences. But when the external air pressure is reduced suddenly, there can be some unhappy consequences, generally referred to as *decompression sickness*. Actually, there are different kinds of physiological reactions. Among these is the formation of nitrogen bubbles in the blood, body tissue, and around the joints; the effect is somewhat like taking the top off a bottle of soda water. The manifestations of the physiological effects include various symptoms, such as those referred to familiarly as the *bends* and the *chokes*, and various skin manifestations. The *bends* consist of generalized pains in the joints and muscles. The *chokes* are characterized by breathing (choking) difficulty, coughing, respiratory distress, and accompanying chest pains; the skin sensations include hot and cold sensations and itching of the skin, and a mottling of the skin surface sometimes occurs. In extreme cases, the above symptoms become more severe, and in some cases shock, delirium, and coma occur; fatalities may occur in such cases.

Protection from Decompression Decompression sickness occurs primarily in underwater construction work (in sealed caissons), in underwater diving operations, in submarines, and in aircraft (as in the case of rapid ascent to, say, 30,000 ft or more, or in rapid decompression from a rupture in a sealed cabin). Where rapid decompression has occurred (say, by accident), it is usually the practice to subject the person to a higher pressure (near that to which he was originally exposed) and then gradually bring the pressure back to that of the earth.

AIR POLLUTION

We humans have inherited from our natural environment a number of environmental hazards to contend with. But our science and technology have resulted in the creation of a whole host of new environmental hazards, especially pollutants, including smoke, exhaust fumes, toxic vapors and gases, insecticides, herbicides, and ionizing radiation. Some of these contaminants are the by-products of industrial processes and tend to be confined to their industrial environments, whereas others escape into the general atmosphere; and our automobiles, household furnaces, and other nonindustrial sources add their bit to the ever-increasing level of pollution in the air we breathe. The constraint of pages precludes a thorough discussion of this topic.[1] However, Figure 14-21 is presented for the purpose of illustrating the effects of atmospheric pollutants; this particular example deals with the effects of carbon monoxide on man as a function of the degree of concentration and exposure time. For some other types of pollutants, exposure limits are set in terms of the amount or concentration, and in some instances the exposure times, that should not be exceeded.

[1] The interested reader is referred to such sources as Roth (1968, sec. 13) and NASA (1973, chap. 3).

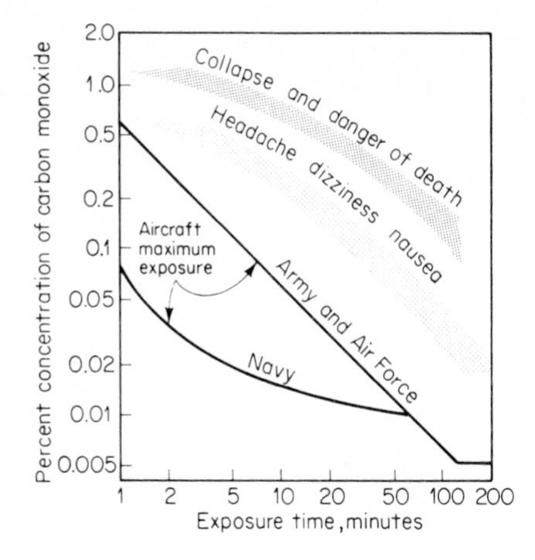

FIGURE 14-21
Effects of carbon monoxide on humans as functions of concentration and exposure time. Shaded areas show conditions that are dangerous and lethal (heavy shading) and milder (light shading). The solid lines represent exposure limits set by the military services for aircraft. (*Source: Adapted from Webb, 1964, fig. 10-2, p. 468, as based on various sources; for original sources, see Webb.*)

REFERENCES

American Society of Heating, Refrigerating, and Air-Conditioning Engineers. *ASHRAE handbook and product directory* (1977 fundamentals vol.). New York: American Society of Heating, Refrigerating and Air-Conditioning Engineers, Inc., 1977.

Azer, N. Z., & Hsu, S. The use of modeling human responses in the analysis of thermal comfort of indoor environments. In B. W. Mangum & J. E. Mill (Eds.), *Thermal analysis–human comfort-indoor environments* (Proceedings of a symposium). Washington, D. C.: National Bureau of Standards, September 1977.

Balke, B. *Human tolerances* (Report 62-6). Oklahoma City: Civil Aeromedical Research Institute, Federal Aviation Agency, Aeronautical Center, April 1962.

Belding, H. S., & Hatch, T. F. Index for evaluating heat stress in terms of resulting physiological strains. *Heating, Piping and Air Conditioning*, August 1955, pp. 129–136.

Bell, P. A. Effects of noise and heat stress on primary and subsidiary task performance. *Human Factors,* 1978, *20*(6), 749–752.

Burton, A. C., & Edholm, O. G. *Man in a cold environment: physiological and pathological effects of exposure to low temperatures.* London: Edward Arnold (Publishers), Ltd., 1955.

Fanger, P. O. *Thermal comfort, analysis and applications in environmental engineering.* Copenhagen: Danish Technical Press, 1970.

Hardy, J. D. (Ed.). *Temperature–its measurement and control in science and industry.* New York: Reinhold, 1963.

Hatch, T. F. Assessment of heat stress. In J. D. Hardy (Ed.), *Temperature–its measurement and control in science and industry* (Vol. 3, Pt. 3, pp. 307–318). New York: Reinhold, 1963.

Hertig, B. A., & Belding, H. S. Evaluation and control of heat hazards. In J. D. Hardy (Ed.), *Temperature–its measurement and control in science and industry* (Vol. 3, Pt. 3, pp. 347–355). New York: Reinhold, 1963.

Kamon, E. & Belding, H. S. The physiological cost of carrying loads in temperate and hot environments. *Human Factors,* 1971, *13*(2), 153-161.

Krajewski, J. T., Kamon, E., & Avellini, B. Scheduling rest for consecutive light and heavy workloads under hot ambient conditions. *Ergonomics,* 1979, *22*(8), 975-987.

Leithead, C. S., & Lind, A. R. *Heat stress and heat disorders.* London: Cassell & Co., Ltd., 1964.

Lind, A. R. A physiological criterion for setting thermal environmental limits for everyday work. *Journal of Applied Physiology,* 1963, *18*, 51-56.

Lind, A. R. Tolerable limits for prolonged and intermittent exposures to heat. In J. D. Hardy (Ed.), *Temperature–its measurement and control in science and industry* (Vol. 3, Pt. 3, pp. 337-345). New York: Reinhold, 1963.

Lind, A. R., & Bass, D. E. The optimal exposure time for the development of acclimatization to heat. *Federal Proceedings,* 1963, *22*(3), 704-708.

Lockhart, J. M., & Kiess, H. O. Auxiliary heating of the hands during cold exposure and manual performance. *Human Factors,* 1971, *13*(6), 457-465.

Lockhart, J. M., Kiess, H. O., & Clegg, T. J. Effect of rate and level of lowered finger-surface temperature on manual performance. *Journal of Applied Psychology,* 1975, *60*(1), 106-113.

McIntyre, D. A., & Griffiths, I. D. The effects of added clothing on warmth and comfort in cool conditions. *Ergonomics,* 1975, *18*(2), 205-211.

McKarnes, J. S., & Brief, R. S. Nomographs give refined estimate of heat stress. *Heating, Piping and Air Conditioning,* January 1966, *38*(1), 113-116.

National Aeronautics and Space Administration. *Bioastronautics data book* (2d ed.) (NASA SP-3006). Washington, D. C.: U.S. Government Printing Office, 1973.

New York State Commission on Ventilation. *Report of the New York State Commission on Ventilation.* New York: Dutton, 1923.

Occupational Safety and Health Administration. Recommendation for a standard for work in hot environments (Draft No. 5). Washington, D. C.: Department of Labor, January 9, 1974.

Pandolf, K. B., Burse, R. L., & Goldman, R. F. Role of physical fitness in heat acclimatization, decay and reinduction. *Ergonomics,* 1977, *20*(4), 399-408.

Rohles, F. H., Jr. The modal comfort envelope and its use in current standards. *Human Factors,* 1974, *64*(3), 314-323.

Roth, E. M. (Ed.). *Compendium of human responses to the aerospace environment* (Vols. 1-4) (NASA CR-1205). November 1968.

Shapiro, Y., Pandolf, K. B., Avellini, B. A., Pimental, N. A., & Goldman, R. F. Physiological response of men and women to humid and dry heat. *Journal of Applied Physiology,* 1980, *49*(1), 1-8.

Shapiro, Y., Pandolf, K. B., & Goldman, R. F. Sex differences in acclimation to a hot-dry environment. *Ergonomics,* 1980, *23*(7), 635-642.

Shvartz, E. The application of conductive cooling to human operators. *Human Factors,* 1975, *17*(5), 438-445.

Siple, P. A., & Passel, C. F. Movement of dry atmospheric cooling in subfreezing temperatures. *Proceedings of the American Philosophical Society,* 1945, *89,* 177-199.

System Safety Design Handbook (AFSC DH 1-6, U.S. Air Force Systems Command Design Handbook 1-6). Wright Patterson Air Force Base, Ohio: Aeronautical Systems Division, December 20, 1978.

Tichauer, E. R. The effects of climate on working efficiency. *Impetus,* Australia, July 1962, *1*(5), 24-31.

Tufts University. *Handbook of human engineering data* (2d ed.). Medford, Mass.: Tufts University, 1952.

United States Air Force. *Flight surgeons manual* (U.S. Air Force Manual 1960-5). July 1954.

United States Air Force. *Your body in flight* (U.S. Air Force Pamphlet 160-10-3). January 1, 1960.

Vogt, J. J., Foehr, R., Kuntzinger, E., Seywert, L., Libert, J. P, Candas, V., & Van Peteghem, Th. Improvement of the working conditions at blast furnaces. *Ergonomics,* 1977, *20*(2), 167-180.

Webb, P. (Ed.). *Bioastronautics data book* (NASA SP-3006). 1964.

Wing, J. F. *A review of the effects of high ambient temperature on mental performance* (TR 65-102). U.S. Air Force, AMRL, September 1965.

Winslow, C. E. A., et al. Physiological influence of atmospheric humidity (2d Report ASHVE Technical Advisory Committee on Physiological Reactions). *Transactions of the ASHVE,* 1942, *48*, 317-326.

Winslow, C. E. A., & Herrington, L. P. *Temperature and human life.* Princeton, N. J.: Princeton University Press, 1949.

Wyon, D. P. The effects of moderate heat stress on typewriting performance. *Ergonomics,* 1974, *17*(3), 309-318.

Wyon, D. P., Fanger, P. O., Olesen, B. W., & Pedersen, C. J. K. The mental performance of subjects clothed for comfort at two different air temperatures. *Ergonomics,* 1975, *18*(4), 359-374.

CHAPTER **15**

NOISE

Before the days of machines and mechanical transportation equipment, our noise environment consisted of noises such as those of household activities, animals (including a few blood-curdling wild ones), horse-drawn vehicles, hand tools, and weather. But human ingenuity changed all this by creating machines, motor vehicles, subways, radios, guns, bombs, fire sirens, jet aircraft, and New Year's Eve horns. Noise has become such a pervasive aspect of working situations and community life as to be referred to as *noise pollution* and to be considered a health hazard.

Although noise has commonly been referred to as *unwanted sound*, a somewhat more definitive concept is the one proposed by Burrows (1960), in which noise is considered in an information-theory context, as follows: Noise is "that auditory stimulus or stimuli bearing no informational relationship to the presence or completion of the immediate task." This concept applies equally well to attributes of task-related sounds that are informationally useless, as well as to sounds that are not task-related.

HOW LOUD IS IT?

We touched briefly in Chapter 5 on the fact that the human ear is not equally sensitive to all frequencies of sound. In general we are less sensitive to low frequencies (below 1000 Hz) and more sensitive to higher frequencies. Thus a low-frequency tone will not sound as loud to us as a high-frequency tone of equal intensity (i.e., sound pressure). Put another way, a low-frequency tone must have more intensity than a higher-frequency tone to be of equal loudness. This particular fact has led investigators to search for a metric to measure the subjective quality of sound. We will discuss several basic measures here. When we discuss the annoyance quality of noise we will introduce several other measures that are derived from these basic measures.

Sound Level Meter Scales

As we indicated in Chapter 5, sound pressure meters built to American National Standards Institute (ANSI) standards contain frequency-response weighting networks (designated A, B, and C). Each of these networks electronically attenuates sounds of certain frequencies and produces a weighted total sound-pressure level. Figure 15-1 shows the relative response curves of the A, B, and C scales and the response characteristics of the human ear at threshold. As can be seen, the C scale weights all frequencies almost equally. The B scale, originally intended to represent how people might respond to sounds of moderate intensity, is rarely if ever used. The most commonly used scale is the A scale. The Occupational Safety and Health Administration (OSHA) standards for daily occupational noise limits are specified in terms of this measure, and the U.S. Environmental Protection Agency (1974) has selected the A scale as the appropriate measure of environmental noise. As we will see, the many indices of loudness, noisiness, and annoyance are all based on the A scale (i.e., dBA). Of the three scales, the A scale comes closest to approximating the response characteristics of the human ear.

As if the A, B, and C scales were not enough, there exist on some meters D scales (there is more than one D scale). The scales were designed primarily to provide a measure of aircraft noise but have yet to gain complete universal acceptance, and are currently used only rarely and for very specific measurement applications.

FIGURE 15-1
Relative response characteristics of the A, B, and C sound-level meter scales and the human ear at threshold. (*Source: Jensen, Jokel, & Miller, 1978, fig. 2.4.*)

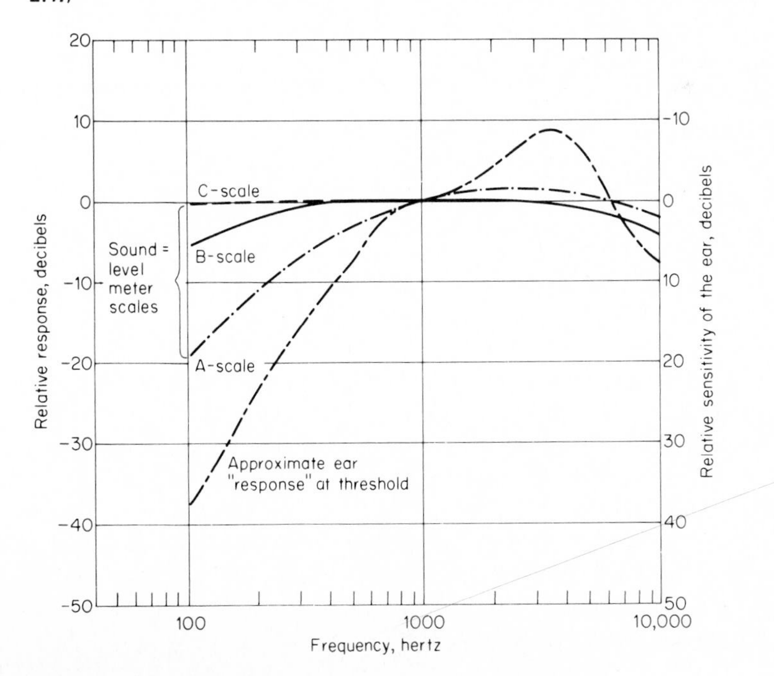

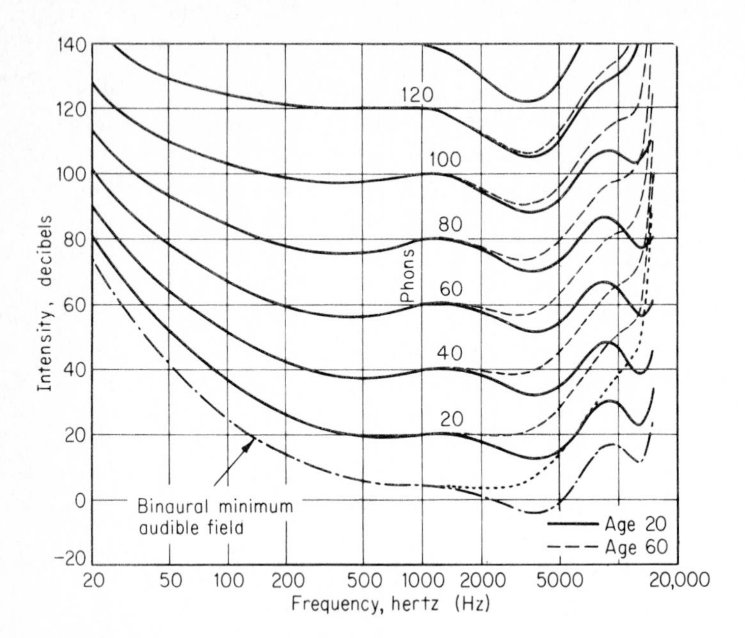

FIGURE 15-2
Equal-loudness curves of pure tones. Each curve represents intensity levels of various frequencies that are judged to be equally loud. The lowest curve shows the minimum intensities of various frequencies that typically can be heard. (*Source: Robinson & Dadson, 1957. Crown copyright reserved. Courtesy National Physical Laboratory, Teddington, Middlesex, England.*)

Loudness Level in Phons

Some years ago Fletcher and Munson (1933) developed what have become known as *equal-loudness* contours. They presented subjects with sounds of different combinations of frequencies and intensities to determine which combinations were judged to have equal loudness.

More recently Robinson and Dadson (1957) in Great Britain have developed somewhat more refined equal-loudness curves, these being shown in Figure 15-2. Each contour shows the decibel intensities of different frequencies that were judged to be equal in loudness to that of a 1000-Hz tone of the specified decibel level of intensity. To illustrate the curves, one can see that a tone of 50 Hz of about 62 dB is judged equal in loudness to a 1000-Hz tone of only 40 dB.

The unit *phon* (of German origin) is used to indicate the loudness level of sounds. Thus, any point along a given contour in Figure 15-2 represents sounds of the same number of phons. The loudness level in phons, then, is numerically equal to the decibel level of a tone of 1000 Hz, which is judged equivalent in loudness. So our 50 Hz, 62 dB tone would have a loudness level of 40 phons.

Loudness in Sones

The phon tells us only about subjective *equality* of various sounds, but it tells us nothing of the *relative subjective loudness* of different sounds. For such comparative purposes we need still another yardstick. Fletcher and Munson developed such a scale—a

TABLE 15-1
EXAMPLES OF LOUDNESS LEVELS IN SONES

Noise source	Decibels	Loudness, sones
Residential inside, quiet	42	1
Household ventilating fan	56	7
Automobile, 50 ft	68	14
"Quiet" factory area	76	54
18-in automatic lathe	89	127
Punch press, 3 ft	103	350
Nail-making machine, 6 ft	111	800
Pneumatic riveter, 4 ft	128	3000

Source: Bonvallet, 1952, p. 43.

ratio scale of loudness. Stevens (1936), in turn, labeled the scale, using the term *sone*. In developing this scale (as with the phon) a reference sound was used. One *sone* is defined as the loudness of a 1000-Hz tone of 40 dB (i.e., 40 phons). A sound that is judged to be twice as loud as the reference sound has a loudness of 2 sones, a sound that is judged to be three times as loud as the reference sound has a loudness of 3 sones, etc. In turn, a sound that is judged to be half as loud has a loudness of ½ sone. To provide some basis for relating sones to our own experiences, consider the examples given in Table 15-1.

A procedure for estimating the loudness of complex sounds has been developed and revised by Stevens (1956, 1961). The procedure uses a table (see Peterson & Gross, 1978, table 3-1) for converting octave-band dBs to loudness indices.[1] These loudness indices are then combined according to a formula (see Peterson & Gross, pp. 24–27).

There is a relationship between phons and sones. Forty phons equals one sone and every ten additional phons doubles the number of sones. For example, 50 phons = 2 sones, 60 phons = 4 sones, and 70 phons = 8 sones. In like manner 30 phons = 0.5 sones, and 20 phons = 0.25 sones.

More recently Stevens (1972) has suggested a further refinement in the measurement of sones, and has introduced new measures called *perceived level of noise* (measured in PLdB) and *Mark VII* sones. The Mark VII sone is related to the original sone (which Stevens called Mark VI) and is based on a ⅓-octave band of noise centered at 3150 Hz as the reference sound (rather than a 1000 Hz tone). This reference noise is at a level of 32 dB (rather than 40 dB) and is assigned a perceived magnitude of 1 sone (that is, a Mark VII sone). The Mark VII sone is interpreted the same as a Mark VI sone, i.e., a 2-sone sound is twice as loud as a 1-sone sound. Mark VII sones can be converted into a perceived level of noise, PLdB, which is analogous to a phon. One Mark VII sone equals 32 PLdB and each increase of 9 PLdB doubles the number of Mark VII sones. Presumably, these changes from Mark VI to Mark VII and phons to PLdB bring about an improvement in accuracy and measurement.

The procedure and tables required for computing Mark VII sones and converting to PLdB can be found in Stevens (1972) or Peterson and Gross (1978, pp. 28-33).

[1] Octave-band midpoints are 31.5, 63, 125, 250, 500, 1000, 2000, 4000, 8000 Hz.

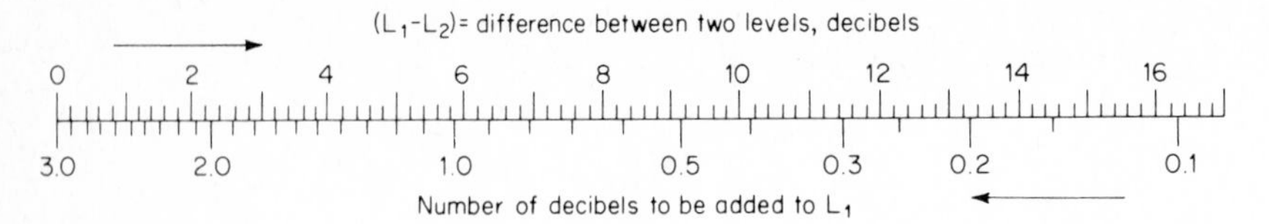

FIGURE 15-3
Nomograph for combining two sound-pressure levels, L_1 and L_2; where L_1 is the higher of the two. The left scale shows the number of decibels to be added to the higher level (L_1) to obtain the level of the combination $L_1 + L_2$.

Noisiness and Perceived Noise Level

If you think you understand phons, sones (Mark VI and VII), and PLdB, wait–there's more to come. Kryter (1970) has developed another set of metrics analogous to the phon and sone, except related to the "noisiness" of a sound rather than the "loudness." The metric which is analogous to the sone is–are you ready for this?–the *noy*. One *noy* is a sound judged to be subjectively equal in *noisiness* to an octave band of random noise centered at 1000 Hz at a sound pressure level of 40 dB. A sound of 2 noys is twice as noisy as a sound of 1 noy. The metric analogous to the phon is *perceived noise level* (PNL) measured in PNdB. Be sure to differentiate this from Stevens's perceived level of noise and PLdB. The relationship between PNdB and noys is the same as between phons and sones (Mark VI, that is); 40 PNdB equals 1 noy, and each additional 10 PNdB doubles the number of noys.

The procedure for calculating noys and PNL is similar to computing sones and phons in the Mark VI system, except that different tables are used. Since PNL has been found useful for assessing aircraft noise, and is the basis for several other measures of annoyance of noise (which we will discuss later), it will be worthwhile to present the procedure for calculating noys and PNL. First measure the sound intensity (dB) at each octave band (centers at 31.5, 63, 125, 250, 500, 1000, 2000, 4000, 8000 Hz). Given these values one proceeds through the following seven steps:

1 (This is tricky.) Combine the levels in the 63-Hz and 125-Hz bands using the nomograph in Figure 15-3.[2] Replace the 125-Hz or the 63-Hz level, whichever is higher, with the new combined level, and ignore the other band level.

2 In the frequency range from 500 Hz and higher, note if any band level projects above the level in adjacent bands. If the difference between this projecting level and the average level in the two adjacent bands is 3 dB or greater, a *tone correction* is required. This correction usually adds between 2 and 5 dBs to the band level. (The exact correction involves lengthy calculations.)

3 From Table 15-2, find the appropriate perceived noise index (noys) for each octave-band level (N).

[2] Figure 15-3 can also be used to determine the combined dB level of two individual sounds. For example, 78 dB + 82 dB = 83.4 dB. The difference between 78 and 82 dB is 4 dB; from Figure 15-3 we find that 1.45 dB must be added to the larger value, hence 82 + 1.45 is 83.4 dB (rounded off).

TABLE 15-2

EXCERPTS FROM TABLE FOR USE IN CALCULATING PERCEIVED NOISINESS (NOYS) AND PERCEIVED NOISE LEVEL, PNdB, FOR COMPLEX SOUNDS

Band level,	Perceived noisiness index, N, for octave bands Midpoint of octave bands									Perceived noisiness,	Perceived noise level,
dB	31.5	63	125	250	500	1000	2000	4000	8000	noys	PNdB
20							0.3	0.4			
30				0.2	0.4	0.4	0.9	1.1	0.4	0.4	30
40			0.3	0.8	1.0	1.0	1.8	2.2	1.4	1.0	40
50		0.2	1.0	1.6	2.0	2.0	3.5	4.3	3.1	2.0	50
60		1.0	2.2	3.2	4.0	4.0	7.0	8.7	6.1	4.0	60
70	0.3	2.6	4.6	6.9	8.0	8.0	14.1	17.3	12.2	8.0	70
75	0.8	4.1	7.0	9.8	11.3	11.3	19.8	24.3	17.3	11.3	75
80	1.7	6.4	10.5	13.9	16.0	16.0	27.9	34.2	24.3	16.0	80
85	3.1	9.9	14.9	19.7	22.6	22.6	39.1	48.0	34.2	22.6	85
90	5.5	14.7	21.1	27.9	32.0	32.0	55.0	67.0	48.0	32.0	90
95	9.5	21.1	29.9	39.4	45.3	45.3	76.0	94.0	67.0	45.3	95
100	15.5	29.9	42.0	56.0	64.0	64.0	107.0	132.0	94.0	64.0	100
105	23.2	42.0	60.0	79.0	91.0	91.0	152.0	187.0	132.0	91.0	105
110	33.6	60.0	84.0	111.0	128.0	128.0	215.0	264.0	187.0	128.0	110
115	49.0	84.0	119.0	158.0	181.0	181.0	304.0	374.0	264.0	181.0	115
120	71.0	119.0	169.0	223.0	256.0	256.0	429.0	529.0	374.0	256.0	120
125	101.0	169.0	239.0	315.0	362.0	362.0	607.0	748.0	529.0	362.0	125
130	145.0	239.0	338.0	446.0	512.0	512.0	859.0	1057.0	748.0	512.0	130

Source: Williams, 1978, table 2-3, pp. 50–53.

4 Add all the noys indices (ΣN).

5 Multiply this sum by 0.3.

6 Add this product to 0.7 of the noys index with the largest value (N_{max}). The total noisiness level in noys, then, is ($0.3\Sigma N + 0.7N_{max}$).

7 This total noisiness (noys) can be converted to a perceived noise level (PNdB) by using the two columns at the right of Table 15-2.

Equivalent Sound Level

Noise often varies in intensity over time. Fortunately, a baby's scream will change to a whimper and with luck to a yawn. Over the years many single-number measures have been proposed for time-varying sound levels. We will discuss a few when we talk about the annoyance of noise. The Environmental Protection Agency (1974), however, concluded that, insofar as cumulative noise effects are concerned, the long-term average sound level was the best measure for the magnitude of environmental noise. This long-term average is designated the *equivalent sound level* (L_{eq}) and is equal to the sound-pressure level (usually measured in dBA) of a *constant* noise that, over a given time period, transmits to the receiver the same amount of acoustic energy as the actual time-varying sound.

L_{eq} depends on the time interval and acoustic events occurring during that period. For example, if a 100-dBA noise occurred for 1 h, the L_{eq} for that hour would be 100 dBA. This is expected because when the sound level is constant, the level that produces the same energy is the identical constant level. Consider, however, a situation in which it was quiet during the next 4 h. The L_{eq} for the total 5-h period would now be less than 100 dB; in fact it would be 94 dBA. What this says is that 5 h of 94-dBA noise is equivalent in acoustic energy to 1 h of 100-dBA noise and 4 h of quiet.

A. Taylor and Lipscomb (1978) present a simplified method for determining L_{eq}. Using a sound-pressure meter set on the A scale and "fast" meter response, the decibel reading on the meter is noted every 5 (or 10) s. A work sheet such as that shown in Table 15-3 can be used to make tally marks adjacent to the appropriate decibel level. If the sound level changes rapidly, shorter sampling intervals will increase the accuracy of the measure. This should be continued throughout the time period of interest (unless the sound levels are stable). As Taylor and Lipscomb point out, the upper limit of samples is determined by the endurance of the person taking the measures. If possible, use automated equipment. After completing the tally, proceed with the following eight steps, using the worksheet in Table 15-3.

1 Sum the tally marks for each discrete sound level (enter in column 3).

2 Decide on the size of the sound level intervals to be used in the calculations. (A 5-dB interval was chosen in Table 15-3; using 1-dB intervals would change the result by only a fraction of a dBA.)

3 Record the midpoint of the intervals in column 4.

4 Count the tally marks for the entire interval (enter in column 5).

5 Compute the fraction of total time the sound level was in each interval (i.e., divide each entry in column 5 by the total number of tally marks—200 in our case).

TABLE 15-3

A WORKSHEET FOR DATA COLLECTION AND CALCULATION OF L_{EQ}

(1) Sound pressure level, dBA	(2) Field measurements counts, 5-s intervals	(3) Counts per dB	(4) Midpoint of interval, dBA	(5) Counts per interval	(6) Fraction of total time	(7) Correction factor	(8) Partial L_{eq}	(9) Combining partial L_{eq}
100								
99	I	1						
98	II	2	98	6	$\frac{6}{200} = 0.03$	−15.0	83	
97								
96	III	3						
95	II	2						86
94	III	3						
93	𝍸 𝍸	10	93	21	$\frac{21}{200} = 0.10$	−10.0	83	
92	𝍸	5						
91	I	1						88.3
90	𝍸 𝍸 I	11						
89	𝍸 𝍸 𝍸 III	18						
88	𝍸 𝍸 𝍸 𝍸 I	21	88	88	$\frac{88}{200} = 0.44$	− 3.3	84.7	
87	𝍸 𝍸 𝍸 II	17						88.8
86	𝍸 𝍸 𝍸 𝍸 I	21						
85	𝍸 𝍸 𝍸 I	16						
84	𝍸 𝍸 IIII	14						
83	𝍸 𝍸 𝍸 𝍸 I	21	83	71	$\frac{71}{200} = 0.36$	− 4.2	78.8	
82	𝍸 III	8						88.8
81	𝍸 𝍸 II	12						
80	𝍸 III	8						
79	II	2						
78	II	2	78	14	$\frac{14}{200} = 0.07$	−12.0	66	
77								
76	II	2						
	Total	200						

Source: Taylor & Lipscomb, 1978.

6 Find the correction term for each interval using the following nomograph (enter the correction factor in column 7):

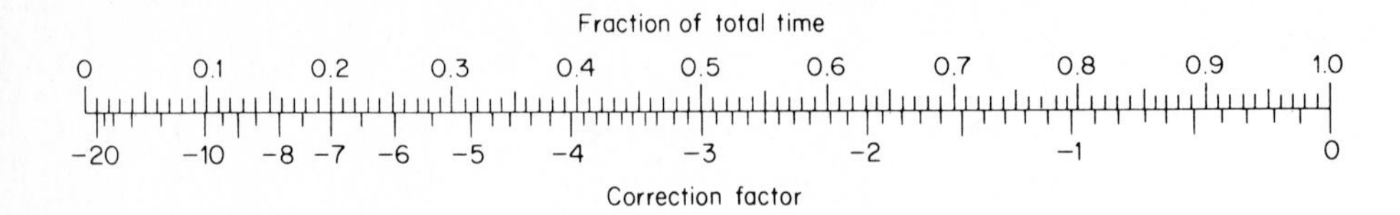

7 Calculate partial L_{eq} values by adding each entry in column 7 to its counterpart in column 4 (enter the sum in column 8).

8 Add the partial L_{eq} values by using the nomograph in Figure 15-3. Add the first two partial L_{eq} values, then add the third to the sum of the first two, and so on. The result is the L_{eq} value for the exposure being measured (see column 9).

The example worked out in Table 15-3 can be interpreted to mean that the sound during those approximately 17 min (200 5-s samples) delivered energy to the ear which would be equivalent to a steady-state sound of 88.8 dBA over the same period of time.

NOISE AND LOSS OF HEARING

Of the different possible effects of noise, one of the most important and clearly established is hearing loss. There are really two primary types of deafness, *nerve* deafness and *conduction* deafness. Nerve deafness usually results from damage or degeneration of the hair cells of the organ of Corti in the cochlea of the ear (see Chapter 5). Conduction deafness is caused by some condition of the outer or middle ear that affects the transmission of sound waves to the inner ear.

The hearing loss in nerve deafness is typically uneven; usually the hearing loss is greater in the higher frequencies than in the lower ones. Normal deterioration of hearing through aging is usually of the nerve type, and continuous exposure to high noise levels also typically results in nerve deafness. Once nerve degeneration has occurred, it can rarely be corrected. Conduction deafness, on the other hand, is only partial, since airborne sound waves strike the skull and may be transmitted to the inner ear by conduction through the bone. It may be caused by different conditions, such as adhesions in the middle ear that prevent the vibration of the ossicles, infection of the middle ear, wax or some other substance in the outer ear, or scars resulting from a perforated eardrum.

People with this type of damage sometimes are able to hear reasonably well, even in noisy places, if the sounds to which they are listening (for example, conversation) are at intensities above the background noise. This type of deafness can sometimes be arrested, or even improved. Hearing aids are more frequently useful in this type of deafness than they are in nerve deafness.

Measuring Hearing

In order to review the effects of noise on hearing, we should first see how hearing (or hearing loss) is measured. There are two basic methods of measurement, namely, the use of simple tests of hearing and the use of an audiometer. Each of these methods is discussed below.

Simple Hearing Tests For some purposes simple hearing tests are used. These include a voice test, a whisper test, a coin-click test, and a watch-tick test.

In the voice and whisper tests, for example, the tester (out of sight) speaks or whispers to the testee, and the testee is asked to repeat what was said. This may be done at different distances and with different voice intensities. The primary shortcoming of such tests is that they usually lack standardization. Even if reasonable standardization can be achieved (such as using a particular person's voice or a particular watch), such tests would still serve only as a gross test of hearing.

Audiometer Tests *Audiometers* are of two types, the most common being an instrument that is used to measure hearing at various frequencies. It reproduces, through earphones, pure tones of different frequencies and intensities. As the intensity is increased or decreased, the people being tested are asked to indicate when they can hear the tone or when it ceases to be audible. It is then possible to determine for each frequency tested the lowest intensity that can just barely be heard; this is the *threshold* for the frequency.

Another type of audiometer is a speech audiometer. Direct speech, or a recording of speech, is reproduced to earphones or to a loudspeaker, and intensity is controlled. Various types of speech intelligibility tests were discussed earlier in the text, in Chapter 6.

Normal Hearing and Hearing Loss

Before we see what effect noise has on hearing, we should first see what normal hearing is like.

Surveys of Hearing Loss Surveys have been made to determine the hearing abilities of people. In such surveys individuals are tested at various frequencies to determine their loss of hearing at each of the tested frequencies. The average hearing loss at each frequency is then determined.

An analysis of such data by Spoor (1967) has given the results shown in Figure 15-4. The two sets of curves show, for men and for women, the average shifts in the threshold of hearing for pure tones that occur with age. It is clear that hearing loss typically becomes increasingly severe at the higher frequencies. The hearing loss shown in Figure 15-4 generally represents the effects of two factors, one being aging (*presbycusis*), and the other being the normal stresses and nonoccupational noises of our current civilization.

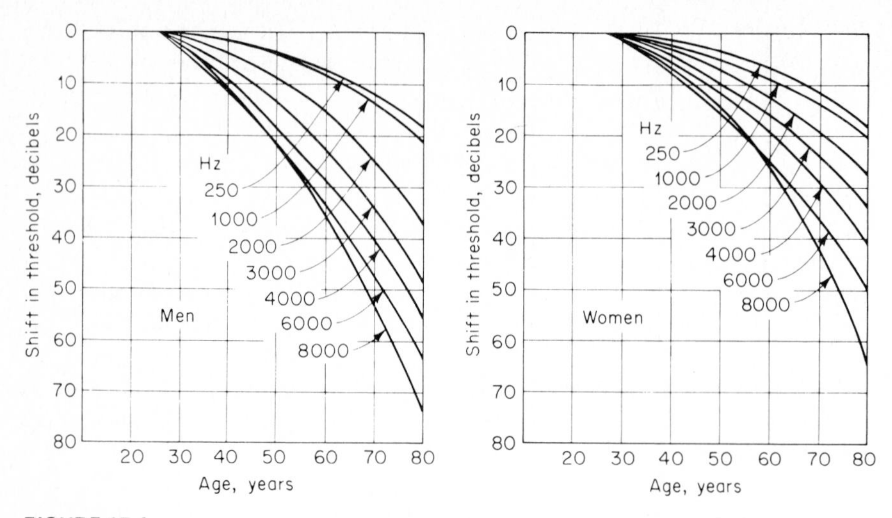

FIGURE 15-4
The average shifts with age of the threshold of hearing for pure tones of persons with "normal" hearing, using a 25-year-old group as a reference group. (*Source: Spoor, 1967, as presented by Peterson & Gross, 1978.*)

Hearing Loss from Continuous Noise Exposure

After exposure to continuous noise of sufficient intensity there is some temporary hearing loss which usually is recovered a few hours or days after exposure. However, with additional exposure the amount of recovery gradually becomes less and less and the individual is left with some permanent loss. The implications of permanent hearing loss are obvious. Temporary hearing loss, however, can also have serious consequences if a person depends on auditory information in the performance of a job or task. We will discuss permanent and temporary hearing loss and the effects that characteristics of the exposure have on them.

Temporary Hearing Loss Since hearing generally recovers with time after exposure, the measurement of hearing loss must take place at a fixed time after exposure to be comparable. Traditionally, this has been done 2 min after the end of exposure. Any shift in threshold (from preexposure levels) is called the *temporary threshold shift at 2 min* (abbreviated TTS_2).

The relationship between TTS_2 and the sound level of the noise to which one is exposed is not a simple one. Some sound levels will not produce any measurable TTS_2 regardless of the duration of exposure. These sound pressure levels define *effective quiet* where the hazardous effects of noise on hearing are concerned. This lower limit depends somewhat on frequency but is generally considered to be around 70 to 75 dBA. For exposures to noises of moderate sound levels (80 to 105 dBA) and for durations of less than 8 h, TTS_2 shows a linear increase with increasing sound level of the exposure noise. Under these same conditions, TTS_2 increases proportionally to the logarithm of exposure duration. Thus, TTS_2 builds up quickly during the first few minutes of exposure and then increases more slowly as exposure time increases.

There is a curious relationship between the frequency of the exposure noise and the TTS_2 produced. The maximum threshold shift is produced not at the frequency of the exposure noise, but rather at frequencies well *above* the exposure noise; e.g., exposure to a pure tone of 700 Hz will produce a maximum TTS_2 at 1000 Hz or higher.

Recovery from temporary threshold shifts is also no simple matter and seems to depend on the magnitude of TTS_2 itself. If TTS_2 is under 40 dBA, recovery is proportional to the logarithm of time after exposure (fast recovery followed by more gradual recovery). If TTS_2 is over 40 dBA (which is quite severe) then recovery is linear with time, i.e., a constant amount of recovery during each unit of time since exposure.

There are of course marked individual differences in TTS_2 that result from a given noise. Some people may experience considerable TTS_2 while others may experience hardly any. Further, some people are more sensitive to high-frequency sounds, others to low-frequency sounds. These individual differences must be considered when setting criteria to protect ears from damage.

Permanent Hearing Loss With repeated exposure to noise of sufficient intensity, *permanent threshold shifts* (PTS, or NIPTS for noise-induced permanent threshold shifts) will gradually appear. Usually the PTS occurs first at 4000 Hz. As the number of years of noise exposure increases, the hearing loss around 4000 Hz becomes more pronounced, but is generally restricted to a frequency range of 3000 to 6000 Hz. With further noise exposure, the hearing loss at 4000 Hz continues and spreads over a wider frequency range (Melnick, 1979). (It might be noted that 4000 Hz is in that region of frequencies to which the human ear is most sensitive.) This can be seen in Figure 15-5,

FIGURE 15-5
Median noise-induced permanent threshold shifts at various frequencies as a function of increasing exposure to noise among female jute weavers. The noise was wide-band and continuous with a spectrum that peaked in the octave bands centered at 1000 and 2000 Hz, with overall sound level varying from 99 to 102 dBA. (*Source: W. Taylor, Pearson, Mair, & Burns, 1965, as presented by Melnick, 1979.*)

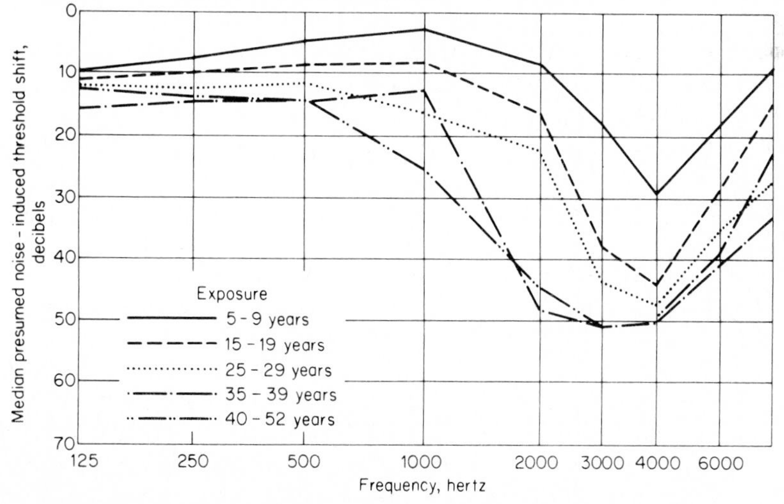

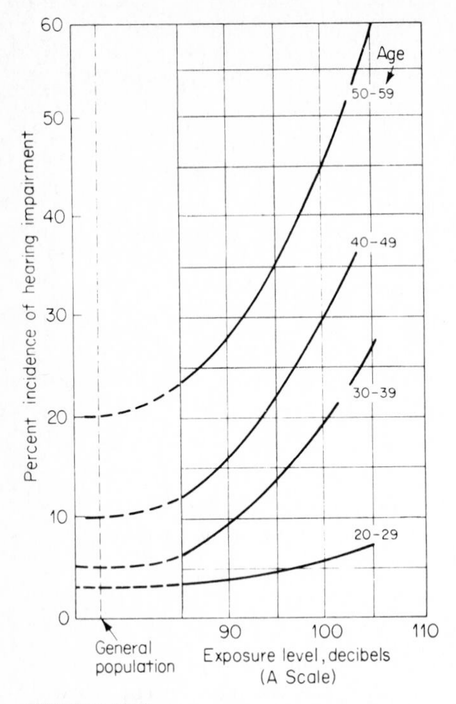

FIGURE 15-6
Incidence of hearing impairment in the general population and in selected populations by age group and occupational noise exposure; *impairment* is defined as a hearing threshold level in excess of an average of 15 dB at 500, 1000, and 2000 Hz. (*Source: Industrial Noise Manual, 1966, fig. 1, p. 420.*)

which shows hearing loss curves for jute weavers exposed to wide-band continuous noise with a noise spectrum that peaked in the octave bands centered at 1000 and 2000 Hz. The overall sound level varied from 99 dBA to 102 dBA.

A committee under the aegis of the American Industrial Hygiene Association has teased out and condensed, from many sources, data that show the incidence of hearing impairment, in a consolidated figure, for each of several age groups of individuals who have been exposed (in their work) to noise intensities of different levels. This consolidation is shown in Figure 15-6. Actually this figure shows, for any group, the probabilities of individuals having hearing impairment (impairment being defined as an average hearing threshold shift in excess of 15 dB at 500, 1000, and 2000 Hz). Although the probabilities of impairment (so defined) are not much above those of the general population for individuals exposed to 85 dB, the curves shoot up sharply at higher levels, except for the youngest group (but their time will come!).

Must we wait 10 or 20 years to discover if a noise environment is potentially harmful? Fortunately not. It is widely accepted that temporary threshold shift (TTS_2) is a good indicator of the permanent threshold shift which will result from repeated exposure (on a daily or work-week basis) to that noise.

Hearing Loss from Noncontinuous Noise

The gamut of noncontinuous noise includes intermittent (but steady) noise (such as machines that operate for short, interrupted periods of time), impact noise (such as that from a drop forge), and impulsive noise (such as from gunfire). In heavy doses, such noise levies its toll in hearing loss, but the combinations and permutations of intensity, noise spectrum, frequency, duration of exposure, and other parameters preclude any simple, pat descriptions of the effects of such noise.[3] In the case of impact and impulsive noise, however, it might be noted that the toll sometimes is levied fairly promptly. For example, 35 drop-forge operators showed a noticeable increase in hearing threshold within as little as 2 years, and 45 gunnery instructors averaged 10 percent hearing loss over only 9 months, even though most of them had used hearing-protection devices (Machle, 1945).

PHYSIOLOGICAL EFFECTS OF NOISE

Permanent hearing loss is of course the consequence of physiological damage to the mechanisms of the ear. Aside from possible damage to the ear itself, one would wonder whether continued exposure to noise might induce any other temporary or permanent physiological effects.

The onset of a loud noise will cause a *startle response*, characterized by muscle contractions, blink, and a head-jerk movement. In addition, larger and slower breathing movements, small changes in heart rate, and dilation of the pupils occur. There is also a moderate reduction in the diameter of blood vessels in the peripheral regions, particularly the skin (Burns, 1979). All these responses are relatively transient, and settle back to normal or near normal levels very quickly. With repeated exposure to the noise, the magnitude of the initial response is also diminished.

The physiological reactions to noise, however, usually would not be regarded as being of a pathological nature if the noise occurred only a few times. However, there is an accumulating body of evidence that indicates that exposure to high noise levels (such as 95 dB or more) acts as a stressor and that over a period of years it may produce pathological side effects and thus may constitute a health hazard (Jansen, 1969; Burns; Gulian, 1974).

Unfortunately, in long-term noise exposure studies it is not possible to separate the effects of noise from the effects of other factors such as heat, vibration, and physical job requirements. These long-term effects are most often associated with the cardiovascular and gastrointestinal systems. Symptoms range from hypertension (high blood pressure) to hypotension (low blood pressure) and from gastroduodenitis to duodenal ulcers. Gulian reports on several European studies which show that people working in high levels of noise have more somatic complaints than workers in low levels of noise. "High-noise" workers complain more of irritability, headaches, tiredness, bad sleep, and heart pains. Here again, the noise levels involved have typically been high, over 95 dBA, and exposure has usually been for more than 10 years.

[3] A thorough treatment of exposure to intermittent noise is presented by Kryter, Ward, Miller, and Eldredge (1966).

EFFECTS OF NOISE ON PERFORMANCE

The effects of noise on performance are not simple. By choosing the right studies, one can show that noise produces either a decrement, no effect, or an improvement in performance (Broadbent, 1979; Hockey, 1978). The particular effect depends on, among other things, the characteristics of the noise itself and the nature of the task being performed. However, we can make a few generalizations about the effects of noise on performance. With the possible exception of some memory tasks (Hockey, 1978), the level of noise required to obtain reliable performance effects is quite high, generally over 95 dBA. Performance of simple, routine tasks may show no effect and often will even show an improvement as a result of noise. The detrimental effects of noise are usually associated with difficult tasks which require high levels of perceptual or information processing capacity, or both (Eschenbrenner, 1971).

Broadbent (1976) identifies three detrimental effects of noise which he has gleaned from the research available. In the first place he reports that certain decisions become more confident in the presence of noise. For example, if a person is presented a visual display and is asked to report signals that occur, that person will be more confident of the responses (whether correct or incorrect) when the task is performed in high levels of noise. Easily detected signals are usually not missed any more often in high-noise conditions than in quiet conditions. Doubtful and uncertain signals, however, may be more often missed in noisy conditions because the person sometimes is "confident" that they did not occur. In quiet conditions, the person may reconsider such signals and report their presence.

Broadbent also reports that there is a "funneling of attention" on the task. At high noise levels, a person typically will focus attention on the most important aspects of a task or the most probable sources of information. If relevant task information is missed owing to this funneling phenomenon, then performance will suffer. This may explain why noise improves performance on simple boring tasks. The person's attention is funneled onto the task rather than wandering because of possible boredom.

Broadbent also reports that continuous work in which there is no opportunity for relaxation will often show occasional moments of low performance and gaps in performance where no recorded response is made. The overall, average performance may not suffer, but rather the variability in performance increases.

These detrimental effects (increased confidence, funneling, and gaps), as we said, have typically been associated with high levels of noise. An apparent exception is performance on cognitive tasks (Hockey, 1978). Here we often find a disruptive effect of noise at levels of 70 or 80 dBA. This is especially true of tasks involving short-term memory requirements. Weinstein (1974, 1977), for example, reported that 68 to 70 dBA noise significantly impaired the detection of grammatical errors in a proofreading task (which requires short-term memory), but did not adversely affect the detection of spelling errors.

Although there is beginning to be some consensus on the effects of noise, there still remains considerable controversy over why, or how, noise has the effects it does. Poulton (1976, 1977, 1978) and Broadbent (1976, 1978) have carried on a lively "discussion" of the mechanisms underlying the effects of noise on performance. Poulton

(1978) contends that all the known effects of noise on performance can be explained by four determinants which combine to affect performance: (1) masking of acoustic task-related cues and inner speech; (2) distraction; (3) a beneficial increase in arousal when noise is first introduced, which gradually lessens and falls below normal when the noise is first switched off; and (4) positive and negative transfer from performance in noise to performance in quiet.[4]

Broadbent (1976, 1978, 1979), on the other hand, rejects the notion that detrimental effects of noise are caused by masking of acoustic task cues or inner speech. Rather, Broadbent attributes most of the detrimental effects of noise to "over-arousal." It is a long standing observation that there exists an inverted U relationship between arousal and performance. Too little or too much arousal results in lower performance than does a moderate level of arousal. Poulton, on the other hand, rejects the "overstimulation" hypothesis when applied to noise.

Whether masking plays a major role in determining the effects of noise remains to be seen. Probably, as is so often the case, both sides are right–to a degree. Undoubtedly if task-related acoustic cues are present and noise masks them, a decrement may well result. The question is whether noise can have a detrimental effect on a task in which no acoustic cue, inner speech, distraction, or transfer is present. Poulton would probably say no; Broadbent probably would contend that it would still be possible owing to overstimulation.

It should be stated that most of the evidence relating to either the degrading or enhancing effects of noise on performance is based on experimental studies and not on actual work situations. And extrapolation from such studies to actual work situations is probably a bit risky. Further, the level of noise required to exert measurable degrading effects on performance is, with the exception of short-term memory tasks, considerably higher than the highest levels that are acceptable by other criteria, such as hearing loss and effects on speech communications. Thus, if noise levels are kept within reasonable bounds in terms of, say, hearing loss considerations, the probabilities of serious effects on performance probably would be relatively nominal.

THE ANNOYANCE OF NOISE

No one need tell us that noise can be annoying; we have all had the experience sometime or another in our lives. Annoyance is not the same as loudness. Loud noises are usually more annoying than soft noises, but there are exceptions. Consider, for example, the slow rhythmic drip of a water faucet versus the roar of the ocean surf–which would annoy you more?

Annoyance is measured by having subjects rate noises on a verbal scale, such as: *noticeable–intrusive–annoying–very annoying–unbearable.* There are a host of factors, both acoustic and nonacoustic, that influence the annoying quality of a noise. A list of some of these factors is presented in Table 15-4.

[4] Positive transfer results from the better learning of the task in noise under the influence of the increased arousal. Negative transfer results from techniques of performance in noise used to counteract the masking or distraction, being used in quiet where they are not appropriate.

TABLE 15-4
SOME FACTORS THAT INFLUENCE THE ANNOYANCE QUALITY OF NOISE

Acoustic factors	Sound level
	Frequency
	Duration
	Spectral complexity
	Fluctuations in sound level
	Fluctuations in frequency
	Rise-time of the noise
Nonacoustic factors	Past experience with the noise
	Listener's activity
	Predictability of noise occurrence
	Necessity of the noise
	Listener's personality
	Attitudes toward the source of the noise
	Time of the year
	Time of the day
	Type of locale

Measures of Noise Exposure

Considerable work has been done to develop a measure of noise exposure that would represent, in a single number, many of the important acoustic factors and some of the nonacoustic factors influencing the annoyance of noise. Sperry (1978) lists 13 different measures which are currently used around the world to assess community exposure to noise. Figure 15-7 lists these measures and shows the relationships among them. Most of these measures were developed in the context of community exposure to air-

FIGURE 15-7
Various measures of exposure to noise. Many were developed in the context of community exposure to aircraft noise. Measures on the left of the figure are primary measures, those on the right are derived from, or based on, those to the left.

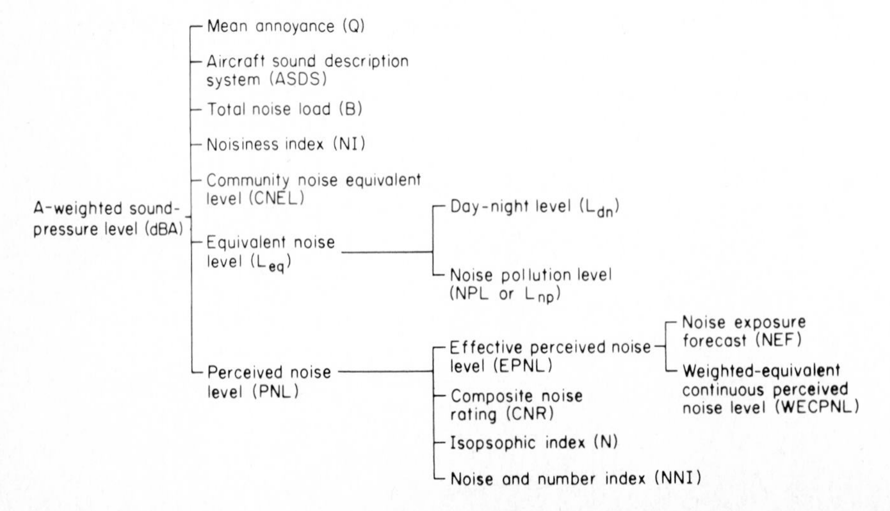

craft noise. From Figure 15-7 it can be seen that all the measures are based on the A-weighted sound level (dBA). Equivalent sound level (L_{eq}) and perceived noise level (PNL), which we discussed previously, have formed the bases for a variety of other measures. The various measures make corrections for such factors as the time of day, season of year, variability in the noise, and number of aircraft flyovers.

We will not take the time to describe each of the measures in Figure 15-7. However, one measure has been related to annoyance and community action and deserves further attention. The day-night level (L_{dn}) is used by the U.S. Environmental Protection Agency to rate community exposure to noise. The day-night level is the equivalent sound level (L_{eq}) for a 24-h period with a correction of 10 dB added to noise levels occurring in the nighttime (2200-0700 hours).

It may be encouraging to note that many of the measures shown in Figure 15-7, when measured over a 24-h period, correlate almost perfectly with one another. For example, L_{eq}, L_{dn}, and CNEL rarely differ by more than ±1 dB (Fidell, 1979). This should not be too surprising since they are all measuring the same noise environment.

Annoyance and Community Response

Whatever these indices of noise exposure measure, they are not highly correlated to whatever gives rise to community response (Fidell, 1979). From Table 15-4 it can be seen that a host of nonacoustical factors influence the annoyance quality of noise. Most of these are not taken into consideration by the various noise exposure measures. Nevertheless, the U.S. Environmental Protection Agency, from a review of British and American community noise surveys, found a linear relationship between L_{dn} and the percentage of people in the survey highly annoyed. This relationship is shown in Figure 15-8.

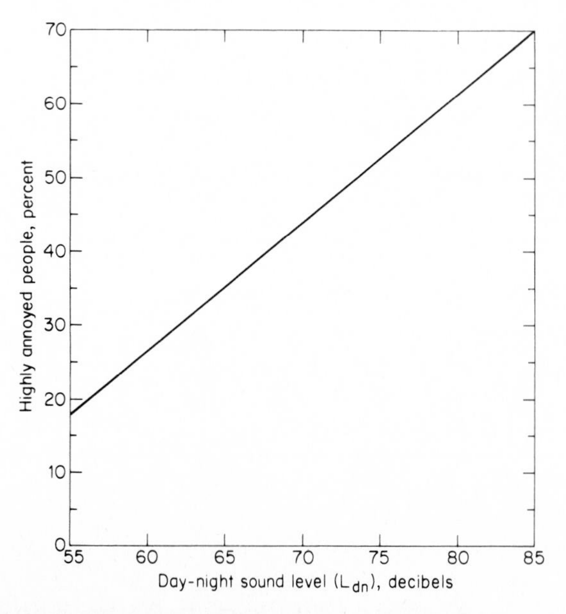

FIGURE 15-8
Relationship between noise exposure and percentage of community highly annoyed. (*Source: U.S. Environmental Protection Agency, 1974.*)

Taking it a step further, attempts have been made to relate noise exposure to community reactions such as complaints, threats, and legal action. There exists a giant step between being annoyed and taking action. A host of sociological factors, political factors, and psychological factors intervene in such a decision. Fidell indicates that the noise exposure itself usually does not account for even half the variance in community reactions.

The U.S. Environmental Protection Agency, using L_{dn}, however, has devised a procedure for predicting community reactions to noise. The procedure attempts to take into account some of the sociopolitical factors that influence community reactions. Table 15-5 lists corrections that are to be added to the L_{dn} to obtain a *normalized* L_{dn}. The normalized L_{dn} is then used with Table 15-6 to predict community reactions.

Generally, a normalized L_{dn} of 55 dBA or lower will not result in complaints. It must be remembered, however, that these predictions are not precise, but rather are only rough indications of the probable community reaction.

TABLE 15-5
CORRECTION FACTORS TO BE ADDED TO THE MEASURED DAY-NIGHT LEVEL (L_{dn}) TO OBTAIN NORMALIZED L_{dn}

Type of correction	Description	Correction added to measured L_{dn} in dB
Seasonal correction	Summer (or year-round operation)	0
	Winter only (or windows always closed)	−5
Correction for outdoor residual noise level	Quiet suburban or rural community (away from large cities, industrial activity, and trucking)	+10
	Normal suburban community (away from industrial activity)	+5
	Urban residential community (not near heavily traveled roads or industrial areas)	0
	Noisy urban residential community (near relatively busy roads or industrial areas)	−5
	Very noisy urban residential community	−10
Correction for previous exposure and community attitudes	No prior experience with intruding noise	+5
	Community has had some exposure to introducing noise; little effort is being made to control noise. This correction may also be applied to a community which has not been exposed previously to noise, but the people are aware that bona fide efforts are being made to control it	0
	Community has had considerable exposure to introducing noise; noisemaker's relations with community are good	−5
	Community aware that operation causing noise is necessary but will not continue indefinitely. This correction may be applied on a limited basis and under emergency conditions	−10
Pure tone or impulse	No pure tone or impulsive character	0
	Pure tone or impulsive character present	+5

Source: U. S. Environmental Protection Agency, 1974.

TABLE 15-6
EXPECTED COMMUNITY RESPONSE BASED ON NORMALIZED L_{DN}

Community response	Normalized L_{dn}, dBA
No reaction or sporadic complaints	50–60
Widespread complaints	60–70
Severe threats of legal action or strong appeals to local officials	70–75
Vigorous action	75–80

Source: U.S. Environmental Protection Agency data as presented by R. Taylor, 1978, table 8-4.

ACCEPTABLE LIMITS OF NOISE

In trying to figure out the upper ceiling of noise that would be acceptable in a given situation, the question of the criterion of acceptability immediately bobs up. Criteria in terms of speech communications were discussed in Chapter 6 [such as the speech interference level (SIL) and noise criteria (NC) curves], and the above discussion of residential noise levels dealt in part with the criterion of annoyance.

Hearing Loss Criteria

In typical work situations, hearing loss is perhaps the prime criterion for acceptable noise levels. Damage-risk criteria and other standards have been set forth by various organizations, differentiating between continuous and noncontinuous noise.

Continuous Noise In the United States the Occupational Safety and Health Administration (OSHA) of the Department of Labor has established a set of permissible noise exposures for persons working on jobs in industry, the permissible levels depending on the duration of exposure, as shown in Table 15-7.

TABLE 15-7
OCCUPATIONAL SAFETY AND HEALTH ADMINISTRATION'S PERMISSIBLE NOISE EXPOSURES

Duration per day, h	Sound level, dBA
8	90
6	92
4	95
3	97
2	100
1½	102
1	105
½	110
¼ or less	115

Note: dB levels given are based on the A scale.
Source: Federal Register, 1971.

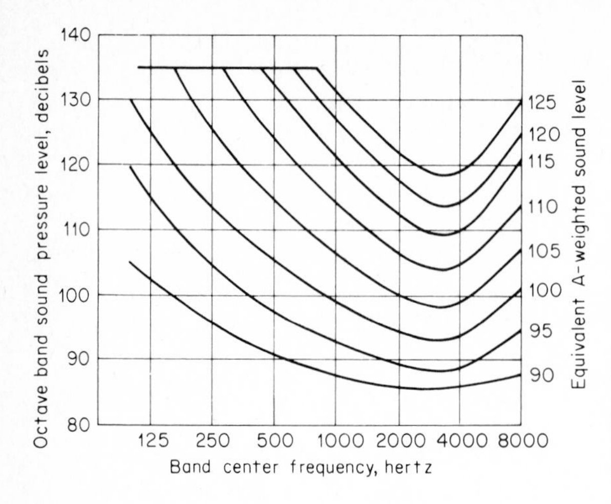

FIGURE 15-9
Equivalent sound-level contours used in determining the A-weighted sound level on the basis of an octave-band sound analysis. The curve at the point of the highest penetration of the noise spectrum reflects the A-weighted sound level. (*Source: Federal Register, 1971, p. 10518.*)

When the noise level is determined by octave-band analysis, the equivalent A-weighted sound level may be determined by the use of Figure 15-9. That figure shows several equivalent sound-level contours. In using this procedure, the spectrum of the noise is plotted over that figure, and the point of the highest penetration of that spectrum is the sound-level contour that is noted. That contour is the one that reflects the equivalent A-weighted sound level. That level, in turn, is used in determining the permissible duration of exposure as discussed above. However, this procedure offers only a rough approximation to dBA, especially in instances of marked discontinuities in octave-band spectra.

The OSHA regulations deal with exposure to changing sound levels by the use of the *noise dose* concept. Exposure to any sound level at or above 90 dBA results in the person incurring a partial dose of noise. (Exposures to sound levels less than 90 dBA are ignored in calculating doses.) A partial dose is calculated for each specified sound-pressure level above 90 dBA as follows:

$$\frac{\text{Time actually spent at the sound level}}{\text{Maximum duration allowed at that sound level}}$$

The total or daily noise dose is equal to the sum of the partial doses. If the daily noise dose exceeds 1.0, the exposure is in violation of the OSHA regulations.

Based on this concept, a curious situation can exist. Let us consider a worker who, during a workday, is exposed to the following noise levels:

95 dBA for 4 h
105 dBA for 1 h
85 dBA for 3 h

Such exposure is within the OSHA permissible limits for each of the levels *individually*, but the daily noise dose exceeds 1.0 and thus would be in violation of the OSHA regulations.

The U.S. Air Force has established a simple standard, using only the four octave-band pressure levels having center frequencies of 500, 1000, 2000, and 4000 Hz; this

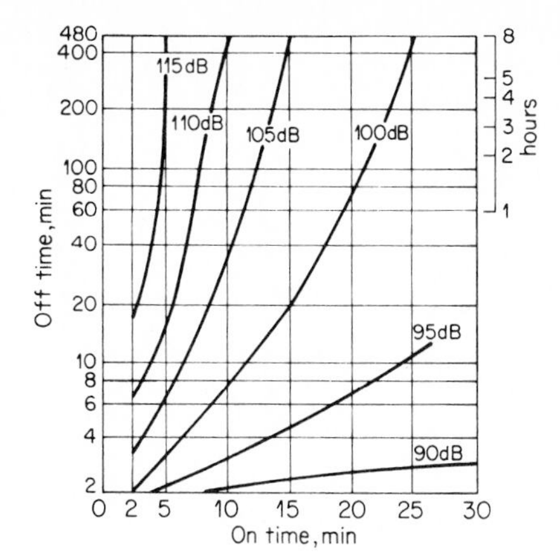

FIGURE 15-10
Guide to allowable exposure times for intermittent noise. (Each curve is labeled with the average value for octave bands with center-line frequencies of 500, 1000, and 2000 Hz.) The vertical scale shows the off time which must follow noise exposure of the duration shown on the horizontal scale (on time) to avoid temporary threshold shifts greater than 12 dB at 2000 Hz. (*Source: Guide for conservation of hearing in noise, 1964.*)

standard provides that the sound-pressure level should not exceed 85 dB for *any* of these four octave bands, for conventional daily exposure of 8 h (although there is provision for permitting 3 dB increases for each *halving* of the daily exposure duration).

Intermittent Noise In the case of intermittent noise, the tolerance limits depend on a trade-off between intensity and the relationship between the duration of exposure and the duration of subsequent nonexposure. An example of such trade-off is given in Figure 15-10.

THE HANDLING OF NOISE PROBLEMS

When a noise problem exists, or might be anticipated, there is no substitute for having good, solid information to bring to bear and for attacking the problem in a systematic manner. The manner of working out such problems is, of course, for the acoustical engineers, and we shall not go into this topic except in a very cursory way.

Defining the Noise Problem

Defining a possible noise problem consists of essentially two phases. The first of these is the measurement of the noise itself. The overall sound-pressure level (e.g., dBA) will give a gross indication of a potential noise problem. An octave-band analysis of the noise, however, gives a more detailed and useful picture of the noise situation. The second is to determine what noise level would be acceptable, in terms of hearing loss, annoyance, communications, etc. Such limits normally would be those adapted from relevant criteria such as discussed above. An example is shown in Figure 15-11. This figure shows the spectrum of the original noise of a foundry cleaning room, and a tentative-design baseline that was derived from a set of relevant noise standards. Incidentally, part of the original spectrum was below this level, but the high frequencies were not; the difference represents the amount of reduction that should be achieved (in this case, the shaded area). The third line shows the noise level after abatement.

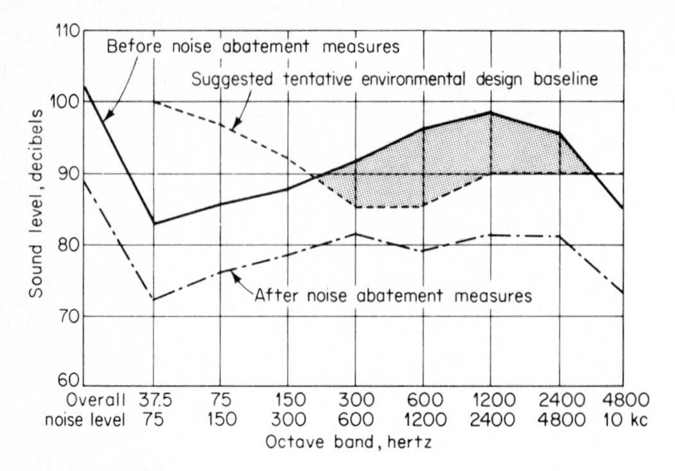

FIGURE 15-11
Spectrum of noise of foundry cleaning room before abatement, the baseline that represents a desired upper ceiling, and the spectrum after abatement. The shaded area represents the desired reduction. The abatement consisted primarily in spraying a heavy coat of *deadener* on the tumbling barrels and surfaces of tote boxes. (*Source: Foundry Noise Manual, 1966, p. 52.*)

Noise Control

Controlling a noise problem can be accomplished by attacking the noise at the source, along its path from the source to the receiver, and at the receiver. Control at the source includes proper design and maintenance of the machines, use of vibration-absorbing mountings and mufflers, and use of sound absorption materials on the inside and outside surfaces of the machine. Controlling noise along its path includes use of barriers, enclosures, acoustical treatment, and baffles.

Actually, there are many ingenious variations of these and other means that have aided in noise reduction. Figure 15-12 illustrates the possible effects of various noise-control measures. As another example, Figure 15-13 shows the spectrum of noise in one situation before and after the use of acoustical treatment.

Ear Protection

Where the noise level cannot reasonably be reduced to "safe" limits, some form of ear protection should be considered for those people who are exposed to the noise. The two types of ear-protection devices that are reusable are earplugs and earmuffs. Disposable forms include dry cotton, waxed cotton, and glass down. *Glass down* is a form of glass wool in which the fibers are so fine that they form a material of downlike softness which has been reported to be quite harmless to the delicate skin of the ear canal (Coles, 1969).

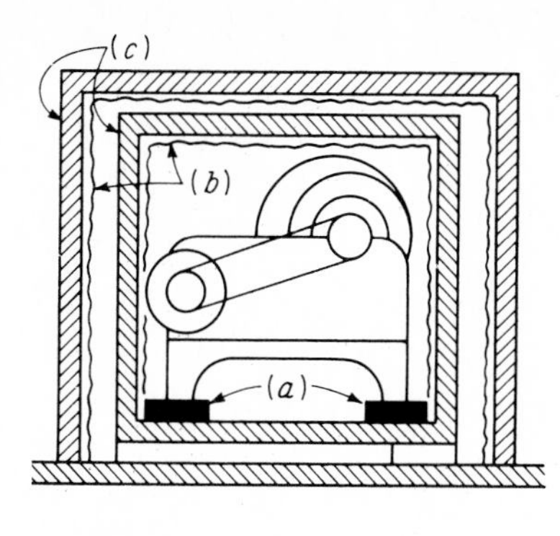

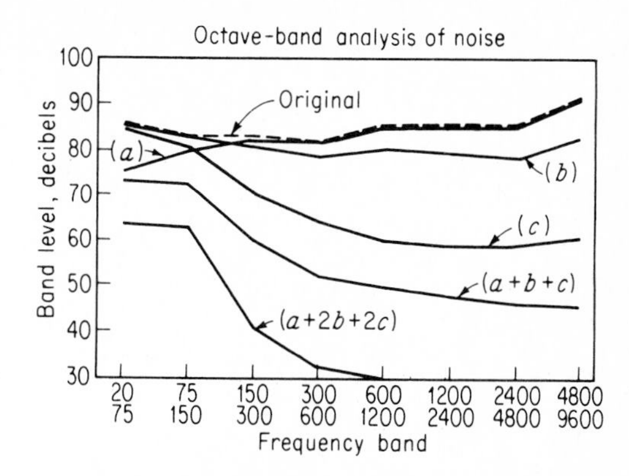

FIGURE 15-12
Illustrations of the possible effects of some noise-control measures. The lines on the graph show the possible reductions in noise (from the original level) that might be expected by vibration insulation, ***a***; an enclosure of acoustical absorbing material, ***b***; a rigid, sealed enclosure, ***c***; a single combined enclosure plus vibration insulation, ***a + b + c***; and a double combined enclosure plus vibration insulation, ***a + 2b + 2c***. (*Source: Adapted from Peterson & Gross, 1978.*)

FIGURE 15-13
Spectrum of noise before and after acoustical treatment. (*Source: Jensen, Jokel, & Miller, 1978, fig. 6.2.1.*)

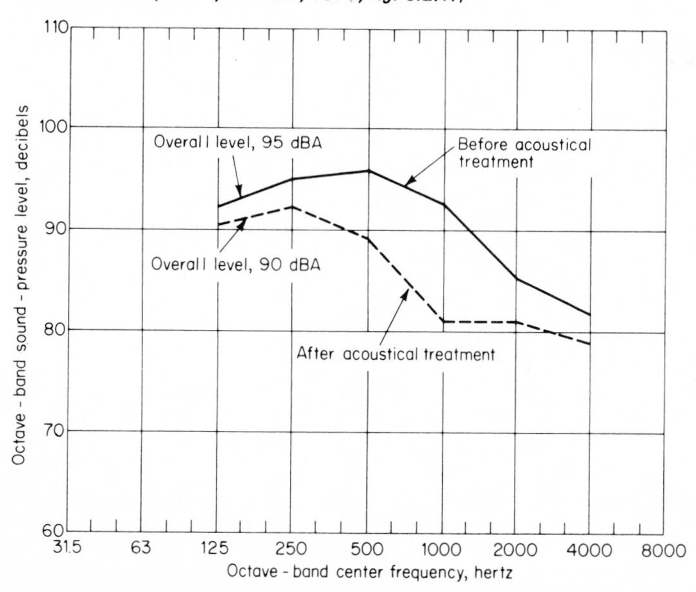

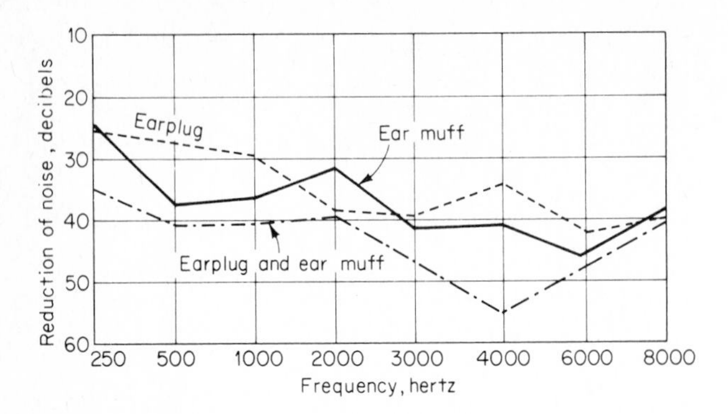

FIGURE 15-14
Protection provided by the use of earplugs, earmuffs, and a combination of the two. (*Source: USAF, Wright-Patterson AFB.*)

Effectiveness of Ear-Protection Devices The effectiveness of ear-protection devices is somewhat variable, depending on the nature of the noise, the duration of exposure, the fit of the device, its attenuating characteristics, and possibly other variables. By and large, however, such devices provide a worthwhile degree of protection from high noise levels.

An example of such protection is shown in Figure 15-14 by the use of earplugs, earmuffs, and a combination. Coles expresses the opinion that the fluid-seal type of earmuff comes close to being the most efficient ear protectors that can be made. These have a plastic ring containing fluid (such as glycerin) that fits around the ear, thereby minimizing sound leakage. He also believes that the maximum amount by which an ear protector can reduce the sound reaching the ear is limited to about 35 dB at 250 Hz, up to about 60 dB at the higher frequencies. In practice, however, the typical attenuation is substantially lower. Different types and models of ear protectors have different attenuation characteristics. It is important, therefore, to select the device that is best for the particular noise conditions of concern.

Sometimes people do not wear ear protection because they believe it will make it harder to hear someone talk in a noisy situation. Howell and Martin (1975), however, found that ear protectors do not degrade speech intelligibility for the listener and may even effect a slight improvement. If worn by the talker, however, speech becomes reduced in level and quality, and intelligibility for the listener is actually reduced. When ear protectors were worn by both talker and listener, the composite effect was an overall *reduction* in speech intelligibility compared to the open-ear condition. Therefore, in a noise situation in which people are wearing ear-protection devices, extra effort must be made to speak loudly and clearly.

REFERENCES

Bonvallet, S. *Noise.* Lectures presented at the Inservice Training Course on the Acoustical Spectrum. Ann Arbor: University of Michigan Press, February 1952.

Broadbent, D. Noise and the details of experiments: A reply to Poulton. *Applied Ergonomics,* 1976, *7,* 231–235.

Broadbent, D. The current state of noise research: Reply to Poulton. *Psychological Bulletin,* 1978, 85, 1052–1067.

Broadbent, D. Human performance and noise. In C. Harris (Ed.), *Handbook of noise control.* New York: McGraw-Hill, 1979.

Burns, W. Physiological effects of noise. In C. Harris (Ed.), *Handbook of noise control.* New York: McGraw-Hill, 1979.

Burrows, A. A. Acoustic noise, an informational definition. *Human Factors,* 1960, *2*(3), 163–168.

Coles, R. R. A. Control of industrial noise through personal protection. In W. D. Ward & J. E. Frick (Eds.), *Noise as a public health hazard.* Washington, D. C.: American Speech and Hearing Association, 1969.

Eschenbrenner, A. J., Jr. Effects of intermittent noise on the performance of a complex psychomotor task. *Human Factors,* 1971, *13*(1), 59–63.

Federal Register, May 29, 1971, *36*(105).

Fidell, S. Community response to noise. In C. Harris (Ed.), *Handbook of noise control.* New York: McGraw-Hill, 1979.

Fletcher, H., & Munson, W. A. Loudness, its definition, measurement, and calculation. *Journal of the Acoustical Society of America,* 1933, *5*, 82–108.

Foundry noise manual (2d ed.). Des Plaines, Ill.: American Foundryman's Society, 1966.

Gulian, E. *Noise as an occupational hazard: Effects on performance level and health. A Survey of the European literature.* Cincinnati, Ohio: National Institute for Occupational Safety and Health, 1974.

Guide for conservation of hearing in noise. Prepared by Subcommittee on Noise, American Academy of Ophthalmology and Otolaryngology, revised 1964.

Hockey, G. Effects of noise on human work efficiency. In D. May (Ed.), *Handbook of noise assessment.* New York: Van Nostrand Reinhold, 1978.

Howell, K., & Martin, A. An investigation of the effects of hearing protectors on vocal communication in noise. *Journal of Sound and Vibration,* 1975, *41*, 181–196.

Industrial noise manual (2d ed.). Detroit: American Industrial Hygiene Association, 1966.

Jansen, G. Effects of noise on physiological state. In W. D. Ward & J. E. Frick (Eds.), *Noise as a public health hazard* (ASHA Reports 4). Washington, D. C.: The American Speech and Hearing Association, February 1969.

Jensen, P., Jokel, C., & Miller, L. *Industrial noise control manual* (Rev. ed.). Cincinnati, Ohio: National Institute for Occupational Safety and Health, 1978.

Kryter, K. *The effects of noise on man.* New York: Academic Press, 1970.

Kryter, K., Ward, W., Miller, J., & Eldredge, D. Hazardous exposure to intermittent and steady-state noise. *Journal of the Acoustical Society of America,* 1966, *39*, 451–463.

Machle, W. The effect of gun blast on hearing. *Archives of Otolaryngology,* 1945, *42,* 164–168.

Melnick, W. Hearing loss from noise exposure. In C. Harris (Ed.), *Handbook of noise control.* New York: McGraw-Hill, 1979.

Peterson, A., & Gross, E., Jr. *Handbook of noise measurement* (8th ed.). New Concord, Mass.: General Radio Co., 1978.

Poulton, E. Continuous noise interferes with work by masking auditory feedback and inner speech. *Applied Ergonomics,* 1976, *7,* 79–84.

Poulton, E. Continuous intense noise masks auditory feedback and inner speech. *Psychological Bulletin,* 1977, *84,* 977–1001.

Poulton, E. A new look at the effects of noise: A rejoinder. *Psychological Bulletin,* 1978, *85,* 1068–1079.

Robinson, D., & Dadson, R. Threshold of hearing and equal-loudness relations for pure tones, and the loudness function. *Journal of the Acoustical Society of America,* 1957, *29*(12), 1284–1288.

Sperry, W. Aircraft and airport noise. In D. Lipscomb and A. Taylor (Eds.), *Noise control: Handbook of principles and practices.* New York: Van Nostrand Reinhold, 1978.

Spoor, A. Presbycusis values in relation to noise-induced hearing loss. *International Audiology,* 1967, *6*(1), 48–57.

Stevens, S. S. A scale for the measurement of a psychological magnitude: loudness. *Psychological Review,* 1936, *43,* 405–416.

Stevens, S. S. Calculation of the loudness of complex noise. *Journal of the Acoustical Society of America,* 1956, *28,* 807–832.

Stevens, S. S. Procedure for calculating loudness: Mark VI. *Journal of the Acoustical Society of America,* 1961, *33*(11), 1577–1585.

Stevens, S. S. Perceived level of noise by Mark VII and decibels (E). *Journal of the Acoustical Society of America,* 1972, *51*(2, Pt. 2), 575–601.

Taylor, A., & Lipscomb, D. The use and measurement of equivalent sound level. In D. Lipscomb and A. Taylor (Eds.), *Noise control: Handbook of principles and practices.* New York: Van Nostrand Reinhold, 1978.

Taylor, R. Exterior industrial and commercial noise. In D. May (Ed.), *Handbook of noise assessment.* New York: Van Nostrand Reinhold, 1978.

Taylor, W., Pearson, J., Mair, A., & Burns, W. Study of noise and hearing in jute weavers. *Journal Acoustical Society of America,* 1965, *38,* 113–120.

U. S. Environmental Protection Agency. *Information on levels of environmental noise requisite to protect public health and welfare with an adequate margin of safety* (EPA 550/9-74-004). Washington, D. C.: U.S. Protection Agency, March 1974.

Weinstein, N. Effects of noise on intellectual performance. *Journal of Applied Psychology,* 1974, *59*(5), 548–554.

Weinstein, N. Noise and intellectual performance: A confirmation and extension. *Journal of Applied Psychology,* 1977, *62,* 104–107.

Williams, K. An introduction to the assessment and measurement of sound. In D. Lipscomb and A. Taylor (Eds.), *Noise control: Handbook of principles and practices,* New York: Van Nostrand Reinhold, 1978.

CHAPTER 16

MOTION

Technological ingenuity in recent times has resulted in the creation of methods of travel that our ancestors probably never dreamed about. These include space capsules, aircraft, zero ground-pressure vehicles, rockets strapped to one's back, and of course various earthbound vehicles such as automobiles, buses, and trucks. Many of these make it possible for people to move at speeds and in environments never before experienced, to which they are not biologically adapted. The disparity between people's biological and physical nature, on the one hand, and on the other, the environmental factors imposed on them by their "exotic" modes of travel, defines a domain within which human factors can make a contribution.

The variables imposed by our increased mobility include vibration, acceleration and deceleration, weightlessness, and an assortment of more strictly psychological phenomena associated with these factors, such as disorientation and other illusions. All of these involve, to one degree or another, the sense of motion and orientation. Therefore, before discussing vibration, acceleration, and other motion topics, it will serve us well to review briefly the sensory receptors associated with motion and body orientation.

MOTION AND ORIENTATION SENSES

The *five senses*, in the Aristotelian tradition (vision, audition, smell, taste, and touch), basically deal with stimuli external to the body. The sensory receptors involved (the eyes, ears, etc.) are referred to as *exteroceptors.* There are, however, a number of other sensory receptors related to motion and body orientation.

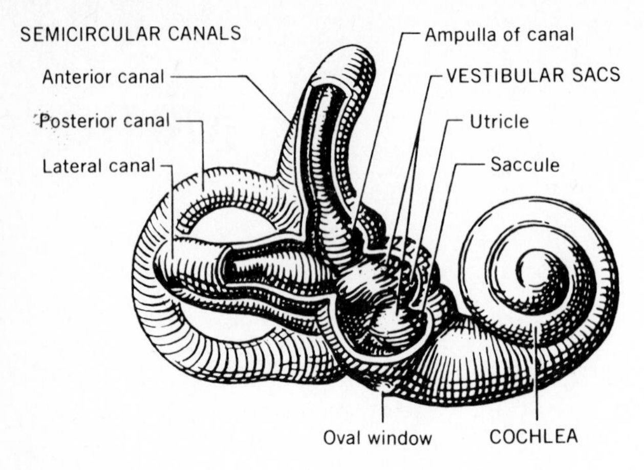

FIGURE 16-1
The body orientation organs. The semicircular canals form roughly a three-coordinate system that provides information about body movement. The vestibular sacs (the utricle and saccule) respond to the forces of gravity and provide information about body position in relation to the vertical.

Proprioceptors

The *proprioceptors* are sensory receptors of various kinds that are embedded within the subcutaneous tissues, such as in the muscles and tendons, in the coverings of the bones, and in the musculature surrounding certain of the internal organs. These receptors are stimulated primarily by the actions of the body itself. A special class of these are the *kinesthetic receptors*, which are concentrated around the joints and which are primarily used to tell us where our limbs are at any given moment and to coordinate our movements.

Semicircular Canals

The three semicircular canals in each ear are interconnected, doughnut-shaped tubes that form, roughly, a three-coordinate system, as shown in Figure 16-1 (along with the vestibular sacs). With changes in acceleration or deceleration, the fluid shifts its position in these tubes, which stimulates nerve endings that then transmit nerve impulses to the brain. It should be pointed out that movement of the body at a constant rate does not cause any stimulation of these canals. Rather, they are sensitive only to a *change* in rate (acceleration or deceleration).

Vestibular Sacs: Utricle and Saccule

The *vestibular sacs* (also called the *otolith organs*) are two organs with interior hair cells and containing a gelatinous substance. The *utricle* is generally positioned in a horizontal plane, and the *saccule* is more in a vertical plane. As the body changes position, the gelatinous substance is affected by gravity, setting up nerve impulses via the hair cells. The utricle apparently is the more important of the two organs. The primary function of these organs is that of sensing body posture in relation to the vertical and thus serving as something of a gyroscope that helps to keep us on an even keel. While their dominant role is that of aiding in sensing postural conditions of the body, they are also somewhat sensitive to acceleration and deceleration and presumably supplement the semicircular canals in sensing such changes.

Interdependence of Motion and Orientation Senses

In the maintenance of equilibrium and body orientation, or in reliably sensing motion and posture, all of these senses play a role, along with the skin senses, vision, and sometimes audition. The importance of vision in orientation was illustrated, for example, in experiments with people in a "tilting room," in some cases with a chair that could be tilted within the room (Witkin, 1959). Subjects seated in the room were tilted to various angles and asked to indicate what direction they considered vertical. When blindfolded, they were able to indicate the vertical more accurately than when they could see the inside of the room; when they could see, they tended to indicate that the ceiling of the room was upright even if the room itself was tilted and the chair was actually vertical. The implication of such investigations is that misperceptions of the *true* upright direction may occur when there is a *conflict* between the sensations of gravity and visual perceptions; in such a case one's visual perceptions may dominate, even when they are erroneous.

WHOLE-BODY VIBRATION

"If it moves, it probably vibrates." This is especially true in the field of transportation; be it car, truck, train, airplane, or boat, all subject occupants to vibration to some degree. Vibration is not a new topic for us. In Chapter 15 we discussed noise, which is a form of vibration that is audible, and in Chapter 10 we discussed vibration induced from hand tools. Our concern in this chapter, however, is with low-frequency (generally less than 100 Hz), whole-body vibration typically encountered in trucks, tractors, airplanes, etc.

Vibration Terminology

Vibration is primarily of two types, *sinusoidal* and *random. Sinusoidal vibration* may be a single sine wave of some particular frequency or it may contain combinations of sine waves of different frequencies. Its principal characteristic is its regularity, that is, the wave form repeats itself at regular intervals. This type of vibration is most often encountered in laboratory studies. *Random vibration* is, as the name implies, irregular and unpredictable. This is the most common type of vibration encountered in the real world.

Vibration, be it sinusoidal or random, occurs in one or more directional planes. Table 16-1 lists these and indicates the terminology and symbols used to designate the direction of the vibration. The terminology relates to the direction of the vibration

TABLE 16-1
VIBRATION TERMINOLOGY WITH RESPECT TO DIRECTION OF VIBRATION

Direction of motion	Heart motion	Other description	Symbol
Forward-backward	Spine-sternum-spine	Fore-aft	$\pm g_x$
To left-to right	Left-right-left	Side-to-side	$\pm g_y$
Headward-footward	Head-feet-head	Head-tail	$\pm g_z$

Source: Adapted from Hornick, 1973, table 7-1, p. 229.

relative to the human body. Thus, a person standing on a platform which is vibrating up-and-down would be experiencing head-tail vibration ($\pm g_z$). If, however, the person lies down on his or her back, the vibration would then be considered fore-aft ($\pm g_x$).

In addition to type and direction, vibration is described in terms of *frequency* and *intensity*. Frequency is measured in hertz (Hz), i.e., cycles per second, as described in Chapter 15. Intensity is measured in a variety of ways such as peak or maximum: (1) *amplitude* (in or cm); (2) *displacement* (in or cm); (3) *velocity*, the first derivative of displacement (in/s or cm/s); (4) *acceleration*, the second derivative of displacement (in/s^2 or cm/s^2) (sometimes acceleration is expressed in terms of numbers of gravities[1] and is labeled in terms of lowercase *g* units as $\pm g_x$, $\pm g_y$, or $\pm g_z$, depending upon the direction of the oscillation as given in Table 16-1); or (5) *rate of change of acceleration*, also called *jerk*, the third derivative of displacement (in/s^3 or cm/s^3).

In the case of random vibration, maximum amplitude or acceleration is somewhat inappropriate because the oscillations vary randomly in terms of frequency and intensity. To handle this variability, the frequency spectrum is usually indicated by *mean-square spectral density* and expressed as *power spectral density* (PSD) in g^2/Hz units. The PSD defines the power at discrete frequencies within the selected bandwidth. A plot of PSD (g^2/Hz) versus frequency illustrates the power distribution of the vibration environment. Figure 16-2, for example, shows PSD plots for an automobile and an aircraft.

Intensity is usually expressed as a root-mean-square (rms) value of acceleration (in/s^2, cm/s^2, or *g*).[2] *Root-mean-square acceleration* defines the total energy across the entire frequency range. Figure 16-3 shows rms acceleration values (actually ranges of rms *g*) for various air and surface vehicles during cruise.

One last point should be made about vibration measurement: acceleration, displacement, and frequency are all related. If, for example, acceleration is held constant and frequency is varied, displacement must also vary. It is impossible to hold two of the quantities constant and vary only the third. This makes it difficult to isolate the specific parameter that is affecting either performance, subjective reactions, or physiological reactions.

[1] 1 g = 386 in/s^2 or 980 cm/s^2.

[2] *Root mean square* (rms) is the square root of the arithmetic mean of instantaneous values (amplitude or acceleration) squared. In the case of a simple sine wave, rms $g = 0.707 \times$ peak g.

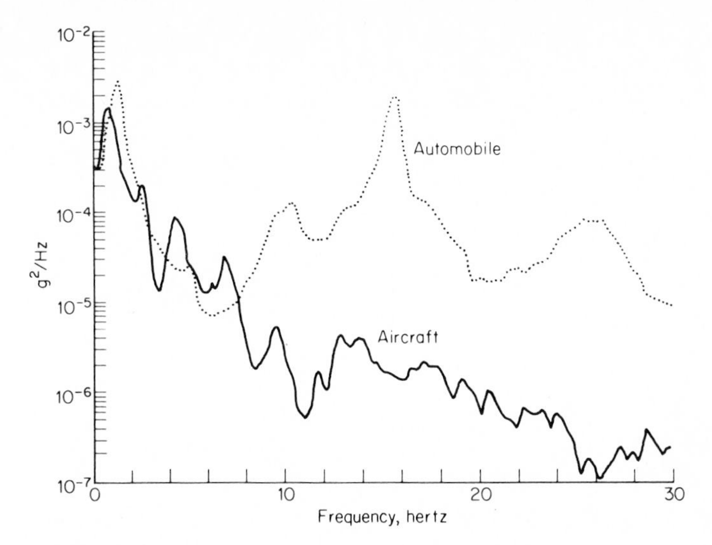

FIGURE 16-2
Power spectral density (PSD) plots for an automobile and an aircraft in cruise. (*Source: Stephens, fig. 2.*)

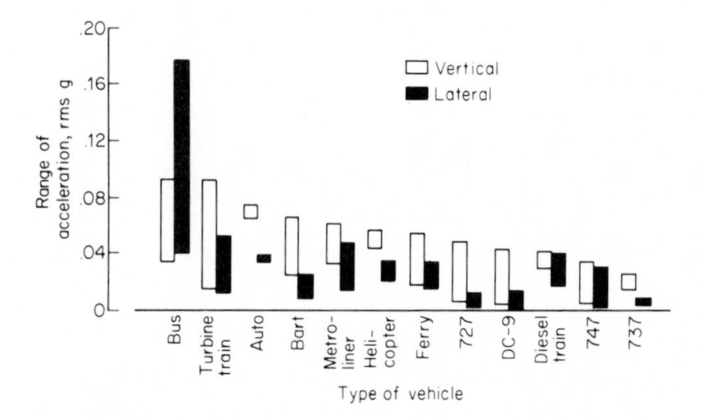

FIGURE 16-3
Root-mean-square (rms) acceleration ranges of various air and surface vehicles during cruise. The variability within a vehicle type is due to such factors as road quality, air turbulence, and age and general condition of the test vehicles. (*Source: Stephens, 1979, fig. 4b.*)

Attenuation, Amplification, and Resonance

As vibration is transmitted to the body, it can be amplified or attenuated as a consequence of body posture (e.g., standing or sitting), the type of seating, and the frequency of the vibration. Every object (or *mass*) has a resonant frequency, in somewhat the same sense that a pendulum has a natural frequency. When an object is vibrated at its resonant frequency, the object will vibrate at a maximum amplitude which is larger than the amplitude of the original vibration. This phenomenon is called *resonance.*

The human body and the individual body members and organs have their own resonant frequencies. As the human body is caused to vibrate (as in a vehicle), the effect upon the body and upon the body members and organs is the consequence of the interaction between the frequency of the vibrating source and the resonant frequencies of the individual *masses*. Since the body members and organs have different resonant frequencies, and since they are not attached rigidly to the body structure, they tend to vibrate at different frequencies, rather than in unison.

The degree to which a part of the body will amplify the input vibration is a function of the ratio of the input frequency to the resonant frequency of the body part. *Amplification* will occur if this ratio is 1.414 or less (Radke, 1957) with, of course, maximum amplification occurring at the resonant frequency (i.e., when the ratio equals 1.0). When the ratio is greater than about 1.414, there is *attenuation*, i.e., the displacement amplitude of the body part is less than that of the input vibration.

The amplification and attenuation that would result from the ratio mentioned above, however, would be expected with no *damping*. In practice, some damping usually would be provided, either by intent or otherwise. Examples of the actual amplification and attenuation of the human body while standing and seated are illustrated in Figure 16-4. With the vibration of the vibrating table used as a base (expressed as 100 percent), the amplification, or attenuation, is measured by the use of accelerometers positioned at different locations, such as at the belt, the neck, and the head. For a person standing, there is a typical attenuation effect; this is because the legs serve to absorb the effects as the individual bends and straightens them in response to the movement. In the case of a seated individual, however, there is an amplification effect (at the head, neck, and belt) especially in the range of frequencies from about 3.5 to 4.5 Hz. Figure 16-5 shows the resonant frequency at the shoulder of a seated man in a relaxed posture and with the trunk muscles tensed. The resonant frequency here is approximately 4.5 to 6.0 Hz depending on the state of the trunk muscles.

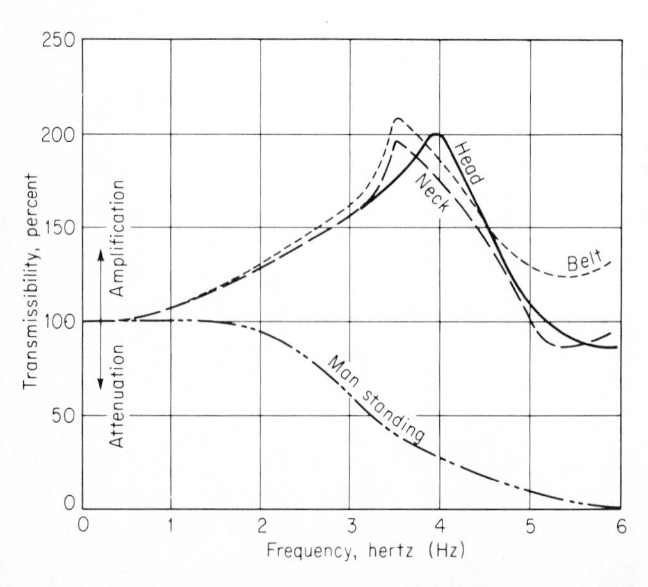

FIGURE 16-4
Mechanical response of person to vertical vibration, showing amplification and attenuation, by frequency, for seated and standing positions. (*Source: Radke, 1957.*)

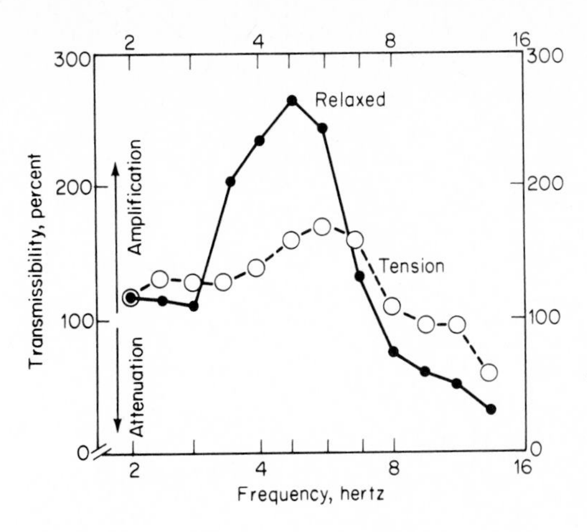

FIGURE 16-5
The effect of tensed and relaxed trunk muscles upon the amplitude of vertical vibration of the shoulders of a seated man. (*Source: Adapted from Guignard, 1965, fig. 380.*)

Damping can be controlled to some degree by appropriate seat design, spring design, cushioning, etc. Figure 16-6 illustrates the amplification and attenuation effects of various seats. It can be seen that, in this particular comparison, the suspension seat resulted in substantial attenuation over the critical frequencies of 3 to 6 Hz.

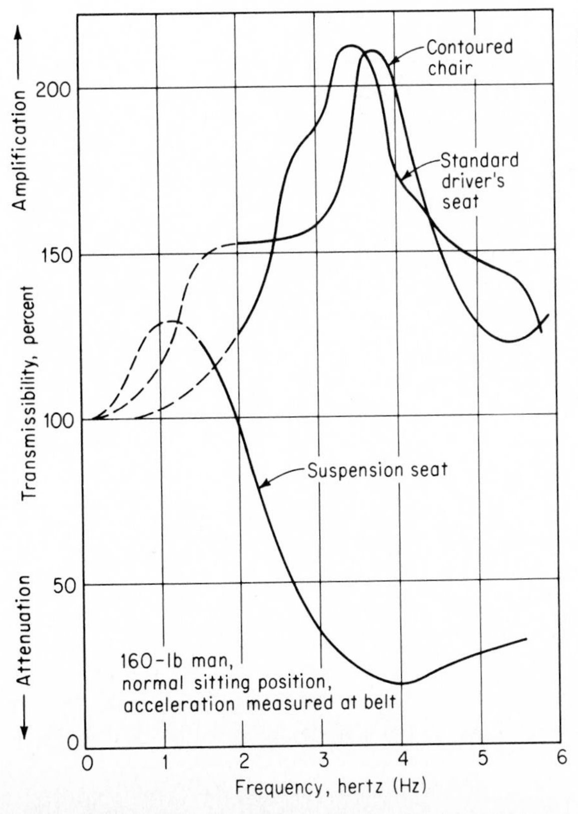

FIGURE 16-6
Mechanical response of a person's body to vibrations, when seated in different seats. (*Source: Simons, Radke, & Oswald, 1956.*)

Physiological Effects of Vibration

The evidence suggests that short-term exposure to vibration causes only very small physiological effects which are of little practical significance (Hornick, 1973). A slight degree of hyperventilation has been reported and increases in heart rate are found during the early periods of exposure. The elevated heart rate appears to be an anticipatory general stress response (Hornick). No significant changes are found in blood chemistry or endocrine chemical composition (Gell & Moeller, 1972).

The physiological effects of long-term exposure to whole-body vibration are not clear. Most of the evidence comes from epidemiological investigations of truck drivers and heavy-equipment operators (Troup, 1978). These workers have disproportional incidences of lumbar spinal disorders, hemorrhoids, hernias, and digestive and urinary problems (Wasserman, 1975). The problem with these sorts of studies is that it is difficult to separate the effects due to vibration from the effects of sitting all day, or manually loading and unloading a vehicle, all of which are part of operating trucks and heavy equipment.

Performance Effects of Vibration

Considerable research has demonstrated the effects of vibration on the tracking performance of seated subjects. Hornick (1973) has reviewed much of this body of literature and has gleaned a few generalizations from it. The effects of vibration are somewhat dependent on the difficulty of the tracking task, the type of display, and the type of controller used. For example, the use of side-stick and arm support can reduce vibration-induced error by as much as 50 percent as compared with conventional center-mounted joysticks. Detrimental effects of vertical sinusoidal vibration generally occur in the 4 to 20 Hz range with accelerations exceeding 0.20*g*. These conditions can produce tracking errors up to 40 percent greater than in nonvibratory control conditions. The effect of sinusoidal vibration, however, is not the same as the effect of random vibration. Random vibration does not seem to be as detrimental to tracking performance as does sinusoidal. Vertical vibration is generally more debilitating than lateral or fore-aft vibration. Further, the effects of vibration on tracking performance do not seem to end with the cessation of the vibration, but residual effects may last up to ½ h after exposure.

Visual performance is generally impaired by vibration and is greatest in the range of 10 to 25 Hz. The amplitude of vibration seems to be a key factor. The effect, of course, depends greatly on the visual task and the display characteristics (e.g., size of letters, larger being better; or clutter, less being better).

Tasks that require steadiness or precision of muscular control are likely to show decrements from vibration. On the other hand, tasks that measure primarily central neural processes, such as reaction time, monitoring, and pattern recognition, appear to be highly resistant to degradation during vibration. In fact, Poulton (1978) points out that vibration between 3.5 and 6 Hz can have an alerting effect on subjects engaged in "boring" vigilance tasks. Figure 16-3 indicated that, within this frequency range, tensing the trunk muscles attenuates the amplitude of shoulder vibration. Tensing the muscles is a good method for maintaining alertness. Outside of the 3.5- to 6-Hz range the subjects can attenuate shoulder vibration more by relaxing the trunk muscles—a good way to fall asleep.

Subjective Responses to Whole-Body Vibration

The subjective response most often assessed in vibration studies is comfort. *Comfort,* of course, is a state of feeling and thus depends in part on the person experiencing the situation. We cannot know directly or by observation the level of comfort being experienced by another person; we must ask people to report to us how comfortable they are. This is usually done using adjective phrases such as "mildly uncomfortable," "annoying," "very uncomfortable," or "alarming." Investigators have tried to link the physical characteristics of vibration, most notably frequency and acceleration, to subjective evaluations of comfort. Their studies usually result in *equal-comfort contours* for combinations of frequency and acceleration. Unfortunately, the resulting contours differ widely from study to study (e.g., see Oborne, 1976 for a review). This is due to different methodologies, subject populations, vibration environments, and semantic comfort descriptors.

Probably the most comprehensive research on ride comfort (or more properly *discomfort*) has been conducted at the National Aeronautics and Space Administration (NASA), Langley Research Center, using over 2200 test subjects (Leatherwood, Dempsey, & Clevenson, 1980). To avoid semantic confusion, NASA developed a ratio scale of discomfort, measured in DISC units. The scale is anchored (1 DISC) at the discomfort threshold (i.e., where 50 percent of the passengers feel uncomfortable). A vibration rated 2 DISC is considered twice as uncomfortable as a vibration rated 1 DISC. The percentage of people feeling uncomfortable rises rapidly with each unit of DISC, as shown in Table 16-2.

The NASA experiments were carried out in a simulator with an interior configuration of a modern jet airliner (including actual airplane cushioned seats). Many studies were conducted to establish a model to predict subjective comfort ratings (in DISC units) for various complex ride environments. The methodology and resulting formulas and tables are too complex to present here. The approach, however, consists of three elements: (1) estimation of discomfort due to sinusoidal or random vibration, or both, within single axes; (2) estimation of the discomfort due to vibration in combined axes; and (3) application of corrections for the effects of interior noise and duration of exposure. Incidentally, it was found that perceived discomfort *decreased* with increasing exposure time.

Although we cannot present the details of the model, Figure 16-7 illustrates the equal-discomfort contours for seated subjects experiencing sinusoidal vertical vibration. As Leatherwood et al. (1980) point out, the model is useful for estimating the average level of discomfort experienced by a group of passengers in a combined noise

TABLE 16-2
PASSENGERS FEELING UNCOMFORTABLE AT VARIOUS LEVELS OF DISC

Discomfort scale, DISC	Percentage uncomfortable
1	50
2	90
3	100

Source: Leatherwood, Dempsey, & Clevenson, 1980.

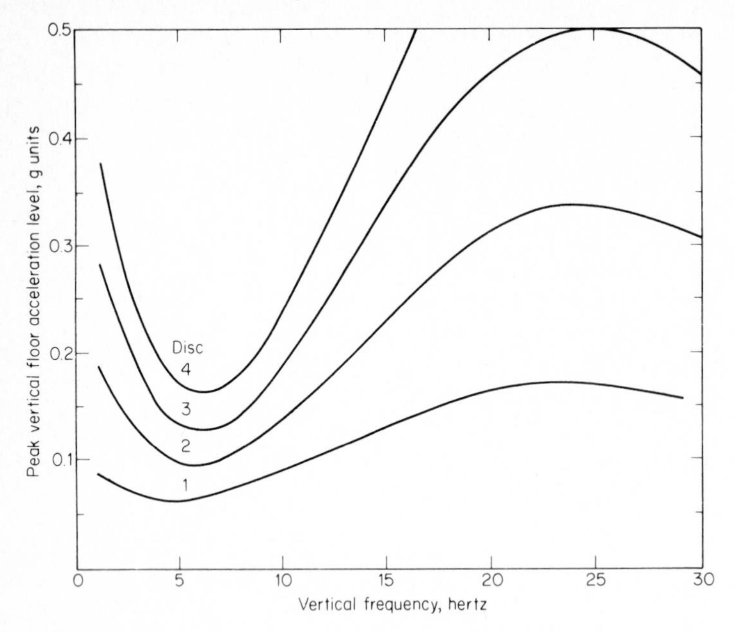

FIGURE 16-7
Equal discomfort curves for seated subjects experiencing sinusoidal vertical vibration. Discomfort is expressed in DISC units (see text for explanation). (*Source: Leatherwood, Dempsey, & Clevenson, 1980, fig. 4a.*)

and vibration environment, and not for predicting the discomfort of an individual passenger.

It is not easy to identify the bodily sensations that form the basis for comfort or discomfort judgments. Whitham and Griffin (1978) asked subjects to indicate which specific body locations were uncomfortable at various vibration frequencies (acceleration was held constant at 1.0 m/s^2 rms). In general, most responses of seated subjects implicated the lower abdomen at 2 Hz, moving up the body at 4 and 8 Hz, with most responses implicating the head at 16 Hz. At 32 Hz the responses were divided between the head and lower abdomen, while at 64 Hz they were mostly located near the principal vibration input site, i.e., the seat of the pants.

In actual vehicle environments, subjective evaluations of comfort depend on the expectations, anxiety, past experiences, and other psychological factors of the passengers. In addition to vibration, other physical environmental factors also impact on comfort evaluations. Richards and Jacobson (1977), for example, found that among airline passengers the presence of smoke, lighting, and work space were important factors in their comfort ratings.

Limits for Exposure to Whole-Body Vibration

Specifying the limits for exposure to whole-body vibration depends on the criterion adopted. The criterion can be based on comfort, task performance, or physiological response. The equal-comfort contours discussed above are, in essence, limits based on the criterion of comfort.

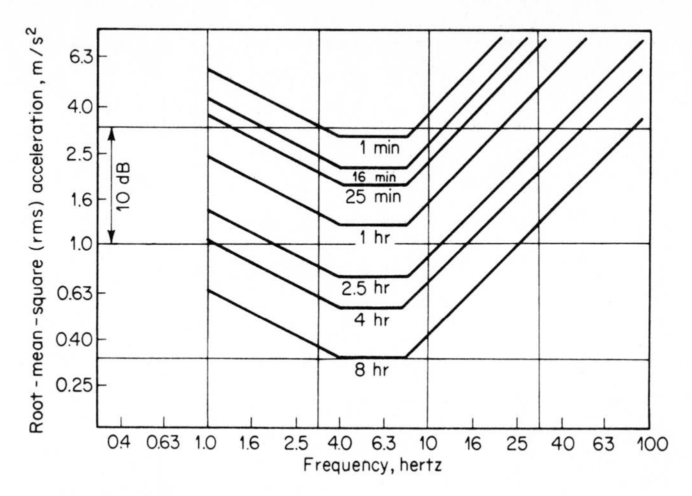

FIGURE 16-8
The fatigue-decreased proficiency boundary for vertical vibration contained in International Organization for Standardization (ISO) 2631. To obtain boundary of reduced comfort, subtract 10 dB; and to obtain boundary for safe exposure, add 6 dB. (*Source: International Organization for Standardization, 1974.*)

The International Organization for Standardization (ISO), after wrestling with the problem for 7 years, has developed a standard for exposure to whole-body vibration (ISO, 1974). The standard specifies the limits in terms of acceleration, frequency, and exposure duration. Limits are specified for comfort, task proficiency, and physiological safety. Figure 16-8 illustrates the *fatigue-decreased proficiency* boundaries for vertical vibration. The trough between 4 and 8 Hz corresponds to the resonance frequencies of the body (see Figures 16-4 and 16-5). Each line in Figure 16-8 represents an estimate of the upper limits that people generally can tolerate before fatigue effects would catch up with them. Amplitudes of about 10 dB less than those indicated tend to characterize the upper boundary of comfort, and the safe exposure limits are about 6 dB higher than the values shown (ISO).

These standards, although heralded as a major contribution, have nonetheless been criticized (Allen, 1971; Morrissey & Bittner, 1975; Sandover, 1979). Some of the criticisms are that (1) the comfort and fatigue-decreased proficiency limits for short time exposures, less than 1 h, may be too high; (2) the limits appear to be based on mean results and do not take into account variability in the population, i.e., what percent of the exposed population will experience fatigue? 50 percent? 100 percent?; and (3) the proposals imply that the effects of combinations of single-axis vibration are additive, yet there is strong evidence to the contrary (e.g., Leatherwood et al., 1980). The standards are a major step in the right direction, but will need additional modification as new research findings accumulate.

ACCELERATION

Acceleration of the human body is part and parcel of riding in any moving vehicle. In most vehicles, such as automobiles, trains, buses, or commercial aircraft, the levels of acceleration are moderate and the effects nominal. There are, however, vehicles that produce very high levels of acceleration for their occupants, such as high-performance jet aircraft and space rockets, with effects that can be of some consequence. Before discussing these, however, we should discuss certain basic concepts and terminology used to characterize acceleration.

Acceleration Terminology

Acceleration is a rate of change of motion of an object having some mass. Rate of change of motion is expressed as feet per second per second (ft/s^2) or as meters per second per second (m/s^2). The basic unit of acceleration, G, is derived from the force of gravity in our earthbound environment. The acceleration of a body in free fall is 32.2 ft/s^2 (9.81 m/s^2); this being $1G$. If a person is accelerated at $2G$, the body effectively weighs twice its normal amount.

Acceleration forces applied to a mass can be either linear or rotational. *Linear* acceleration is the rate of change of *velocity* of a mass, the direction of movement being kept constant. In turn, *rotational* acceleration is the rate of change of *direction* of a mass, the velocity of which is kept constant. There are two forms of rotational acceleration. One form, *radial* acceleration, is that in which the axis of rotation is external to the body (as in an aircraft turn). The other, commonly referred to as *angular* acceleration, is that in which the axis of rotation passes through the body (as when a ballet dancer is twirling around on a toe). In some circumstances people are subjected to an admixture of these.

Although rotational acceleration is important in some circumstances (as in certain aircraft maneuvers), we will here discuss only linear acceleration. Such acceleration can of course occur in any direction with respect to the human body. In this regard, if the human body is caused forcibly to change velocity, that is, to accelerate or decelerate, as in a vehicle that is changing its velocity, the change in velocity causes a physiological reactive force that is opposite the direction of movement. This reactive force is manifested by a displacement of the heart and other organs, body tissues, and blood, since these body components are not rigid. Thus, as an automobile accelerates rapidly, the internal organs and the blood tend to "lag behind" the structure of the body as the body is propelled forward at increasing velocity.

The three possible directions of linear motion are depicted in Figure 16-9, these being called x (forward-backward), y (left and right), and z (upward-downward). Acceleration forces in these directions are expressed in terms of $\pm G$ units, as shown in column 4 of Table 16-3. Column 2 shows, for each direction of motion of the body, the (opposite) direction of motion of the heart and other internal organs. In turn, column 3 gives a vernacular label for each direction of motion as described by the sensation of movement of the eyeballs, such as "eyeballs in" or "eyeballs down."

Rarely, if ever, is someone exposed to a simple unvarying linear acceleration in one direction. Instead, acceleration may vary in magnitude or direction, and may be accompanied by complex oscillations and vibration. For purposes of summary, however, it is simpler to consider the effects of each direction of acceleration separately without adding additional complexities.

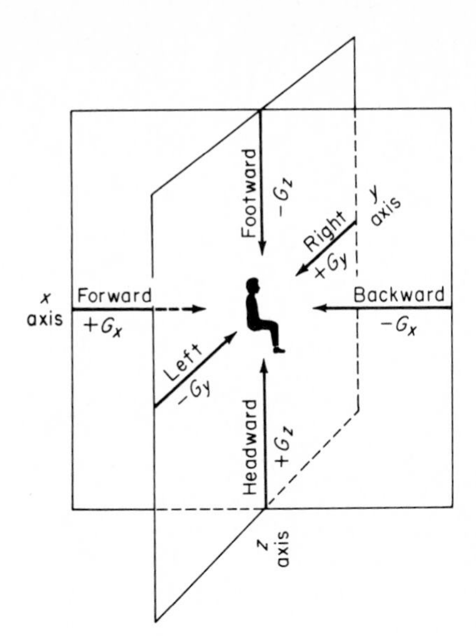

FIGURE 16-9
Illustration of three directions of linear acceleration. The direction of displacement of the heart and other organs is opposite that of the motion of the body.

TABLE 16-3
ACCELERATION TERMINOLOGY WITH RESPECT TO DIRECTION OF ACCELERATION

Acceleration			
1	2	3	4
Direction of motion	**Heart motion toward**	**Motion of eyeballs**	**Symbol**
Forward	Spine	In	$+G_x$
Backward	Sternum	Out	$-G_x$
To right	Left	Left	$+G_y$
To left	Right	Right	$-G_y$
Headward	Feet	Down	$+G_z$
Footward	Head	Up	$-G_z$

Source: Adapted from Hornick, 1973, table 7-1, p. 229.

Effects of Headward ($+G_z$) Acceleration

In headward acceleration, the force is acting down in a footward direction on the body. Thus, there is an increase in apparent body weight, a tendency for soft tissue to droop, and for blood to pool in the lower parts of the body. Figure 16-10 summarizes

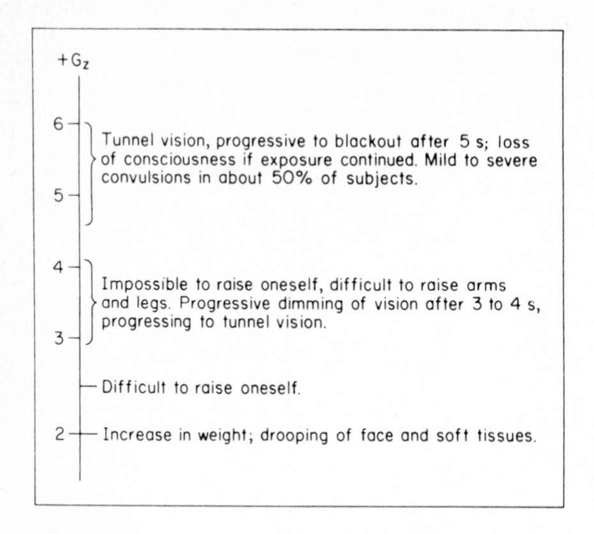

FIGURE 16-10
Summary of effects of headward, $+G_z$, acceleration. (*Source: Fraser, 1973, p. 150.*)

these effects as a function of acceleration magnitude. The predominant effects of headward acceleration are on gross body movement and vision.

Headward acceleration results in decreased ability to detect targets, especially in the peripheral visual field. Figure 16-11, for example, illustrates the change in threshold luminance, i.e., the minimum light intensity at which a stimulus can be perceived, for foveal and peripheral vision, as a function of $+G_z$ acceleration. Complete loss of peripheral vision generally occurs at about $+4.1G_z$, and loss of central vision at about $+5.3G_z$ (Zarriello, Norsworthy, & Bower, 1958).

FIGURE 16-11
Threshold of foveal and peripheral vision under $+G_z$ acceleration. (*Source: White, 1960, as shown in Fraser, 1973, figs. 4-5 and 4-6.*)

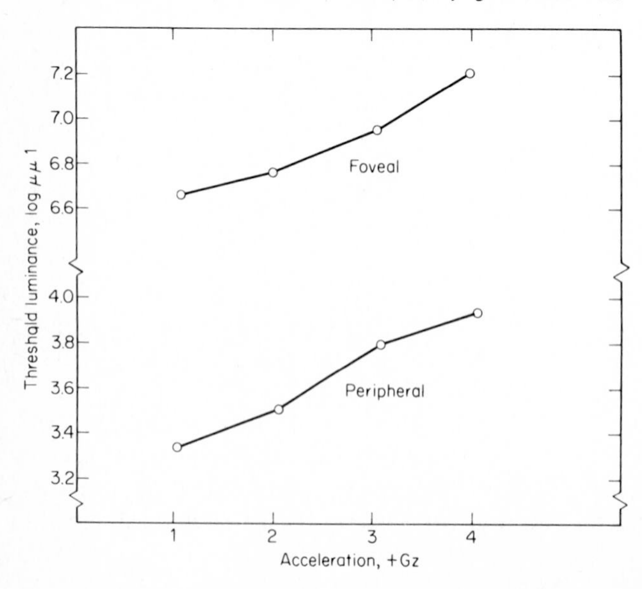

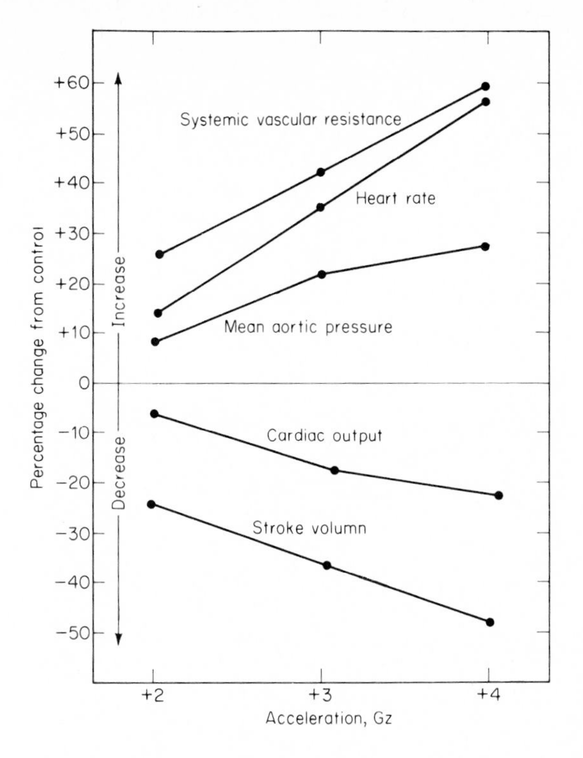

FIGURE 16-12
Cardiovascular response to headward acceleration, $+G_z$. (*Source: Wood, Sutterer, Marshall, Lindberg, & Headley, 1961.*)

These effects of $+G_z$ acceleration have consequences for performing tasks which involve movement, vision, or both, including such tasks as visual reaction time, reading, tracking, and certain higher mental processes (Fraser, 1973).

In addition to these effects, various cardiovascular and respiratory effects of headward acceleration have been noted (Fraser, 1973). For example, Figure 16-12 illustrates some cardiovascular changes occurring with various levels of $+G_z$ acceleration. Cardiac output and stroke volume decrease, while heart rate, aortic pressure, and systematic vascular resistance increase (Wood, Sutterer, Marshall, Lindberg, & Headley, 1961).

Effects of Footward ($-G_z$) Acceleration

Far less research has been conducted on the effects of footward acceleration, undoubtedly because it is far less common than headward acceleration. In footward acceleration, the force is acting up toward the head.

At $-1G_z$ there is unpleasant, but tolerable, facial congestion. At -2 to $-3G_z$, the facial congestion becomes severe. At these levels one experiences throbbing headaches, progressive blurring, graying, or occasionally reddening of vision after about 5 s. The limit of tolerance at $-5G_z$ is about 5 s, reached only by the most exceptional subjects (Fraser, 1973).

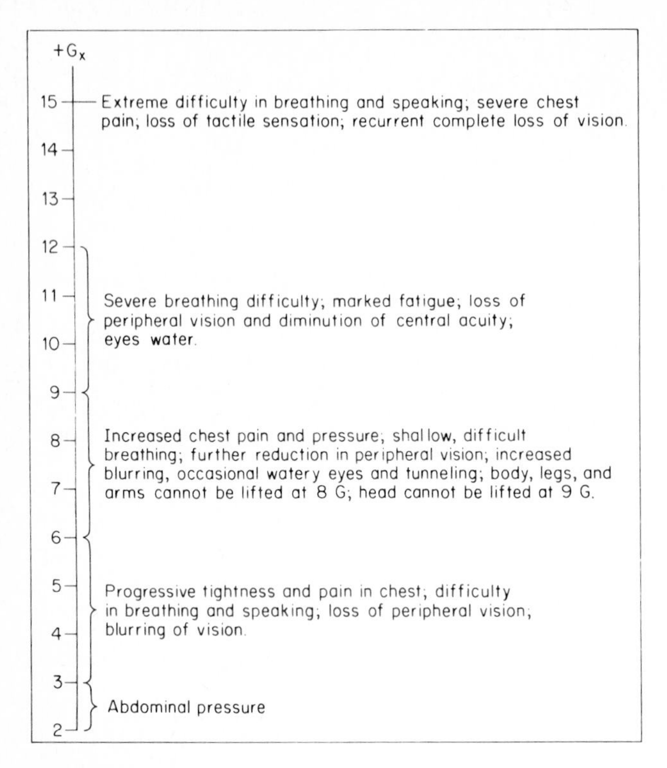

FIGURE 16-13
Summary of the effects of forward, $+G_x$, acceleration. (*Source: Fraser, 1973, p. 151.*)

Effects of Forward ($+G_x$) Acceleration

Seated subjects experiencing forward, $+G_x$, acceleration are pushed back in their chairs. Figure 16-13 summarizes some of the known effects of forward acceleration. A more detailed portrayal of the effects of vehicular (forward) acceleration $+G_x$ on gross body movements is shown in Figure 16-14, on the opposite page. This figure shows the typical levels of G that are near the threshold of ability to perform the acts indicated. Some of the implications for design decisions are fairly obvious; it would seem unwise, for example, to place the ejection-control lever of an airplane over the pilot's head.

In addition to the effects on gross body movement and vision, there is a marked effect on respiration. Figure 16-15 illustrates the effect of 120-s exposure to $+G_x$ acceleration on arterial oxygen saturation, i.e., the degree to which the blood leaves the lungs saturated with oxygen. At approximately $+6G_x$, one begins to experience sensory impairment, with mental impairment due to lack of oxygen occurring around $+8G_x$.

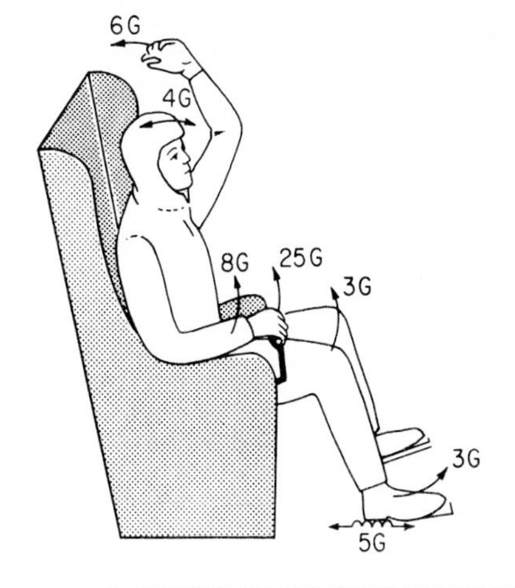

FIGURE 16-14
The forward, $+G_X$, forces that are near the threshold of various body movements. For any given motion indicated, the movement is just possible at the G_X forces indicated; greater G_X forces usually would make it impossible to perform the act. (*Source: Chambers & Brown, 1959.*)

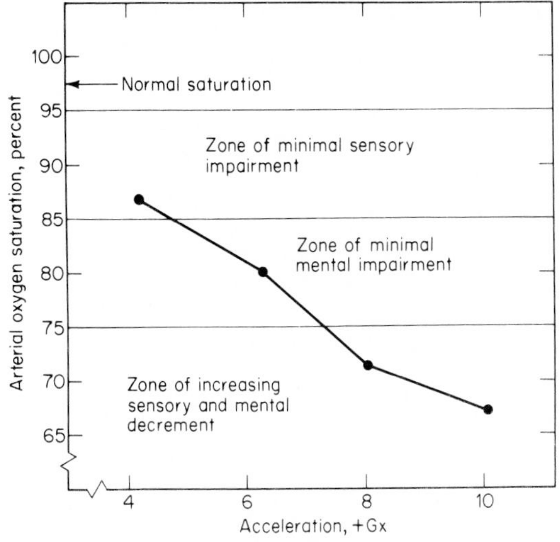

FIGURE 16-15
Arterial oxygen saturation after 120-s exposure to forward, $+G_X$, acceleration while breathing air. (*Source: Adapted from Alexander, Sever, & Hoppin, 1965.*)

Effects of Backward ($-G_X$) Acceleration

The effects of backward acceleration are similar to those experienced in forward acceleration. Chest pressure is reversed, making breathing easier. There still exists, however, pain and discomfort from outward pressure against the restraint harness being worn (and you don't want to be in such an environment without some type of upper body restraint). The effects on vision are similar for both forward and backward acceleration, but the reduction in arterial oxygen saturation is less in backward acceleration than in forward acceleration (Fraser, 1973).

Effects of Lateral ($\pm G_Y$) Acceleration

There is little information available on the effects of lateral acceleration (Fraser, 1973); but high levels of lateral acceleration are relatively rare in operational environments.

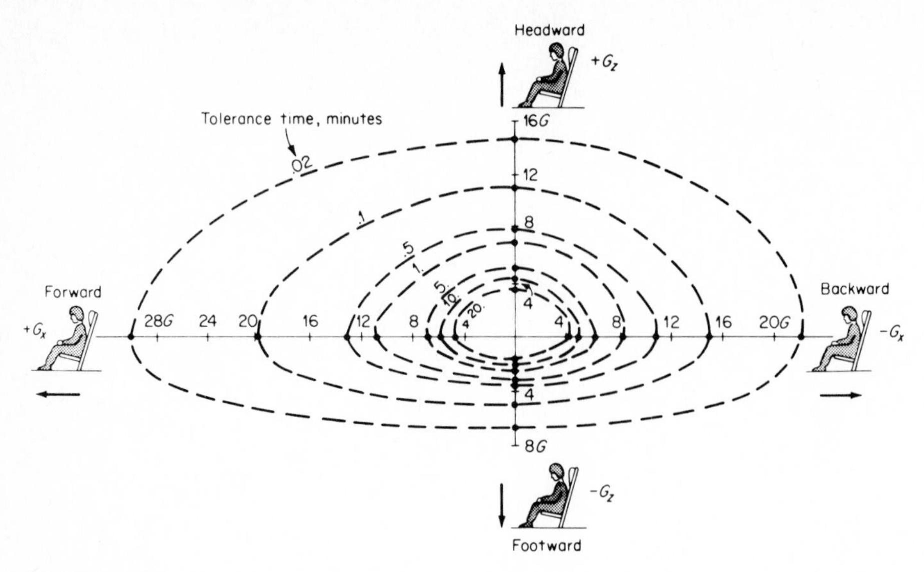

FIGURE 16-16
Average levels of linear acceleration, in different directions, that can be tolerated on a voluntary basis for specified periods of time. Each curve shows the average G load that can be tolerated for the time indicated. The data points obtained were actually those on the axes; the lines as such are extrapolated from the data points to form the concentric figures. (*Source: Adapted from Chambers, 1963, fig. 6., pp. 193–320.*)

Voluntary Tolerance to Acceleration

The physiological symptoms brought about by acceleration, of course, have a pretty direct relationship to the levels that people are willing to tolerate voluntarily. The physiological effects of differential levels of linear acceleration in various directions mentioned above are paralleled by somewhat corresponding differences in average voluntary tolerance levels, such as shown in Figure 16-16. This figure shows the average levels for the various directions that can be tolerated for specific times. It can be seen that tolerance to footward acceleration, $-G_z$, is least, followed by headward acceleration, $+G_z$, with forward acceleration, $+G_x$, being the most tolerable. Individual differences are of considerable magnitude; trained and highly motivated personnel frequently can endure substantially higher levels than the average.

Protection from Acceleration Effects

Although minor doses of acceleration (as in most land vehicles and most commercial airplanes) pose no serious problem either in safety and welfare or in performance, the effects of higher levels (especially with long exposures as in space travel) usually require some protective measures. One way to provide protection is to assume a posture which increases the tolerance to the direction of acceleration being experienced. Figure 16-17, for example, shows various body positions which resulted in the greatest tolerance to various directions of acceleration, as well as similar, but less advantageous, postures. Incidentally, with water immersion conditions the same acceleration forces

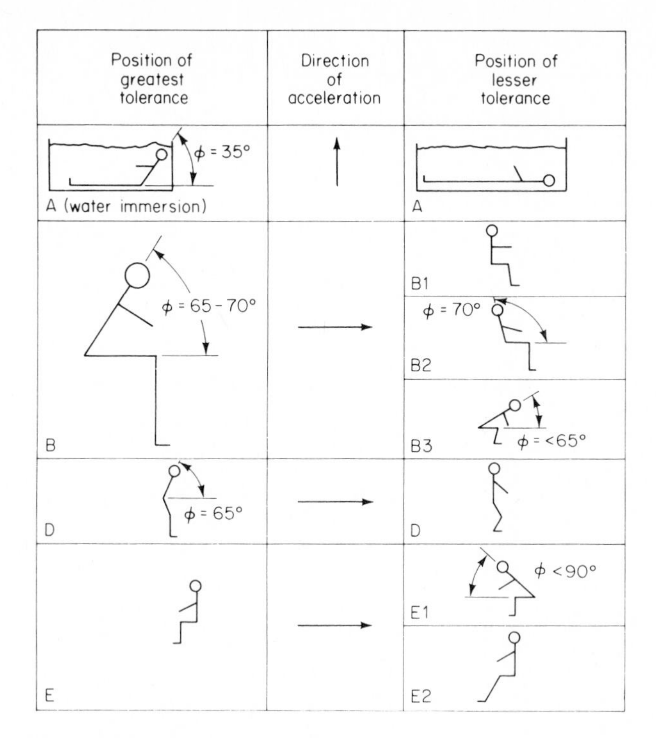

FIGURE 16-17
Comparison of various body postures which influence tolerance to acceleration. (*Source: Bondurant, Clark, Blanchard, et al., 1958, as adapted by Fraser, 1973, fig. 4-30.*)

affect the water as well as the person immersed in it. This has the effect of equalizing the forces over the parts of the body. Individuals so immersed can even move their body members under acceleration and can tolerate more acceleration than under any other known condition, especially in a supine position with about a 35° angle of the trunk to the line of force.

Immersion, although providing considerable protection, is generally not considered feasible in space flight because of the problem of lugging around a private swimming pool. Aside from the use of optimum postures, the presently available protection schemes include restraining devices, anti-*G* suits, and special body supports such as contour and net couches.

For high, sustained *G* loads in a forward direction, $+G_x$, the use of contour couches (sometimes molded for the individual) are reasonably satisfactory. Such couches, however, do not provide adequate support for the rearward direction, $-G_x$. For such purposes additional contraptions are required to provide adequate restraint, such as a restraint helmet and supporting face and chin pieces, and frontal supports (such as nylon netting) for the chest, torso, and body members; to further hem in the individual, an anti-*G* suit may be in order. Anti-*G* suits, however, generally have their greatest utility in connection with headward, $+G_z$, acceleration, with some designs being more effective than others (Nicholson & Franks, 1966). Anti-*G* suits typically apply pressure to the lower abdominal area and legs, thus reducing the tendency of the blood to drain to those areas during headward, $+G_z$, acceleration. There is no particularly effective scheme for providing protection against footward acceleration, $-G_z$, so if you insist on being shot out of a cannon, have them shoot you out headward, and don't forget your anti-*G* suit!

Deceleration and Impact

The normal deceleration of a vehicle imposes forces upon people that are essentially the same as those of actual acceleration, but in reverse. But acceleration usually is a gradual affair (except in unusual instances, such as with astronauts during blast-off), whereas deceleration can be extremely abrupt, especially in the case of vehicle accidents. When a vehicle hits a solid object or another vehicle head on, an unrestrained occupant will continue forward at the initial velocity until the body strikes some part of the interior of the vehicle, or is thrown clear from the vehicle. This is sometimes called the *second collision*. The deceleration of the occupant is a function of the deformation of the part of the interior which is hit by the occupant's body, if any. To be protected from this "second" collision, the occupant must be "tied" to the vehicle (as with restraining devices) and decelerate with the vehicle (which may be through a distance of a couple of feet or more), or be provided with some energy-absorbing object which results in more gradual deceleration by increasing the travel distance of the occupant during deceleration (air bags, collapsible steering wheels, etc.).

Restraining Devices

An occupant can be "tied" to a vehicle by quite a few types of restraining devices, in automobiles and commercial aircraft these nearly always being seat (lap) belts. In some automobiles shoulder harnesses are used by themselves or in combination with seat belts, and in military aircraft more complex restraints are used, individually or in combination. A major deficiency of automobile seat belts is, of course, the possibility of head damage caused by the torso and head being propelled forward with the body hinged at the hips. Considerably greater protection is provided by shoulder-lap belts, which aid the occupant in "riding down" the vehicle as it impacts. A general comparison of some of these effects is shown in Figure 16-18. (However, there is some suspicion that, under impact conditions, shoulder belts might apply excessive force on the spine.) Some indication of the protection provided by seat belts comes from a survey in California (Tourin & Garrett, 1960). The use of seat belts reduced the incidence of fatalities by 35 percent, largely because they prevented the ejection of the individual. Although the seat-belt wearers were injured with the same frequency as nonwearers, the degree of injury was lower for those who used seat belts.

Another method of "tying" passengers to a vehicle is by the use of rear-facing seats. It has been estimated that properly designed aft-facing seats could protect occupants against as much as 40-*G* impacts with little or no injury (Eiband, 1959). Such seats would be particularly feasible in aircraft, trains, etc., but are not completely unrealistic for use in automobiles (if we could adapt ourselves to them). In the case of vehicles with forward-facing seats involved in rear-end crashes, the use of a full back support with headrest minimizes the possibility of whiplash.

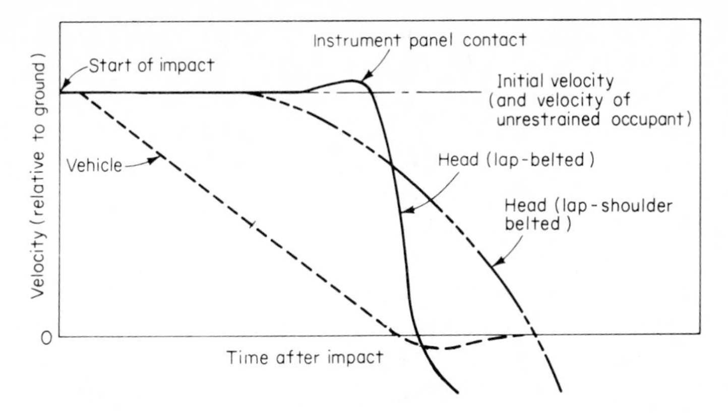

FIGURE 16-18
Velocity of vehicle occupants following impact with lap belt and shoulder-lap belt. The shoulder-lap restraint causes the upper torso and head to follow more closely the vehicle velocity change than does the seat belt alone. On impact, an unrestrained occupant would be propelled forward at the impact velocity until hitting some part of the interior. (*Source: Cichowski & Silver, 1968, fig. 1.*)

Energy-Absorbing Objects

The air bag seems destined to replace seat belts as the primary device for protection from vehicular head-on accidents. The air bag system is so designed that when the vehicle hits the back of another vehicle, the air bag is immediately inflated in front of the occupant and is then deflated. The effect is that of absorbing the energy of the occupant being catapulted forward upon the impact. In other words, the deceleration takes place over a longer travel distance than if the occupant were propelled forward at the vehicle velocity at the time of impact until striking a rigid object such as the windshield or the steering wheel.

WEIGHTLESSNESS

Since human beings have evolved in an earthbound environment the force of gravity has been our constant companion. Its constant presence has influenced our physiological makeup and is basic to all of our activities. Except for the case of a few roller coasters, it is only in space flight and in certain aspects of aircraft flight that the "natural" phenomenon of weightlessness or reduced gravity is experienced. For those few who do venture into outer space, there are two aspects of the weightlessness or reduced-weight state that are particularly important (Berry, 1973). The first of these is the absence of weight itself; the removal from the normal gravitational environment could be expected to have an impact on the human organism, such as on its physiological functioning and on perceptual-motor performance and sensory performance. In the second place the weightless condition is accompanied by a tractionless condition when moving and working.

Since the weightless condition does not exist in our earthbound environment, various schemes have been developed for creating such a condition, or simulating it, in order to study its effects. True weightlessness can be achieved within the earth's gravitational field by the use of aircraft in a parabolic flight maneuver. During the maneuver (up and down) a weightless condition can be achieved for upwards of 1 min (Moran, 1969).

Various forms of partial simulation of the weightless state include cable suspension rigs (which reduce the friction between persons and the surface on which they are operating), gimbals, air bearings, and water immersion (to achieve neutral buoyancy) (Deutsch, 1969). Although some such methods have provided some inklings about the effects of weightlessness, the most reliable basis for estimating the long-term effects is undoubtedly actual experience in orbital space flights. We are accumulating a wealth of such information, as American and Russian space travelers have already spent tens of thousands of hours in space. To date, the longest flight has been 185 days in the Soviet Salyut 6 space station.

Physiological Effects of Weightlessness

In summarizing the early space flight experience of both the United States and the Soviet Union, Berry (1973) indicates that although some physiological changes have been consistently noted, none of them have been permanently debilitating. Some of the temporary effects that have been observed include aberrations in cardiac electrical activity; changes in the number of red and white blood cells; loss of muscle tone; and loss of weight.

Reason (1974), summarizing more recent space flight experience, reports cases of *space sickness* ranging from mild sensations of tumbling to serious cases of prolonged nausea and vomiting. Homick (1979), however, reports that the symptoms generally disappear after 3 to 5 days, but can inhibit task performance during that time.

Thornton (1978) reports various anthropometric changes that occur in a weightless environment. Space travelers grow in height by approximately 3 percent (approximately 2 in or 5 cm). This has implications for, among other things, pressure-suit design and control stations with critical eye level requirements. Even more significant is the natural relaxed posture assumed in a weightless environment as shown in Figure 16-19. Notice the lower line of sight under zero-*G* conditions, the angled foot, and the height of the arms. Seats, work stations, etc. must be designed to accommodate this unique posture. Thornton reports that the body "rebels" with fatigue and discomfort against any attempt to force it into a more normal earthbound posture.

Performance Effects of Weightlessness

Locomotion within a spacecraft adds a third dimension to our usual two-dimensional movements. Apparently no serious problems have occurred in normal locomotion within spacecraft (Berry, 1973). Although there has been serious consideration given to generating artificial gravity in spacecraft (by rotating the entire space vehicle or station, or by having an on-board centrifuge), the experience of United States astronauts has so far suggested that this would not be necessary.

Performance outside a spacecraft in extravehicular activities (EVA) is a bit of a different story, since some astronauts have experienced considerable exhaustion in such

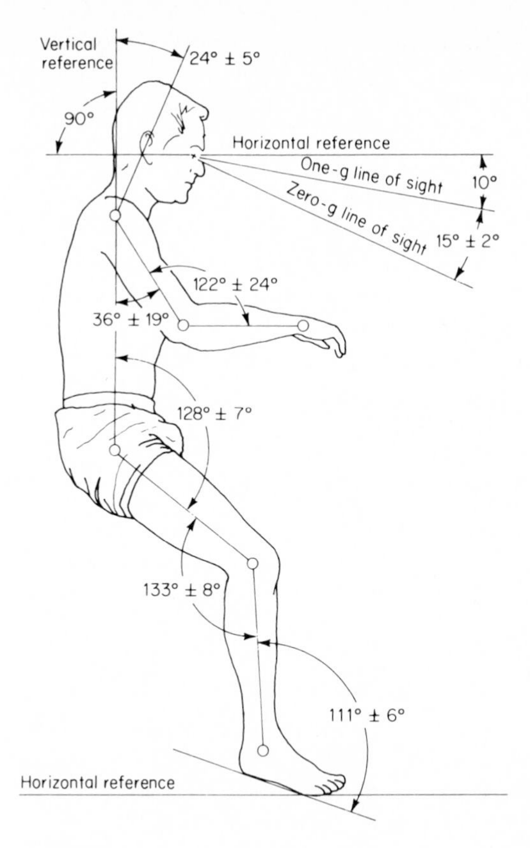

FIGURE 16-19
Typical relaxed posture assumed by people in weightless conditions. (*Source: Thornton, 1978, fig. 16.*)

activities. As Berry points out, this probably argues for careful consideration of work load in planning EVA missions.

ILLUSIONS DURING MOTION

When humans are in motion, they receive cues regarding their whereabouts and motion from sense organs, especially the semicircular canals, the vestibular sacs (the otolith organs), the eyes, the kinesthetic receptors, and the cutaneous senses. These sense organs and the intricate interactions among them were specifically designed for a self-propelled terrestrial animal. They were not intended to cope with the unusual and prolonged forces encountered in three-dimensional flight. Such forces can push our sense organs beyond their capabilities to function accurately and cause them to signal erroneous information concerning the position and motion of the body.

Pilots in airplanes sometimes experience *disorientation* regarding their position or motion with respect to the earth. Actually there are numerous phenomena which cause disorientation; some give rise to sensations that the person is spinning (true vertigo) and some cause dizziness (a feeling of movement within the head). Indeed, the three terms *disorientation, vertigo,* and *dizziness* are usually, though inaccurately, used interchangeably to describe a variety of symptoms such as false sensations of turning, linear velocity, or tilt (Kirkham, Collins, Grape, Simpson, & Wallace, 1978).

Disorientation can be the basis for fatalities in an aircraft environment. Kirkham et al. (1978), for example, report that disorientation was the cause of 16 percent of all fatal general aviation aircraft accidents in the United States. Reason (1974) reports that virtually all pilots experience some form of disorientation at least once in their flying careers—for some, it is their last.

There are a multitude of disorienting effects, and we cannot hope to present or discuss them all. To understand some of them requires a knowledge of physics and physiology that is more than we wish to tackle here.

Reason (1974) distinguishes two types of disorientation effects: those arising from false sensations and those arising from misperception.

Disorientation from False Sensations

The mental confusion or misjudgment resulting from false sensations is due to inaccurate or inappropriate sensory information coming from the semicircular canals, otolith organs, or both. Those that result from false sensations from the semicircular canals involve angular acceleration cues. One such situation occurs when a pilot executes a roll at an angular acceleration that is below the threshold of perception of the semicircular canals. The pilot is not aware of the full extent of the roll and so overcorrects when attempting to reestablish straight and level flight. The aircraft ends up in a bank turned in the opposite direction, but the pilot thinks it is flying straight and level.

Another example occurs after recovering from a spin; a pilot will often think the aircraft is actually spinning in the opposite direction. Attempting to recover from this false spin will often send the airplane into another real spin. Pilots call this the *graveyard spin*, and for good reason.

Another form of disorientation arising from inaccurate semicircular canal information is called the *Coriolis illusion* or *cross-coupling effect*. This can occur when the head is tilted during a long-established turning or circling maneuver. The head must be tilted in a different plane from the plane of the aircraft's turn. The experience is an illusion of roll, often accompanied by dizziness.

False sensations from the otoliths often give rise to false sensations of tilt. The *oculogravic illusion* is one example. Whenever an aircraft's forward speed is suddenly increased (i.e., accelerated), the acceleration force vector combines with the gravity force vector to fool the otoliths into signaling that the body (and hence the aircraft) is tilted in a "nose-up" attitude. The eyes also roll back, thus adding to the impression of being tilted. Attempting to level the aircraft (when in fact it is already level) can send it into the ground rather abruptly. Incidentally, decelerating often gives rise to the sensation of tilting nose-down.

It should be pointed out that these false-sensation forms of disorientation tend to occur under conditions of poor visibility, e.g., at night, in clouds, or in fog. It is for this reason that pilots are trained to trust their instruments rather than their senses. Usually a visual reference can reduce or eliminate disorientation, but not always.

Disorientation from Misperception

These forms of disorientation arise because the brain misinterprets or misclassifies perfectly accurate sensory information, usually provided by the visual sense. One of these is *autokinesis*, in which a fixed light appears to move against its dark background. It has been reported that pilots have attempted to "join up" in formations with stars, buoys, and street lights which appear to be moving (Clark & Graybiel, 1955). Another common visual misperception occurs under conditions of limited visibility wherein a pilot will accept a sloping cloud bank as an indication of the horizontal and, as a result, align the wings with the cloud bank and fly along at a "rakish" tilt.

Discussion

The reduction of disorientation and illusions generally depends more on procedural practices and training than upon the engineering design of aircraft. Among such practices are the following (Clark & Graybiel, 1975): understanding the nature of various illusions and the circumstances under which they tend to occur, maintaining either instrument *or* contact flight, avoiding night aerobatics, shifting attention to different features of the environment, learning to depend upon the correct cue to orientation (such as using visual cues or instruments), avoiding sudden accelerations and decelerations at night, and avoiding prolonged constant speed turns at night.

MOTION SICKNESS

We would be remiss if we failed to include *motion sickness* in a chapter on humans in motion.[3] Although motion sickness may never actually kill us, there are times when we wish it would! Motion sickness, of course, is not really a sickness in the pathological sense at all; it only makes us "feel sick." Motion sickness is associated with most forms of travel—cars, boats, trains, and even camels, but not horses. (We will explain the camel-horse paradox shortly.) People differ in their susceptibility to motion sickness, but that susceptibility is a relatively stable and enduring characteristic of the individual (Reason, 1974). Further, if you are susceptible to one type of motion sickness you are probably susceptible to all types except, for some unexplained reason, *space sickness*.

The symptoms of motion sickness are familiar to us all. Reason (1974) categorizes them into two classes, head and gut. *Head symptoms* include drowsiness and a general apathy, together called the *sopite syndrome* by Graybiel and Krepton (1976). Headaches are also experienced. *Gut symptoms* range from a disconcerting awareness of the stomach region to acute nausea and vomiting (or, more politely, *emesis*).

The most widely accepted theory of motion sickness, called *sensory rearrangement theory* (Reason & Brand, 1975), considers motion sickness as a consequence of incongruities among the spatial senses, i.e., the organs of balance, the eyes, and the nonvesti-

[3] See Money (1970) for an excellent review of the motion sickness literature published before 1970.

bular position senses (joints, muscles, and tendons). The incongruity among the senses is incompatible with what we have come to expect on the basis of our past experience. Reason (1974, 1978) points out that the vestibular system, i.e., the semicircular canals and otolith organs, must be implicated for motion sickness to be an outcome. And, since the vestibular system responds only to accelerations (linear and angular), acceleration (or apparent acceleration) must always be involved in motion sickness.

Reason (1974, 1978) identifies two classes of sensory rearrangement: *visual-inertia rearrangements*, and *canal-otolith rearrangements*. Further, these rearrangements, or incongruities, are of two types: (1) both systems (e.g., canal and otolith) simultaneously signal contradictory information; and (2) one system signals in the absence of an expected signal from the other. For example, some people can get motion sickness from watching a movie filmed from inside a roller-coaster. In this case, the visual system senses accelerations, but the semicircular canals and otoliths sense no motion.

Low-frequency oscillations of less than 1 Hz, as in the up-and-down motion of a ship, usually cause motion sickness. Reason and Brand (1975) believe this is because the otolith signals are out of phase with the semicircular canal signals; however, the exact mechanism is unclear. McCauley and Kennedy (1976) summarized the results of exposing over 500 subjects to 2 h of vertical sinusoidal vibration of various combinations of frequency (<1 Hz) and acceleration. The dependent variable was the percentage of subjects that vomited; not the kind of research you dream about doing after graduate school! Figure 16-20 shows the percentage of subjects vomiting at various

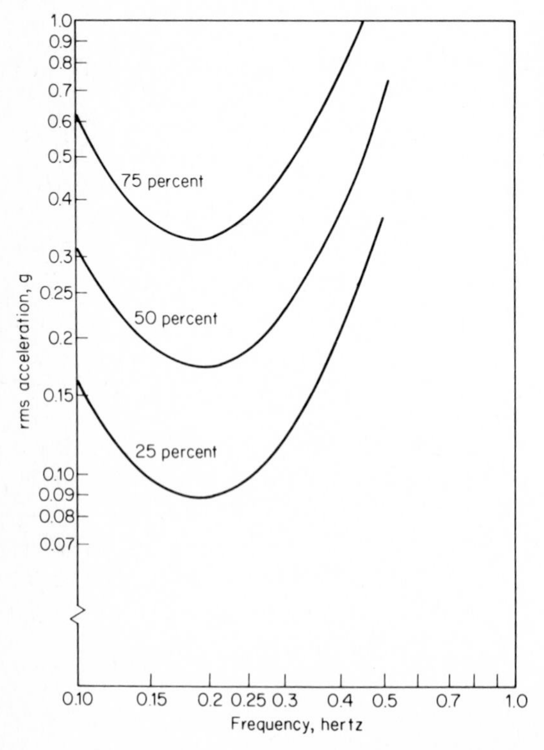

FIGURE 16-20
Equal motion sickness contours, based on percent emesis within 2 h for 500 male subjects exposed to vertical sinusoidal motion. (*Source: McCauley & Kennedy, 1976, fig. 4.*)

frequency-acceleration combinations. Vibration in the 0.15 to 0.25 Hz (i.e., 9 to 15 cycles per minute) range is, for some reason, especially effective for inducing gut symptoms of motion sickness. Incidentally, McCauley and Kennedy indicate that those who are going to get sick generally do so within the first hour when exposed to very regular (sinusoidal) motion.

Were you wondering about the camel-horse paradox? It has nothing to do with whether horses smell better than camels; actually, the answer lies in Figure 16-20. The characteristic frequency of a camel's gait happens to be in the 0.2 Hz range, while a horse tends to oscillate at higher frequencies.

You will recall that frequency, amplitude, and acceleration are all related in sinusoidal vibration. For very low-frequency vibration relatively large amplitudes are required to achieve relatively low levels of acceleration. For example, a 0.2 Hz vibration, having a 0.09 rms *g*, would have a peak-to-peak displacement of 5.2 ft (1.58 m).[4]

Now that we know what causes motion sickness, what can we do to reduce its effects? Reason (1974) lists several practical suggestions, including getting off the boat. In the case of seasickness, lying down on your back will often help; holding your head still will also reduce the symptoms. There are several drugs available which also seem to be effective against motion sickness. Most people will eventually adapt to the motion environment and "get their sea legs" in a few days. NASA uses *adaptation schedules* in which space crews are exposed to sickness-inducing stimulation in graded doses, the idea being to build up tolerance (Reason, 1974). We can all add a few "home grown" preventative measures to the list.

One last point worth mentioning involves motion sickness and ride quality. Obviously a major source of ride discomfort is motion sickness, yet it is rarely addressed in the "ride quality" literature. Look back at the equal-discomfort curves in Figure 16-7 and the ISO recommended criteria shown in Figure 16-8; in both cases the curves start at 1 Hz with no mention of discomfort below 1 Hz. Future research in ride quality will have to address the region below 1 Hz to provide comprehensive data on ride comfort.

DISCUSSION

The common denominator of the exotic experiences of astronauts in outer space, of the more mundane processes of moving around in automobiles or other vehicles, and of the operation of tractors and other related items of equipment is the fact that the human body is being moved somehow, e.g., accelerated, decelerated, transported, jostled, or shaken. Although some forms and degrees of such shuffling are not of any particular moment, other forms or degrees obviously have potentially adverse effects in terms of physiological, performance, or subjective criteria. The challenge to the human factors disciplines is that of so designing the physical systems or protective devices in question as to minimize such effects, and where they cannot be eliminated, to design the system to match the reduced capabilities of the operator and to make whatever motion is involved more tolerable for passengers.

[4] Peak-to-peak displacement (in) for a sine wave = $(27.675 \times \text{rms}\, g)/[\text{frequency (Hz)}^2]$.

REFERENCES

Alexander, W., Sever, R., & Hoppin, F. *Hypoxemia induced by sustained forward acceleration in pilots breathing pure oxygen in a five psi absolute environment* (NASA-TX-X-51649). Washington, D. C.: National Aeronautics and Space Administration, 1965.

Allen, G. Human reaction to vibration. *Journal of Environmental Science,* September/December 1971, pp. 10–15.

Berry, C. A. Weightlessness. In *Bioastronautics data book* (2d ed.) (NASA SP-3006). Washington, D. C.: National Aeronautics and Space Administration, U. S. Government Printing Office, 1973.

Bondurant, S., Clark, N. P., Blanchard, W. G., et al. *Human tolerance to some of the accelerations anticipated in space flight* (TR 58-156). Ohio: Wright-Patterson Air Force Base, Wright Air Development Center, 1958.

Chambers, R. M. Operator performance in acceleration environments. In N. M. Burns, R. M Chambers, & E. Hendler (Eds.), *Unusual environments and human behavior.* New York: The Free Press, 1963.

Chambers, R. M., & Brown, J. L. *Acceleration.* Paper presented at Symposium on Environmental Stress and Human Performance. American Psychological Association, September 1959.

Cichowski, W. G., & Silver, J. N. *Effective use of restraint systems in passenger cars* (SAE 680032). Detroit, Michigan: Paper presented at Automotive Engineering Congress, Society of Automotive Engineers, January 8–12, 1968.

Clark, B., & Graybiel, A. *Disorientation: a cause of pilot error* (Research Project NM 001 110 100.39). U. S. Navy School of Aviation Medicine, March 2, 1955.

Deutsch, S. Preface to special issue on reduced-gravity simulation. *Human Factors,* 1969, *5*(11), 415–418.

Eiband, A. M. *Human tolerance to rapidly applied acceleration* (NASA Memo 5-19-59E). Washington, D. C.: National Aeronautics and Space Administration, 1959.

Fraser, T. Sustained linear acceleration. In *Bioastronautics data book* (2d. ed.) (NASA-SP-3006). Washington, D. C.: National Aeronautics and Space Administration, U. S. Government Printing Office, 1973.

Gell, C., & Moeller, G. The biodynamics aspects of low altitude high speed flight. *Ergonomics,* 1972, *15,* 655–670.

Graybiel, A., & Krepton, J. Sopite syndrome: A sometimes sole manifestation of motion sickness. *Aviation, Space, and Environmental Medicine,* 1976, *47,* 873–882.

Guignard, J. Vibration. In J. Gillies (Ed.), *A textbook of aviation physiology.* Oxford, England: Pergamon Press, 1965.

Homick, J. Space motion sickness. *Acta Astronautica,* 1979, *6,* 1259–1272.

Hornick, R. Vibration. In *Bioastronautics data book* (2d ed.) (NASA Sp-3006). Washington, D. C.: National Aeronautics and Space Administration, U. S. Government Printing Office, 1973.

International Organization for Standardization. *Guide for the evaluation of human exposure to whole-body vibration* (ISO 2631). 1974.

Kirkham, W., Collins, W., Grape, P., Simpson, J., & Wallace, T. Spatial disorientation in general aviation accidents. *Aviation, Space, and Environmental Medicine,* 1978, *49,* 1080–1086.

Leatherwood, J., Dempsey, T., & Clevenson, S. A design tool for estimating passenger ride discomfort within complex ride environments. *Human Factors,* 1980, *22,* 291–312.

McCauley, M., & Kennedy, R. *Recommended human exposure limits for very low frequency vibration* [TP-76-36(U)]. Point Mugu, Calif.: Pacific Missile Test Center, 1976.

Money, K. Motion sickness. *Physiological Reviews,* 1970, *50*(1), 1-38.

Moran, M. J. Reduced-gravity human factors research with aircraft. *Human Factors,* 1969, *11*(5), 463-472.

Morrissey, S., & Bittner, A. *Effects of vibration on humans: Performance decrements and limits* (TP-75-37U). Point Mugu, Calif.: Pacific Missile Test Center, 1975.

Nicholson, A. N., & Franks, W. R. Devices for protection against positive (long axis) acceleration. In P. I. Altman & D. S. Dittmer (Eds.), *Environmental biology* (Tech. Rep. 66-194). Aerospace Medical Research Laboratory, November 1966, pp. 259-260.

Oborne, D. A critical assessment of studies relating whole-body vibration to passenger comfort. *Ergonomics,* 1976, *19*, 751-774.

Poulton, E. Increased vigilance with vertical vibration at 5 Hz: An alerting mechanism. *Applied Ergonomics,* 1978, *9*, 73-76.

Radke, A. O. *Vehicle vibration: man's new environment* (paper 57-A-54). American Society of Mechanical Engineers, December 3, 1974.

Reason, J. *Man in motion: The psychology of travel.* London: Weidenfeld and Nicolson, 1974.

Reason, J. Motion sickness: Some theoretical and practical considerations. *Applied Ergonomics,* 1978, *9*, 163-167.

Reason, J., & Brand, J. *Motion sickness.* London: Academic Press, 1975.

Richards, L., & Jacobson, I. Ride quality assessment III: Questionnaire results of a second flight programme. *Ergonomics,* 1977, *20*, 499-519.

Sandover, J. A standard on human response to vibration—one of a new breed? *Applied Ergonomics.* 1979, *10*, 33-37.

Simons, A. K., Radke, A. O., & Oswald, W. C. *A study of truck ride characteristics in military vehicles* (Report 118). Milwaukee: Bostrom Research Laboratories, March 15, 1956.

Stephens, D. Developments in ride quality criteria. *Noise Control Engineering,* 1979, *12*, 6-14.

Thornton, W. Anthropometric changes in weightlessness. In Anthropology Research Staff (Eds.), *Anthropometric source book, Vol. I: Anthropometry for designers* (NASA RP-1024). Houston, Tex.: National Aeronautics and Space Administration, 1978.

Tourin, B., & Garrett, J. W. *Safety belt effectiveness in rural California automobile accidents.* Automotive Crash Injury Research of Cornell University, February 1960.

Troup, J. Driver's back pain and its prevention. *Applied Ergonomics,* 1978, *9*, 207-214.

Wasserman, D. Bumps, grinds take toll on bones, muscles, mind. *Health and Safety,* 1975, *45*, 19-21.

White, W. *Variations in absolute visual threshold during acceleration stress* (WADC-Tech. Rep.-60-34). Ohio: Wright-Patterson Air Force Base, 1960.

Whitham, E., & Griffin, M. The effects of vibration frequency and direction on the location of areas of discomfort caused by whole-body vibration. *Applied Ergonomics,* 1978, *9*, 231-239.

Witkin, H. A. The perception of the upright. *Scientific American,* February 1959.

Wood, E., Sutterer, W., Marshall, H., Lindberg, E., & Headley, R. *Effects of headward and forward accelerations on the cardiovascular system* (WADC-Tech. Rep.-60-634). Ohio: Wright-Patterson Air Force Base, 1961.

Zarriello, J., Norsworthy, M., & Bower, H. *A study of early grayout thresholds as an indicator of human tolerance to positive radial acceleratory force* (Project NM-11-02-11). Pensacola, Fla.: Naval School of Aviation Medicine, 1958.

PART SIX

HUMAN FACTORS: SELECTED TOPICS

CHAPTER **17**

APPLICATION OF HUMAN FACTORS DATA

Some of the previous chapters have dealt in part with the human factors aspects of the design of certain types of things people use (displays, controls, hand tools and devices, etc.), and of the environments in which people work and live. This chapter deals with some of the procedures and problems involved in the actual application of human factors data to the design of the things people use and to their environments.[1] The following chapters will include some examples of human factors research and the application of human factors in certain specific areas.

THE NATURE OF HUMAN FACTORS DATA

The human factors data that are available are based on varying combinations of research results, experience, and expert judgments. (Such data can be found in various types of sources, such as some of the references given in this book, especially those in Appendix C. Some illustrative data are also given in this book.)

TYPES OF HUMAN FACTORS DATA

The data relevant to various human factors design problems cover a wide span (with many gaps yet to be filled) and exist in many forms, such as the following:

[1] For a more thorough treatment of the processes of applying human factors data to practical engineering and design problems the reader is referred to Meister (1971), and Meister and Rabideau (1965).

- Common sense and experience (such as the designer has in memory, some of which may be valid, and some not)
- Comparative quantitative data (such as relative accuracy in reading two types of visual instruments, as illustrated in Chapter 4)
- Sets of quantitative data (such as anthropometric measures of samples of people, as given in Chapter 11, and error rates in performing various tasks, as illustrated in Chapter 3)
- Principles (based on substantial experience and research, that provide guidelines for design, such as the principle of avoiding or minimizing glare when possible)
- Mathematical functions and equations (that describe certain basic relationships with human performance, such as in certain types of simulation models)
- Graphic representations (nomographs or other representations, such as tolerance to acceleration of various intensities and durations)
- Judgment of experts
- Design standards or criteria (consisting of specifications for specific areas of application, such as displays, controls, work areas, and noise levels)

Much of the research in human factors is directed toward the development of design standards or criteria (sometimes called *engineering standards*) that can be used directly in the design of relevant items. The recommendations regarding the design of control devices in Table B-2 of Appendix B represent examples of such standards. Such design specifications sometimes are incorporated into the form of check lists; in the case of military services, they are incorporated into military specifications called MIL-SPECS.

CONSIDERATIONS IN APPLYING HUMAN FACTORS DATA

The application of relevant human factors data to some design problems is a fairly straightforward proposition and in some circumstances can even be done by computers. More generally, however, it is necessary to exercise judgment in this process–in particular, in evaluating the applicability of potentially relevant data to specific design problems. In making such evaluations there are at least four considerations that should be taken into account.

Practical Significance

One of these deals with the practical significance of the application of some relevant human factors information. For example, although the time required to use control device A might be significantly less than for device B, the difference might be so slight, and of such nominal utility, that it would not be worthwhile to use device A, especially if other factors (such as cost) argued against it.

Extrapolation to Different Situations

Much of the available human factors data has been based on research findings or experience in certain specific settings. The second consideration in the application of human factors data deals with possible extrapolation of such data to other settings. For example, could one assume that the reaction time of astronauts in a space capsule would be the same as in an earthbound environment? Referring specifically to laboratory studies, Chapanis (1967) urges extreme caution in applying the results of such studies to real-world problems.

Perhaps three points might be made about these sobering reflections. In the first place, despite Chapanis's words of caution, there are some design problems for which available research data or experience are fully adequate to resolve the design problem. In the second place, in certain areas of investigation it may be possible to "simulate" the real world with sufficient fidelity to derive research findings that can be used with reasonable confidence. In the third place, there are undoubtedly circumstances in which one might extrapolate from data that are not uniquely appropriate to the design problem at hand, on the expectation that this would increase the odds of achieving a better design than if such data were not used. Clearly this strategy involves risks, and should be used only in the case of noncritical design problems.

Consideration of Risks

A third consideration in applying human factors data is associated with the seriousness of the risks in making bad guesses. Bad guesses in designing a capsule for shipping an astronaut to the moon obviously would be a matter of greater concern than those in designing a hat rack. The more serious the risks, the greater is the need for relevant, high-quality data.

Consideration of Trade-Off Function

Still a fourth consideration is trade-off values. Since it frequently is not possible to achieve an optimum in all possible criteria for a given design of a man-machine system, some give-and-take must be accepted. The possible payoff of one feature (as suggested by the results of research) may have to be sacrificed, at least in part, for some other more desirable payoff.

HUMAN FACTORS IN THE DESIGN PROCESS

In the consideration of human factors in the design of equipment, facilities, and other physical items that people use, there are certain basic stages or processes that typically have to be carried out. If the item or system is very complex (such as a new aircraft or a petroleum refinery), these processes usually are highly organized and elaborate; whereas in the design of very simple items (such as a new type of egg beater or hedge trimmer), these processes may be very informal, and in some instances certain stages may be completely irrelevant.

A representation of certain of the human-factors-related functions in system design is given in Figure 17-1. This representation would apply in the case of a complex system, such as certain military systems. In the development of certain systems (especially in the military services) the human factors functions extend into various personnel-

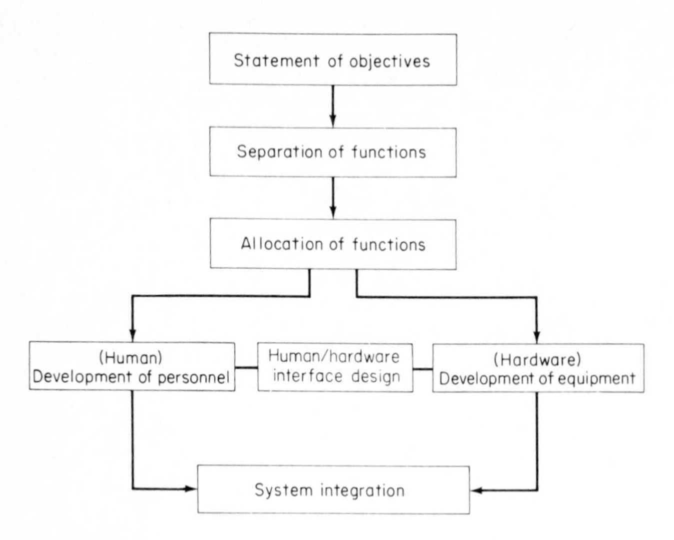

FIGURE 17-1
Representation of certain human-factors-related functions involved in systems design. (*Source: Applied Ergonomics Handbook, chap. 15, fig. 15.5, p. 156.*)

related areas such as job design and task analysis, job specifications, selection criteria, training, and staff planning.

We will touch on certain of the human-factors-related functions involved in design processes, recognizing that all of these would not apply on an across-the-board basis.

Specification of System Functions

A system is designed to serve some purpose. In some instances—especially in the case of complex systems—the performance requirements of the system should be formally set forth before the design process begins. For example, in the design of an urban transportation system there might be some specification of the number of passengers to be served, the maximum desired travel time, and the locations of loading and unloading stations. Such specifications need to take into account current technical feasibility as well as the desired objectives.

Given the objectives of a system (especially a complex one), one usually then determines the functions that need to be performed in order to achieve the objectives. In the case of an urban transportation system, for example, some of the functions that would need to be performed include the sale of tickets, the collection of tickets, the movement of passengers to loading platforms, opening and closing of vehicle doors to permit entrance and exit of passengers (possibly taking into account handicapped passengers), vehicle control, and movement of passengers to exits of stations, along with providing relevant information and directions for passengers and a communication system for use in the control and spacing of vehicles. This function analysis initially should be concerned with *what* functions need to be performed to fulfill the objectives and not with the *way* in which they are to be performed (such as whether they are to be performed by individuals or by machine components).

Allocation of Functions

Given functions that have to be performed, in some instances there may be an option as to whether any particular function should be allocated to a human being or to some physical (machine) component or components. In this process, the allocation of certain functions to human beings, and of others to physical components, is virtually predetermined by certain manifest considerations, such as obvious superiority of one over the other or economic considerations. Between these two extremes, however, may be a range of functions that are within the reasonable repertoire of both human beings and physical components. (The control of vehicles in an urban transportation system, for example, could be by individual operators or by a central computer system.) In this allocation process, however, we should keep in mind the qualms expressed by Jones (1967) that there are circumstances in which the notion of a one-to-one correspondence between functions and components may be severely strained. In such instances one might view the functions (on one side of the fence) and the components (on the other side) in more of a "collective" frame of reference than a one-to-one relationship.

Because of the importance of these decisions for some systems, one would hope that there would be available some guidelines to aid the system designer in allocating specific functions to human beings versus physical machine components. The most common types of guidelines that heretofore have been proposed consist of general statements about the kinds of things human beings can do better than machines and vice versa. As Chapanis (1965) points out, however, such comparisons serve a useful function in only the most elementary kind of way. He points out that such generalizations are useful primarily in directing one's thinking toward man-machine problems and in reminding one of some of the general characteristics that humans and machines have as system components. Aside from the potential practical utility of such comparisons for use in the allocation process, however, Jordan (1963) and others raise the more basic question of the appropriate role of human beings in systems, especially as we continue the trend toward automation. Before discussing such views, however, let us present a set of such generalizations, first to demonstrate what such "lists" are like, and second to serve what Chapanis refers to as the "elementary" purposes for which they may have some utility.

Relative Capabilities of Human Beings and of Machines The following generalizations about the relative capabilities of human beings and of machine components are drawn from various sources (Chapanis, 1960, p. 543; Fitts, 1951, 1962; Meister, 1971; Meister & Rabideau, 1965; and others), plus some additional items and variations on previously expressed themes.

Humans are generally better in their abilities to

- Sense very low levels of certain kinds of stimuli: visual, auditory, tactual, olfactory, and taste.
- Detect stimuli against high-"noise"-level background, such as blips on cathode-ray-tube (CRT) displays with poor reception.
- Recognize patterns of complex stimuli which may vary from situation to situation, such as objects in aerial photographs and speech sounds.
- Sense unusual and unexpected events in the environment.

- Store (remember) large amounts of information over long periods of time (better for remembering principles and strategies than masses of detailed information).
- Retrieve pertinent information from storage (recall), frequently retrieving many related items of information; but reliability of recall is low.
- Draw upon varied experience in making decisions; adapt decisions to situational requirements; act in emergencies. (Do not require previous "programming" for all situations.)
- Select alternative modes of operation, if certain modes fail.
- Reason inductively, generalizing from observations.
- Apply principles to solutions of varied problems.
- Make subjective estimates and evaluations.
- Develop entirely new solutions.
- Concentrate on most important activities, when overload conditions require.
- Adapt physical response (within reason) to variations in operational requirements.

Machines are generally better in their abilities to

- Sense stimuli that are outside the normal range of human sensitivity, such as x-rays, radar wavelengths, and ultrasonic vibrations.
- Apply deductive reasoning, such as recognizing stimuli as belonging to a general class (but the characteristics of the class need to be specified).
- Monitor for prespecified events, especially when infrequent (but machines cannot improvise in case of unanticipated types of events).
- Store coded information quickly and in substantial quantity (for example, large sets of numerical values can be stored very quickly).
- Retrieve coded information quickly and accurately when specifically requested (although specific instructions need to be provided on the type of information that is to be recalled).
- Process quantitative information following specified programs.
- Make rapid and consistent responses to input signals.
- Perform repetitive activities reliably.
- Exert considerable physical force in a highly controlled manner.
- Maintain performance over extended periods of time (machines typically do not "fatigue" as rapidly as humans).
- Count or measure physical quantities.
- Perform several programmed activities simultaneously.
- Maintain efficient operations under conditions of heavy load (humans have relatively limited channel capacity).
- Maintain efficient operations under distractions.

In discussing such "lists" of relative advantages of humans and machines, Jordan boils the assortment down to a nub, as follows: "Men are flexible but cannot be depended upon to perform in a consistent manner, whereas machines can be depended upon to perform consistently but have no flexibility whatsoever."

Limitations of Man-Machine Comparisons The previously implied limitations regarding the practical utility of general comparisons of human and machine capabilities stem from various factors. Some of these have been pointed up by Chapanis (1965) and Corkindale (1967) and are discussed in the following (with a fair portion of editorial license):

1 General man-machine comparisons are not always applicable. Given some general superiority of humans or machines, there are circumstances in which it would be inappropriate to apply the dictates of that generality. For example, the amazing computational abilities of computers do not imply that one should use a computer whenever computations are required.

2 Lack of adequate data on which to base function allocations.

3 Relative comparisons are subject to continual change. For example, machines are not very effective in pattern recognition, but this may change in the years to come.

4 It is not always important to provide for the "best" performance. For example, although human beings who serve as toll collectors on superhighways offer some advantage over mechanical collectors, the mechanized devices do the job well enough to be acceptable.

5 Function performance is not the only criterion. One also has to consider the trade-off values of other criteria such as availability, cost, weight, power, reliability, and cost of maintenance. For example, although remote-control devices for the family TV and garage are available, because of costs they are not yet used universally. At the present stage of affairs, there are very few systematic guidelines to follow in figuring out the relative trade-off values of various criteria.

6 Function allocation should take social and related values into account. The process of allocation of functions to humans versus machine components directly predetermines the role of human beings in systems and thereby raises important questions of a social, cultural, economic, and even political nature. The basic roles of human beings in the production of the goods and services of the economy have a direct bearing upon such factors as job satisfaction, human motivation, and the value systems of individuals and of the culture. Since our culture places a premium on certain human values, the system should not require human work activities that are incompatible with such values. In this vein, Jordan postulates the premise that humans and machines should not be considered comparable, but rather should be considered complementary. Whether one would agree with his conclusion that the allocation concept becomes entirely meaningless, the fact does remain that decisions (if not allocations) need to be made concerning the relative roles of humans and machines. In this context, although the objectives of most systems are not the entertainment of the operators (pinball machines and gambling devices excepted), it would seem, for example, that, within reason, the human work activities that are generated by a system preferably should provide the opportunity for reasonable intrinsic satisfaction to those who perform them. That is, we must design jobs that people can do and that they want to do.

A Strategy for Allocating Functions The discussion above implies that there are no sets of clear-cut guidelines available for use in deciding what system functions should be performed by people and by machines. Rather, one needs to pursue a general strategy, bringing to bear at various phases the most adequate data that are available and exercising the best (well-informed) judgment possible. As proposed by Chapanis (1965), this strategy should *not* be directed toward the allocation of functions as though each function were in vacuum-packed isolation from the others. Rather, the strategy should be directed toward making decisions about functions in such a manner as to enhance the operation of the system as a whole, and toward the creation of jobs that are interesting, motivating, and challenging to the human operator.

Design of Physical Components

Although the actual design of the physical components is dominantly an engineering chore, this phase represents a second stage at which human factors inputs usually are of considerable moment. The specific nature of the design decisions made during this phase can (if inappropriate in terms of human considerations) forever plague the user and cause decrement in system performance or, conversely (if appropriately designed), facilitate the user's use of the equipment and bring about better system performance. In a sense, the design of physical components embraces two phases, the development of a design *concept* (i.e., the basic conceptual formulation of the component) and the composition of design *detail* (Meister, 1971, p. 273), the latter being an elaboration of the first.

The human factors design phase consists of tracking down sources of human factors data that are potentially relevant to the design problem and evaluating their possible applicability, taking into account the various considerations discussed earlier in this chapter. Such sources include various sets of design standards and specifications and the other types of human factors data referred to above.

Problems in Applying Human Factors Data

It would be very satisfying to say that available human factors data are being extensively applied to the design of the things people use in their work and daily living. But such is not the case. Meister (1971), for example, reported that (unfortunately) many design engineers did *not* consider human factors in their design procedures. And Rogers and Armstrong (1977), on the basis of a study dealing with the use of human factors design standards in engineering design, came to the dismal conclusion that such standards have little, if any, effect on product design. They suggest that the apparent reasons for this are varied and complex, involving resistance on the part of designers and managers, education of human factors specialists and designers, the standards themselves, and interdisciplinary communication.

The human factors research community probably has contributed to this state of affairs in at least a couple of ways, in particular in failing to "communicate" adequately with people in other disciplines such as engineering, architecture, and industrial design, and in failing to present human factors data in a reasonably useful form.

Presentation of Human Factors Data With regard to the presentation of human factors data, Blanchard (1975) proposes a set of guidelines for the development of human factors data banks that would contribute more adequately to the use of such data. In connection with the development of human factors design standards, Rogers and Armstrong (1977) state that such standards would be enhanced if the following suggestions are incorporated:

1 Eliminate general or ambiguous terms such as "proper feel" or "high torque."

2 Present quantitative data in a manner consistent with designer preference, i.e., graphical or pictorial first, followed by tabulations.

3 Eliminate the use of narrative statements when data can be presented quantitatively.

4 Eliminate inconsistencies in data within and among standards.

5 Provide revisions and updating of standards on a more timely basis.

Rogers and Pegden (1977) elaborate on certain of these recommendations, and add certain others such as providing a quick access system, including relevant definitions, and providing adequate cross referencing.

JOB DESIGN

To a very considerable degree the design of the equipment and other facilities that people use in their jobs predetermines the nature of the jobs they perform. Thus, the designers of some types of equipment are also, in effect, designing the jobs of people who use the equipment. They may, or may not, be aware of this, but they certainly should be.

Human Values in Job Design

As discussed above in the frame of reference of function allocation, there are certain philosophical considerations in the design of jobs. In this regard there has been considerable interest in the notion of job enlargement and job enrichment, this concern being based on the assumption that enlarged or enriched jobs generally bring about higher levels of job satisfaction. There are a number of variations on this theme, such as increasing the number of activities to be performed, giving the worker responsibility for inspection of the employee's own work, delegating responsibility for a complete unit (rather than for a specific part), providing opportunity to the worker to select the work methods to be used, job rotation, and placing greater responsibility on work groups for production processes.

On the other hand, the human factors approach has placed primary emphasis on efficiency or productivity, and it has then tended toward the creation of jobs that are more specialized and that require less skill than jobs that are enlarged. In discussing these somewhat disparate objectives, McCormick and Ilgen (1980) emphasize the fact of individual differences in value systems, indicating that some people do not like "enlarged" or "enriched" jobs.

It should be noted that the degree of possible conflict between these two approaches is not yet clear, and that there are as yet no clear-cut guidelines for designing jobs that would be compatible with the dual objectives of work efficiency and providing the opportunity for job satisfaction on the part of workers. Although we cannot here come forth with any pat solutions to this problem, it is to the credit of the human factors clan that increased attention is being given to this issue.

Task Analysis

In the case of some systems, especially in the military services, some form of task analysis is carried out. It is not appropriate here to include an extensive discussion of various task analysis procedures,[2] but we will mention certain objectives of such procedures. In the actual development of certain systems the task analysis is based on inferences from the tentative system design features of the tasks that ultimately would be required in the operation and maintenance of the system. Such task analyses can serve to identify human factors problems in the tentative design so that modifications in the design can be made. In addition it can serve as the basis for predicting what the jobs will be like when the system is actually produced, in order to begin the processes of personnel selection and training before the system is available. In the case of systems in operation, task analysis may be carried out as a form of job analysis, as for use in personnel training.

Another form of task analysis involves the development of task inventories that include the various tasks that may be performed in an occupational area. These usually are completed by job incumbents who are asked to indicate what tasks they perform and usually such other information as the estimated time spent on the various tasks. Such data from many job incumbents can be used to identify the groups of tasks that typically occur in combination and that then result in the identification of *job types*. The results of such analyses may then be used for developing training programs for the individual job types, or for other purposes such as job restructuring.

PERSONNEL SUBSYSTEM

We can see that, for at least some systems, one has to start worrying about personnel affairs during most of the design and development process; such affairs include task and function analysis, job design, operating procedures, personnel selection, training, the development of training aids, and staff planning. Undoubtedly the most comprehensive program of this type is that described in U.S. Air Force *Human factors engineering* (1980), and referred to as the *personnel subsystem* (PSS). (In fact, this term is also being used somewhat outside the Air Force.) As developed by the Air Force, this program embodies a systematic set of procedures and guidelines to be followed during the development of at least major systems, for which trained personnel must be available at the time the final system is produced. Although many (perhaps most) systems do not require an elaborate personnel subsystem in this sense, there may be somewhat corresponding functions that require attention during system development for the benefit of the ultimate user of the system. Such functions might include actual training of people (such as mechanics for a new model of automobile), and more likely the

[2] For a further discussion of task analysis the reader is referred to E. J. McCormick. *Job analysis: Methods and applications.* New York: Amacom, the American Management Associations, 1979, especially chap. 5.

preparation of instructional manuals for self-training of users (such as how the householder can sharpen the blade of a new lawn mower).

HUMAN FACTORS EVALUATION[3]

Evaluation in the context of system development has been defined as the measurement of system-development products (hardware, procedures, and personnel) to verify that they will do what they are supposed to do (Meister & Rabideau, 1965, p. 13), and, in turn, *human factors evaluation* is the examination of these products to ensure the adequacy of attributes that have implications for human performance. Actually, almost every decision during the design of a system includes some evaluation, such as deciding whether to use a visual signal or an auditory signal. Although many such evaluative decisions need to be made as part of the ongoing development cycle, in the case of most items that are developed for human use there should be some systematic evaluation upon the completion of the development stage and before the item goes into production. Although such evaluation should be carried out, it is a sad commentary to report that many of the things people use in their work and everyday lives do not benefit from such evaluation, with the result that many things do not serve their purposes effectively or safely, and in some instances have to be recalled for expensive retrofitting or replacement.

The Nature of Human Factors Evaluation

The evaluation of some items in terms of human factors considerations should be carried out systematically, in much the same manner as an experiment. In many instances, however, it is probable that the evaluation procedures have left much to be desired in terms of acceptable procedures.

In connection with experimental procedures, Meister (1978) differentiated between what he called controlled experimentation (CE) and personnel subsystem measurement (PSM). The CE experimenter deals largely with the *explanation* of relationships and phenomena, such as explaining *why* reaction time is a function of the numbers of stimuli and the possible responses to them. On the other hand, the PSM investigator must be concerned with the measurement of human performance (and related criteria) *in operational terms*, that is, in terms that are relevant to the system, subsystem, or item in question, specifically to the functions of the personnel who are involved in the operation and maintenance of the system in question.

Special Problems in Human Factors Evaluation

The above comments may give some inklings about the problems of human factors evaluation. We can touch here on only a few of the specific problems that haunt those responsible for human factors evaluation in actual operational situations or in circum-

[3] For a thorough treatment of human factors evaluation in system development, the reader is referred to D. Meister and G. F. Rabideau. *Human factors evaluation in system development.* New York: Wiley, 1965.

stances that approximate operational conditions. Although much of the research on which human factors data and principles are developed is carried out in laboratory settings, DeGreene (1977) makes the point that the real laboratory for human factors research actually should be the real world. This is indeed so in the final evaluation and testing of things to be used by people.

The overview of research methods discussed in Chapter 2 is as applicable to the final evaluation of the things people are to use as it is to the research aimed toward developing human factors data and principles. However, the problems involved in evaluation in the real world usually are much more complex than those within the pristine walls of a laboratory. In the design of evaluation procedures consideration must be given to three factors: subjects, criteria, and experimental procedures.

Subjects The evaluation should be carried out with the same types of individuals who are expected to use the item in question, such as homemakers, trained pilots, persons with driver's licenses, factory assemblers, handicapped persons, or the public at large.

Criteria The criteria (the dependent variables) should be those that are in some manner relevant to the operational use of the item, and depending on the item can include work performance, physiological effects, accidents, effects on health, learning time, job satisfaction, attitudes and opinions, and economic considerations, to mention a few. And the problems of measuring some of these can drive an experimenter up the wall. The use of multiple criteria is characteristic of most evaluation research. Rarely, if ever, is a single criterion sufficient for use in basing an evaluation.

Experimental Procedures and Controls The problem of carrying out appropriate experimental procedures, including adequate controls (involving in some instances control groups), is much greater in the real world than in a laboratory. We cannot discuss this problem in detail here, but it should be recognized by the investigator.[4]

E. M. Johnson and Baker (1974) point out that in many field studies it is not possible to replicate the study, control groups cannot always be used, and the physical environment cannot be completely controlled. Recognizing these and other limitations of evaluation in real-world circumstances, the investigator should of course make every reasonable effort to follow appropriate experimental procedures.

An Example of a Field Evaluation

The human factors literature includes very few examples of systematic final evaluations of items produced for human use in work and everyday living, probably reflecting the fact that (unfortunately) such evaluation is not a common practice. One evaluation reported by Hicks (1977) dealt with operational field tests of army vehicles. This particular evaluation was based largely on obtaining drivers' judgments of the hu-

[4] For a discussion of experimental methods the reader is referred to A. Chapanis. *Research Techniques in Human Engineering*. Baltimore, Md.: Johns Hopkins, 1959.

TABLE 17-1
BASIS OF EVALUATION OF VEHICLE CHARACTERISTICS USED IN STUDY BY HICKS

Vehicle characteristic	Basis of evaluation
Driver's compartment	Effort required in entering
Driver's seat	Comfort
Driver's front vision	Visibility
Accelerator	Design, location and effort required to operate
Speedometer	Readability
Handling characteristics	Ability to corner at low speeds

Source: Hicks, 1977.

TABLE 17-2
RATINGS OF THREE TRUCKS IN TERMS OF SPECIFIED CHARACTERISTICS

Category of characteristics	Mean vehicle rating		
	Truck A	Truck B	Truck C
Driver's compartment	2.7	1.8	2.6
Handling characteristics	2.8	1.7	3.1
Ride characteristics	2.7	2.0	2.5

Source: Hicks, 1977.

man factors characteristics of the vehicles being included. In particular the drivers were asked to judge the vehicles on 85 characteristics such as those shown in Table 17-1. The 85 judgments so obtained were grouped into a few major categories, and average ratings derived for the several characteristics in each category, as illustrated in Table 17-2 for three trucks. (High values indicate favorable ratings.) These (and other) results indicated quite clearly that trucks A and C were superior to truck B on virtually all characteristics.

Discussion

It is almost inevitable that each of you has had the occasion to use some item about which you could ask the question, "Why did they design it this way?" This could apply to the controls on the kitchen stove, the controls on an elevator, the design of a chair, the location of the speedometer on a car, a road sign, the arrangement of an office or factory building, or the labels of TV channels that are hard to read. Such examples probably represent failure to evaluate the items before they were produced. Guidelines for carrying out evaluations are few and far between. In this regard, however, the military services probably have done a much better job than civilian industries. Malone (1976), for example, describes the human factors test and evaluation manual developed for use by the U.S. Navy. Among other features, the manual includes guidance on how to plan and carry out tests for various types of equipment.

REMINDERS ABOUT HUMAN FACTORS DESIGN

It is probably fairly obvious that the application of human factors data to design processes does not (at least yet) lend itself to the formulation of a completely routine, objective set of procedures and solutions. However, a systematic consideration of the human factors aspects of a system would at least focus attention on features that should be designed with human beings in mind. In this connection, therefore, it might be useful to list at least some reminders that are appropriate when approaching a design problem. These reminders are presented in the form of a series of questions (with occasional supplementary comments). Two points should be made about these questions. In the first place some of these would not be pertinent in the design of some items; and, in turn, this is not intended as an all-inclusive list of questions. In the second place (and as indicated frequently before) the fulfillment of one objective may of necessity be at the cost of another. Nevertheless, this list of questions should serve as a good start in the design process, deleting and adding questions to fit the specific situation. The questions follow:

1 What are the functions that need to be carried out to fulfill the system objective?

2 If there are any reasonable options available, which of these should be performed by human beings?

3 For a given function, what information external to the individual is required? Of such information, what information can be adequately received directly from the environment, and what information should be presented through the use of displays?

4 For information to be presented by displays, what sensory modality should be used? Consideration should be given to the relative advantages and disadvantages of the various sensory modalities for receiving the type of information in question.

5 For any given type of information, what type of display should be used? The display generally should provide the information when and where it is needed. These considerations may take into account the general type of display, the stimulus dimension and codes to be used, and the specific features of the display. The display should provide for adequate sensory discrimination of the minimum differences that are required.

6 Are the various visual displays arranged for optimum use?

7 Are the information inputs collectively within reasonable bounds of human information-receiving capacities?

8 Do the various information sources avoid excessive time sharing?

9 Are the decision-making and adaptive abilities of human beings appropriately utilized?

10 Are the decisions to be made at any given time within the reasonable capability limits of human beings?

11 In the case of automated systems or components, do the individuals have basic *control*, so that they do not feel that their behavior is being controlled by the system?

12 When physical control is to be exercised by an individual, what type of control device should be used?

13 Is each control device easily identifiable?

14 Are the controls properly designed in terms of shape, size, and other relevant considerations?

15 Are the operational requirements of any given control (as well as of the controls generally) within reasonable bounds? The requirements for force, speed, precision, etc., should be within limits of virtually all persons who are to use the system. The man-machine dynamics should so capitalize on human abilities that, in operation, the devices meet the specified system requirements.

16 Is the operation of each control device compatible with any corresponding display, and with common human response tendencies?

17 Are the control devices arranged conveniently and for reasonably optimum use?

18 Is the work space suitable for the range of individuals who will use the facility?

19 Are the various components and other features of the facility arranged in a satisfactory manner for ease of use and safety?

20 When relevant, is the visibility from the work station satisfactory?

21 If there is a communication network, will the communication flow avoid overburdening the individuals involved?

22 Are the various tasks to be done grouped appropriately into *jobs*?

23 Do the tasks which require time sharing avoid overburdening any individual or the system? Particular attention needs to be given to the possibility of overburdening in emergencies.

24 Is there provision for adequate redundancy in the system, especially of critical functions? Redundancy can be provided in the form of backup or parallel components (either persons or machines).

25 Are the jobs of such a nature that the personnel to perform them can be trained to do them?

26 If so, is the training period expected to be within reasonable time limits?

27 Do the work aids and training complement each other?

28 If training simulators are used, do they achieve a reasonable balance between transfer of training and costs?

29 Is the system or item adequately designed for convenient maintenance and repair, including any individual components? For example, is there adequate clearance space for reaching parts that need to be repaired or replaced? Can individual parts be repaired or replaced easily? Are proper tools and adequate trouble-shooting aids available? Are adequate instructions for maintenance and repair available?

30 Are the environmental conditions (temperature, illumination, noise, etc.) such that they permit satisfactory levels of human performance and provide for the physical well-being of individuals?

31 In any evaluation or test of the system (or components) does the system performance meet the desired performance requirements?

32 Does the system in its entirety provide reasonable opportunity for the individuals involved to experience some form and degree of self-fulfillment and to fulfill some of the human values that we should all like to have the opportunity to fulfill in our daily lives?

33 Does the system in its entirety contribute generally to the fulfillment of reasonable human values? In the case of systems with identifiable outputs of goods and services, this consideration would apply to those goods and services. In the case of systems that relate to our life space and everyday living, this consideration would apply to the potential fulfillment of those human values that are within the reasonable bounds of our civilization.

In the resolution of these and other kinds of human factors considerations, one should draw upon whatever relevant information is available. This information can be of different types, including principles that have been developed through experience or research, sets of normative data (such as frequency distributions of, say, body size), sets of factual data of a probability nature (such as percentage of signals that are detected under specified conditions), mathematical formulas, tentative theories of behavior, hypotheses that have been suggested by research investigations, and even the general knowledge acquired through everyday experience.

With respect to information that would have to be generated through research (as opposed to experience), while there is very comprehensive information available in certain areas of knowledge, in others it is pretty skimpy, and there are some areas in which one draws virtually a complete blank. Where adequate information is not available, the opinion is expressed that considered judgments based on partial information will, in the long run, result in better design decisions than those that are pulled out of a hat. But let us reinforce the point that such judgments usually should be made by those whose professional training and experience put them in the class of *experts*, whether in the field of night vision, physical anthropometry, hearing disorders, perception, heat stress, acceleration, learning, decision making, or otherwise.

We should here, again, mention the almost inevitability of having to trade off certain advantages for others. The balancing out of advantages and disadvantages generally needs to take into account various types of considerations—engineering feasibility, human considerations, economic considerations, and others. Granting that there probably are few guidelines to follow, nonetheless the general objective of this horse trading is fairly clear. This basically goes back to the stated or implied system objectives and the accompanying performance requirements. In other words, any trade-offs should be made on the basis of the considerations of their relative effects in terms of system objectives.

PRODUCTS LIABILITY

The production and sale of products (be they hair dryers or nuclear power plants) is not necessarily the end of the line for the producer or seller. In the case of certain items the producer or seller maintains relationships with the buyers for service, maintenance, or repair, for example. But in recent years in the United States the matter of products liability has taken on increasing importance, to the point that producers and sellers must give advance attention to such liability in designing, producing, and selling products. The liability issue applies equally to products that are used in industry (machines, tools, etc.) and to those that are purchased by consumers (household appliances, utensils, etc.).

Products liability is the legal term used to describe an action in which an injured party (the *plaintiff*) seeks to recover damages for personal injury or loss of property from a manufacturer or seller (the *defendant*) because the plaintiff believes that the injuries or damages resulted from a defective product. Today, more than ever, people look to the courts for redress when they suffer injury or damage. This attitude has been acutely felt in the area of products liability. Each year more and more cases are brought to court for adjudication. This growth in products liability cases has created a greater demand for human factors experts, both in the initial design of products to make them safer, and in the courtroom as expert witnesses.

In this section we will present some basic concepts in products liability that are relevant to human factors people. We will not dwell on the innumerable nuances of the law involved in products liability cases; our intention here is not to make the reader a lawyer (God knows we have enough lawyers already). Products liability is case law, that is, each new court decision adds, changes, clarifies, or sometimes clouds the accumulated legal precedents. Each state has different legal precedents and, hence, cases may be tried and decided differently in different states. Even as you read this, a case may be decided that drastically alters the nature and course of products liability cases.

Products liability cases are usually tried under one of the following bodies of law: (1) *negligence*, which tests the conduct of the defendant; (2) *strict liability*, which tests the quality of the product; (3) *implied warranty*, which also tests the quality of the product; and (4) *express warranty* and *misrepresentation*, which tests the performance of a product against the explicit representations made about it by the manufacturer or sellers.

Products liability cases typically involve three types of defects: manufacturing defects, design defects, and warning defects. The Interagency Task Force on Products Liability (1977) reported that 35 percent of all products liability cases involve manufacturing defects, 37 percent involve design defects, and 18 percent involve warning defects. The absence of a proper warning can easily be seen as a kind of design defect. Weinstein, Twerski, Piehler, and Donaher (1978), for example, repeatedly point out the trade-off between providing warning to decrease the danger of a product and designing to reduce the danger. They rightly stress the desirability of designing dangers out of the product rather than warning of their presence.

Making a Case

There are certain assertions that must be established to make a products liability case, regardless of the body of law under which the case is tried (Weinstein et al., 1978). It must be established that the product was defective in manufacture or design. This requires a definition of when a product is defective. We will discuss this further in the next section. Suffice it to say, however, that a product can be dangerous without being defective; a knife is a good example. It must be established that the product was defective at the time the product left the defendant's hands. Products are often abused by users, and through abuse the product becomes defective. One might think that in such a case the manufacturer would not be held liable; actually the manufacturer could still be held liable if the abuse was foreseeable. The mere presence of a defect in a product at the time of injury is not enough to make a case. It must be established that the defect was involved in the injury. The injury, after all, may have had no relationship whatever to the defect. A closely related, and more difficult, question that must also be established is whether the defect actually caused the harm. Here the *"but for" test* is applied; "but for the presence of the defect, product failure, or malfunction, would the injury have occurred?" Imagine, for example, that a car is sold with defective brakes. The plaintiff was driving down a street, encountered a patch of ice, slammed on the brakes, which failed, and hit the car in front. A products liability case? Not necessarily; the court could find that even with no defect, slamming on the brakes would not have stopped the car under the icy conditions.

Let us now turn to a discussion of how to determine when a product is defective. The rules here have changed over the years and will undoubtedly continue to do so. There do seem to be, however, a few common threads emerging from recent decisions that probably will set the trend for the near future at least.

When Is a Product Defective?

We said that a product can be dangerous without being defective in design. In fact, for many years the legal precedent, established in *Campo v. Scofield,*[5] held that a manufacturer was not responsible for dangers in a product that were open and obvious. This came to be called the *patent-danger rule*, i.e., a patently obvious danger. It was not until 26 years later that the patent-danger rule was rejected in the landmark case of *Micallef v. Miehle Company.*[6] The plaintiff was employed as an operator on a huge photo-offset printing press. One day he discovered a foreign object on the printing plate, called a *hickie* in the trade, which causes a blemish on the printed page. To correct the situation, the plaintiff informed his supervisor that he intended to "chase the hickie," a common practice wherein a piece of plastic is inserted against the plate which is wrapped around a cylinder that spins at high speed. The plastic caught and drew the plaintiff's hand into the unit between the cylinder and an ink roller. The machine had no safety guards to prevent such an occurrence, and the plaintiff was unable to stop the machine quickly because the shut-off button was distant from his position at the machine.

The plaintiff was fully aware of the obvious danger, but it was the custom to "chase hickies on the run" because once the machine was stopped, it required 3 h to start it back up again. The court, if it maintained the patent-danger rule, would have been forced to deny payment to the plaintiff. The court, however, rejected the rule and stated that it would judge a product for *reasonableness*. Therefore, a product is defective if it presents an *unreasonable danger* to the user.

The question, then, becomes "What is unreasonable?" or "What is reasonable?" First, it must be recognized that most products present some risks to the user. These risks, however, must be balanced against the functions the product performs and the cost of providing for greater safety. The California Supreme Court, in *Barker v. Lull Engineering Company,*[7] went even further in specifying the conditions under which a

[5] *Campo v. Scofield*, New York, Vol. 301, pp. 468, Northeastern 2d, Vol. 95, pp. 802 (1950).

[6] *Micallef v. Miehle Company*, New York 2d, Vol. 39, pp. 376, Northeastern 2d, Vol. 348, pp. 571, New York Supplement 2d, Vol. 384, pp. 115 (1976).

[7] *Barker v. Lull Engineering Company*, California 3d, Vol. 20, pp. 413, California P. 2d, Vol. 573, pp. 431, California Reporter Vol. 143, pp. 225 (1978).

product would be found defective in design. The court set out a two-pronged test. A product would be found defective (1) if it fails to perform safely as an ordinary user would expect when used in an intended or *reasonably foreseeable manner* (emphasis added); or (2) if the risks inherent in that design are not justified by the design's intrinsic benefits.

The idea that a manufacturer can be held liable even if a user abuses or misuses a product, so long as the abuse or misuse was reasonably foreseeable, has considerable implications for human factors. Human factors deals with human behavior and how people respond in certain situations. The courts have now extended our concern to include consideration of what unusual, perhaps even harebrained, things people might do with a product for which it was not intended. Weinstein et al. (1978) present the example of a hair dryer. Is it reasonable to expect that a hair dryer might be used to defrost a freezer, or thaw frozen water pipes, as well as dry things other than hair? It will thus be used in humid environments and near water. Even in drying hair, it may be reasonable to foresee the user sitting in a bathtub when he or she turns the unit on. Surely it is foreseeable that users will be male and female, young and old, and have a broad range of manual dexterity and levels of understanding and awareness.

Ritter v. Narragansett Electric Company[8] illustrates a case that hinged on the issue of foreseeable use. The defendant manufactured a small free-standing 30 in gas range. The plaintiff, a 4-year-old girl, opened the oven door and used it as a step stool to look into a pot on top of the stove. The stove tipped forward, seriously injuring the plaintiff. Expert testimony concluded that the oven door could not hold a weight of 30 lb without tipping. The issue was whether the use of the open door as a step stool was so unforeseeable that the manufacturer should not be held liable. Note, however, that had the stove tipped because a homemaker used the open door as a shelf for a heavy turkey, there probably would have been no question of the manufacturer's liability since it could be argued that one of the intended uses of the door was as a shelf for checking food during preparation. But, as a step stool, was that foreseeable? The jury said "yes" and attached liability to the manufacturer.

One last point before we leave the question of when a product is defective. Designing a product to meet government or industry standards does not guarantee that it will not be found unreasonably dangerous. Weinstein et al. (1978) point out that traditionally courts have taken the position that all standards provide, *at best*, lower limits for product acceptability. Consider a case in point, *Berkebile v. Brantly Helicopter Corporation.*[9] The plaintiff took off in a helicopter with a nearly empty gas tank. Shortly after takeoff the helicopter crashed. The charge of defect was that the helicopter system of autorotation (wherein the aircraft "glides" to the ground) required that the pilot be able to throw the helicopter into autorotation within 1 s. The plaintiff contended that this was too short a time period and that this was the cause of the crash. The defendant argued that the 1-s time met Federal Aviation Administration (FAA) regulations. The court agreed with the defendant, but the appeals court stated that

[8] *Ritter v. Narragansett Electric Company*, Rhode Island, Vol. 109, pp. 176, Atlantic 2d, Vol. 285, pp. 255 (1971).

[9] *Berkebile v. Brantly Helicopter Corporation,* Pennsylvania Superior, Vol. 219, pp. 479, Atlantic 2d, Vol. 281, pp. 707 (1971).

"such compliance (with regulations) does not prevent a finding of negligence where a reasonable man [and we assume woman] would take additional precautions."

Designing a Reasonably Safe Product

Products must be designed for reasonably foreseeable use, not solely intended use. This requires an analysis be made to determine the types of use and misuse a product could be subjected to; this may require a survey of users of similar products or potential users of a new product. Laboratory simulation tests might also be conducted on product prototypes to gain insight into user behaviors and product performance. Weinstein et al. (1978) list seven steps that should be included in the design process of a product to help ensure that a reasonably safe product evolves:

1 Delineate the scope of product uses.
2 Identify the environments within which the product will be used.
3 Describe the user population.
4 Postulate all possible hazards, including estimates of probability of occurrence and seriousness of resulting harm.
5 Delineate alternative design features or production techniques, including warnings and instructions, that can be expected to effectively mitigate or eliminate the hazards.
6 Evaluate such alternatives relative to the expected performance standards of the product, including the following:
  a Other hazards that may be introduced by the alternatives.
  b Their effect on the subsequent usefulness of the product.
  c Their effect on the ultimate cost of the product.
  d A comparison to similar products.
7 Decide which features to include in the final design.

WARNINGS AND INSTRUCTIONS

Closely allied to the issue of whether a product is unreasonably dangerous are the warnings and instructions that accompany it. A distinction can be made between warnings and instructions, although the line between these is rather fuzzy. *Warnings* inform the user of the dangers of improper use and tell how to guard against those dangers, if possible, whereas *instructions* tell the user how to use the product effectively.

Warnings

In the design of a product, Fowler (1980) points out that there are three basic approaches in making a product or substance safe for use:

- Design the dangerous features out of the product.
- Protect against the hazards by guarding or shielding.
- Provide adequate warnings and instruction for proper use and foreseeable misuse.

The use of warnings is inexpensive, but their use does not ensure safety since people may ignore them. By and large, however, effectively worded and prominently displayed warnings have been recognized by the courts as an effective method of reducing the liability of manufacturers (Weinstein et al., 1978). The designing of dangerous features out of some products can be much more expensive, but if this can be done the safety of the product is of course enhanced, usually more than would be the case with warnings.

Legal Requirements for Warnings The legal requirements for an adequate warning, as set forth by the court in the case of *Muncy v. Magnolia Chemical Company,*[10] are as follows:

1 It must be in such a form that it could reasonably be expected to catch the attention of the reasonably prudent man in the circumstance of its use.
2 The content of the warning must be of such a nature as to be comprehensible to the average user and convey a fair indication of the nature and extent of the danger to the mind of a reasonably prudent person.
3 Implicit in the duty to warn is the duty to warn with a degree of intensity that would cause a reasonable man to exercise the caution commensurate with the potential danger.

Weinstein et al. (1978) add a fourth requirement based on their analysis of the *Tucson Industries v. Schwartz*[11] case; i.e., an adequate warning must tell the consumer *how* to act to avoid the danger.

In *Tucson Industries v. Schwartz*, the plaintiff suffered severe eye injuries when fumes from an adhesive being used were circulated by an air conditioning system. The label on the adhesive container read:

> DANGER, extremely flammable, read the instructions, be sure to provide adequate ventilation and safety first.

The air conditioning system for the building was turned on in an attempt to remove the fumes. The system, however, only cooled and recirculated the air without bringing in any fresh air. The issue was whether foreseeable users would understand the term *adequate ventilation* and know how to provide it. The warning did not adequately describe the actions the user was to take to eliminate the hazard, or reduce it to an acceptable level. Weinstein et al. (1978) suggest that the following warning would have been judged adequate:

> Fumes are dangerous. They *must* be exhausted directly to the *outside air*. Use as near as possible to open window or outside door. Fans or blowers can be used *only* if they exhaust to *outside air*. Exposure to fumes can cause blindness or other serious injury.

[10] *Muncy v. Magnolia Chemical Company,* Southwestern 2d, Vol. 437, pp. 15, Court of Civil Appeals of Texas (1968).

[11] *Tucson Industries v. Schwartz*, Arizona Appeals, Vol. 15, pp. 166, Pacific 2d, Vol. 487, pp. 12 (1971).

We might take exception to the specific wording of this warning, but it is obvious that providing the "how to" information definitely improves it. Human factors specialists have been designing displays, warnings, and instructions for years. It is not surprising, therefore, that they are often called on as expert witnesses in products liability cases in which warnings or instructions are an issue.

The Design of Warnings Granting that the legal requirements must be fulfilled in the use of warnings, Fowler (1980), in discussing the hazards associated with products, argues that the type of warning should be commensurate with the dangers inherent in using the product, and differentiates among the following:

- *Danger* is used where there is an immediate hazard which, if encountered, will result in severe personal injury or death.
- *Warning* is the signal word for hazards or unsafe practices which *could* result in severe personal injury or death if encountered.
- *Caution* is for hazards or unsafe practices which could usually result in minor personal injury, product damage, or property damage.

Although we will not elaborate on these points, Fowler provides certain guidelines for use in the design of warnings on products. Aside from the use of any guidelines, however, it is possible to compare the effectiveness of different warning signs experimentally. The best method of doing so (but a method that involves serious ethical aspects) is to use various signs in actual situations and then later compare the accidents or injuries resulting from the use of the different designs. However, some indication of the relative effectiveness of different designs can be obtained by asking a sample of products users (or potential users) their opinions of various alternative designs. This method was used by McGuinness (1977) with five designs of warnings for lawn mowers. These five designs are shown in Figure 17-2 along with the percent of 621 respondents who were asked to state which of the designs they considered to be most effective. In this study the first two were found to be greatly preferred over the last three. However, the primary purpose of referring to this study is to demonstrate the method used.

Instructions and Manuals

You probably have encountered a set of instructions or a manual relating to the use of a particular product that you simply could not understand. (The instructions regarding the timing mechanism of the stove of one of the authors are so intricate that the timing mechanism is simply not used.) The preparation of clear-cut, simple instructions is an art, and we cannot here go into the details of their preparation. In a detailed discussion of instruction manuals, however, Coskuntuna and Mauro (1980) offer these overly simplified rules of thumb:

- Less is more: avoid informational overload like the plague.
- Avoid abstract information; use only concrete information.
- Forget why's; concentrate on how's.

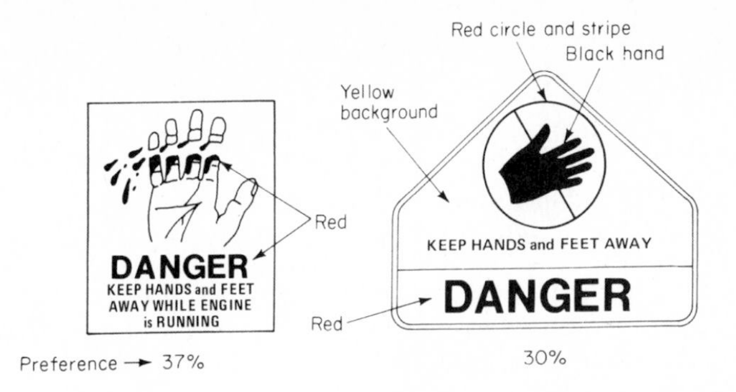

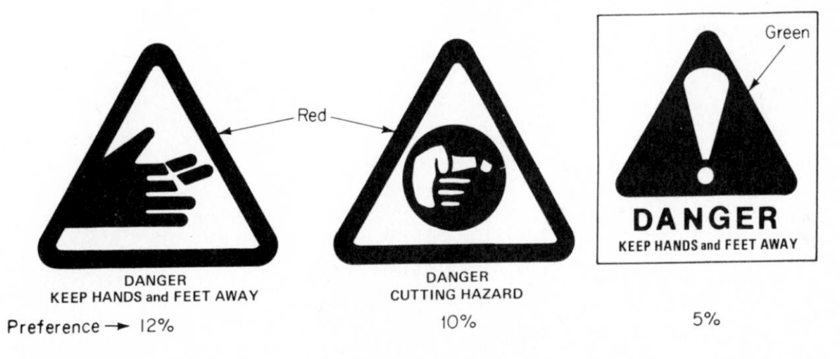

FIGURE 17-2
Lawn mower warning signs used in obtaining opinions of subjects regarding effectiveness of signs, and percents of subjects who selected each design as most effective. (*Source: Adapted from McGuinness, fig. 1, p. 293.*)

- Remember that learning will come from doing.
- Forget the *hype* (the sales pitch): the users already have the product; what they want to do now is set it up or use it with minimum hassle.

As an aside, they urge putting important material at the front, or in a prominent position.

The Use of Illustrations Verbal material certainly has its place in instructions and manuals, and on some labels. However, illustrations of some type have their place either in connection with manuals or instructions, or simply by themselves either to give instructions or to serve as warnings. In this regard D. A. Johnson (1980) differentiates between pictures and symbols, as follows: a *picture* is a realistic photograph or drawing or an object about which information is to be conveyed; a *symbol* is a photograph or drawing that represents something *else*, as the Red Cross symbol. In turn he describes a *pictogram* as a series of associated pictures intended to give information about performing a series of actions.

In discussing the use of illustrations to convey information to users, he makes the point that concepts about which information is to be conveyed range along a continuum from simple to complex, as indicated by the following "ends" of the continuum:

- Simple concept: instructions are given to perform a single action (why and when are obvious).
- Complex concept: multiple actions are required, and multiple results can occur (when and why an action is to be performed may not be obvious).

He also makes the point that the concepts in question can range from *concrete* (such as an emergency door), to *abstract* (with no physical referent, such as the general notion of *danger*). The various combinations of these distinctions, along with proposed methods of conveying information about them, are given in Table 17-3. Examples of displays to represent combinations *a, b,* and *c* are given in Figure 17-3. The procedure for developing relevant illustrations to convey various concepts is discussed by D. A. Johnson and is not repeated here, but the examples in Figure 17-3 show how appropriate illustrations can convey the intended meaning in many situations.

DISCUSSION

It has not been the intent of this book to compile and organize a tremendous amount of available human factors data or to provide how-to-do-it procedures and guidelines for applying such data to the design of the things people use. Rather, one of the objectives has been to present at least some information about certain human characteristics (such as sensory and motor processes) that might contribute to greater understanding of human performance and behavior that might, in turn, have some relevance for those who are concerned with designing things for people to use. In addition it had been the intent to include discussions of at least certain methods and techniques that are relevant to the human factors field and to discuss at least some of the aspects of the application of human factors data and principles to practical design problems.

But perhaps the more important underlying objective has been to develop an increased sensitivity to, or awareness of, the many human aspects of the systems and situations that abound in our current civilization. Such awareness is at least the first prerequisite to the subsequent processes of creating those systems and situations in which human talents can be most effectively utilized in the furtherance of human welfare.

It must be granted that, to date, the systematic consideration of human factors in the design of things people use has not taken on epidemic proportions, but rather has been limited to certain types and classes of systems and facilities. In this regard, however, we would like to propose that virtually any kind of equipment, facility, gadget, "thing," process, service, or environment that is created or influenced by humans—presently in existence or to be created—be viewed as a prime candidate for human factors attention. Such systematic concern should of course be directed toward the identification of any existing or potential human factors problems, with the objective of designing the product or conditions for more effective human use and the improvement of the quality of life.

The remaining chapters of this book deal with examples of human factors research,

TABLE 17-3
PROPOSED METHODS FOR CONVEYING INSTRUCTIONAL INFORMATION OF VARIOUS TYPES

	Concrete	Abstract
Simple	*a* use pictures/words	*b* use symbols/words
Complex	*c* use pictograms/words	*d* use words

Source: D. A. Johnson, 1980.

(a) Simple/concrete concept: fastening seat belt

(b) Simple/abstract concept: emergency oxygen supply

(c) Complex/concrete concept: installing ribbon cartridge

FIGURE 17-3
Illustrations representing three combinations of "simple/complex" and "concrete/abstract" concepts and appropriate methods of illustration, as follows: (*a*) simple/concrete (picture); (*b*) simple/abstract (symbol); (*c*) complex/concrete (pictogram). (*Source: a and b from D. A. Johnson, 1980, figs. 1 and 2; c from Boyer, 1980, fig. 8, p. 350, and reproduced through the courtesy of the International Business Machines Corporation.*)

and in certain instances of the application of human factors data and principles, as related to the following areas: the "built" environment (buildings, communities, etc.); transportation and related facilities; and certain work-related situations that have not been dealt with extensively in other chapters. In presenting such examples it should not be assumed that they all represent exceptionally good examples of research or applications. Rather, they are intended to reflect some of the types of human factors design problems that people have worked on in these areas.

REFERENCES

Applied ergonomics handbook (Pt. 1), A first introduction. Systems design. *Applied Ergonomics,* 1971, *2*(3), 150–158.

Blanchard, R. E. Human performance and personnel resource data store design guidelines. *Human Factors,* 1975, *17*(1), 25–34.

Boyer, H. Do it yourself IBM products and the role of the technical illustrator in human factors testing. In *Proceedings of the symposium: Human factors and industrial design in consumer products.* Medford, Mass.: Tufts University, 1980, pp. 342–351.

Chapanis, A. Human engineering. In C. D. Flagle, W. H. Huggins, & R. H. Roy (Eds.), *Operations research and systems engineering.* Baltimore: Johns Hopkins, 1960.

Chapanis, A. On the allocation of functions between men and machines. *Occupational Psychology,* 1965, *39*, 1–11.

Chapanis, A. The relevance of laboratory studies to practical situations. *Ergonomics,* 1967, *10*(5), 557–577.

Corkindale, K. G. Man-machine allocation in military systems. *Ergonomics,* 1967, *10* (2), 161–166.

Coskuntuna, S., & Mauro, C. L. Human factors and industrial design in consumer products. In *Proceedings of the symposium: Human factors and industrial design in consumer products.* Medford, Mass.: Tufts University, 1980, pp. 300–313.

DeGreene, K. B. Has human factors come of age? In *Proceedings of the Human Factors Society.* Santa Monica, Calif.: Human Factors Society, 1977, pp. 457–461.

Fitts, P. M. (Ed.). *Human engineering for an effective air-navigation and traffic-control system.* Washington, D. C.: NRC, 1951.

Fitts, P. M. Functions of men in complex systems. *Aerospace Engineering,* 1962, *21* (1), 34–39.

Fowler, F. D. Failure to warn: A product design problem. In *Proceedings of the symposium: Human factors and industrial design in consumer products.* Medford, Mass.: Tufts University, 1980, pp. 241–250.

Hicks, J. A., III. A methodology for conducting human factors evaluations of vehicles in operational field tests. In *Proceedings of the Human Factors Society.* Santa Monica, Calif.: Human Factors Society, 1977, pp. 211–215.

Human factors engineering (AFSC DH 1-3). Air Force Systems Command Design Handbook 1-3, 3d ed., rev. 1. Ohio: Wright Patterson Air Force Base, June 25, 1980.

Interagency Task Force on Products Liability (Final Report II-54). Washington, D. C.: Department of Commerce, 1977.

Johnson, D. A. The design of effective safety information displays. In *Proceedings of the symposium: Human factors and industrial design in consumer products.* Medford, Mass.: Tufts University, 1980, pp. 314–328.

Johnson, E. M., & Baker, J. D. Field testing: The delicate compromise. *Human Factors,* 1974, *16*(3), 203–214.

Jones, J. C. The designing of man-machine systems. *Ergonomics,* 1967, *10*(2), 101–111.

Jordan, N. Allocation of functions between man and machines in automated systems. *Journal of Applied Psychology,* 1963, *47*(3), 161–165.

Malone, T. B. The Navy's human factors test and evaluation manual: HFTEMAN. In *Technical program of the 20th annual meeting of the Human Factors Society.* Santa Monica, Calif.: Human Factors Society, 1976, pp. 178–182.

McCormick, E. J., & Ilgen, D. R. *Industrial psychology* (7th ed.). Englewood Cliffs, N. J.: Prentice-Hall, 1980.

McGuinness, J. Human factors in consumer product safety. In *Proceedings of the Human Factors Society.* Santa Monica, Calif.: Human Factors Society, 1977, pp. 292–294.

Meister, D. *Human factors: Theory and practice.* New York: Wiley, 1971.

Meister, D. A theoretical structure for personnel subsystem management. In *Proceedings of the Human Factors Society.* Santa Monica, Calif.: Human Factors Society, 1978, pp. 474–478.

Meister, D., & Rabideau, G. F. *Human factors evaluation in system development.* New York: Wiley, 1965.

Rogers, J. G., & Armstrong, R. Use of human engineering standards in design. *Human Factors,* 1977, *19*(1), 15–23.

Rogers, J. G., & Pegden, C. D. Formatting and organization of a human engineering standard. *Human Factors,* 1977, *19*(1), 55–61.

Weinstein, A., Twerski, A., Piehler, H., & Donaher, W. *Products liability and the reasonably safe product.* New York: Wiley, 1978.

CHAPTER **18**

THE BUILT ENVIRONMENT

Although the human factors discipline has traditionally been primarily concerned with the design of equipment, facilities, and environments as related to work activities, the basic approach of human factors is equally applicable to a wide spectrum of other areas, including virtually all man-made or -influenced features of our total life space. Our life space can be viewed as encompassing our cities and communities (including their physical and cultural aspects), our neighborhoods, the buildings in which we live and work and that we use for other purposes, and the natural outdoor environment. This chapter deals largely with some of the human factors aspects of what has been called the *built environment*, particularly the buildings and related facilities that people use for various purposes.

In recent years there has been considerable research and discussion relating to what is sometimes referred to as *environmental psychology*, and to the more specific area of *architectural psychology*. In this regard Heimstra and McFarling (1978) refer to *built environments* as including various "levels" starting with the geographical regions in which people live and work, their cities or communities, their neighborhoods, and finally the buildings and rooms they use.

PEOPLE AND PHYSICAL SPACE

As we consider the physical space of the built environment we must remember that there are different "contexts" with respect to the physical and social features of that environment within which people interact. We will touch briefly on certain concepts that have been discussed in connection with these interactions.

Density and Crowding

These terms sometimes are used interchangeably and sometimes are used to mean quite different things. Heimstra and McFarling (1978), however, following the practice of certain investigators, consider *density* to be a physical concept and *crowding* to be a psychological concept. Thus, *density* can be viewed as characterizing the relationship between the number of people and some unit of space, as the number of people per square mile or (within buildings) the number of square feet or square meters per person. (It should be added that some investigators differentiate between *spatial* and *social* density, social density being based on the ratio of number of people per unit of space.) In turn (following Heimstra and McFarling) we will consider *crowding* to represent a psychological state in which individuals have personal, subjective reactions based on the feelings of too little space. As Loo (1977) points out, feelings of crowding can be influenced by various factors, including: (1) inside physical factors (such as room dimensions and furniture); (2) outside physical factors (such as number of apartments in a building and populaton size and density of the community); and (3) various social factors (such as number of people in a room and the cultural values and norms of people who are associated in interpersonal relationships or in the community).

Personal Space

Somewhat related to the notion of crowding is the concept of *personal space*. This can be thought of as an envelope or bubble surrounding an individual. If "intruders" invade this space the individual usually experiences a feeling of unease and may wish to withdraw. Such space is, in a sense, portable, in that individuals take it along with them wherever they go. However, Hall (1976) emphasizes the fact that amount of personal space varies with the situation. It has some of the flexibility of an accordion in that people will tolerate being closer to others in some circumstances (such as in a subway or at a football game) than in other circumstances (such as in a conversation at home or in an office). The amount of space also shrinks or expands depending on whether an invader is a close acquaintance or a stranger, and on the differential status of individuals (for example, a lowly subordinate of a high-ranking executive typically will remain farther away from the executive in the executive's offices than, let us say, an individual with more nearly equal status). The amount of personal space is also a function of cultural background.

There are many manifestations of the existence of personal space, such as in seating oneself in a conversational group; in looking for a seat in a library, on an airplane, or in a theater (when it is usual for one to try to find a seat separated from others); or in "keeping one's distance" when walking along a sidewalk. People use various schemes (some rather ingenious) to define their space, such as glaring at a stranger who sits too close on a park bench (or even getting up and leaving, taking one's portable space along), or by putting a coat or briefcase on a chair beside oneself in a public place.

Territoriality Although personal space is *portable* (following people like their shadow), *territoriality* tends to characterize a piece of "real estate" to which an individual makes some claim. (This is a very common phenomenon in the case of some animals. Intruders will be attacked by the resident.) *Territoriality* as related to human beings actually has been defined in various ways, but as Edney (1974) concludes, taken together the term includes such concepts as space, defense, possession, identity, markers, personalization, control, and exclusiveness of control. In the most typical circumstances a person's *territory* consists of his or her place of abode along with any land area associated with it. As related to human dwellings Newman (1972) defines it as "the capacity of the physical environment to create perceived zones of territorial influences." Newman points out that, by its very nature, a single-family house is its own statement of territorial claim in that it is on an integral piece of land that buffers the house from neighbors and from the public street. Shrubs, fences, walls, and gates sometimes serve as symbolic reinforcers of the house. The territoriality of semidetached or row houses is sometimes characterized by stoops, porches, or fenced-in front yards. In the case of apartments the territoriality depends very much upon the physical relationship of individual apartments to other areas. Often the door of the apartment separates the territory of the dwelling unit from the *public* area—the hallway.

Defensible Space Closely related to the territoriality of the dwelling unit is the concept of *defensible space*, which is characterized by Newman (p. 3) as a model for residential environments which tends to inhibit crime by creating the physical expression of a social fabric that *defends* itself. It is, in essence, an area with features which form an environment in which latent territoriality and sense of community in the inhabitants can be translated into responsibility for ensuring a safe, productive, and well-maintained living space.

Privacy

The concept of *privacy* is another complex concept, but in general has links with the notions of crowding, personal space, and territoriality described above. In everyday language *privacy* generally is considered to reflect the need for people to be by themselves or to get away from others. Ittelson, Proshansky, Rivlin, and Winkel (1974) define it more generally as individuals' freedom to choose what they will communicate about themselves and to whom they will communicate it in a given circumstance.

Discussion

These concepts relating to the interactions of individuals with the *space* in which they live and work are related both to the physical features of the built environment and to the social and personal situations that take place in such environments. From the point of view of design of the built environment our primary focus is on the physical features of such environments as they affect the people who use them.

INDEPENDENT AND DEPENDENT VARIABLES

The varied features of the built environment can have profound effects on people. Aside from the obvious physical features of such environments, people are also influenced by such nonphysical features as social, cultural, technological, economic, and political factors that characterize the environment. The built environment and its associated nonphysical factors affect a wide gamut of human experience, including life style, performance in work and other aspects of life, social behavior, attitudes, satisfaction and dissatisfaction, mental health, and physical health and well-being.

Dependent Variables

If we were to ask a number of people to give their opinions about the standards (i.e., the criteria or dependent variables) by which they would evaluate the features of the built environment they use, we would of course receive a wide variety of responses, but such an assortment probably would include the following:

- Physical convenience (convenience of location, rooms, spaces, and facilities)
- Effects on activities (on a job, doing housework, recreation, etc.)
- Physical comfort (temperature, avoidance of excessive noise, comfort as in seating, etc.)
- Health, safety, and security
- Opportunity for social interaction
- Opportunity for privacy when desired
- Pleasing esthetics
- Subjective reactions (attitudes, satisfaction or dissatisfaction, "feeling tone," etc.)

Independent Variables

In turn, the features of the environment (that is, the independent variables) that might influence such criteria (the dependent variables) include those that characterize the physical features of both the outside and inside built environment and the varied assortment of features of the life space of people (mostly nonphysical) that influence the quality of life. The generally physical features of the built environment outside one's dwelling or place of work could include such factors as location; the nature of the neighborhood; population density of the neighborhood; esthetics of the neighborhood and community; proximity to stores, transportation, and other services; closeness of dwelling units to each other; and availability of facilities for recreation and entertainment. The inside features of buildings include such factors as size and shape of rooms, including arrangement of doors and windows, height, and other design features; arrangement of facilities within rooms (facilities for work, furniture in homes, kitchen cabinets, etc.); arrangement of rooms and spaces relative to each other; size and arrangement of halls, passageways, entrances, etc.; utilities (as heating, plumbing, and electrical facilities); architectural style and decor.

Human Factors Information Taxonomy The practices of architects and city planners in designing the built environment generally should be aimed toward the fulfillment of reasonable human needs, such as independent variables (the criteria) referred to above. In reflecting about this objective, Harrigan (1974) states that the most difficult aspect of this problem is to determine how the various relevant behavioral and human factors considerations can serve as the basis for professional application within architectural programs. Since the tie-in between the behavioral sciences and the architectural profession depends upon information interchange, he proposes the use of a human factors information taxonomy of 27 "informational objectives," these being grouped into seven broad categories as follows:

1 Facility description
2 Sociocultural character
3 User activity support (furnishings, hardware, environmental control features, etc.)
4 Surfaces (including effects of surfaces on user activity; color, texture, and pattern; and durability and maintenance)
5 Circulation (of information, people, materials, etc.)
6 Spatial configurations and arrangements (spatial requirements, unit adjacencies, etc.)
7 Location (relative to services, transportation, etc.)

Psychological Dimensions of the Built Environment People tend to react to virtually any stimulus object with some sort of attitude or attitudes based on their perceptions. In the case of many stimuli the attitudes of people can be characterized in terms of each of several "dimensions." In the case of buildings and environments there have been a few investigations directed toward identifying the psychological dimensions that represent people's perceptions, such as pleasantness or comfort.

In this regard Wools and Canter (1970), following up some previous research by Canter (1968), have identified certain of the dimensions that people generally reflect when appraising their physical environments. In their study they used a semantic differential scale (Osgood, Suci, & Tannenbaum, 1957), which consists of several or many pairs of opposing adjectives (such as *pleasant–unpleasant*) with a scale (usually a 7-point scale) between them that can be used to indicate a person's assessment of a stimulus object. In this study 49 pairs of adjectives were used, these being adjectives that could be considered as relevant in describing rooms in a building.

The subjects in this study (nonarchitectural students) were presented with 24 monochromatic line drawings such as the ones shown in Figure 18-1, and then "described" each room, using the 49 scales. The responses were then subjected to a statistical procedure (a principal-components analysis) for identifying the sets of adjectives that tended to "go together" or to form groups. Eight such groupings were identified, these being referred to as *attitudinal* dimensions. Of these, three seemed to be of particular importance. These are listed in Table 18-1, along with an illustrative pair of adjectives. For any given room the responses of people to the several pairs of adjectives that form each dimension serve as the basis for deriving a score for the dimension as it applies to that room.

In a subsequent phase of the study, other subjects were asked to judge the drawings of eight rooms that had various combinations of windows, desk arrangements, and ceilings. Figure 18-1 shows the combinations that were judged to be most friendly and least friendly. The features which were associated with the most friendly room were a sloping ceiling; a floor-to-ceiling window; and a grouping of chairs without an intervening table.

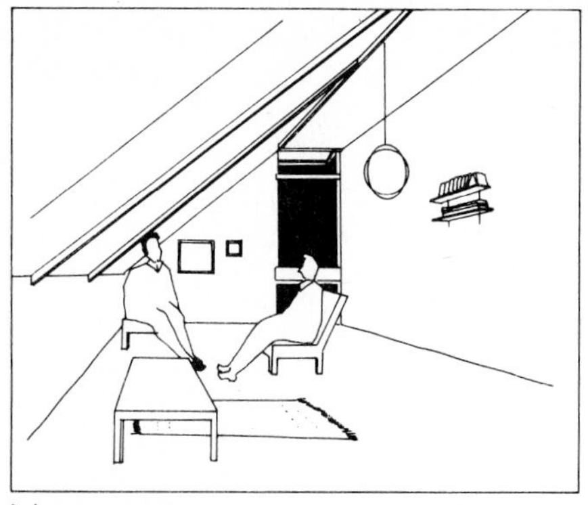

(*a*) Most friendly room

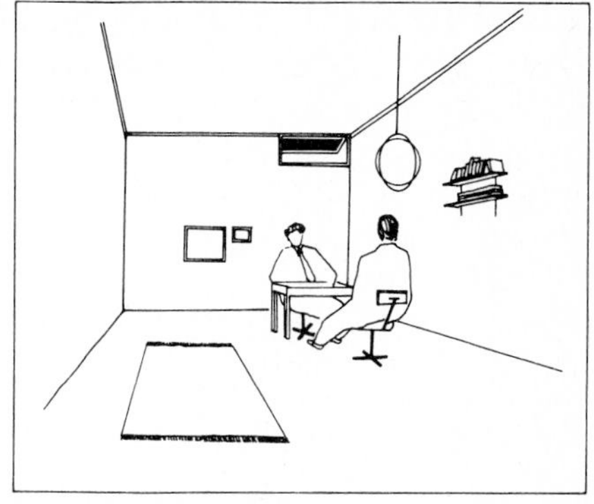

(*b*) Least friendly room

FIGURE 18-1
Examples of the drawings of rooms used in a study relating to the perceived dimensions of rooms. These examples are of rooms that were appraised to be most friendly and least friendly. (*Source: Adapted from Wools & Canter, 1970, figs. 4 and 5.*)

TABLE 18-1
PSYCHOLOGICAL DIMENSIONS OF THE BUILT ENVIRONMENT REPORTED BY WOOLS AND CANTER

Dimension	Illustrative adjectives
Activity	fast–slow
Harmony	clear–obscure
Friendliness	welcoming–unwelcoming

Source: Wools and Canter, 1970.

Somewhat more meaningful psychological dimensions of buildings (and environments in general) were identified in a similar manner by Küller (1972). These are listed below, with brief descriptions of certain ones.

1 Pleasantness (amount of pleasantness and security that an individual perceives)
2 Complexity
3 Unity
4 Enclosedness (closedness or its obverse, spaciousness)
5 Potency (potential "powerfulness" of the environment and also a clear aspect of sex, which means that the environment is more or less associated with one sex or the other)
6 Social status
7 Affection (age of an environment but also a feeling for the old and genuine)
8 Originality

When using the rating scales on which these dimensions are based it is possible to derive ratings for any given environment. Ratings on such dimensions, of course, could range from the concept implied by the title of a dimension to its opposite concept.

Discussion

The design of buildings and other features of the built environment is a difficult process because good design should be directed toward satisfying several goals (i.e., criteria) simultaneously. However, it usually is not feasible to fulfill all the desired objectives. In this regard, Bennett (1977) makes the point that various design objectives are not equal, and tend to form a hierarchy. In planning a private home, for example, a person might consider such factors as location, convenience to transportation, type of neighborhood, size and number of rooms, style of architecture, and type of building materials, along with cost. The relative values the individual places on these and other features usually would place them in some rough hierarchy, so that if all desirable features cannot be fulfilled those low in the hierarchy usually would be dropped or modified. The decisions made under such circumstances represent the trade-offs that plague human factors designers in the design of many items.

It is not feasible here to discuss the many relevant human factors features of the built environment, although the later discussion will at least touch on certain of these in specific contexts. Although our primary interest is with regard to the design of the physical features of the built environment, we should keep in mind the fact that such features interact with various "nonphysical" features of everyday life that influence the quality of life, for better or worse. Certain negative features can be thought of as environmental *stressors*. Current folklore, for example, includes the common assumption that a hectic, hurly-burly pace of life causes stress. There probably are some grains of truth in this belief. In any event, Heimstra and McFarling (1978) cite evidence that such factors as city living, population density and crowding, information overload (as in work and other aspects of life), and noise tend to serve as stressors. Such generalizations, however, should be regarded with two qualifications, namely that there are marked individual differences in susceptibility to stress, and that some people become adapted to environmental conditions.

WHAT KIND OF OFFICE?

Arguments have been bandied about regarding the pros and cons of various types of offices—large and small, and landscaped and conventional. There have been a few studies dealing with this matter, although it must be said that the results of these have been somewhat ambiguous if not conflicting.

Office Size and Social Behavior

There are some indications that small-office environments are more conducive to the development of social affinity for others than large offices. This was reflected, for example, by the results of a sociometric study by Wells (1965a) in which office personnel in large and small offices were each asked to indicate their choices of individuals beside whom they would like to work. The data summarized in Table 18-2 show the percents of the choices made by individuals (in open and small areas) of other persons who were in their own section or department. Although there was greater internal cohesion among personnel working in the smaller areas than in the open areas, it should also be added that there were more isolates (individuals not chosen by anyone).

TABLE 18-2
SOCIOMETRIC CHOICES OF OFFICE WORKERS IN OPEN AND SMALL OFFICE AREAS

	Percent of choices of members of own section	Percent of reciprocal choices within own section
Open areas	64	38
Small areas	81	66

Source: Wells, 1965a.

Office Size and Expressed Preference

Although smaller offices tend to result in more work-group cohesion than open areas, there apparently is no widely pervasive preference for small offices. In one study Manning (1965) asked a group of office workers in Great Britain to express their preferences for five different office arrangements varying in the degree of openness and found that the workers ranked the most closed arrangement first, with the most open last (and the others in between). On the other hand, Nemecek and Grandjean (1973), on the basis of a survey of 519 workers in 15 large-space offices in Switzerland, found no such pattern of preferences. For example, one of the questions asked was: "Would you accept another job in a large-space office?" In reply to this 59 percent said "Yes" and 37 percent said "No," with 4 percent expressing no opinion. However, it might be added that although 59 percent said they would accept a job in a large office, this does not necessarily mean that they would prefer such offices.

Granting that there are differences in opinions of people regarding office size, it is probable that such differences reflect in part the ever-present fact that individuals do differ in their reactions to almost every aspect of our environment, and also to the fact that there are both advantages and disadvantages to large, open offices and to small offices. These advantages and disadvantages of large offices were reflected by certain questions in the survey by Nemecek and Grandjean. The primary factors referred to were as follows (expressed as percents of all responses):

- *Advantages:* better communications (40 percent); personal contacts (28 percent); work flow, supervision, discipline (15 percent)
- *Disadvantages:* disturbances in concentration (69 percent); confidential conversations impossible (11 percent)

Assuming a bit of editorial license, it is the opinion of the authors that, by and large, most people would tend to prefer small offices except in those circumstances in which the work activities involve considerable communications and personal contact, or require a flow of work that is most effectively achieved when people are in the same work space.

Landscaped Offices

In recent years there has been a flurry of interest in the office landscape concept (*Bürolandschaft*), which originated in Germany (Brookes, 1972). Such an office con-

FIGURE 18-2
Example of a landscaped office in which individual offices and work groups are separated by plants, screens, cabinets, and shelves. This office is in the Administration Services Building at Purdue University.

sists of one large, open–but "landscaped"–area which is planned and designed about the organizational processes that are to take place within it. The people who work together are physically located together, the geometry of the layout reflecting the pattern of the work groups. The areas of the various work groups are separated by plants, low, movable screens, cabinets, shelves, etc., as shown in Figure 18-2.

The attitudes and opinions of people about landscaped offices represent something of a mixed bag. People who work in such offices have been reported by certain investigators (Brookes, 1972; Brookes & Kaplan, 1972; McCarrey, Peterson, Edwards, & Von Kulmiz, 1974) as complaining of noise, visual bustle, lack of privacy and confidentiality of communications, and lack of "territory definition." On the other hand, such offices are generally viewed as developing greater solidarity and as offering greater personal contact. In addition, most people react favorably to the esthetic aspects of such offices. The implications regarding productive efficiency are ambiguous. Brookes (1972), for example, states that "It looks better but it works worse," whereas McCarrey et al. (1974) report opinions tending to imply increased productivity.

At the present time it is probably not feasible to say that the reported advantages of landscaped offices do, or do not, outweigh the disadvantages. However, it probably can be said that they would tend to be preferred over large, open, bullpen types of offices.

Windows or No Windows?

Although windows are no longer necessary to provide light and ventilation, the question has arisen as to whether they have value such as in fulfilling what Manning (1967) refers to as a "psychological need" for some "contact with the outside world" or for actual daylight. There are not many data bearing on this, but one study by Wells (1965b) sheds a bit of light on this topic. As one phase of his study he asked clerical workers to estimate, for the total illumination at their work stations, the percentage of that illumination that was from daylight (from the windows). He found that workers whose stations were quite removed from the windows markedly overestimated the amount of daylight at their stations. These results, along with the dominant opinion expressed that daylight is "better" for one's eyes than artificial light, add up to the impression that people generally seem to want "daylight" at their work (even though they seem to overestimate the amount they do have). At the same time, there are many people who work in windowless offices who presumably have not been driven up the walls because of that. (As an aside, the use of windowless buildings can conserve energy for heating.)

Office Furnishings and Arrangements

Some fairly definite indications of the subjective reactions of people to office furnishings and arrangements can be reported. For one thing, people tend to report feelings of spaciousness as being greatest in offices that are moderately furnished (in terms of number of chairs, desks, etc.), as contrasted with those that are more empty or overfurnished (Imamoglu).

In addition, people react more favorably to offices that have living things (as plants) and esthetic objects (as posters) and that are tidy, than to those that lack these features (Campbell, 1979). These features tend to cause visitors to feel more welcome and comfortable, as reflected by the responses of subjects presented with slides of offices with various combinations of features (see Table 18-3).

TABLE 18-3
RATINGS OF RESPONDENTS REGARDING SLIDES OF OFFICES WITH VARIOUS FEATURES

	Mean ratings (9 = favorable; 1 = unfavorable)					
Question	**Plants**	**No plants**	**Posters**	**No posters**	**Tidy**	**Messy**
Comfort of visitor	6.0	4.6	5.7	4.9	6.1	4.5
Invitingness of office	5.9	4.3	5.6	4.7	6.1	4.1
How welcome visitor feels	6.0	4.7	5.8	5.0	6.2	4.5

Source: Campbell, 1979.

Another feature of offices that influences the reaction of visitors and interpersonal relationships is the arrangement of chairs for visitors. When a visitor is seated on the opposite side of a desk from the person being visited, the desk tends to serve as a barrier. In general, visitors have a greater sense of friendliness when both chairs are on the same side of the desk, as illustrated in Figure 18-1*a* as contrasted with 18-1*b* (Campbell, 1979; Zweigenhaft, 1976).

Discussion

Large or small offices? Landscaped offices? Windows or no windows? Although research data about these and other aspects of offices are still quite skimpy, the research that is available (such as that discussed above) gives an impression of ambiguity, inconsistency, and lack of support for at least certain expectations or hypotheses. In reflecting about this disturbing state of affairs one needs to keep in mind the fact that in part the measures of the effects of some design features consist of subjective reactions of people, such as preferences, attitudes, and esthetic impressions. Overt manifestations of these in terms of behavioral criteria such as work performance are difficult to document, but at the same time the favorable disposition of workers to certain types of working situations indicates that they probably have some long-range hidden values. Further, it probably can be said that, although people might prefer a particular environmental situation, they have a fair quota of resiliency or adaptability that makes it possible for them to adjust to a variety of circumstances.

DWELLING UNITS

There are many aspects of dwelling units that have human factors implications. For our illustrative purposes in the sections below we will touch briefly on only certain such aspects.

Room Usage in Dwellings

One human factors aspect of dwelling units relates to the usage of rooms, including indexes of the spatial adequacy of dwelling units in relation to the number of occupants. One index of spatial adequacy is the number of *persons per room* (PPR), which is the simple ratio of the total number of occupants divided by the number of rooms. As pointed out by Black (1968, p. 58), the upper limits of what are usually considered acceptable PPRs are somewhere around 1.00 or 1.20, with a national average of about 0.69. Another index that is sometimes used is the *square feet per person* (SFPP). As pointed out by Black, there are no United States norms, or standards, for the SFPP, although Chombart de Lauwe (1961) has proposed the categories given in Table 18-4; the last column shows the percent in each category resulting from a survey of 121 houses in Salt Lake City. It might be added that, of the home owners surveyed, 7 out of 10 were satisfied with their present houses, and in the case of those who were not, house size had no apparent relation to their dissatisfaction.

TABLE 18-4
RESULTS OF SURVEY OF ADEQUACY OF 3 CATEGORIES OF SQUARE FEET PER PERSON (SFPP) OF LIVING AREA

SFPP	Category	Salt Lake City survey, percent
Less than 130	Poor housing	2
131-215	Adequate	21
Over 215	Very good	77

Source: Black, 1968, pp. 62-63.

TABLE 18-5
RESULTS OF SURVEY REGARDING FREQUENCY OF USE OF VARIOUS ROOMS FOR READING

	Type of reading		
Space	Books, percent	Magazines, percent	Newspapers, percent
Living room	50	62	50
Recreation room	24	25	21
Bedroom	24	12	2
Kitchen-dining area	5	7	32
Other (and "never")	6	3	2

Source: Black, 1968.

Although Black's survey resulted in a correlation of −.77 between the PPR and SFPP values for the 121 houses,[1] he expresses the general opinion that the SFPP is the more discriminating of the two indexes (except in the case of overcrowded houses).

But, aside from deriving these gross indexes of spatial adequacy of the houses in the survey, Black was more concerned with usage of rooms, and in this connection used a questionnaire in which home owners reported the frequency of use of various rooms for such purposes as studying, TV viewing, family activities, and seeking privacy. Summaries of the responses to three such questions are given in Table 18-5, specifically questions regarding rooms in which reading usually was done.

Data on room usage could ultimately aid in designing dwellings which would be more useful to the occupants.

Room Arrangement

Dwelling units (be they houses or apartments) come in many arrangements and sizes. In this regard an extensive survey by Becker (1974) of apartment occupants of public

[1] The negative correlation is brought about by the fact that PPR values *decrease* with house size, whereas the SFPP values *increase*.

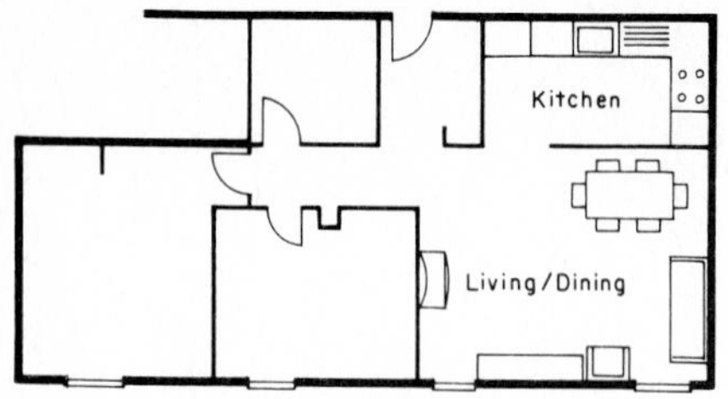

(*a*) One of least preferred arrangements (in convenience of carrying food, food odors in living area, limited hobby space)

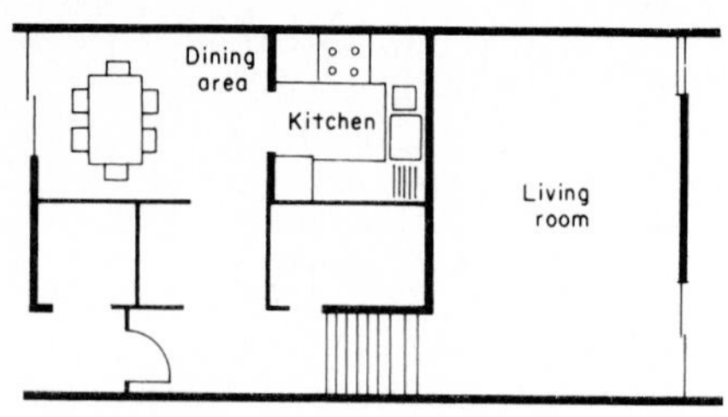

(*b*) More preferred arrangement (particularly because of separated eating and living areas)

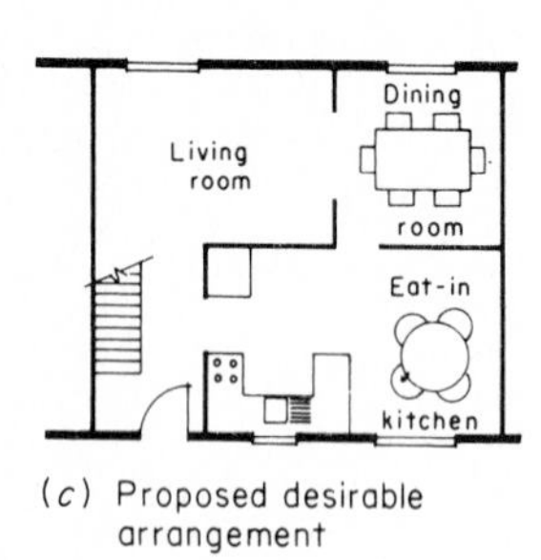

(*c*) Proposed desirable arrangement

(*d*) More flexible proposed arrangement (with movable "wall" unit to permit flexibility)

FIGURE 18-3
Order of preference of residents of a couple of multifamily housing units regarding dining and eating arrangements, and a couple of proposed designs, one of which (*d*) would permit several options for eating and arrangement of living space (this one having a movable wall unit which could be placed in various locations, such as at *A* or *B*). (*Source: Adapted from Becker, 1974, figs. 5a, 5b, 5c, and 5d.*)

housing developments included data obtained by interviews with 257 residents and questionnaire checklists from 591 residents. Certain questions dealt with room size and arrangement of the living-dining-eating spaces. The results of the responses to these questions are summarized below:

- Many variations of room size and arrangement were considered to be equally satisfactory by the residents, but (as would be expected) satisfaction tended to be greater with larger rooms.
- More residents preferred a separate dining area (39 percent) or separate dining area and large eat-in kitchen (28 percent) to combined living-dining area (19 percent) or living room with large eat-in kitchen (18 percent). A couple of these arrangements are illustrated in Figure 18-3 along with a couple of proposed designs that would permit several options for arrangement of eating and living space. Although most residents preferred a separate dining area, 83 percent of the residents said they would not be willing to give up any or all of their living room space for a separate dining area or an all-purpose room. In other words, residents presumably found living space so essential (whatever the size) that they could not conceive of reducing it.

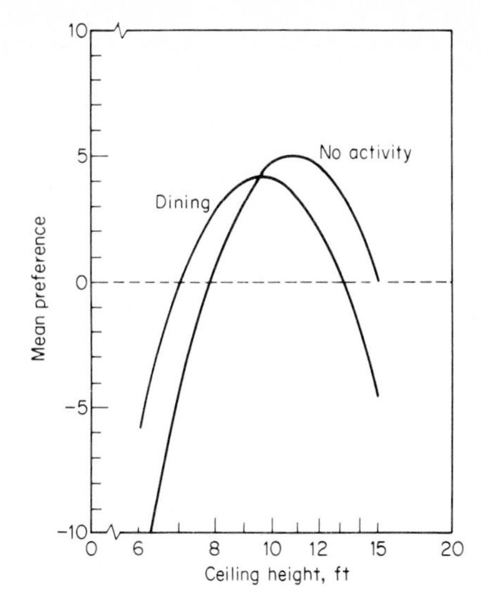

FIGURE 18-4
Mean preferences for ceiling height of rooms for "no activity" and for dining. Note that the ceiling height is shown on a logarithmic scale. (*Source: Adapted from Baird, Cassidy, & Kurr, 1978, fig. 3, p. 725. Copyright 1978 by the American Psychological Association. Reprinted by permission.*)

Room Height and Related Features

The usual height for rooms in dwelling units is 8 ft (2.4 m). In a study by Baird, Cassidy, and Kurr (1978) subjects were placed in a room that had an adjustable ceiling. After the ceiling was set at any given height the subject was asked to give a preference rating on a scale ranging from −10 to +10, these ratings being given in two hypothetical frames of reference, one in which they would be involved in no activity, and the other in which they would be dining. The mean ratings, which are shown in Figure 18-4, indicate that the preferred heights were about 2 ft (60 cm) above the conventional height.

In another phase of this study they found that subjects tended to prefer sloping ceilings (in particular those with $\frac{3}{12}$ and $\frac{5}{12}$ pitches) and wall corners that were greater than 90° rather than less than 90°.

Housing Features Associated with "Difficulty" of Homemaking Tasks

As one phase of a broader study dealing with the "difficulty" of homemaking tasks by Steidl (1972), 208 women described as "housewives" were asked to rate the housing and equipment factors that they perceived as being associated with the difficulty of various tasks around the home, these being separated into the so-called *high-cognitive* tasks (those involving primarily mental activities) and *low-cognitive* tasks (those involving primarily physical activities). A number of these features were associated with the design features of the house and of the associated equipment, as shown in Table 18-6. Although some of these factors are associated with food preparation (such as the level of work surfaces), many of them deal with many other features of the house, a number of which are a function of the design characteristics.

TABLE 18-6
HOUSING AND EQUIPMENT FACTORS CONSIDERED TO MAKE HOMEMAKING TASKS "LESS DIFFICULT" OR "MORE DIFFICULT"

	Satisfactory (that make work less difficult)		Unsatisfactory (that make work more difficult)	
Feature or factor	**High-cognitive tasks**	**Low-cognitive tasks**	**High-cognitive tasks**	**Low-cognitive tasks**
Work surface	28	17	80	21
Storage space	32	16	58	28
Quality of equipment, supplies	105	109	91	69
Availability of equipment, supplies	80	50	39	39
Location of equipment, task	24	23	22	28
Furniture, furnishings	15	35	22	32
Amount of space, number of rooms	92	76	135	53
Arrangement of rooms	56	23	25	10
Temperature, light, ventilation, sound, safety	35	18	40	24
Quality of housing structure, age	15	24	20	21
Other	24	14	48	36
Total	506	405	580	361

Source: Steidl, 1972, table 3, p. 476.

MULTIFAMILY HOUSING

Multifamily housing developments involve human factors considerations over and above those associated with the individual dwelling units within them. These factors are associated with such features as the height, size, and arrangement of apartment buildings, and with certain associated outdoor features. The survey referred to above by Becker included additional data from three high-rise and four low-rise housing developments in urban and suburban areas throughout New York State. Aside from obtaining responses by interviews and questionnaires from residents, the survey procedures also included systematic observations (nearly 100 observations) and interviews with managers of the developments. A few of the relevant findings of the survey are given below in summarized form:

- Low-rise developments were preferred slightly to high-rise developments ("satisfaction" averaged 93 percent for the low-rise as compared with 87 percent for the high-rise).
- The exterior appearance of a development was "very important" to most residents (67 percent). Variation in the shape, pattern, and form of buildings which increased their "individuality" were much appreciated; straight, rectilinear, and symmetrical forms were strongly disliked.
- Residents liked lobbies that were visually pleasing (in the survey, the lobby of only one of the seven developments received a very high rating from the residents).

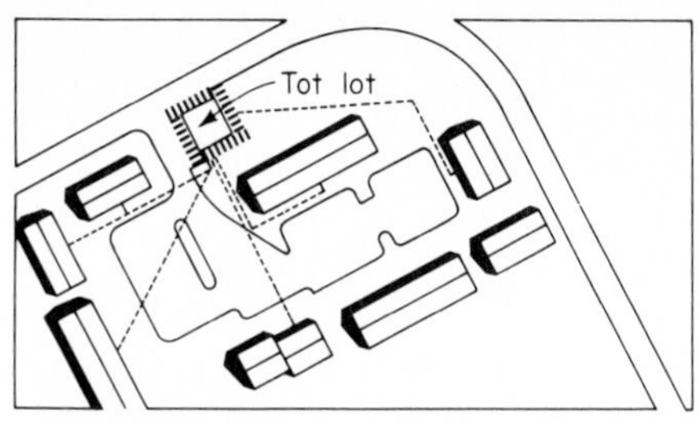

(*a*) Tot lot with negative characteristics

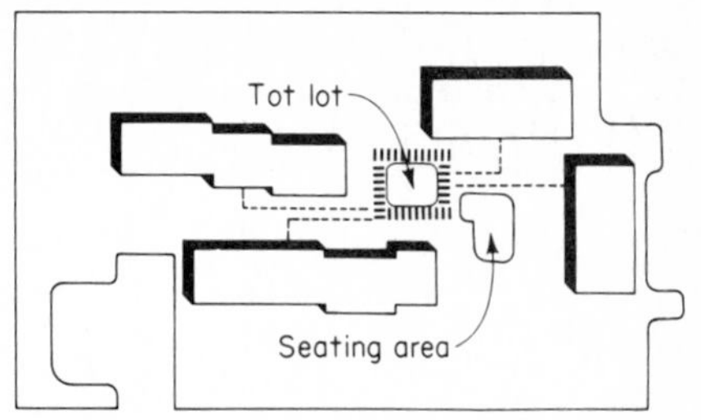

(*b*) Tot lot with more positive characteristics

FIGURE 18-5
Illustration of two *tot lots* (playgrounds for small children) in a multifamily housing development, one with negative and the other with positive characteristics. The one at the right provides more convenient access, better opportunity for surveillance from the adjacent buildings, and greater safety and avoids crossing the parking lot. (*Source: Adapted from Becker, 1974, figs. 9d and 9a.*)

Aside from presenting the results of the opinion survey, Becker also depicts illustrative contrasting designs of certain specific features of multifamily housing developments, in each instance illustrating characteristics that can be considered as positive or negative in terms of human factors considerations. One of these features is the *tot lot* (a playground for small children), as illustrated in Figure 18-5. The arrangement at the right has a clear superiority in terms of ease of access and of increased opportunity for surveillance from the adjacent buildings. The other feature is a barrier around a development, as illustrated in Figure 18-6. The wire fence at the left gives an institutional impression, yet probably would not serve as a completely impenetrable barrier for someone intent on breaching it. On the other hand, the hedgelike fence at the right creates a psychological boundary and a feeling of enclosure, and would tend to discourage strangers from entering while not communicating rejection to nonresidents.

FIGURE 18-6
Illustration of two types of barriers for a multifamily housing development. The one at the left gives an institutional feeling, whereas the one at the right has a more pleasing appearance yet still serves as a psychological boundary and gives a feeling of enclosure. (*Source: Adapted from Becker, 1974, figs. 7 and 8.*)

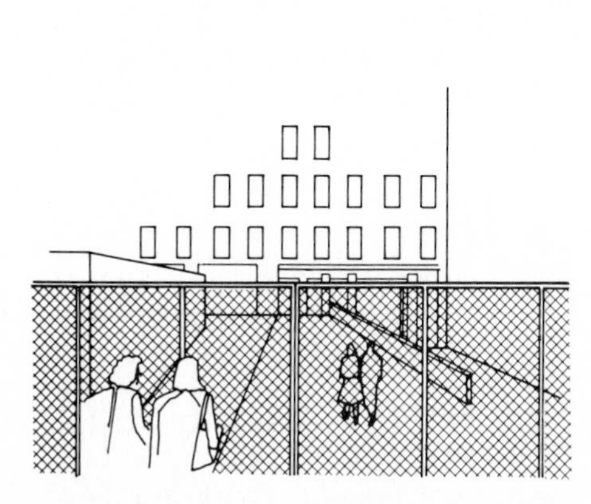

Chain-like fence gives "institutional" impression.

Hedge-like fence creates psychological boundary, with "soft" territorial definition.

Social Behavior as Related to Building Design

The study by Becker and data from other studies start to add up to a fairly consistent pattern regarding the effects of the design features of multifamily housing. For example, it seems that the social behavior of residents of typical high-rise buildings with long corridors is somewhat different from that of residents in buildings in which there are small clusters of apartments. Mehrabian (1976), for instance, reports that studies have shown that if the number of apartments on a floor (or presumably with a common entrance way) is limited to six, the corridor becomes a common space which residents can identify as being partly theirs and for which they are inclined to take some responsibilities. They tend to become more friendly with each other than in buildings with many apartments on the same corridor. Ittelson et al. (1974), referring to low-rise buildings (that tend to have small clusters of apartments), echo this point of view, stating that such residents have a greater territorial sense and that they are more likely to know their neighbors and exercise a proprietary attitude toward their living space. And Newman (1972) notes that many high-rise structures lack "defensible space."

In addition, there are indications that crime and vandalism tend to be more common in high-rise structures, especially those with many apartments on the same corridor. Such effects probably are especially accentuated in public, as opposed to private, housing projects. The classic example of the failure of a public housing project of this type is the Pruitt-Igoe project in St. Louis. The building plan did not provide adequate space for the development of social networks, there were few places where tenants could meet and form friendships, the living hallways became a "no man's land" and areas for vandalism and misbehavior, and there was little opportunity for supervision of children by parents. The physical condition of the buildings deteriorated and the project was such a social disaster that most of the buildings were demolished after about 20 years.

Although some public housing projects have been constructed to replace slum areas (as in the case of the Pruitt-Igoe project), some have resulted in a loss of a pattern of social relationships that is characteristic of some slums. In fact, as Heimstra and McFarling (1978) point out, residents of slum areas often staunchly defend them as suitable places to live. In addition, they express the opinion that at least some public housing projects have provided less satisfaction to residents than slum neighborhoods from which they moved. However, as discussed below, proper design of such developments presumably can make them satisfying places to live.

Lack of Privacy in Multifamily Structures

Another feature of multifamily housing developments that has been viewed as a disadvantage is the lack of privacy, especially because of transmission of noise and voices through walls and ceilings. This disadvantage can be overcome with appropriate construction, but the costs involved frequently preclude such soundproof construction.

Discussion

In general terms, then, it seems that multifamily residence structures should be constructed so as to enhance the possibility of greater social interchange and surveillance. Thus, apartments should be in small clusters, preferably in low-rise buildings, with adequate opportunity for surveillance and spaces for social interchange; they should have soundproofed walls, and should have pleasing esthetics—all of this to contribute to a sense of identity and community, and at the same time to provide the opportunity for privacy.

SINGLE-FAMILY DWELLING COMMUNITIES

The social relations among neighbors in single-family dwellings vary considerably, as with the size of the community and the nature of the people in a given neighborhood. However, Heimstra and McFarling (1978), in summarizing some research on such behavior, report that two features of residential environments that affect social behavior are spatial: distance between houses and relative locations of houses. In general, friendships between families tend to increase as the distances between houses is decreased. Further, families living in culs-de-sac or dead-end streets tend to be more friendly with each other than those living on main through streets; this may be due in part to the possibility that through streets serve as a barrier to interpersonal relationships.

COLLEGE DORMITORIES

College dormitories represent another example of the fact that the design of buildings influences the behavior of people.

Dormitory Design and Social Behavior

As one aspect of such effects let us consider the effects of dormitory design on social behavior. As one phase of a study reported by Baum and Valins (1977), for example, the investigators compared residents of suite-type and corridor-type dormitories in terms of the residential locations of their friends. (The individual rooms in both types of dormitories were nearly the same size, and the average square feet per person, including lounges and bathrooms, was about the same.) The comparison showed quite clearly that more of the friends of residents of suite-type dormitories lived within the same dormitory (an average of 61 percent) as contrasted with friends of residents of corridor-type dormitories (an average of 27 percent). These results strongly suggest that suite-type dormitories are more conducive to the formation of friendships.

Another phase of this study dealt with residents of long versus short corridors. In general, those living on short corridors had a sense of greater privacy and of less crowding and experienced less aggressiveness on the part of other residents than did those living on long corridors. These and other results from this and other studies rather consistently reflect differences in the social behavior and attitudes of residents of different types of dormitories.

HOUSING FOR THE ELDERLY

For some elderly persons the human factors aspects of housing are basically the same as for other persons. There are, however, certain circumstances related to many elderly persons that argue for special attention to housing features for them. These circumstances include the following: physical and health conditions; the fact that many such persons are retired and spend more time in their housing units; and cost constraints for many such persons.

Parsons (1978) refers to numerous publications dealing with housing for the aged, and it is not feasible here to discuss in detail the many facets of such housing. Rather, the intent here is simply to highlight a few of the major aspects of such housing that have important human factors implications.

Specific Housing Features for the Elderly

Some of the specific features of housing facilities that have been cited as being especially relevant for the aged are the following (Browning & Saran, 1978; Lang, 1978; Parsons, 1978; Rohles, 1978): safety and security; bathing facilities; handles and railings; warning devices; wide doorways and halls (for possible wheelchairs); adequate provision for mobility in general; furniture; facilities for recreation and entertainment (including TV, hobby facilities, etc.). Adequate provision for such features would apply equally to private homes as well as to various types of facilities for the elderly and retired such as retirement housing developments, retirement homes, public housing, and convalescent homes.

Multiple-Unit Facilities for the Elderly

In connection with multiple-unit facilities for the elderly, however, there can be additional features that would be relevant. In this regard, for example, Lang reports the experiences resulting from the presentation of plans for a large residential complex in California that included provision for rental apartments for senior citizens. At a conference with many senior citizens, the original plans were sharply criticized, and the persons attending the conference were subsequently divided into nine groups for discussion purposes; questionnaires and interviews were also used to elicit reactions that could be incorporated into a "consensus" of the participants. Certain of the findings are summarized below:

- Cost: For the group in question this was the most basic important concern, even if it resulted in "basic" housing without luxuries. (For other groups this cost constraint would not apply.)
- Location of units: There was a preference for senior citizens to be separated from general family housing units, and to be close to markets and medical and other services.
- Size of units: In part because of cost considerations, many senior citizens were willing to have small apartments (with no more than 500 ft^2), consisting of a single bedroom with a large living room.
- Area around units: Many individuals expressed a desire for walking paths and green areas, patios and balconies, and convenient parking facilities.

Discussion

There are many specific aspects of housing for the elderly that cannot be discussed here. (For example, in certain retirement facilities there is the matter of single rooms versus rooms for two or more persons; with two or more persons in a room interpersonal conflicts can erupt.) However, the discussion above will at least reflect some of the features of housing for the elderly that have human factors implications.

OTHER TYPES OF BUILDINGS AND FACILITIES

We have discussed above the human factors aspects of certain types of buildings and facilities people use. Although it is not the intent here to discuss the human factors implications of all aspects of the built environment, we should keep in mind that the design of other types of buildings and facilities can similarly affect the people who use them. A partial inventory of other types of buildings and facilities would include factories and other production and service facilities, commercial establishments, hospitals, institutions, prisons, schools, entertainment and recreation facilities, and museums.

AUXILIARY FEATURES AND FURNISHINGS OF STRUCTURES

Aside from the fact that human factors are relevant to the designs of structures as such, the design of the many auxiliary features and facilities of structures, and of furniture and other related items, also has human factors implications. We will mention a couple of such features as illustrations.

Ramps and Stairs

Simple answers to human factors problems are few and far between, and this generalization applies to the question as to whether ramps or stairs should be used in specific situations, and the matter of their specific design when used. Certainly ramps are preferable to stairs for use with wheelchairs. In connection with circumstances in which there might be an option, Corlett, Hutcheson, DeLugan, and Rogozenski (1972) carried out an experiment in which the subjects climbed and descended a height of 704 in (1690 cm) using at various times steps with risers of 4 in (10 cm) and 6 in (15 cm) with various tread lengths that resulted in the slopes shown in Table 18-7. The ramp conditions corresponded to these slopes. The eight subjects were young males, four being short and lightweight and the other four being tall and heavyweight. The results are a bit complicating and will be summarized only briefly, as follows:

TABLE 18-7
SLOPES OF STAIR RISERS USED IN STUDY BY CORLETT ET AL.

Riser		Slope	
4 in (10 cm)	10.5°	13.6°	19.5°
6 in (15 cm)	15.8°	20.0°	30°

Source: Corlett, Hutcheson, DeLugan, & Rogozenski, 1972.

- Where joint rotation and muscle strength are not limiting factors, stairs are more efficient than ramps from a physiological point of view (in terms of oxygen consumption and heart rate).
- The higher 6-in (15-cm) step resulted in less physiological cost than the lower 4-in (10-cm) step.
- Where knee angle (and probably ankle angle) are important (as with old or lame people), it is probable that a ramp would be easier to negotiate for any given slope, although the maximum ramp angle requires further study to specify.

It might be added that the American National Standards Institute (ANSI) specifies that for physically handicapped persons ramps shall not have a slope greater than 1 ft (25 cm) rise in 12 ft (300 cm), or 8.33 percent, and that steps should have risers that do not exceed 7 in (14.5 cm) (ANSI, 1961).

Hardware and Related Items

Other auxiliary features of structures include a wide range of hardware and related items. Woodson (1976) has presented quite a laundry list of such items with at least general comments about some of them. Although we will not repeat that material here, a very few examples are given below to illustrate the kinds of considerations that should be given to such items from the human factors point of view.

- Windows: ease of cleaning both sides, ease of opening, closing, and locking.
- Stairs: handrails to fit both children and adults; ramps for wheelchairs.
- Cabinetry: drawer slides for heavy loads; doors that stay open or closed; design to fit articles to be stored.
- Plumbing: handles that fit, operate logically, easily, and accurately; handrails and handholds for bathtubs.
- Electrical switches and conveniences: locate switches and outlets where they can be used conveniently; make outlets childproof.
- Heating and cooling systems: use understandable controls; minimize noise produced by system.

Specific standards for the design of many such items in terms of human factors are not yet available, but reference to some of these can at least serve as reminders to designers that people have to use them.

FACILITIES FOR THE HANDICAPPED

In recent years there has been increased interest in the design of facilities and other things for the handicapped, including features of the built environment such as homes, ramps for public buildings and as replacements for sidewalk curbs, restroom facilities, and convenient access to public transportation facilities. It is not feasible here to deal extensively with this topic, but human factors designers should be fully aware of the need to design facilities for such persons.[2]

[2] A book by G. Hale (Hale, G. *The Source Book for the Disabled.* New York: Paddington Press, 1979) includes discussion of certain design features relevant for the handicapped.

COLOR IN OUR ENVIRONMENT

There have been many speculations about human reactions to color. Although we will discuss certain aspects of color as related to human beings, we should precede any such discussion with a warning that this domain is dominated more by opinions of people than by supporting research evidence.

Color Preferences

People differ markedly in their preferences for color, but in general terms, blue, green, and red hues seem to be most commonly preferred. However, the context in which colors are used can have a bearing on preferences. (Although a woman might like lavender clothing, such a color fortunately is not often seen on houses or automobiles.) Aside from hue, Bennett (1977) reports research by others that indicates that people tend to prefer light colors to dark colors and saturated colors to unsaturated colors. When objects are placed against a background, people tend to prefer high-contrast combinations, i.e., light-colored objects with dark background colors or vice versa.

In all of this, however, we need to remember that the esthetics of interior design, with various combinations of colors in rooms and furnishings, are essentially based on subjective preferences, and that one person's meat is another person's poison.

Color and Perception

There are some cues that colors influence certain perceptions of people. For example, Acking and Küller (1972) had 27 college students and nine hospital patients and personnel describe three hospital rooms that were identical except for color. The "descriptions" were in terms of five previously identified dimensions. The colors were as shown in Table 18-8. The results indicated that the impressions of the three rooms differed on two of the five dimensions, as follows:

- Spatial enclosedness (*C* being most "open," with *B* and *A* following in that order)
- Complexity (with *A* and *B* being considered more "motley" than *C*)

Granting that these results are not profoundly earthshaking, they do reflect in quantitative terms the fact that people do have different perceptions of rooms of various colors, there being reasonable consistency across people in these perceptions.

TABLE 18-8
CHARACTERISTICS OF ROOM COLORS USED IN SURVEY BY ACKING AND KÜLLER

Room	Hue	Lightness	Chromatic strength
A	green	low	medium
B	green	medium	high
C	white	high	low

Source: Acking & Kuller, 1972.

TABLE 18-9
RECORDED BEHAVIOR OF SUBJECTS IN BEIGE AND BROWN ROOM OF MUSEUM OF ART

Room	Number of steps	Area covered, ft^2	Time spent, s
Beige	42.7	9.0	38.1
Brown	46.2	17.9	26.4

Source: Srivastava & Peel, 1968.

Physiological Effects of Color

On the basis of a review of various studies on the physiological effects of color, Acking and Küller state that it seems to be beyond doubt that color has some direct physiological influence on people, such as that reflected by blood pressure, respiratory rate, and reaction time; however, the mechanisms that cause this are not known.

Effects of Color on Behavior

There have been occasional cues indicating that, in certain specific circumstances, color does influence the nature of the behavior of people. For example, Srivastava and Peel (1968) report a study carried out in a room of the museum of art at the University of Kansas, with Japanese paintings displayed on the walls. Paid (volunteer) subjects were informed that the experiment was intended to determine their reactions to the paintings (and this was done). But for the first half of the subjects (301) the room was painted light beige and had a corresponding rug, whereas for the second half the room was painted dark brown and had a matching rug. A special sensing device under the rug (a *hodometer*) made it possible to record each square foot of area over which the subjects walked, and a watch was used to measure the time spent in the room. Some of the comparisons of movements of the two groups are given in Table 18-9. It can be seen that the subjects in the brown room took more steps, covered more area (almost twice as much), and spent less time than those in the beige room. Although one might be hard pressed to explain why the differences in color affected the movements of people so markedly, one must conclude that the effect must be the indirect consequence of some subjective reaction to the environmental color.

Although color apparently had some effect on the behavior of people in the museum, it is very risky to speculate about the effects of color on the behavior of people in their many everyday activities on the job and in their living environments.

Discussion

In assessing the influence of color in our lives, one has the impression of being caught between Scylla (accepting the many common beliefs and pronouncements about the effects of color) and Charybdis (rejecting such beliefs and notions, and assuming that color is of only nominal consequence in human life—except possibly for the fate of paint stores). An in-between frame of reference actually seems to be warranted, but it must be acknowledged that hard data about the various aspects of color are still limited.

URBAN COMMUNITIES

The burgeoning problems of urban centers are undoubtedly one of the major challenges of current life. The many facets of these problems leave few inhabitants unscathed. A partial inventory would include problems associated with health, recreation, mobility, segregation, education, congestion, physical housing, crime, and loss of individuality. The current manifestations of these problems lend some validity to the forebodings of Ralph Waldo Emerson and Henry Thoreau, who viewed with deep misgivings the encroachment of civilization on human life, especially in the form of large population centers. The tremendous population growth, however, makes it inevitable that many people must live in close proximity to others (and thus requires the existence of urban centers); accepting this inevitability, however, it is proposed that one should operate on the hypothesis that, by proper design, urban centers can be created which might make it possible to achieve the fulfillment of a wide spectrum of reasonable human values—perhaps even those that Emerson and Thoreau, and maybe that you and we, might esteem.

As Proshansky, Ittelson, and Rivlin (1976) point out, cities are not just social, political, economic, and cultural systems, but geographical and physical systems as well. The interaction of these various features of cities with the inhabitants clearly influences the quality of life of the inhabitants, but these influences are of course not the same for all inhabitants. The crowding and density in slum areas is of course in marked contrast to those features in spacious luxury apartments or in posh suburban areas.

It is not feasible within the limits of these pages to discuss the problems of urban communities and to speculate about their possible resolution. The sociologists, social psychologists, and clinical psychologists clearly could contribute through research to the understanding of the impact of such communities on the lives of inhabitants. In addition, however, it is believed that the human factors disciplines have much to contribute to the resolution of some of the human problems of urban communities.

DISCUSSION

It is probable that Utopia in the built environment that forms our living space is a will-o'-the-wisp, an unattainable goal. In other words, it is doubtful if human beings' living space in the buildings and other structures they use and in the communities they build can ever provide for the broad-scale fulfillment of the various criteria that are relevant to each of us—physical and mental health and welfare, esthetic values, opportunity for social interchange or privacy, recreation, entertainment, culture, convenience, mobility, safety and security, psychological identity, optimum facilities for contribution to efficient production of goods and services, or whatever. Although perfection in our built environment is not realistic, we should never cease trying to achieve it in the matters of rehabilitating our existing built environment, and ensuring that newly constructed components (buildings, cities, etc.) are so designed as to be reasonably optimum in terms of fulfilling human needs.

REFERENCES

Acking, C. A., & Küller, R. The perception of an interior as a function of its colour. *Ergonomics,* 1972, *15*(6), 645-654.

American National Standards Institute, Inc. American National Standard specifications for making buildings and facilities accessible to, and usable by, the physically handicapped [ANSI A117.1-1961 (R1971)]. New York: American National Standards Institute, 1961.

Baird, J., Cassidy, B., & Kurr, J. Room preference as a function of architectural features and user activities. *Journal of Applied Psychology,* 1978, *63*(6), 719-727.

Baum, A., & Valins, S. *Architecture and social behavior.* Hillsdale, N. J.: Lawrence Erlbaum Associates, 1977.

Becker, F. D. *Design for living: The resident's view of multi-family living.* Ithaca, N. Y.: Center for Urban Development Research, Cornell University, May 1974.

Bennett, C. *Spaces for people: Human factors in design.* Englewood Cliffs, N. J.: Prentice-Hall, 1977.

Black, J. C. *Uses made of spaces in owner-occupied houses* (Ph.D. thesis). Salt Lake City: University of Utah, April 1968.

Brookes, M. J. Office landscape: does it work? *Applied Ergonomics,* 1972, *3*(4), 224-236.

Brookes, M. J., & Kaplan, A. The office environment: Space planning and affective behavior. *Human Factors,* 1972, *14*(5), 373-391.

Browning, H. W., & Saran, C. A proposed plan of home safety for the older adult. In *Proceedings of the Human Factors Society.* Santa Monica, Calif.: Human Factors Society, 1978, pp. 592-596.

Campbell, D. E. Interior office design and visitor response. *Journal of Applied Psychology,* 1979, *64*(6), 648-653.

Canter, D. *The study of meaning in architecture* (GD/16/DC/A). Glasgow, Scotland: Building Performance Research Unit, University of Strathclyde, April 25, 1968.

Chombart de Lauwe, P. The sociology of housing methods and prospects of research. *International Journal of Comparative Sociology,* March 1961, *2*(1), 23-41.

Corlett, E. N., Hutcheson, C., DeLugan, M. A., & Rogozenski, J. Ramps or stairs. *Applied Ergonomics,* 1972, *3*(4), 195-201.

Edney, J. J. Human territoriality. *Psychological Bulletin,* 1974, *81*(12), 959-973.

Hall, E. T. The anthropology of space: An organizing model. In H. M. Proshansky, W. H. Ittelson, & L. G. Rivlin (Eds.). *Environmental Psychology.* New York: Holt, Rinehart and Winston, 1976.

Harrigan, J. E. Human factors information taxonomy: Fundamental human factors applications for architectural programs. *Human Factors,* 1974, *16*(4), 432-440.

Heimstra, N. W., & McFarling, L. H. *Environmental psychology.* Monterey, Calif.: Brooks/Cole, 1978.

Imamoglu, V. The effect of furniture density on the subjective evaluation of spaciousness and estimation of size of rooms. In R. Küller (Ed.), *Architectural psychology: Proceedings of the Lund conference.* Stroudsburg, Pa.: Dowden, Hutchinson & Ross, Inc., 1973, pp. 341-352.

Ittelson, W. H., Proshansky, H. M., Rivlin, L. G., & Winkel, G. H. *An introduction to environmental psychology.* New York: Holt, Rinehart and Winston, 1974.

Küller, R. *A semantic model for describing perceived environments* (Document D12). Sweden: National Swedish Institute for Building Research, 1972.

Lang, C. L. Seniors plan senior housing. In *Proceedings of the Human Factors Society.* Santa Monica, Calif.: Human Factors Society, 1978, pp. 545-549.

Loo, C. Beyond the effects of crowding: Situational and individual differences. In D. Stukols (Ed.)., *Perspectives on environment and behavior.* New York: Plenum Press, 1977.

Manning, P. (Ed.). *Office design: a study of environment* [SfB (92): UDC 725.23]. Liverpool, England: Pilkington Research Unit, Department of Building Science, University of Liverpool, 1965.

Manning, P. Windows, environment and people. *Interbuild/Arena,* October 1967.

McCarrey, M. W., Peterson, L., Edwards, S., & Von Kulmiz, P. Landscape office attitudes. *Journal of Applied Psychology,* 1974, *59*(3), 401-403.

Mehrabian, A. *Public places and private spaces.* New York: Basic Books, 1976.

Nemecek, J., & Grandjean, E. Results of an ergonomic investigation of large-space offices. *Human Factors,* 1973, *15*(2), 111-124.

Newman, O. *Defensible space.* New York: Macmillan, 1972.

Osgood, C. E., Suci, G. J., & Tannenbaum, P. H. *Measurement of meaning.* Urbana: The University of Illinois Press, 1957.

Parsons, H. M. Bedrooms for the aged. In *Proceedings of the Human Factors Society.* Santa Monica, Calif.: Human Factors Society, 1978, pp. 550-557.

Proshansky, H. M., Ittelson, W. H., & Rivlin, L. G. *Environmental psychology.* New York: Holt, 1976.

Rohles, F. H., Jr. Habitability of the elderly in public housing. In *Proceedings of the Human Factors Society.* Santa Monica, Calif.: Human Factors Society, 1978, pp. 693-697.

Srivastava, R. K., & Peel, T. S. *Human movement as a function of color stimulation.* Topeka, Kan.: Environmental Research Foundation, April 1968.

Steidl, R. E. Difficulty factors in homemaking tasks: Implications for environmental design. *Human Factors,* 1972, *14*(5), 471-482.

Wells, B. W. P. The psycho-social influence of building environment: sociometric findings in large and small office spaces [SfB (92): UDC 301.15]. *Building Science,* 1965, *1*, 153-175.

Wells, B. W. P. Subjective responses to the lighting installation in a modern office building and their design implications [SfB:Ab7:UDC 628.9777]. *Building Science,* 1965, *1*, 57-68.

Woodson, W. E. Human factors engineering of architectural hardware. *Environmental Design News,* 1976, 7(3).

Wools, R., & Canter, D. The effect of the meaning of buildings on behavior. *Applied Ergonomics,* 1970, *1*(3), 144-150.

Zweigenhaft, R. L. Personal space in the faculty office: Desk placement and the student-faculty interaction. *Journal of Applied Psychology,* 1976, *61*(4), 524-532.

CHAPTER 19

TRANSPORTATION AND RELATED FACILITIES

One of the features that characterizes the present-day world is the use of transportation—transportation of people and physical materials. The various forms of transportation include those that go by land (automobiles, trucks, buses, trains, etc.), air, and sea. All of these—and their related facilities such as highways—encompass a whole host of human factors problems, some of which we have touched on in previous chapters. Since we can touch on only certain illustrative human factors aspects, and since highway vehicles (especially automobiles) and public transportation systems represent the most common mode of transportation we will deal particularly with them.

VEHICLE ACCIDENTS

The dominant problem with highway vehicles is that of accidents and resulting personal injuries and deaths. As Näätänen and Summala (1976) point out, this is a worldwide problem. Their summary of deaths and injuries for 19 countries over more than 40 years (1930-1973) reflects different patterns for various countries, those patterns being related to the *level of motorization* of the countries. Although the ratio of number of deaths to the number of motor vehicles has tended to go down, the problem is still of such magnitude that it should be given greater attention.

Terminology dealing with *accidents* is used rather loosely, such terms as *accidents, collisions,* and *crashes* sometimes being used interchangeably. For our purpose we will consider *accidents* to be those traffic mishaps that involve some property damage, personal injury or death, or both.

The Causes of Accidents

The *causes* of accidents can be characterized in various ways and at various levels of specificity. At a very general level accidents can be attributed to human behavior, the

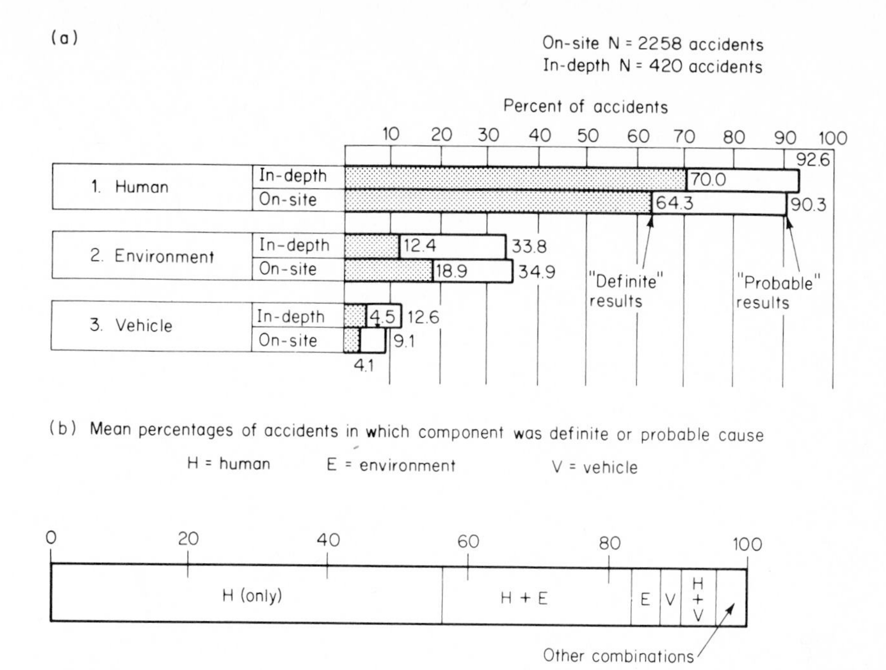

FIGURE 19-1
The percent of accidents caused by human, environmental, and vehicular causes (*a*), and the relative proportion of combinations of these causes (*b*). Because many accidents have multiple causes the sum of the percentages in (*a*) exceeds 100 percent, while the sum in (*b*) adds up to 100 percent. (*Source: Shinar, 1978, fig. 5.2, p. 111.*)

environment, and the vehicle, usually with some interaction among these. In discussing accident causation, Older and Spicer (1976) suggest that accidents can be the consequence of *conflict* situations involving the driver and the environment (and presumably the vehicle) that lead to evasive actions on the part of the driver. Such evasive actions may or may not result in what they call a *collision* (or in our terms an *accident*).

In connection with accident causation, Shinar (1978) presents data on the percentage of accidents from two samples that were attributed to human, vehicle, and environmental causes and their combinations. One sample of 2258 accidents was investigated on-site, and the other sample of 420 was investigated more thoroughly, *in depth*. The results, given in Figure 19-1, show that human behavior was clearly the dominant cause.

We will discuss accident causation in terms of those three categories, and will then elaborate on certain more specific variables that are associated with accidents.

PERSONAL FACTORS IN ACCIDENTS

Discussion of personal factors in accident occurrence frequently involves reference to the notion of *accident proneness*. As usually interpreted this refers to some *attribute* that certain people have that causes them to be accident repeaters. We are disinclined to accept this interpretation and prefer to use this term in a statistical sense—to characterize those individuals who, for whatever reason, tend to have more accidents over time than can be attributed to chance. But it should be noted that the mere fact that an individual has a high accident rate for a while is *not* necessarily indicative of this statistical concept of *accident proneness*—a person who gets a bridge hand with 10 spades should not be called *spades prone.* It is probable that there are some accident repeaters in driving, although collectively such individuals do not account for a very large proportion of all accidents. Further, there are inklings that different types of personal factors, in certain instances, are related to driving behavior. These can be lumped generally into the following categories:

- Personal characteristics
- Driver behavior patterns
- Driver-related capabilities (experience, training, etc.)
- Temporary impairments (Shinar, 1978, refers particularly to fatigue, alcohol, and drugs)

For illustrative purposes we will discuss a few of them that, in certain specific circumstances, have been found to be related to driving behavior.

Personal Characteristics

Various personal characteristics have been investigated as possible contributors to driving behavior; these include personality factors, vision, perception, physical abilities, reaction time, intelligence, age, and sex. Although we will discuss certain of these briefly, it is probable that most of them have significant impact on driving behavior to only a limited degree and with respect to only certain individuals.

Personality There is an increasing body of evidence that manifestations of certain personality characteristics are associated with accident behavior. In summarizing the results of various studies, for example, McGuire (1976) concludes that some highway accidents are just another correlate of being emotionally unstable, unhappy, asocial, antisocial, impulsive, or under stress—or a host of similar conditions referred to by other labels. Such attributes presumably cause some people to be less cautious, less attentive, less responsible, less caring, less knowledgeable, or less capable of driving, and thus increase the risks of accident.

Vision The vision test usually used in issuing driver's licenses is a test of *static* visual acuity, administered when the visual stimuli are stationary and under normal daylight conditions. Although poor static visual acuity has been found to contribute somewhat to accident frequency, Booher (1978) and Shinar (1978) argue for the use of tests that measure more specific driving-related visual skills. One type of test, dynamic visual acuity (DVA), measures the ability to make visual discriminations when the visual targets or the individuals are moving. Shinar presents evidence that indicates that while good static acuity is a prerequisite for good dynamic acuity, having good static acuity is no guarantee of good dynamic acuity. For example, of a total of 356 subjects (16 to 80 years of age), 277 had static acuity of 20/20, but of these more than half had dynamic acuity of 20/40 or poorer.

Shinar also urges the use of measures of visual acuity under low levels of illumination similar to those encountered in nighttime driving. This argument is supported by data that indicate that nighttime accident involvement is related to poor visual acuity under nighttime levels of illumination, but unrelated to acuity under high (daytime) levels of illumination.

In connection with individuals with poor acuity some experimental work has been carried out with the use of bioptic telescope spectacles that have lenses that greatly enhance vision. Booher reports that such individuals tend to have fewer accidents than the general population, but suggests that this may be due to nonvisual factors, in particular self-restrictions such as driving on familiar routes, limited nighttime driving, and driving when traffic is light. Further investigations, therefore, are required to evaluate the utility of such telescope spectacles.

Field Dependence The visual input to drivers is of course influenced in part by the characteristics of the drivers as well as by the nature and location of objects within their environments. One of the personal characteristics that seems to be relevant is a perceptual style called *field dependence–field independence.* Field-independent people are better at distinguishing relevant from irrelevant cues in their environment than are those who are field-dependent. There are various tests that are used for measuring field dependence. Some of these are described by Goodenough (1976). There have been a few studies that have indicated that persons who are field-dependent may be more likely to have accidents than those who are field-independent. One such indication comes from a study by Barrett and Thornton (1968) using an automobile simulator. Figure 19-2 shows the relationship between field dependence–field independence and deceleration rate of subjects when reacting to an emergency in the simulator, specifically the figure of a child appearing on the "road" ahead.

Goodenough summarizes the results of a few investigations that tend to confirm the implications of this perceptual style as contributing to accidents. In a study by Harano (1970), for example, it was found that drivers with accident records (three or more accidents in three years) tended to be more field-dependent than did accident-free drivers. Presumably the field-dependent drivers have higher accident liability because they are more easily influenced by irrelevant cues when driving.

Although Goodenough points out that the reasons for the relationship between this type of perceptual style and driving behavior are not yet clear, he refers to evidence

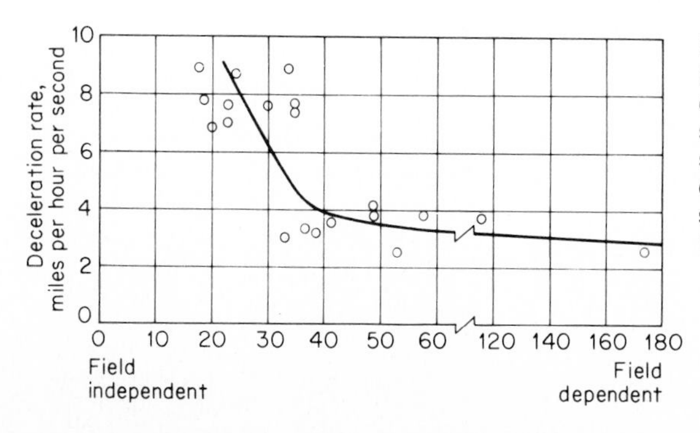

FIGURE 19-2
Relationship between field dependence–field independence style and deceleration rate of subjects reacting to an emergency situation in vehicle simulator. (*Source: Barrett & Thornton, 1968, fig. 3.*)

that suggests that field-dependent drivers do not quickly recognize developing hazards, are slower in responding to embedded road signs (those surrounded by many other stimuli), have difficulty in learning to control a skidding vehicle, and fail to drive defensively in high-speed traffic. In further efforts to "explain" the relationship, Shinar, McDowell, Rackoff, and Rockwell (1978) report data that show a relationship between field dependence and eye-movement behavior. The more field-dependent a driver, the longer his or her eye-fixation duration, thus indicating that it takes longer to pick up relevant information.

In connection with the measurement of field dependence, Williams (1977) suggests the use of a three-dimensional measure rather than the usual two-dimensional test. As evidence for this he presents data from a follow-up study of telephone company truck drivers that indicate that the three-dimensional test was more predictive of subsequent accidents than was the two-dimensional test.

Although it has been suggested that training might help field-dependent individuals to become more field-independent, there is as yet no strong evidence to indicate that this would be possible.

Age and Sex of Drivers Age has been found quite consistently to be one of the higher correlates of accidents. The rate is highest for teenagers, decreases sharply in the twenties, with an increase in the case of older drivers. This basic pattern is shown in Figure 19-3, this being based on data for males in Great Britain for a 2-year period. These data are for fatal and serious injury accidents, but the pattern for slight injury

FIGURE 19-3
Accident rate of males by age for fatal and serious injury accidents in Great Britain for a 2-year period. (*Source: Jones, 1976, fig. 3.4A, p. 73.*)

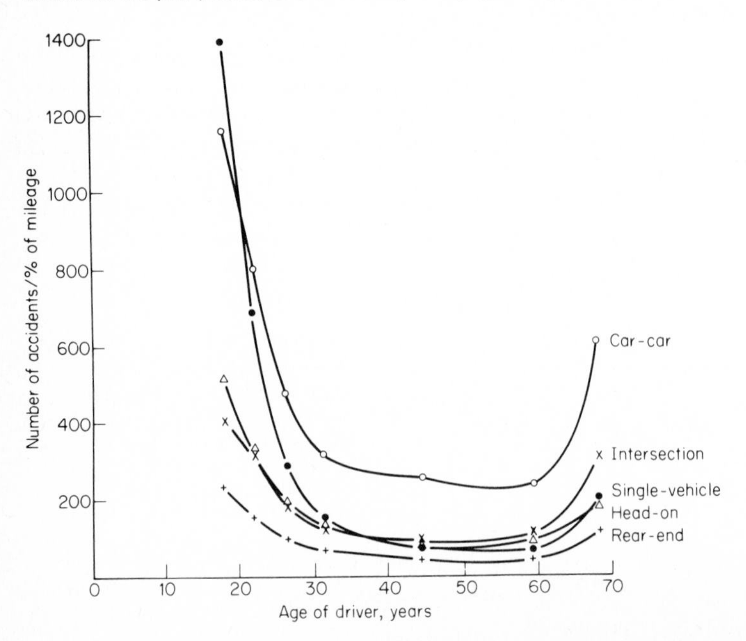

accidents is somewhat similar. Although some differences have been reported by sex, McGuire (1976) states that sex, per se, is probably not an important correlate of accidents, because the differences tend to be very small when age, exposure, etc. are controlled.

Behavior Patterns of Drivers

The behavior of people when driving presumably has its roots in their personal characteristics and experiences. Regardless of the origins, however, various aspects of driver behavior have been investigated toward the objective of determining what relationships there are, if any, between such behaviors and accident occurrence. Such *behaviors* have been characterized in various ways and at various levels of specificity, such as inattention, improper lookout, risk taking, judgment (as of gaps between vehicles and of car-following distances), steering behavior, reaction time, etc. Here (as in previous sections) we cannot deal extensively with all possible variables, but will discuss a few for illustrative purposes.

As something of an overview of human behaviors associated with accident occurrence, some data from a study by Fell (1976) are summarized in Figure 19-4. These data are based on analyses of "definite involvement" and "definite or probable involvement" of four "direct cause" categories. Of these it is clear that decision errors and recognition errors are dominant.

Sensory and Perceptual Input in Driving What Fell (1976) refers to as *recognition errors* probably stem largely from the sensory and perceptual input to drivers, especially through the visual sense. As discussed before, there seems to be evidence that persons who are field-dependent have higher accident rates than those who are field-independent.

FIGURE 19-4
Summary of the causal accident involvement of four human "direct cause" categories, expressed as percentages of accidents in which they were implicated. (*Source: Fell, 1976, fig. 2, p. 92.*)

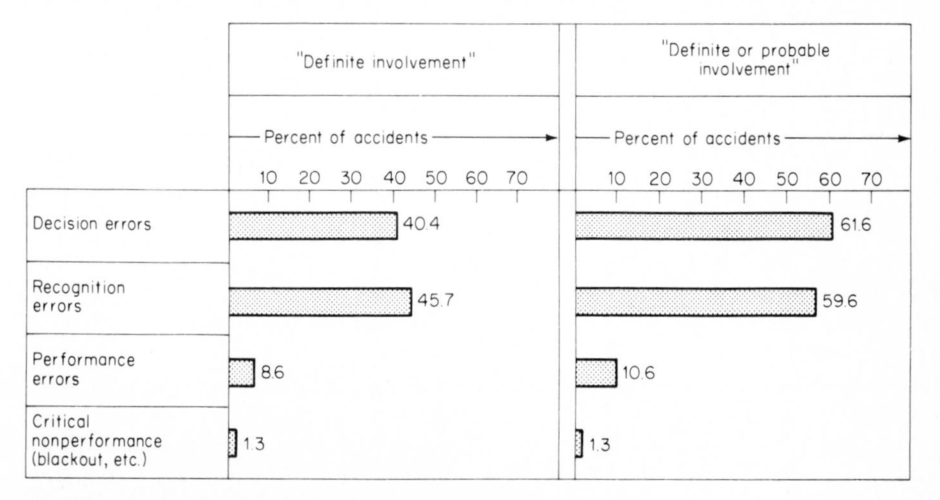

One of the various aspects of sensory and perceptual input is the visual and scan patterns drivers use. Such patterns typically have to be studied under somewhat controlled conditions. In one such study, Mourant and Rockwell (1970) used an eye camera to record the eye movements of drivers who were driving on an expressway at 50 mi/h. One of the implications from the study was that route familiarity plays a role in such patterns. Over unfamiliar routes the drivers typically "sampled" a wide area in front of them, but with increasing familiarity their eye movements tended to be confined to a smaller area. The investigators suggest that, since the unfamiliar driver has to rely on road signs for route guidance information, further investigations should deal with the relationship between density and sequencing of road signs with visual scanning patterns and the required control actions.

The results of this study also lend support to the hypothesis that peripheral vision is used primarily to monitor the road-edge markers. This points out that it is important to have good-quality and easily detected edge lines because of the poor visual acuity of the peripheral portions of the retina of the eye.

The perceptions of individuals of what they sense (as by vision or hearing) serve as the basis for their judgments about things in their environment—what Shinar calls *perceptual judgments*. An example or two will illustrate this point. One example deals with judgments of safe distances in following other vehicles. In this regard there is a "2-s" rule of thumb that has been proposed by a number of officials as a safe practice in following other cars. This practice means that—at the speeds of the cars in question—it would take 2 s for the following car to reach the momentary position of the car ahead. Colbourn, Brown, and Copeman (1978) found in a driving experiment that most drivers did allow this distance.

However, Harte and Harte (1976) summarize studies that indicate that there is a common tendency in actual driving situations (as contrasted with controlled experiments) for many drivers to underestimate following distances. Such underestimates, of course, can be serious when a driver decides to overtake the car ahead. Harte and Harte also report that such judgments are influenced somewhat by the type of road cues available. They found, for example, that underestimation of distances was greater on a segmented-lined road (which is a bit of a surprise) than on unlined roads, and more on unlined roads than on solid-lined roads where such lining is appropriate.

As an aside, Colbourn et al. (1978) point out that the 2-s rule of thumb is inadequate under conditions of poor visibility, bad weather, etc.

In connection with passing other cars, there are inklings that indicate that people tend to underestimate the time required to pass. To add to this potential danger in passing, there is a further tendency on the part of some drivers to make less realistic judgments when passing vehicles traveling at high speeds than at low speeds (Gordon & Mast, 1970). One cannot provide protection from all of the harum-scarum drivers of the world, but drivers who are aware of the human tendency to underestimate the time required for passing can at least increase their own survival probabilities by *not* passing other vehicles if there is the slightest doubt about the time factor.

As an example of another facet of the perceptual judgments of drivers, there is substantial direct and indirect evidence to indicate that the phenomenon of *adaptation* influences the perception of speed. The concept of *adaptation* applies to many circumstances. In the driving situation it is reflected by the tendency to perceive a given speed (say 50 mi/h) to be *less* than that if a person has previously been adapted to a *higher* speed, and to be *higher* than that if a person has previously been adapted to a *lower* speed. Various investigators have tended to confirm this tendency (Denton, 1976; Mathews, 1978). This influence is especially noticeable when changing from a high speed to a lower speed, the estimates of lower speeds usually being underestimates (Schmidt & Tiffin, 1969). Further, the longer the exposure to a high speed, the greater the underestimate of lower speeds. The effects of this would be for people who slow down (as on an exit ramp of a superhighway) to drive at a higher speed than they think they are driving. In this and other driving circumstances a person could then be driving at a higher speed than is warranted, thereby increasing the likelihood of accidents.

Decision Processes A vehicle driver makes many decisions in the course of a trip, and executes these in driving behavior by controlling the steering wheel, accelerator, brake (and sometimes the horn). As an example of the decision processes we will mention briefly the matter of risk taking. Although the genesis of variations in risk taking on the part of different individuals is shrouded, there are indeed differences in the extent to which at least some people are inclined to take risks. And even the same individual varies from time to time in the risks he or she is apparently willing to take.

While granting the fact of rather marked individual differences in risk taking, Näätänēn and Summala (1976) remind us of a fairly universal tendency for people to view themselves as being relatively immune to accidents ("accidents happen to other people—not to me"), and discuss the factors that tend to decrease the subjective dangerousness of traffic for themselves as road users. Some of these are:

- Deluding perceptual and cognitive processes (as sensory adaptation to speed discussed above, underestimation of speed, underestimation of the physical forces at work in a traffic accident, and the interpretation of threatening situations and accidents with only slight consequences).
- Learning (the consequences of observations of many situations that were potentially dangerous, but in which no unfortunate consequences occurred; this is also something of an adaptation process).
- Subjective easiness in driving (the feeling that driving, for most of the time, is quite an easy task).
- The driver as the operator of the vehicle (the driver may feel that the risk of future traffic situations can be reduced because he or she can have some influence on the situation).
- Expectancy (most people have little basis for estimating the probabilities of accidents).
- Little traffic supervision (the risk of apprehension for traffic violation is limited).
- Examples, norms, and images (these are viewed by many individual drivers as being "for other people," not themselves).

The risk-taking tendencies of people are intricately intertwined with people's expectations—that is, the probability of some unhappy consequence resulting from given behaviors—and since accidents are relatively rare (and the odds of accidents therefore low) such limited risks tend to lend a sense of false immunity. As a specific aspect of expectancies, there have been several investigations that have demonstrated that reaction time is increased appreciably if one is not *expecting* to have to make a response (such as with a brake), as contrasted with a situation in which a response might have to be made.

Driver-Related Capabilities

Programs for increasing the capabilities of drivers include driver training programs, licensing of drivers, enforcement practices, and driver improvement programs (especially for those involved in violations or accidents). At various times questions have been raised about the adequacy of all of these types of programs—in some instances with good cause. Although such programs are relevant to vehicle operation, we cannot deal with them in any comprehensive manner here, and will simply state that such programs, when properly established and administered, undoubtedly can contribute somewhat to the improvement of driving and the reduction of accidents.

Temporary Impairment

The primary factors that contribute to temporary impairment of driving are (1) fatigue, and (2) alcohol and drugs. These will be touched on briefly.

Fatigue The primary culprit in generating fatigue is travel time and associated loss of sleep. The degradation in driving performance because of fatigue is reported by Shinar (1978) to account for a small, but significant, percent of highway accidents. However, among long-haul commercial drivers, who often must drive long hours, fatigue can be a significant cause of accidents. The Bureau of Motor Carrier Safety, for example, investigated 286 commercial vehicle accidents and found that 38 percent of them were categorized as attributable to the driver's being either asleep at the wheel or inattentive (Harris & Mackie, 1972).

Some drivers are often aware of the early stages of fatigue and can take steps to minimize it, as by stopping for a snooze or a cup of coffee. Hulbert (1972) lists other behaviors as preceding the final stage of actually falling asleep.

- Longer and delayed deceleration reactions in response to changing road demands
- Fewer steering corrections
- Reduced galvanic skin responses (GSR) to emerging traffic events
- More body movement such as rubbing the face, closing the eyes, and stretching

Presently the primary basis for reducing the risk from fatigue lies with the driver, who must be sensitive to the signs of the onset of fatigue and must stop driving. (An interesting aid to reduce fatigue is suggested by Endou, Ohshima, Watanabe, and Shingo, 1979, namely chewing gum. However, they point out that this is only a palliative, and that the only safe method is to stop driving.) The authors have heard indirectly of the use of corrugated berms as a possible method of rousing a drowsing driver from a snooze. On the freeways in certain locations *Botts dots* are used to demarcate the lanes. These are small raised bumps, and are so arranged that if a driver changes lanes the tires roll over them with a "roar." They are quite effective in alerting drowsy or inattentive drivers. (They cannot be used in snow areas since they interfere with the use of snow-removal equipment.)

Alcohol and Drugs Shinar and other traffic experts have supported the general belief that a major portion of all accidents arise from the use of alcohol and drugs. This is so well documented that we need not discuss it further here.

Discussion

The extent to which corrective action can be taken about personal factors associated with accidents varies markedly with the particular factors in question. Granting this, however, let us mention a few illustrative factors along with possible corrective action. For example, some cases of poor vision can be assisted with properly prescribed glasses or in exceptional cases by bioptic telescope spectacles; further, proper road markings and signs could be useful. To help people who tend to be field-dependent in their perception, road signs might be placed in locations that are not cluttered up, so they tend to be clearly visible. The high incidence of accidents among teenagers might be reduced by more adequate driver training programs and more careful practices in issuing drivers' licenses.

The major problem associated with alcohol and drugs probably needs to be attacked through drivers' training and education programs and enforcement practices. In connection with strictly experimental innovative approaches to the problem of driving while under the influence of alcohol, some exploratory work presumably has been carried out in the development of a computerized digit memory test. A set of digits would be presented to a person before starting a car, and the person would be required to punch them into a keyboard correctly, otherwise the starting mechanism of the car would not operate. The theory behind this is as follows: a person under the influence of alcohol would not be able to remember the digits, and could therefore not start the car.

In connection with driving in general, the decisions and actions involved by virtually all drivers to some degree can be aided and abetted by the appropriate design of highways, road markings, traffic lights, and road signs, and in part by the appropriate design of the vehicles people use.

THE DRIVING ENVIRONMENT

The driving environment includes the roads and highways, street and highway lighting, road markings, road signs, and traffic as well as the *natural* features of the ambient environment such as rain and snow. A few illustrative features of the environment are discussed briefly below.

Road Characteristics

There are various indication regarding the effects of road characteristics on driving performance. One of the most clear-cut factors is with regard to the use of superhighways as contrasted with conventional two-way roads. The accident rate on superhighways is

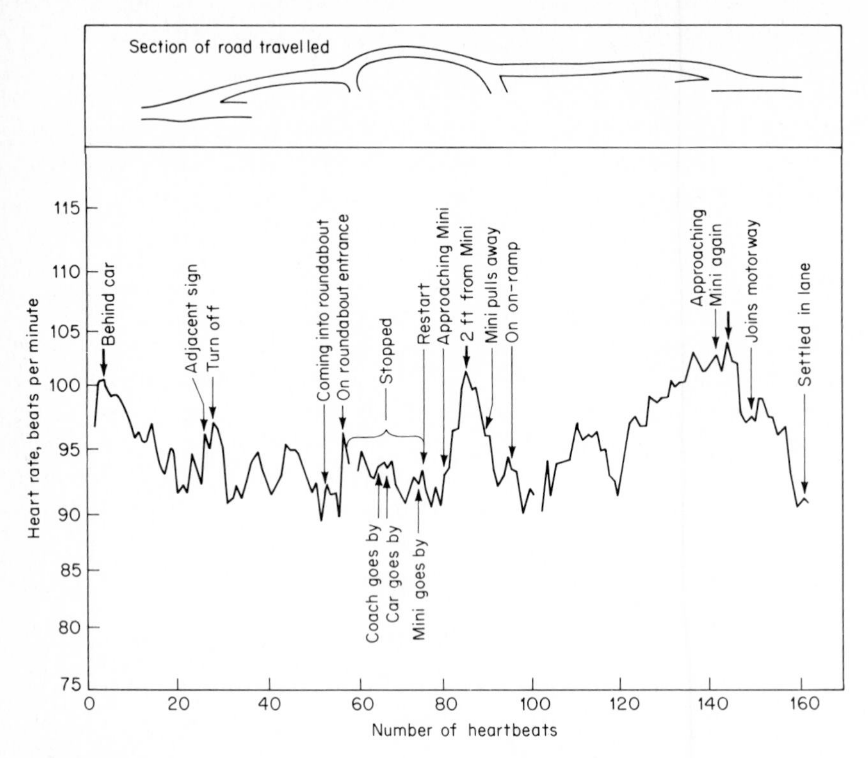

FIGURE 19-5
Heart rate of one driver on a section of a road near the London airport. The horizontal scale represents the time period during which 160 heartbeats were recorded. The features of the road are sketched at the top, and the written entries indicate the "events" occurring during the time period. (*Source: Rutley & Mace, 1972, fig. 3.*)

clearly lower, this lower incidence being attributed to the reduced number of circumstances in which any given vehicle encounters other vehicles moving in other directions.

An example of a different kind of "effect" of road design characteristics is reported by Rutley and Mace (1972), who rigged up electrodes to record the heart rates of a few drivers when driving on an approach to the London airport. An example of these recordings for one driver is shown in Figure 19-5. This figure shows the heart rate in relation to the road being traversed and the "events" that occurred when driving over the section of the road in question. One can consider such physiological responses on the part of the driver (in this instance, changes in heart rate) as indexes of strain that reflect varying levels of stress that are generated by the driving situation.

Road Markings

Everyday experience in driving clearly demonstrates the value of road markings, such as edge markings and center lines, especially in night driving. Aside from such mark-

TABLE 19-1
MEAN SPEEDS OF VEHICLES BEFORE AND AFTER PAINTING OF SPACED TRANSVERSE LINES ON ROADS

	Mean speed	
	km/h	mi/h
Before	57.0	35.3
Immediately after	44.1	27.3
1 year after	52.4	32.5

Source: Rutley, 1975.

ings, however, there are some interesting developments in the value of special markings in specific situations. In Great Britain, for example, Rutley (1975), drawing on Denton's previous research, has experimented with the use of yellow transverse lines on the road starting 0.4 km (about ¼ mi) ahead of traffic circles (called *roundabouts* in Great Britain). The spacing of these lines was reduced the closer they were to the traffic circle. The driver, first becoming "adapted" to lines being crossed at a given (slow) rate, then becomes aware of lines being crossed at ever-increasing rates. This lends an illusion of speed (like telephone poles being passed at a high rate of speed) that typically causes the driver to slow down as he approaches the traffic circle. Rutley presents the data shown in Table 19-1 on the mean approach speed of vehicles at one traffic circle before such markings were painted on the road, immediately after, and 1 year after. At this circle there had been 14 accidents the year before the stripes were painted on the road, and only one accident during the year after, thus strongly suggesting that such stripes can influence driver behavior.

Another road marking scheme was used by Shinar, Rockwell, and Malecki (1980) for marking a road curve. The markings were based on an illusion effect discovered many years ago by a German psychologist, Wundt, who found that V-shaped lines between parallel lines make the parallel lines appear to be closer together at the center than they really are, as illustrated in Figure 19-6. Shinar et al. (1980) painted V-shaped lines in a herringbone pattern on a road starting 97.5 m (about 300 ft) ahead of an obscured curve and on the curve itself. It was hypothesized that the herringbone pattern of lines would cause the illusion of narrowing of the road, and cause drivers to slow down.

FIGURE 19-6
The Wundt illusion that causes parallel lines to appear to be closer together than they are at the center. This illusion was used by Shinar, Rockwell, and Malecki (1980) as the basis for drawing a herringbone pattern of lines approaching, and on, a curve in a road to cause drivers to slow down on the curve.

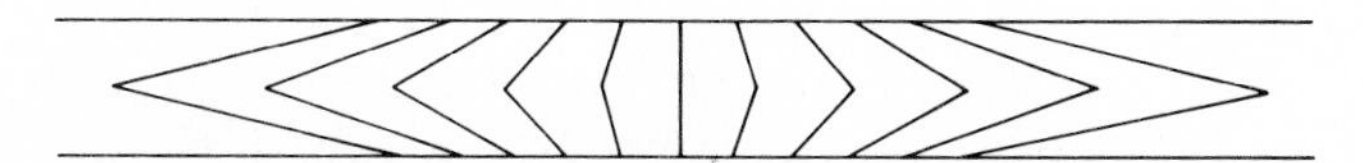

TABLE 19-2
REDUCTION IN SPEEDS OF VEHICLES APPROACHING CURVES BEFORE AND AFTER PAINTING OF HERRINGBONE PATTERN ON ROADS

	Mean reduction in speed (between approach to curve and speed in the curve)
Before	6.4 km/h (4.0 mi/h)
After	11.7 km/h (7.3 mi/h)

Source: Shinar, Rockwell, & Malecki, 1980.

The herringbone pattern presumably did have the intended effect, as illustrated by the *reductions* in mean speeds by which drivers drove around the curves before and after the lines were painted (see Table 19-2).

Road Signs

Chapter 4 included reference to a few studies dealing with the design and effectiveness of certain types of road signs, so we need not refer here to other examples. However, we will mention briefly three aspects of road signs, namely the response of drivers to signs, the size of sign legends (as letters and numerals), and the matter of legibility at night.

Response to Road Signs To serve their intended purposes road signs should be so designed and so located as to catch the driver's eye. In this regard some rather discouraging results of a survey are reported by Johansson and Rumar (1966), who stopped and interviewed 1000 drivers to ascertain how many recalled having seen five specified types of road signs in a previous stretch of road. On the average, the signs were "recalled" by the motorists only 47 percent of the time; however, there were differences in the percentages for the individual signs, as follows: speed limit (78 percent), "police control" sign (63 percent), road condition sign (55 percent), general warning (18 percent), and pedestrian crossing (17 percent).

Size of Letters on Signs Shinar (1978, p. 169) refers to earlier recommendations that letters on road signs be 1 in high (2.54 cm) for every 100 ft (32.8 m) of viewing distance. Thus, an "exit" sign to be seen 500 ft (164 m) ahead of the exit should be 5 in (12.7 cm) high. However, since some states in the United States are issuing licenses to persons with 20/40 vision (as contrasted with 20/20) Shinar recommends a doubling of letter heights over such values.

Legibility of Signs at Night The legibility of road signs at night depends on several factors, such as the size of symbols, the colors of sign legends and of their backgrounds, the luminance of the legends and backgrounds, the type of material used, any ambient illumination, and the lights from approaching cars.

An illustration of the effect of sign brightness on legibility of numerals on signs of high and of low reflectance is based on a controlled experiment by Hicks (1976). In

TABLE 19-3
LEGIBILITY OF NUMERALS BY SOBER AND ALCOHOL-IMPAIRED DRIVERS UNDER SPECIFIED REFLECTANCE AND HEADLIGHT CONDITIONS

		Mean legibility distances			
		Sober		Alcohol-impaired	
Reflectance	Headlights	m	ft	m	ft
Low	Low	147	482	137	450
Low	High	173	568	163	535
High	Low	169	554	155	508
High	High	181	592	173	569

Source: Hicks, 1976.

this study, the drivers used both low headlights and high headlights on different nights. Legibility was measured in terms of the distance at which the numerals could be read by subjects when sober and when alcohol-impaired. For one group of subjects the mean legibility distances under the different experimental conditions were as shown in Table 19-3. Both reflectance of signs and the position of the headlights influenced legibility. And, incidentally, although alcohol impaired the performance of the drivers, high sign reflectance and high headlights made the signs more legible to them than low sign reflectance and low headlights.

Roadway Illumination

Among the postulated advantages of roadway illumination is that of safety. The before-and-after accident records of streets and highways which have been illuminated provide extremely persuasive evidence that this pays off. Table 19-4 is a summary of the data compiled by the Street and Highway Lighting Bureau of Cleveland for 31 thoroughfare locations throughout the country showing traffic deaths for the year before they were illuminated and for the year after. Such evidence is fairly persuasive, and argues for the expansion of highway illumination programs.

Roadway Features That Influence Visibility Several aspects of roadways presumably have some effect upon visibility in night driving, and therefore possibly are related to accidents. One of these is transition lighting. Where roadway luminaires are to be

TABLE 19-4
TRAFFIC DEATHS BEFORE AND AFTER ILLUMINATION OF 31 THOROUGHFARE LOCATIONS

		Reduction	
Year before illumination	Year after	Number	Percent
556	202	354	64

Source: Public lighting needs, 1966.

TABLE 19-5
EFFECTS OF TYPE OF HIGHWAY LUMINAIRE ON GLARE AND VISIBILITY RATINGS

	Type of luminaire	
	Cutoff	Noncutoff
Glare rating, mean	1.8	−0.9
Visibility rating, mean	1.3	0.4

Source: Christie, 1963.

used (such as at an intersection), it is desirable to provide some "transition" on the approaches and exits by graduating the size of lamps used. This helps to facilitate visual adaptation to and from the highly illuminated area. Another factor is what is referred to as *system geometry*, which refers particularly to the mounting height of luminaires; in some installations the mounting height is 35 ft or more, and in some European systems installations at heights of 40 ft or more are used. Such heights (while requiring larger lamps) help to increase the *cutoff* distance, that is, the distance from the light source at which the top of the windshield cuts off the view of the luminaire from the driver's eyes. Under average conditions this cutoff is about 3.5 times the mounting height (MH) of the luminaire.

Still another factor is that of luminaire design, which can control the "cutoff" of luminaire candlepower at greater approach distances, thus improving comfort and visibility. The effects of the "cutoff" type of luminaire are shown in Table 19-5 in terms of glare ratings and visibility ratings of 121 drivers. The rating scales ranged from −2 to +4.

In addition, the reflectivity of roadway surfaces varies substantially, creating a factor which in turn affects visibility. For example, asphalt pavement surface (about 6 years old) reflects approximately 8 percent of the light, whereas concrete pavement has a surface reflectance of about 20 percent. It has been estimated that the amount of illumination required for equal brightness of asphalt pavements, compared with concrete, is of approximately the ratio 1.92:1 (Blackwell, Prichard, & Schwab, 1960); this is roughly a 2:1 ratio.

Vehicle Headlights

Closely related to roadway illumination is the illumination from vehicle headlights. A primary problem arises from glare produced by the headlights of oncoming vehicles. The effects of glare on visibility are shown by the results of a study by Mortimer and Becker (1973) as illustrated in Figure 19-7. This shows the *visibility distance* of targets ahead of drivers (on the left side of the driving lane) as the drivers approached oncoming cars with high-beam or low-beam lights. Glare increases and visibility decreases as the oncoming headlights approach, but as the headlights of the oncoming vehicle start to move away from the center of the visual field (when the two cars are nearing each other) visibility starts to increase. High-beam headlights of oncoming cars clearly reduce visibility more than low-beam headlights, except when the two cars are fairly close to each other.

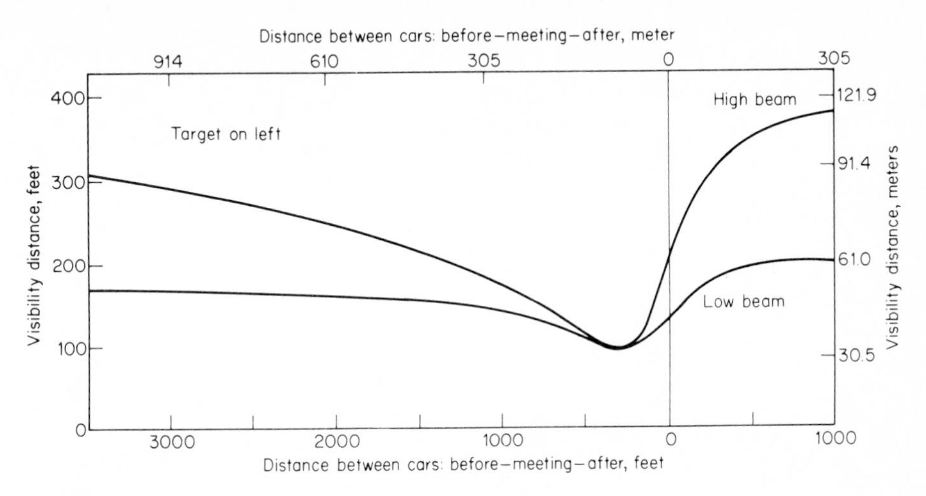

FIGURE 19-7
The effects of glare from oncoming headlights on visibility at varying distances between approaching cars. (*Source: Adapted from Mortimer & Becker, 1973.*)

Making the point that driver behavior is difficult to change, Shinar (p. 146) suggests the use of polarized headlight beams and the possible addition of a third light beam (the *midbeam*) as means of reducing glare from oncoming headlights.

Traffic Events

Although it is not feasible to discuss many implications of the traffic environment in driving behavior, we will mention one interesting procedure used by Helander (1978). He placed electrodes on 60 subjects driving a rural test route, and obtained measures of brake pressure. The electrodes recorded heart rate, electrodermal response (EDR), which is the same as galvanic skin response (GSR), and two electromyograms. The various measures were then related to 15 types of "traffic events" such as "other car passes own car" and "meeting other car." There were 7 million measurement values, corresponding to 120,000 per test run. Without going into details, a primary analysis of the mass of data made it possible to determine the extent to which the various measurements from the electrodes were related to brake pressure. Brake pressure was considered as a criterion of the *mental difficulty* of the various traffic events. In general terms Helander found that the EDR measure had the highest correlation with brake pressure for the various traffic events, this being .65. By eliminating a couple of the events the correlation was .95, indicating that the EDR measure is a potentially very useful measure of the *difficulty* of traffic events. The event that was most difficult was "cycle or pedestrian + meeting other car," and the next one was "other car merges in front of own car."

VEHICLE CHARACTERISTICS

Some of the previous chapters have dealt with topics that have certain implications relating to vehicle design features, such as the discussion of displays, control processes and control mechanisms, seating and space considerations, and vibration and motion. The control characteristics of vehicles are especially important, as discussed by Jones (1976). For illustrative purposes we will discuss here the effects of vehicle mass on deaths and injuries, the physiological effects of the use of manual and automatic transmissions, response time in certain aspects of driving, and the experimental use of *head-up* display speedometers.

Vehicle Mass

It is a commonly accepted notion that heavy cars are "safer" than light cars. Grime and Hutchinson (1979) generally confirmed this belief by the results of an analysis of head-on accidents in rural areas of Great Britain. As they point out, Newton's law, as applied to two-vehicle collisions, would imply that relative velocity change is in inverse proportion to vehicle mass (i.e., weight). Their data showed that, when the larger of two vehicles is twice the mass of the other, the percentage of deaths in the lighter vehicle is about six times that of the heavier. More detailed analyses indicated that the ratio of the masses of the two vehicles is related most to deaths, and least to slight injuries.

The implications of the effects of mass introduce a potentially conflicting problem in this day and age when there is a tendency toward increased use of small cars for energy-conservation purposes. As larger cars are phased out over time, however, this conflict will be minimized.

Manual Versus Automatic Transmissions

The use of manual transmissions involves more control actions than the use of automatic transmissions. In addition, however, the demand for executing more control actions might also add at least some mental stress, especially in heavy traffic situations. Zeier (1979) carried out a study in Zurich, Switzerland, with drivers using both types of transmissions in a 14-km (about 9-mi) route, taking certain physiological measures during and after the driving task (we will not go into the details). He found significant differences when using the two types of transmissions in rate of adrenaline excretion, skin conductance activity (SCR), heart rate, and heart rate variability, and concluded that driving with the manual transmission produces greater activation of the sympathetic nervous system, in effect reflecting a higher level of stress. He suggests that the stress reduction due to the use of an automatic transmission may constitute a valuable measure for improved health and safety since it enables a driver to concentrate more on traffic events.

Response Time in Vehicle Operation

One factor that differentiates between the quick and the dead in traffic is reaction time in operation of controls in emergency conditions. It is impossible to know in advance when an emergency will arise, but it is unfortunately the case that response time is greater under surprise conditions than when one is anticipating the need to make a response. This was demonstrated by Johansson and Rumar (1971), who measured brake reaction time of a sample of 321 drivers under an "anticipation" condition, and then adjusted the values (by a method we need not describe) to estimate the reaction time under a "surprise" condition. The median estimated brake reaction time was 0.9 s as compared with 0.66 s under the "anticipation" condition.

Although one cannot do much to decrease the time required to initiate a response under "surprise" conditions, there are some possible methods of minimizing total response time in activating a brake. One scheme, for example, is to have the brake and accelerator at the same level, this making it possible simply to move the foot laterally to the brake, rather than to raise the foot and then shift it laterally. A different (and promising) twist in the operation of the accelerator and brake is represented by a combined brake-accelerator (Konz, Wadhera, Sathaye, & Chawla, 1971). The combination pedal has two fulcrums, such that when activated by the toe, the pedal serves as an accelerator, and when activated by the heel, it serves as a brake. Some rather persuasive evidence about this is shown in Figure 19-8. This shows the reaction times (i.e., *cycles*) of 50 subjects when using a laboratory model of the dual-function pedal and 50 subjects when using the conventional systems with separate brake and accelerator. The figure shows the distribution of 500 reaction times (10 for each of the 50 subjects). The difference in mean values was 19 ms, this representing a potentially very significant difference in terms of stopping distance. Although the design of vehicular control devices is only one of several factors that influence driving behavior, the design of such devices in terms of human considerations (including any possible innovative features,

FIGURE 19-8
Distribution of 500 reaction times (i.e., *cycles*)—10 for each of 50 subjects using a dual-function brake and accelerator pedal, and using a conventional brake-accelerator system. (*Source: Adapted from Konz, Wadhera, Sathaye, & Chawla, 1971.*)

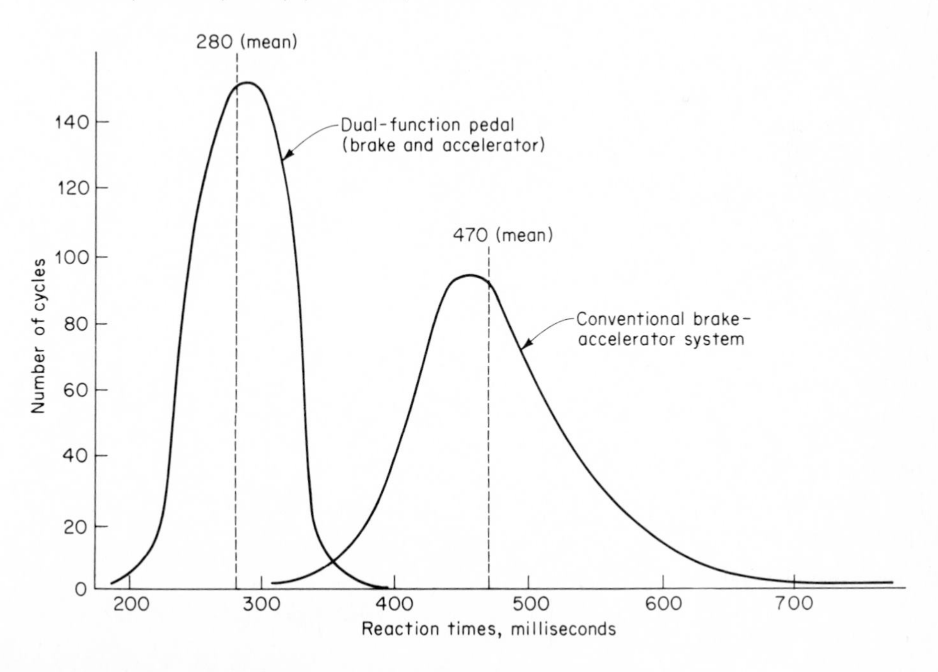

TABLE 19-6
SPEEDS OF DRIVERS USING STANDARD AND HEAD-UP SPEEDOMETERS

	Mean speed, mi/h	
Advised curve speed, mi/h	**Standard speedometer**	**Head-up speedometer**
25	28.1	26.3
30	31.5	30.0
35	35.0	33.6
40	37.6	35.9

Source: Rutley, 1975.

such as the combined brake-accelerator) can, to some degree, contribute to improved driving performance.

Head-Up Display Speedometers

As a possible procedure for making drivers more aware of their speed, Rutley (1975) experimented with a *head-up* display speedometer. This consists of the projection of speed on a partially silvered surface on the windshield directly in front of the driver's eyes, and is common in high performance military aircraft. Rutley suggests that such a speedometer might be particularly useful to drivers when going around curves that have advisory speed limits, and found that, in fact, there was a tendency for drivers to drive a bit more slowly with such a speedometer, as shown in Table 19-6.

PUBLIC TRANSPORTATION SYSTEMS

The previous discussion has dealt largely with the operation of vehicles and some of the individual and situational factors that influence such operation. Let us now turn our attention to some of the human factors aspects of public transportation systems. The various forms of public transportation include airlines, trains, buses, subways and elevated systems, trolleys, and a few new forms such as monorails. The particular forms of course vary with distances to be traveled and the countries in question. Because of the worldwide oil problem there has been some shift from the use of automobiles to the use of public transportation systems, and this trend undoubtedly will continue, especially in connection with urban transportation systems.

Human Factors Aspects of Public Transportation

The human factors overtones of public transportation systems are manifest. From the point of view of the users (or potential users) some of the human factors aspects tend to fall into the following categories:

- Station or terminal facilities and features
- Convenience and mobility of vehicles
- Vehicle interior (convenient entrance and exit, space and seating comfort, envi-

ronmental factors such as minimum noise and vibration, adequate illumination, air quality, etc.)

- Safety and security
- Social factors (an adequate personal space, minimum crowding, etc.)
- Esthetics and other psychological factors

It is of course not feasible here to deal extensively with the many specific human factors aspects of transportation systems, so we will discuss only a few aspects, especially those related to urban transportation systems.

Importance of Design Features to Users

The relative importance of specific design features of transportation systems varies from individual to individual, but some impressions of the judgments of importance across a sample of people comes from a survey by Golob, Canty, Gustafson, and Vitt (1970), based on interviews with, and questionnaires from, 786 individuals. The questionnaires included pairs of items such as the following:

A. Making a trip without changing vehicles
B. Easier entry and exit from the vehicle

The items generally depict something about the service that might be provided by a transportation system. All together there were 32 items. The items were divided into several "blocks," and those within each block were paired together. Any given respondent then indicated a preference for one of the items in each pair. On the basis of these preferences a scale value was derived for each respondent for each item in the block of items in the questionnaire. The mean scale value was then derived for each item, based on the responses of all who had that item in their questionnaires. These scale values are given for all 32 items in Figure 19-9.

A word of caution is in order in evaluating the scale values, since the items were paired within each of nine blocks. In a strict sense, the scale values can be compared with each other only within each such block. Since one item ("lower fare") was common to all nine blocks, it did serve as a common link, but even so some caution in interpretation is in order. Recognizing this possible distortion of the scale values, the figure shows that "arriving when planned" comes out with the highest scale value, followed by "having a seat" and "no-transfer trip." These in turn are followed by a cluster of nine characteristics concerned mainly with the customers' time, fare, and shelters. Lower on the scale is a large cluster of 18 characteristics that are concerned primarily with interior design, esthetic aspects of the trip, and passenger convenience. The items at the bottom of the scale are "coffee, newspapers, etc." on the vehicle, and "stylish vehicle exterior."

As an additional analysis, the characteristics were separated into three subgroupings: levels of service, convenience factors, and vehicle design characteristics. Figure 19-9 represents these groups by different symbols (a solid dot, a triangle, and a square).

The relative scale values of such preferences are of course influenced by the selection of items used. In a somewhat similar survey, for example, Fox (1974, pp. 9–12)

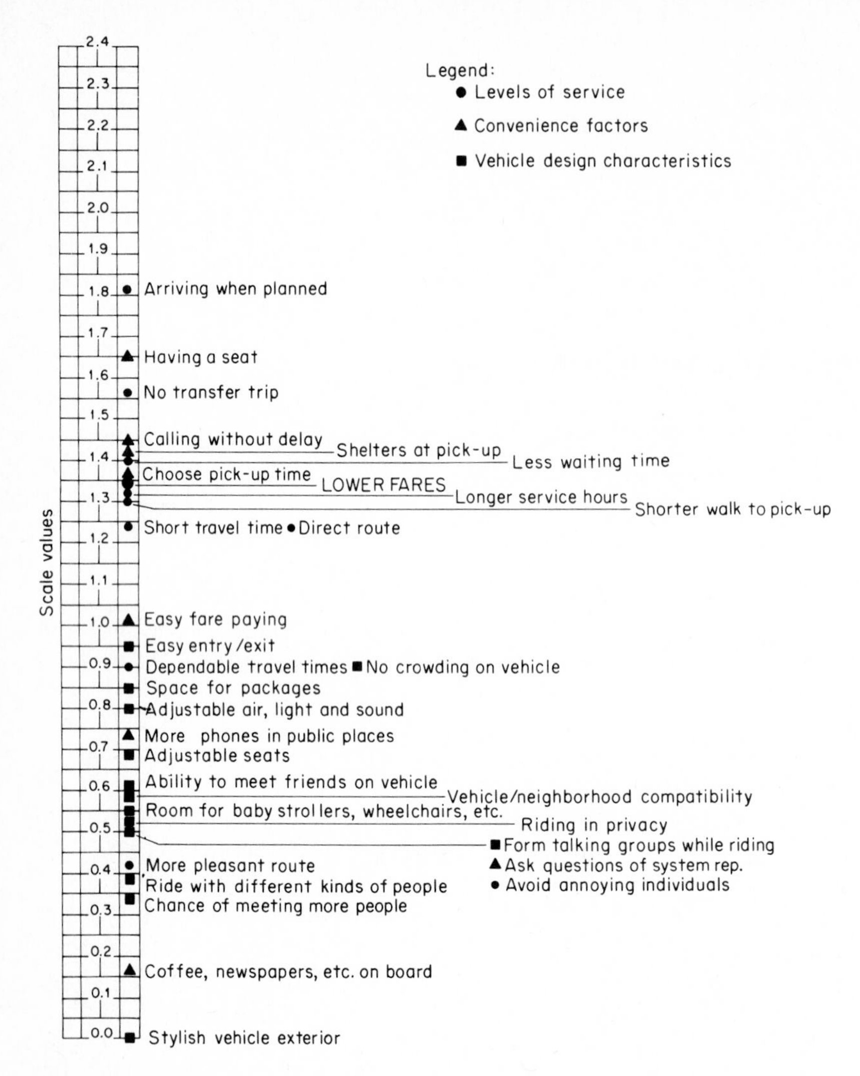

FIGURE 19-9
Scale values of preferences for 32 characteristics of public transportation systems based on the paired-comparison responses of 786 subjects. The different symbols represent three different groups of characteristics. (*Source: Adapted from Golob, Canty, Gustafson, & Vitt, 1970, figs. 6, 8a, 8b, and 8c.*)

found that "vehicle accident safety" was judged as most important by his subjects. Since the survey by Golob et al. (1970) did not include such an item (and since there were numerous other differences in the items in the two studies), comparisons between the two studies are difficult.

The results of surveys such as by Golob et al. (1970) and Fox (1974) give some impression of preferences for various characteristics or types of service of transportation systems. Similar approaches can be used to obtain expressed preferences for alternative features of a specific aspect of a system. An example of this approach is

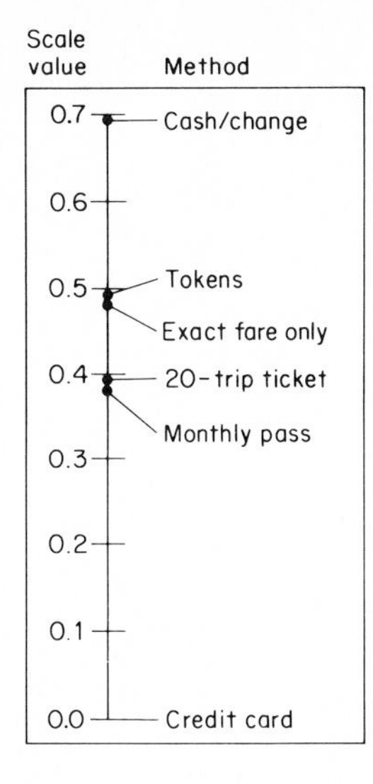

FIGURE 19-10
Scale values of preferences for various methods of fare collection, based on paired-comparison method. (*Source: Gustafson, Curd, & Golob, 1971, fig. 5.*)

found in the results of a survey by Gustafson, Curd, and Golob (1971), this example dealing with alternative methods of payment. Figure 19-10 shows the scale values of the alternative methods (in this instance it was found that the use of cash or change was the preferred method).

Functional Aspects of Transportation Systems

There are a number of functional aspects of transportation systems that can influence overall effectiveness, functions such as ticket selling, baggage handling, and loading and off-loading of passengers. The time for loading and off-loading of passengers, for example, influences the *dwell time* of a system, that is, the time a vehicle (such as a bus) would have to stop at stations or locations, thus of course affecting total trip times. The dwell time itself is in part influenced by the physical features of the doors, platform, and mode of collection of fares (if collected on the vehicle, such as a bus). Some data on such times are reported by Bauer (1970), with examples of average loading times being given in Table 19-7.

It might also be added that the off-loading times for various types of vehicles as reported by Bauer range from 0.8 passengers per second (on a curb-level platform bus in Toronto) to 1.6 passengers per second on Toronto subways, there being a difference of 2 to 1 in this ratio.

In connection with the functional aspects of transportation systems, greater attention is now being paid to features that facilitate the use of systems by the handicapped. In this regard the United States Department of Transportation has developed a procedure for evaluating the suitability of vehicles for the handicapped. In particular this procedure provides for the calculation of the percentages of handicapped who would be able to use a particular transportation mode with the *removal* of various combinations of *barriers*. In the case of buses and trolleys these barriers are sudden movement of the vehicle; riding while standing; rapid self-locomotion; movement in crowds; wait standing; taking short steps; rising from a seat; aisle width (restricted);

TABLE 19-7
ILLUSTRATIONS OF LOADING TIMES OF PASSENGERS ON VARIOUS TRANSPORTATION VEHICLES

Feature	Bus	Streetcar	Subway
Location	Detroit	New Orleans	Toronto
Platform	Street level	Street level	Vehicle-floor level
Door	30 in	50 in	44 in
Steps	3 steps	3 steps	No steps
Fare collection	On board	On board	Prepaid
Passengers per second	0.4	0.7	1.2
Seconds per passenger	2.7	1.4	0.8

Source: Bauer, 1970, table 3, p. 16.

and walking long distances. If all of these *barriers* were removed, 99 percent of the handicapped could use the transportation mode in question, whereas if only the first four were removed, only 35 percent could use it. Although the removal of all such barriers could pose a serious design problem, solutions or partial solutions for at least some of them have been proposed (*Travel barriers*, 1970):

- *Sudden movement:* provide special bus lanes to control traffic; pad hard surfaces to reduce accidental injuries; use vertical floor-to-ceiling stanchions.
- *Crowds:* limit bus seating; use smaller buses with more frequent service; redesign fare turnstiles to eliminate push-bar and widen channel; use pressure mats to open fare gates when coin is deposited; use automatic doors at exits; improve coin receiver to eliminate precision movements; modify buses to lower entrance; use mechanical steps, or add ramps or lift; provide raised platform at bus stops; major redesign of buses.
- *In-vehicle barriers:* pad hard interior surfaces; provide vertical stanchions for all seats; reserve open spaces for wheelchairs; widen aisles.

New Types of Transportation Systems

Aside from the common types of urban systems such as buses and subways, there have been various experimental designs of new types of vehicles, such as a multimodal capsule suggested by Canty and Sobey (1969) and illustrated in Figure 19-11. This type of

FIGURE 19-11
Sketch of a multimodal capsule vehicle suggested for possible use in an urban transportation system. (*Source: Canty & Sobey, 1969.*)

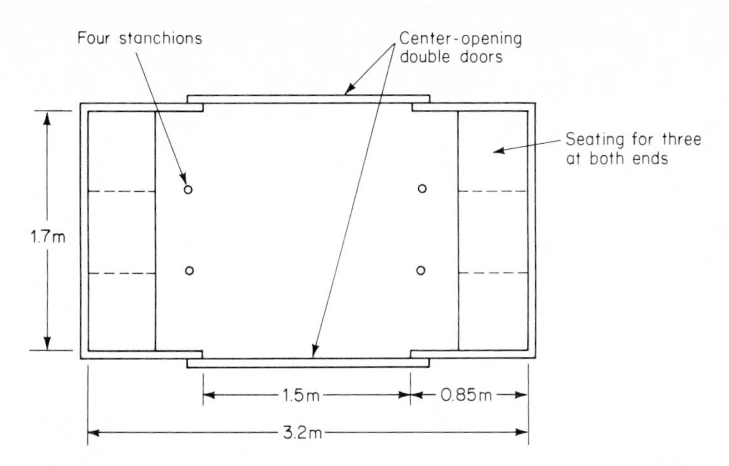

FIGURE 19-12
The dimensions and features of the vehicle recommended for a proposed "minitram" urban transportation system. *(Source: Ashford & Kirk, 1978, fig. 1, p. 476.)*

system would be most useful for transportation between a central business district and outlying residential areas.

In Great Britain, Ashford and Kirk (1978) report a study dealing with the design of a vehicle for possible use in a proposed "minitram" system concept. In particular they experimented with 12 variations in vehicle design varying in width and length, door width, and nominal capacity (ranging from 12 to 20 persons). One phase of the study dealt with variations in design and vehicle population factors on station dwell time. In particular an analysis was made of the number of passenger movements as affecting dwell time. One *passenger movement* was defined as a passage either in or out of the scale model of the vehicle. In general there was a linear relationship between dwell time and passenger movements.

On the basis of the various considerations the "final" vehicle is the one that is shown sketched in Figure 19-12 (above), this vehicle seating six persons and having standing space for six more. The dwell time that should be allowed for such a vehicle was 16 s, this time generally being adequate for use by wheelchair patrons as well as ambulatory riders.

Effects of Commuting

Commuting by train into urban centers is virtually a way of life for millions of people. A rather interesting survey of the subjective feelings of commuters is reported by Singer, Lundberg, and Frankenhaeuser (1974). As one part of their survey they asked a sample of volunteer passengers to rate their subjective comfort, boredom, unpleasantness, and crowdedness after boarding the train at the suburb where they had started, and after stopping at six stations before arriving at Stockholm, the final destination.

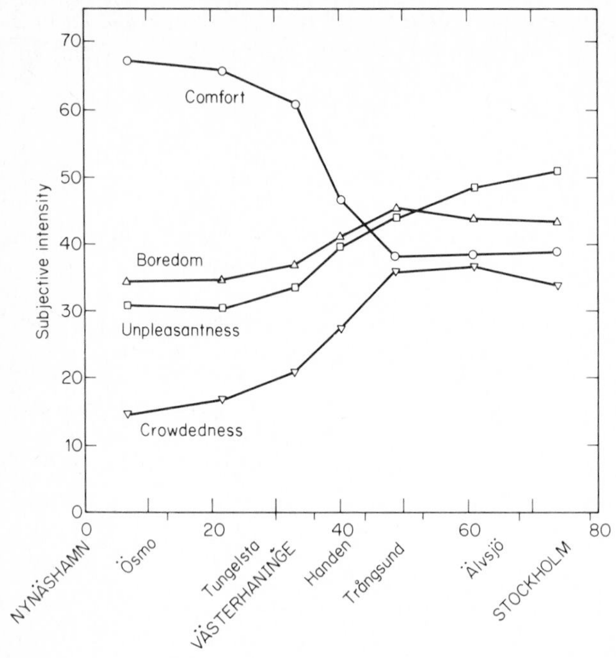

FIGURE 19-13
Results of a survey of commuters in Sweden of their subjective reactions during a commuting trip. (*Source: Singer, Lundberg, & Frankenhaeuser, 1974, fig. 3, p. 9.*)

The results, summarized in Figure 19-13, show a marked decrease in comfort and increased boredom, unpleasantness, and crowdedness during the commuting trip.

Crowding seems to be a particularly important negative aspect of public vehicles. Kogi (1979), for example, reports changes in certain physiological measures (heart rate and palmar skin resistance) with increases in passenger density in an electric railcar in Japan. Although commuting probably never can be an enjoyable experience for many riders, it is hoped that ultimately vehicle design, with minimum crowding, might reduce the unpleasant features as well as minimizing physiological and psychological stress.

DISCUSSION

The sampling of human factors aspects of automobile driving and public transportation systems discussed in this chapter barely touch the surface even of these aspects of transportation. A potpourri of other aspects of transportation that also have numerous human factors implications could include ship design, motorcycle safety, seat belts for adults and children, airport design, public transport station design, airport control tower operations, moving walkways for pedestrians, pedestrian safety, hand signals of bicycle riders, and powerboat design and safety. The list could go on and on, but we hope that the central intent of this chapter has been made, namely that transportation and related facilities are simply permeated with human factors implications, with the hope that such awareness may contribute to their ultimate improvement in this regard.

REFERENCES

Ashford, N., & Kirk, N. S. Designing small vehicles and stations for automatic urban transit systems. *Ergonomics,* 1978, *21*(6), 473–482.

Barrett, G. V., & Thornton, C. L. Relationship between perceptual style and driver reaction to an emergency situation. *Journal of Applied Psychology,* 1968, *52*(2), 169–176.

Bauer, H. J. *Public transportation and human factors engineering* (Research Publication GMR-9982). Warren, Mich.: Research Laboratories, General Motors Corporation, April 3, 1970.

Blackwell, H. R., Prichard, B. S., & Schwab, R. N. *Illumination requirements for roadway visual tasks* (Highway Research Board Bulletin 255, Publication 764), 1960.

Booher, H. R. Effects of visual and auditory impairment in driving performance. *Human Factors,* 1978, *20*(3), 307–320.

Canty, E. T., & Sobey, A. J. *Case studies of seven new systems of urban transportation* (Research Publication GMR-845). Warren, Mich.: Research Laboratories, General Motors Corporation, January 13–17, 1969.

Christie, A. W. Visibility in lighted streets and the effects of the arrangement and light distribution of the lanterns. *Ergonomics,* 1963, *6*(4), 385–391.

Colbourn, C. J., Brown, I. D., & Copeman, A. E. Drivers' judgments of safe distances in vehicle following. *Human Factors,* 1978, *20*(1), 1–11.

Denton, G. G. The influence of adaptation on subjective velocity for an observer in simulated rectilinear motion. *Ergonomics,* 1976, *19*(4), 409–430.

Endou, T., Ohshima, M., Watanabe, T., & Shingo, O. A preventive measure against drowsiness while driving. *Ergonomics,* 1979, *22*(6), 758. (Abstract)

Fell, J. C. A motor vehicle accident causal system: the human element. *Human Factors,* 1976, *18*(1), 85–94.

Fox, J. N. Human factors involved in current public attitudes towards public transit concepts. In *Proceedings of the Human Factors Society.* Santa Monica, Calif.: Human Factors Society, 1974.

Golob, T. F., Canty, E. T., Gustafson, R. L., & Vitt, J. E. Research Publication GMR-1037. Warren, Mich.: Research Laboratories, General Motors Corporation, October 26, 1970.

Goodenough, D. R. A review of individual differences in field dependence as a factor in auto safety. *Human Factors,* 1976, *18*(1), 53–62.

Gordon, D. A. & Mast, J. M. Drivers' judgments in overtaking and passing. *Human Factors,* 1970, *12*(3), 341–346.

Grime, A., & Hutchinson, T. P. Vehicle mass and driver injury. *Ergonomics,* 1979, *22* (1), 93–104.

Gustafson, R. L., Curd, H. V., & Golob, T. F. *User preferences for a demand-responsive transportation system: A case study report* (Research Publication GMR-1047). Warren, Mich.: General Motors Corporation, Research Laboratories, January 1971.

Harano, R. M. Relationship of field dependence and motor-vehicle accident involvement. *Perceptual and Motor Skills,* 1970, *31,* 372–374.

Harris, W., & Mackie, R. A study of the relationships among fatigue, hours of service, and safety of operations of truck and bus drivers (Report No. BMCS-RD-71-2). Washington, D. C.: Bureau of Motor Carrier Safety, November 1972.

Harte, D. B., & Harte, M. R. Estimates of car-following distances on three types of two-laned roads. *Human Factors,* 1976, *18*(4), 393–396.

Helander, M. Applicability of drivers' electrodermal response to the design of the traffic environment. *Journal of Applied Psychology,* 1978, *63*(4), 481–488.

Hicks, J. A., III. The evaluation of the effect of sign brightness on the sign-reading behavior of alcohol-impaired drivers. *Human Factors,* 1976, *18*(1), 45–52.

Hulbert, S. Effect of driver fatigue. In J. W. Forbes (Ed.), *Human factors in highway traffic research.* New York: Wiley, 1972.

Johansson, G., & Rumar, K. Drivers and road signs: a preliminary investigation of the capacity of car drivers to get information from road signs. *Ergonomics,* 1966, *9*(1), 57–62.

Johansson, G., & Rumar, K. Drivers' brake reaction times. *Human Factors,* 1971, *13* (1), 23–27.

Jones, I. S. *The effect of vehicle characteristics on road accidents.* Oxford, England, and New York: Pergamon Press, 1976.

Kogi, K. Passenger requirements and ergonomics in public transport. *Ergonomics,* 1979, *22*(6), 631–639.

Konz, S., Wadhera, N., Sathaye, S., & Chawla, S. Human factors considerations for a combined brake-accelerator pedal. *Ergonomics,* 1971, *14*(2), 279–292.

Mathews, M. L. A field study of the effects of drivers' adaptation to automobile velocity. *Human Factors,* 1978, *20*(6), 709–716.

McGuire, F. L. Personality factors in highway accidents. *Human Factors,* 1976, *18*(5), 433–442.

Mortimer, R. G., & Becker, J. M. Development of a computer simulation to predict the visibility distance provided by headlamp beams (Report No. UM-HSRI-HF-73-15). Ann Arbor, Mich.: Highway Safety Research Institute, University of Michigan, July 1973.

Mourant, R. R., & Rockwell, T. H. Mapping eye-movement patterns to the visual scene in driving: An exploratory study. *Human Factors,* 1970, *12*(1), 81–87.

Näätänen, R., & Summala, H. *Road-user behavior and traffic accidents.* Amsterdam: North-Holland Publishing Company, 1976.

Older, S. J., & Spicer, B. R. Traffic conflicts: A development in accident research. *Human Factors,* 1976, *18*(4), 335–350.

Public lighting needs. *Illuminating Engineering,* 1966, *61*(9), 585–602.

Rutley, K. S. Control of drivers' speed by means other than enforcement. *Ergonomics,* 1975, *18*(1), 89–100.

Rutley, K. S., & Mace, D. G. W. Heart rate as a measure in road layout design. *Ergonomics,* 1972, *15*(2), 165–173.

Schmidt, F., & Tiffin, J. Distortion of drivers' estimates of automobile speed as a function of speed adaptation. *Journal of Applied Psychology,* 1969, *53*(6), 536–539.

Shinar, D. *Psychology on the road: The human factor in traffic safety.* New York: Wiley, 1978.

Shinar, D., McDowell, E. D., Rackoff, N. J., & Rockwell, T. H. Field dependence and driver visual search behavior. *Human Factors,* 1978, *20*(5), 553–559.

Shinar, D., Rockwell, T. H., & Malecki, J. A. The effects of changes in driver perception on road curve negotiation. *Ergonomics,* 1980, *23*(3), 263–275.

Singer, J. E., Lundberg, U., & Frankenhaeuser, M. Stress on the train: A study of urban commuting (Report No. 425). Stockholm: Psychological Laboratories, the University of Stockholm, December 1974.

Travel barriers (PB187-237). U. S. Department of Transportation, National Information Service, 1970.

Williams, J. R. Follow-up study of relationships between perceptual style measures and telephone company vehicle accidents. *Journal of Applied Psychology,* 1977, *67* (6), 751–754.

Zeier, H. Concurrent physiological activity of driver and passenger when driving with and without automatic transmission in heavy city traffic. *Ergonomics,* 1979, *20*(7), 799–810.

CHAPTER 20

SELECTED ASPECTS OF WORK-RELATED HUMAN ERROR

The major thrust of this text has been intended to emphasize the importance of taking human factors into consideration in designing equipment, facilities, procedures, and environments for people involved in the production of goods and services, as in manufacturing industries, service industries, military services, government, and other types of organizations. In large part human factors efforts are directed toward designing things people use in order to enhance the performance, and minimize the errors, of man-machine systems. (This takes us back to the matter of *performance reliability* discussed in Chapter 2, such reliability, of course, being the complement of error.)

This chapter deals with certain human factors topics that have some rather special relevance to human errors and that in part share the common denominator of requiring continual or substantial attention on the part of people. These topics are monitoring activities, inspection processes, and accidents and safety. Design features of equipment and facilities used certainly can aid in minimizing errors in such work-related matters. However, they are all particularly susceptible to human errors of both omission and commission (especially omission), as in identifying relevant stimuli in monitoring activities, detecting defects in inspection processes, and performing in a safe manner in the face of hazards.

MONITORING TASKS

Certain types of work consist basically of monitoring or vigilance functions. In such tasks the monitor's function is that of giving attention to an operation or process to identify circumstances or events that require some action or response. A primary requirement of the monitor is the correct identification of all, or most, of the events that should require action. The input relating to these events may be presented to the monitor by various displays (such as dials, gauges, cathode-ray tubes or other visual displays, or auditory signals of some kind), or they may be observed or detected directly (such as by noticing a change in the sound or the output of a machine or by observing boxes moving off a conveyor after they have been labeled). The indication of the event—whatever it may be—usually is called a *signal*; when a signal occurs the monitor usually is supposed to take some action. Performance on monitoring tasks—whether in actual job situations or in laboratory studies—usually is measured with such criteria as (1) failure to detect relevant stimuli or signals; (2) false detection (*false alarms*); and (3) response lag. Some investigations have suggested that performance on vigilance tasks can be viewed in the framework of signal detection theory (SDT) as discussed in Chapter 3 since the *signals* to be detected can be thought of as occurring with some background *noise* (visual or auditory depending on the nature of the signal). This SDT framework seems to be especially relevant if the signal in question can vary in degree, rather than being clearly present or absent. If signals can "vary in degree" the yes-no decisions can of course be influenced by the standard of acceptability of the monitor, thus influencing the proportions of correct detections versus false alarms. In discussing the relevance of the SDT framework, Loeb (1978) agrees that to some extent SDT may be applied to monitoring tasks, but does suggest some constraints in its applicability.

There are, of course, individual differences in performance of monitoring tasks, but a primary concern is how the situational and task-related aspects of such tasks affect performance.

Duration of Monitoring Periods

A very significant task-related variable is the duration of the monitoring periods. In the case of tasks that consist exclusively of monitoring for signals, the evidence is quite clear-cut—that performance deteriorates with time (especially in terms of failure to detect signals that occur). This is illustrated by some of the results of an analysis of 37 studies by Teichner (1974) as shown in Figure 20-1. This particular figure shows sum-

FIGURE 20-1
Percent of signals detected in monitoring tasks as a function of monitoring time, as based on data from 12 studies. (*Source: Adapted from Teichner, 1974, fig. 10, p. 349.*)

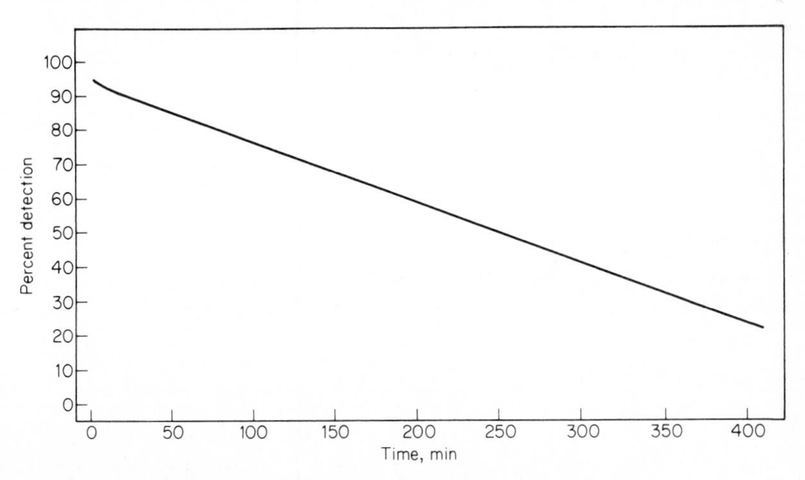

marized data for 12 monitoring tasks for which the initial detection rate was between 90 and 100 percent. For tasks with lower initial detection rates (i.e., the signals being more difficult to detect) the curves start lower and also show deterioration with time.

Some monitoring studies have shown a reasonable level of detection for perhaps ½ h or so before deterioration sets in. (Such a pattern is not revealed in Figure 20-1, perhaps because it summarizes data from several studies.) Because of such a pattern it has become the practice in some circumstances (as in radar watches in the military services) to limit monitoring periods to, say, ½ h. One of the clearest implications of monitoring research is that rest periods should be provided for continuous, demanding monitoring tasks.

Frequency Rate of Signals and Other Stimuli

Another factor that sometimes influences performance on monitoring tasks is the frequency rate of signals. The *percent* of detection of signals that have a very *low* frequency rate tends to be *lower* than if the signal rate is somewhat *higher*. This tendency probably can be attributed to the matter of expectancy. If people are not "expecting" signals very often they tend to miss them more than if the signals occur more frequently.

This effect tends to occur with *all or nothing* displays in which the signals to be detected are the only stimuli presented. A somewhat different pattern occurs if the signals to be detected must be differentiated from other somewhat similar *nonsignal* stimuli. This effect was reflected by the results of a study by Wiener (1977) in which 32 signals were presented in a 48-min vigil, along with other interspersed *nonsignal* stimuli at a "regular" rate of one stimulus per second or at a "slow" rate of one stimulus every 5 s. (The signals to be detected were pairs of dots 13.3 cm apart on a cathode-ray tube, and the *nonsignals* were 10.8 cm apart.) The results are summarized in Figure 20-2, this showing the percent detected for the slow and regular rates for

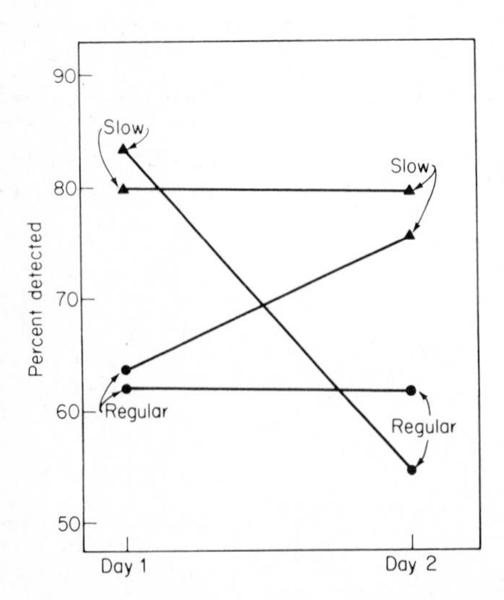

FIGURE 20-2
Detection rates of signals in a monitoring task in which stimuli (signals and somewhat similar stimuli) were presented at a "regular" rate (one every 1 s) and at a "slow" rate (one every 5 s). Data show performance for four groups of subjects that performed under all combinations of regular and slow conditions on 2 days. (*Source: Wiener, 1977, fig. 1, p. 302.*)

four groups of subjects who had performed the tasks on 2 days, with all combinations of slow and regular rates on these 2 days. The poorer performance under the regular rate probably implies that the requirement to differentiate between the two types of stimuli once every second was so demanding that detection of the actual signals was adversely affected. (The comparisons for the 2 days showed no learning effect from 1 day to the next.)

Discussion

Other variables can also affect performance on monitoring tasks, such as the regularity or irregularity of the signals (regularly occurring signals are more likely to be detected), the intensity or clarity of the signals to be detected, feedback (knowledge of results), and the complexity or clutter of the visual background of the signals. Although we will not review all the relevant research here, Bergum and Klein (1961), on the basis of a review of available research, suggested certain principles for more effective design of people-monitored systems; these are given below, with minor adaptations.

1 Visual signals should be as large in magnitude as is reasonably possible. This includes size, intensity, and duration.

2 Visual signals should persist until they are seen (or otherwise detected), or as long as is reasonably possible.

3 In the case of visual signals, the area in which a signal can appear should be as restricted as possible.

4 Although "real" signal frequency often cannot be controlled, where possible it is desirable to maintain signal frequency at a minimum of 20 signals per hour. If necessary, this should be accomplished by introducing artificial (noncritical) signals to which the operator must respond.

5 Where possible, the operator should be provided with anticipatory information. For example, a buzzer might indicate the subsequent appearance of a critical signal.

6 Whenever possible and however possible, the monitor should be given knowledge of results.

7 Noise, temperature, humidity, illumination, and other environmental factors should be maintained at optimal levels.

Aside from these principles, it is important that adequate training be provided to those who are to serve as monitors. This training should make clear the nature of the "signals" that are to be identified and the response that is to be made to each.

INSPECTION PROCESSES

The inspection of parts and products as in manufacturing processes is in certain respects similar to monitoring tasks in that the inspector (usually making visual inspections) must continually be on the lookout for some condition that deviates from the norm–in particular, defective items. Although these two types of tasks do have some differences, much of what has been said about monitoring tasks also applies to inspection tasks.

TABLE 20-1
COMPARISON OF INSPECTION PROFICIENCY UNDER TRADITIONAL AND NEW SCHEDULE SYSTEMS

	Traditional system	New system
Total working time	450 min	395 min
Total break and wash-up time	30 min	85 min
Time working at 0.85 detection proficiency	120 min	395 min
Time working at 0.60 detection proficiency	330 min	0 min

Source: Purswell & Hoag, 1974.

Inspection Periods

As an example of the similarity there are cues that long periods of inspection without breaks usually result in reduced proficiency in terms of defects detected. Purswell and Hoag (1974), for example, compared the proficiency of inspection in an actual production facility under a "traditional" work schedule (a regular 8-h day with a mid-morning and mid-afternoon 10-min break and time out for lunch) as compared with a "new" schedule with a 5-min break every 30 min. Under the traditional schedule the proficiency level for the first 30 min was 85 percent, and after that it dropped to about 60 percent. A comparison of certain data for these two schedules is given in Table 20-1. Despite the reduced working time (and without going into details) the inspection proficiency of the new system was reported to be superior to that of the traditional schedule.

Additional evidence of the improved inspection with frequent breaks comes from an earlier study in Great Britain by Colquhoun (1959), as shown in Figure 20-3. In this

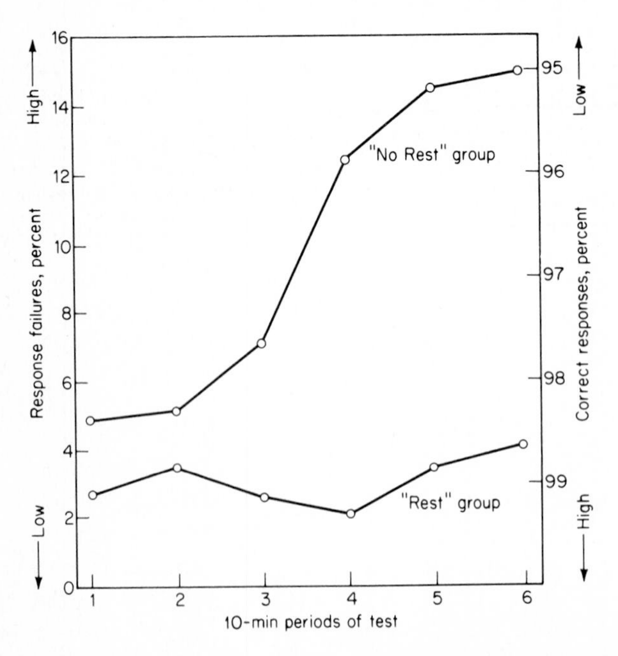

FIGURE 20-3
Percent of failure to detect and correct detections in a simulated inspection task for two groups of inspectors. One group was given a 5-min break midway during the hour, and the other group worked continuously for the full hour. (*Source: Colquhoun, 1959, p. 370.*)

case the subjects inspected defective items for 1 h in a simulated inspection. One group was told that it would have a 5-min break after 30 min; the other group was told that it was to carry on the inspection for a full hour. The performance of the "no rest" group was clearly worse.

Detectability of Defects

Another factor that can affect the performance of inspection is the difficulty of making the visual and perceptual distinctions involved. Difficulty can be influenced by various factors relating to the nature and size of the defects, the total number of stimuli to be scanned, and the illumination available, along with others. Such effects are illustrated by the results of a study by G. L. Smith (1972), who varied the detectability of defects in a simulated inspection task by varying (1) the illumination; (2) the magnification used (that influenced the size of the defects); and (3) the total number of stimuli (the total number of stimuli to be scanned at a given time in order to identify the defective stimuli). The stimuli were circles, and the "defects" were circles with a slight break that made them look like the letter *C* (actually they were Landholdt rings described in Chapter 4). Four combinations of the three variables were ordered in terms of difficulty level of detectability, and the performance of the inspectors for these four conditions is shown in Figure 20-4. That figure shows the performance under two sets of instructions, namely "to avoid misses" and "to avoid false alarms," although the patterns under these two instructions are quite similar. The pattern shows quite clearly the deterioration in inspection performance with increasing level of difficulty of detectability.

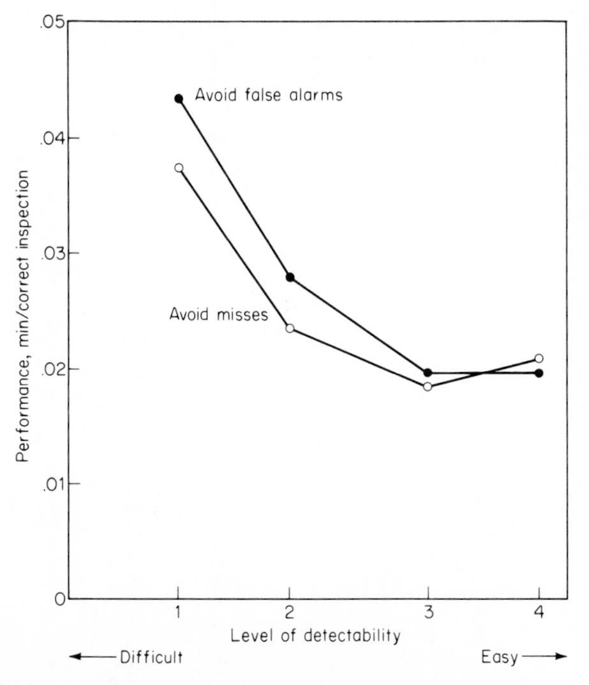

FIGURE 20-4
Inspection performance as related to the difficulty of detectability of defects. Level of detectability was varied by combination of illumination, the magnification used (that affected the apparent size of the defects), and the total number of stimuli to be scanned when identifying defects. (*Source: G. L. Smith, 1972, fig. 2, p. 289.*)

TABLE 20-2
ERROR IN IDENTIFYING DEFECTS AS RELATED TO TOTAL NUMBER OF STIMULI TO BE SCANNED

Grid	Total stimuli to be scanned	Percent error in identifying defects
5 × 5	25	4.7
7 × 7	49	11.2

Source: Purswell, Greenhaw, & Oats, 1972.

Somewhat confirming evidence regarding the effects of the number of stimuli to be scanned comes from a study by Purswell, Greenhaw, and Oats (1972) in which the "defects" (an open circle and a black square) were included in either a 5 × 5 grid (with 25 open and black geometric figures) or a 7 × 7 grid (with 49 figures). The results are summarized in Table 20-2.

The effect of size of defects on inspection was demonstrated by Drury (1975) in the inspection of glass sheets for a single flaw, the flaws varying from 0.19 to 10 mm^2. In inspecting the sheets with a special shadowgraph screen it was found that the time required to identify the flaws varied markedly by size of flaw, with some flaws taking several times as long to detect as others, and one flaw (the smallest) taking over 12 s as contrasted with less than 1 s in the case of the largest flaws.

It is of course frequently not possible to enhance the detectability of defects, but when this is possible it should be done, as by illumination, magnification, minimizing the number of stimuli to be scanned, and otherwise.

Rate of Presentation of Stimuli

Continuing the inventory of variables that can influence inspection efficiency is the rate at which the stimuli to be scanned are presented, as for example if they are presented at a controlled rate on a conveyor belt. Some indication of this effect comes from the study by Purswell et al. (1972), in which the target presentation rate was varied by the velocity of the conveyor and the spacing of the stimuli. These variations were called the *target throughputs* in terms of targets per second. The total errors for the 5 × 5 and 7 × 7 grids are given in Table 20-3.

It is admittedly somewhat risky to extrapolate directly from simulated laboratory studies to actual inspection processes, but such studies as this can illustrate how the rate of presentation of items to be inspected can influence the efficiency of inspection processes. If the operation requires that the rate of presentation in actual inspection situations be controlled, two suggestions are offered: (1) the effect of rate of presentation of items on inspection efficiency should be determined empirically for the particular items or parts in question; and (2) usually some judgment needs to be made

TABLE 20-3
INSPECTION ERRORS AS RELATED TO TARGET PRESENTATION RATE AND SPACING OF TARGET STIMULI

	Throughput, targets per sec			
Grid	**0.70**	**0.50**	**0.35**	**0.25**
5 × 5	18	9	7	1
7 × 7	35	25	11	4

Source: Purswell, Greenhaw, & Oats, 1972.

about the trade-off between speed of inspection and the effects of errors in inspection (either failure to identify defects or false alarms); if failure to identify a defect could have serious consequences, certainly every effort should be made to identify all defects. When feasible, however, it is preferable for the inspectors to control the inspection rate.

Use of Single and Multiple Inspectors

Data on the relative effectiveness of single inspectors versus multiple inspectors come primarily from simulated inspection studies, but indicate that having two or more inspectors tends to increase the effectiveness of inspection processes. Some such evidence comes from Waikar (1973, as reported by Purswell & Hoag, 1974) and is illustrated in Figure 20-5. The results of this study indicated improvement in the percent of defective items rejected. These results indicate improvement from one to two, and from two to three, inspectors, but no further improvement with four. It should be added, however, that if all inspectors must recognize a defective item before it is rejected, the number of defective items rejected decreases as the number of inspectors increases. This implies that each inspector should be able individually to reject an item.

Lion, Richardson, Weightman, and Browne (1975), using a simulated inspection task, also found that inspection was significantly better when pairs of inspectors worked together, rather than each inspector working singly. Paired inspectors viewed discs being moved along six parallel lines, while with single inspectors each inspected those on only three lines. The rate of movement with the single-inspector procedure, however, was twice as fast as when both inspectors scanned all six lines. The advantage of the paired-inspector operation was attributed in part to the stimulus of doing a job in unison, the feeling of competition, and possible reduced boredom because of the opportunity to talk.

Granting that the use of two or more inspectors can increase inspection proficiency, the question arises again as to the trade-off values of additional inspection costs versus the benefits from improved detection of defects.

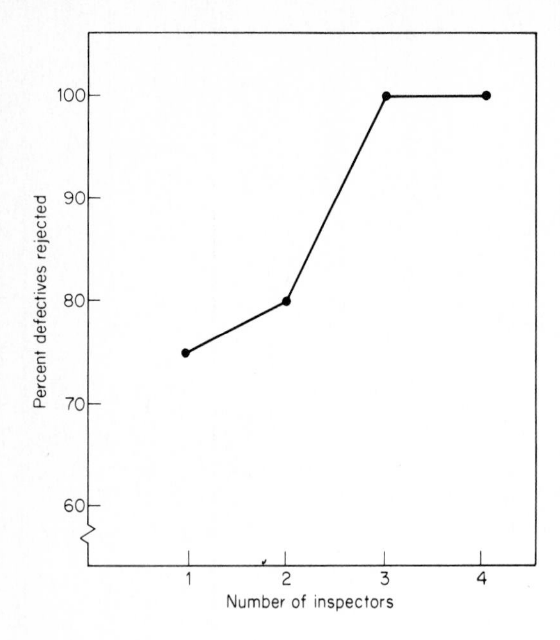

FIGURE 20-5
Percent of defective items rejected when 1, 2, 3, or 4 inspectors made individual decisions about items. (*Source: Waikar, 1973, as presented by Purswell & Hoag, 1974, fig. 2, p. 400.*)

Aids in Inspection Processes

The effectiveness of many inspection tasks leaves much to be desired. Harris and Chaney (1969, p. 169), for example, report that the percents of defects identified by 26 experienced inspectors ranged from 20 to about 40 percent, with a mean of about 40. Certain methods of improvement of inspection operations have been discussed or implied, such as providing rest periods, arranging for inspection of stationary items rather than moving items when feasible, providing adequate illumination, minimizing (when feasible) the total number of items to be scanned simultaneously, using moderate rates of presentation for moving items, and using multiple inspectors when feasible. Certain other methods may be feasible in certain circumstances. A few of these will be mentioned.

Visual Aids One scheme proposed by Chaney and Teel (1967), for example, is by the use of visual aids. In their application of this technique they developed aids for use in inspection of machined parts for such defects as mislocated holes, improper dimensions, and lack of parallelism and concentricity. These aids consisted of simple drawings of the sample parts. The dimensions and tolerances for each feature to be inspected were placed on the drawings to minimize the need for calculation or reference to other materials. The inspection aids were introduced as a part of the regular inspection process in such a way that it was possible to compare the "before" and "after" inspection accuracy over a 6-month period, along with the effects of a specially designed training program. The results of the comparison, shown in Figure 20-6, clearly demonstrated the improvement associated with the visual aids as well as with the training. Needless to say, visual aids would be suitable only in certain special inspection processes.

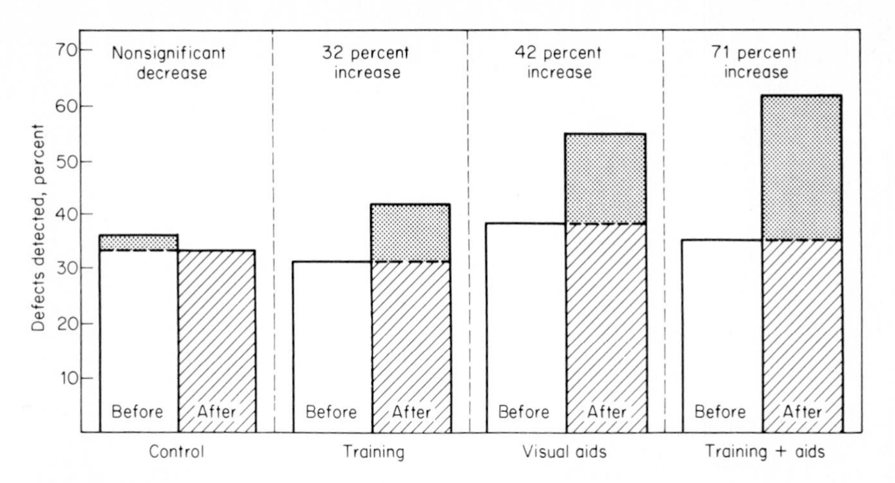

FIGURE 20-6
Percent improvement in detecting specified defects in machined parts following introduction of visual inspection aids, special training, and their combinations. (*Source: Chaney & Teel, 1967, as presented by Harris & Chaney, 1969, fig. 9.20. Copyright 1967 by the American Psychological Association. Reprinted by permission.*)

Limit Samples The psychological literature is sprinkled with references to the fact that people are generally better at making relative judgments than absolute judgments. To take an overly simplified case, it is easier to say that one person is taller than another when seeing them side by side than to judge the actual height of either in absolute terms. As another example, most people can make "absolute" identifications of perhaps a couple of dozen colors varying in hue, saturation, and brightness, but they can differentiate pairs of many thousands if each pair is presented together.

Limit samples are based on this principle. A *limit sample* as used in inspection operations is an example of a product or part that is just barely acceptable in terms of some particular feature, such as a minor defect that still is acceptable. In operational use the limit sample is placed before each inspector with instructions to reject any item that is "worse" than the limit sample.

An example of this approach, as reported by Kelly (1955), involved the inspection of glass panels for the front of TV picture tubes. As a prelude to selecting the limit sample she had a sample of 10 panels rated by the paired-comparison method (in which each item is judged in comparison with every other, as being "better"). The results of this consisted of an ordering of the 10 panels from good to poor in judged appearance. In turn, management representatives made a determination of the panel that they considered to be just barely acceptable, and that panel was then the limit sample used in later actual inspection. It was estimated that inspection accuracy after the introduction of the limit sample improved over 75 percent.

There have been reports that limit samples do not serve a useful purpose in certain inspection processes. However, it is probable that this may have been due to the fact that the inspectors actually were not using them. It is probable that additional training and supervision might help to ensure their continued use. Photographs can also be used in much the same manner as actual limit samples.

TABLE 20-4
EXPERIMENTAL CONDITIONS USED IN STUDY OF TRAINING OF INSPECTORS

Condition	Display mode during orientation training	Diplay mode during experimental trials
1	Stationary	Stationary
2	Stationary	Moving
3	Moving	Stationary
4	Moving	Moving

Source: Williges & Streeter, 1972.

Stationary Versus Moving Inspection Items

Some inspection operations are carried out when the items or parts to be inspected are stationary (static), whereas in other instances the items are moving, such as the inspection of fruit or glassware on a conveyor.

It is generally assumed that inspection is more efficient in detection of defects if the items are stationary rather than moving, and Williges and Streeter (1972) report their earlier research that tended to confirm this assumption. However, the nature of certain inspection operations virtually requires inspection of items while moving on a conveyor belt. In searching for ways and means to improve inspection of moving items, Williges and Streeter performed a study in which the inspectors carried out inspection processes under various combinations of stationary and moving "modes" as shown in Table 20-4.

The details of their experiment and of the results are a bit complex and are not given here, but in general terms they found that inspectors whose *orientation* trials were with moving displays presumably learned visual scanning strategies that transferred to their performance with stationary items in the subsequent *experimental* trials. Thus, an orientation with moving items under controlled speed conditions presumably forces inspectors to develop visual scanning patterns that have a beneficial effect when they are later involved in the inspection of stationary items.

Discussion

The development of inspection processes involves a couple of major decisions. The first of these relates to the development of the procedures and techniques that are reasonably optimum from a human factors frame of reference and thus result in the most efficient method of inspection for the items in question. The second decision that may be required in some circumstances relates to the question of trade-offs that may be necessary between the costs of the "best" inspection method and the consequences of less effective inspection, such as the criticality of the consequences that could accrue from failure to detect defects in the inspection. The consequences of a defect in a Ping-Pong paddle would not be nearly as catastrophic as those of a defect in a nuclear power plant.

ACCIDENTS AND SAFETY

As indicated at the beginning of this text one of the objectives of human factors is to maintain or enhance certain human values, one of these being safety and physical well-being. In this regard the previous pages are sprinkled with references to accidents and injuries, these frequently being used as criteria in human factors research, as to the design of displays, controls, tools, work areas, and vehicles. To a very considerable extent many of the design features discussed in the various chapters (and scads of others not specifically referred to) are directed toward the creation of safer conditions for people in their work and everyday lives.

Since much of the previous material in this text has direct or indirect implications in terms of safety it is not intended here to refer further to such examples or to bring in other illustrations of human factors research or applications to specific design problems. Rather, the intent here is to discuss accidents in a general frame of reference, to illustrate certain techniques and methods that have been developed for use in the analysis of accidents, and to mention certain methods for reducing accidents and injuries. In this regard some of the discussion of vehicular accidents in Chapter 19 will be relevant.

Models of Accident-Related Behavior

Various "models" of accident-related behavior have been proposed, some of these having at least certain common denominators. A model proposed by Drury and Brill (1980), for example, poses the view that accidents occur when the momentary demands of the task to be performed exceed the momentary abilities of the individual in question. In accident analysis they propose a form of task analysis in which the task demands are expressed in the same terms as information about what humans *can* do on the task. Thus the task demands might be specified in terms of the force that must be applied and the direction and distance of the movement involved, and human capabilities in turn could be specified in terms of the proportion of the population that is capable of applying such force. This type of comparison provides a very useful frame of reference in viewing accident occurrence.

A sequential model of the stages leading up to accidents is proposed by Ramsey (1978) as represented in Figure 20-7. Although his model was focused specifically on potentially hazardous consumer products, it is equally relevant to work situations and other circumstances.

His model is intended to trace sequentially the activities that take place within the individual in potentially hazardous circumstances. The first stages deal with the *perception* and the *cognition* (i.e., the recognition) of the hazard. If the hazard is not perceived or recognized as a hazard the likelihood of an accident occurring is of course increased.

If the hazard is perceived and recognized as such the next stage is that of *decision making*. A decision to avoid or not to avoid the hazard is of course influenced by the individual's attitudes and previously acquired behavior patterns. A person's decision not to avoid the hazard would be based on his or her risk-taking proclivities, some people being more willing to take risks than others.

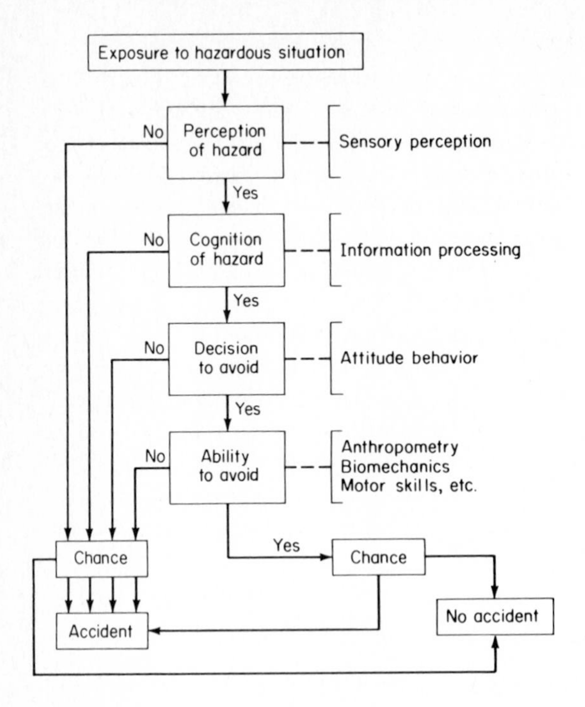

FIGURE 20-7
Accident sequence model representing various stages in the occurrence or avoidance of accidents in a potentially hazardous situation. (*Source: Adapted from Ramsey, 1978.*)

If the individual decides to avoid the hazard the next stage depends on the person's *ability* to do so, which would be based on such factors as anthropometry, biomechanics, and motor skills, along with other human characteristics and skills such as experience, training, and reaction time.

The fulfillment of the sequence of behaviors involved in this model (including the exercise of the abilities to avoid a hazard) does not ensure avoidance of accidents, but certainly reduces such possibilities. Even with the best of intentions and the presence of requisite abilities, "chance" factors bring accidents into human affairs. But the failure to fulfill the indicated sequence certainly increases the chances of accidents.

A model such as this, that crystallizes the behavioral components involved in accidents, can be useful in focusing attention on such human factors aspects as the adequate presentation of relevant information relating to hazards (as in displays), the design of situational features that can be useful in avoiding accidents (control devices, work space, etc.), the provision of adequate information about hazards (including appropriate warnings), and the training of personnel in taking appropriate action in the face of hazards.

Collection and Analysis of Accident and Injury Data

Actions that would be useful for reducing accidents and injuries preferably should be based on the collection and analysis of relevant data. In this regard there are a number of techniques that can be used, although each one usually has some advantages and limitations.

Data Reporting Procedures Data on accidents and injuries are compiled as a matter of routine by various organizations, including insurance companies, the Occupational Safety and Health Administration (OSHA), the Bureau of Labor Statistics (BLS), the National Safety Council, and many trade associations. This is not the place to present the classification systems used by various organizations or the masses of data based on them. However, it may be relevant to illustrate certain of the categories that are included in at least certain classification systems. In discussing the identification of *causal* factors as included in various classification systems, for example, Ramsey (1973) refers to the following (along with a few examples):

- Nature of injury (as amputation, burn, fracture)
- Part of body (head, back, elbow)
- Type of accident (struck by, fall, caught in)
- Unsafe condition (inadequate guarding, placement hazard, defect of agency or equipment involved)
- Source of injury (body motion, air pressure, machine, hand tool, vehicle)
- Activity of injured (using hand tool, body movement, handling, operating power equipment)
- Unsafe act (operating or working at unsafe speed, using unsafe equipment, taking unsafe position or posture, failure to wear protective gear)

Certain sets of injury data consist of frequencies and frequency rates of injuries as related to the above types of categories and their specific subcategories.

Critical Incident Technique As applied to accident research this procedure provides for those who have experienced or have observed unsafe acts or actual accidents to describe those events in detail. Such reports usually provide clues that reflect patterns of behaviors and events that can be useful in developing preventative measures.

Although the critical incident technique probably is used most commonly in reporting unsafe acts (rather than actual accidents), there is evidence from various sources that observed unsafe acts and conditions are definitely related to accidents and injuries. Such confirmation, for example, comes from a survey by Edwards and Hahn (1980) of over 4000 workers in 19 plants in which they used a variation of the critical incident technique for obtaining data on observations by workers of unsafe acts or conditions. They found that the frequency of observations by workers of unsafe acts and conditions correlated, on an across-the-board basis, .61 with accidents and .55 with disabling injuries. Certain of the conditions that were most highly related to accidents were:

Failure to ground (electrical) material or equipment
Handling dangerous material unsafely
Worker did not understand how to do job
Climbing on or over moving equipment
Loading, feeding material too fast

On the basis of their data, Edwards and Hahn suggested that accidents "happen" where they have a chance to happen—where people report the existence of unsafe acts and conditions.

Fault-Tree Analysis (FTA) *Fault-tree analysis* is based on a deductive logic in the analysis of a system or operation that depicts graphically the interrelationship of combinations of events that can result in accidents or injuries. The development of a fault tree can be a rather complicated process, as discussed by Brown (1973), and although the details are not described here an example is given for illustrative purposes. In the diagraming of fault trees certain standard geometric symbols are used to represent certain classes of "events." These symbols are described below.

- Rectangle: used to identify events that will generally be developed further in the analysis
- Circle: used to identify *basic fault events*, that is, events that are in no further need of development because of sufficiency of empirical data
- House: used to identify events that are expected to occur in the normal operation of the system
- Diamond: used for events that will not be developed further in the logic diagram because of insufficient data or because they are inconsequential

In addition, small symbols are used to represent events that must occur *in combination* for some other event to occur (an AND "gate"), or to represent *possible alternative* events (an OR "gate").

Realizing that an illustration cannot be fully comprehended without more detailed description, Figure 20-8 is shown to illustrate a fault-tree analysis of the "event" of a chip in the eye occurring. As shown, this event could occur if (1) an operator fails to wear safety glasses, or (2) a person without safety glasses (other than the operator) gets too close to the operation. The second possibility, in turn, would exist only if other events (those under the AND symbol) would be present. (In the case of one of those there is another OR gate.)

Although fault-tree analysis has a place in accident analysis procedures, Christensen (1980) points out that there is an obvious danger with it (as with many techniques) in that its form and content may blind the user to alternative relationships between events that are not made explicit.

Discussion There are various other methods for collecting and analyzing data relevant to accidents and injuries, including various forms of task analysis and the use of accident review teams (as used in industry and aviation); some of these and other techniques are discussed by Ramsey (1973) and Christensen (1980) and will not be described here. In general terms, however, the various techniques provide for obtaining and analyzing data relating to the behaviors, to the physical conditions, or to both that presumably contribute to accidents or injuries. Such data, when appropriately developed, can serve as the basis for taking remedial action to reduce the incidence of such events.

The Reduction of Accidents and Injuries

As mentioned before in this chapter and elsewhere in this text the application of relevant human factors principles and data to the design of things people use in their jobs and everyday lives can reduce the likelihood of accidents and injuries. Some of the techniques for collection and analysis of data described above can point up some of

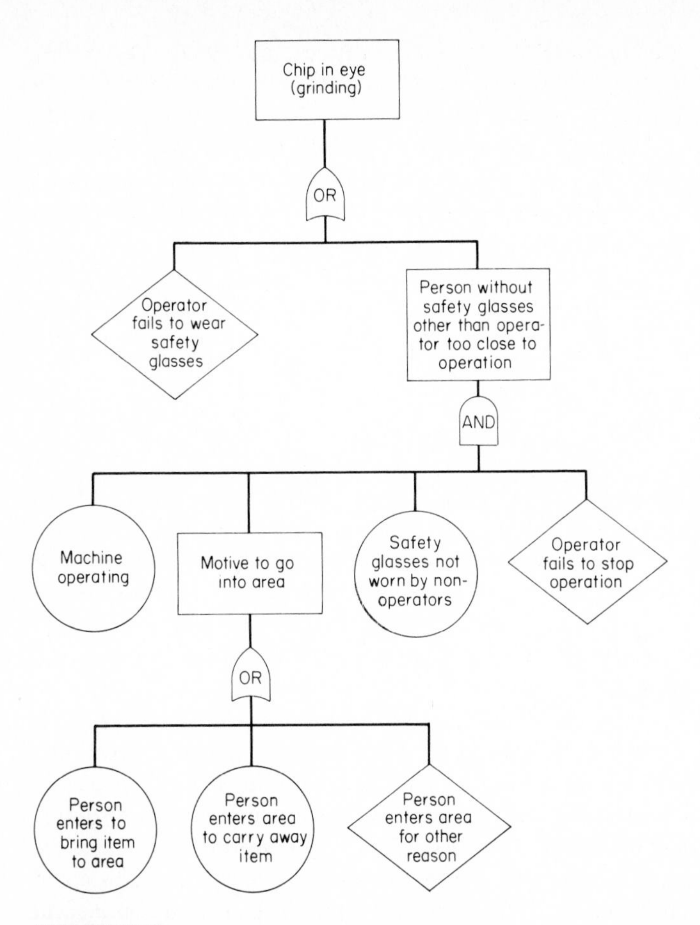

FIGURE 20-8
Example of a fault-tree analysis for use in analyzing the interrelationship between events associated with accidents and injuries. This example is for the event of a chip to the eye from a grinding operation. See text for discussion. (*Source: Brown, 1973, fig. 6, p. 77.*)

the features of the things people use that could be redesigned toward this end. Aside from strictly design matters, however, there are certain other actions that can be taken to reduce accident and injury risks, particularly those that are more directly related to operational procedures or to the behavior of people. Although this is not the place for a "how-to-do-it" or "what-to-do" manual on safety, a few examples may illustrate some such approaches.

Procedural Checklists One technique that has become rather standard practice in aircraft operation, certain military operations, and certain other situations is the use of checklists that are to be followed in executing some specified routine. The individual follows a list of steps or functions that are to be executed, and checks off each one as it is completed. This is, in effect, a substitute for memory in carrying out the stated activities.

Training Training is of course one of the standard procedures used by safety directors to aid people in acquiring safe behavior practices. In this regard Cohen, Smith, and Anger (1979), on the basis of a review of approaches for fostering self-protective measures against workplace hazards, concluded that training remains the fundamental method for effecting such self-protection. However, they point out that the success of such training depends on (1) positive approaches that stress the learning of safe behaviors (not the avoidance of unsafe acts); (2) suitable conditions for practice that ensure the transferability of these learned behaviors to the real settings and their resistance to stress or other interferences; and (3) the inclusion of means for evaluating their effectiveness in reaching specified protection goals with frequent feedback to mark progress.

Feedback Closely related to training is the matter of feedback, specifically feedback following desirable work behaviors on the part of workers. Cohen et al. (1979), for example, report the effects of feedback following training in practices that would reduce exposure to styrene (a suspected carcinogen) in a work situation. After initial training the instructor visited each worker once or twice each day to provide encouragement or feedback in observing the work procedures in question. Observers, from remote vantage points, logged the number of times that the behaviors in question occurred both before and after training, and found reductions of from 36 to 57 percent in exposure levels to styrene.

In another situation, in two departments of a large bakery, a brief safety training program was instituted. This was followed by an "intervention" period in which there was an updated posting of the percentage of behavioral incidents that were performed safely (as based on the observations of an observer). As another form of feedback the supervisors took occasion to recognize workers when they performed certain selected incidents safely.

The results are summarized in Figure 20-9, this showing the percent of incidents performed safely for "baseline" observation sessions, for "intervention" sessions, and a "reversal" period at the end of the 25-week study period, the four reversal observation sessions being those after the feedback of the "intervention" sessions was terminated. The immediate drop following the intervention sessions was so pronounced that within 2 weeks the management assigned and trained an employee to reinstitute the posting of data on a weekly basis. Within a year the injury frequency rate had stabilized at less than 10 lost-time accidents per million hours worked, as contrasted with the previous rates of 35 and above. In connection with feedback regarding safety practices it is evident that this must be a continuing process and not a "one-shot" proposition.

Contingency Reinforcement Strategies Contingency reinforcement has its roots in the early work of Skinner, whose extensive research demonstrated that the probability of repetition of specific behaviors may be increased by providing rewards (i.e., *reinforcement*) when the specific behaviors occur. In discussing such reinforcement in the safety context, Cohen et al. (1979) suggest that the reinforcement rewards for safe behaviors might include bonuses, promotions, informal feedback (as discussed in the previous section), social recognition, and special privileges (time off, preferred parking locations, etc.).

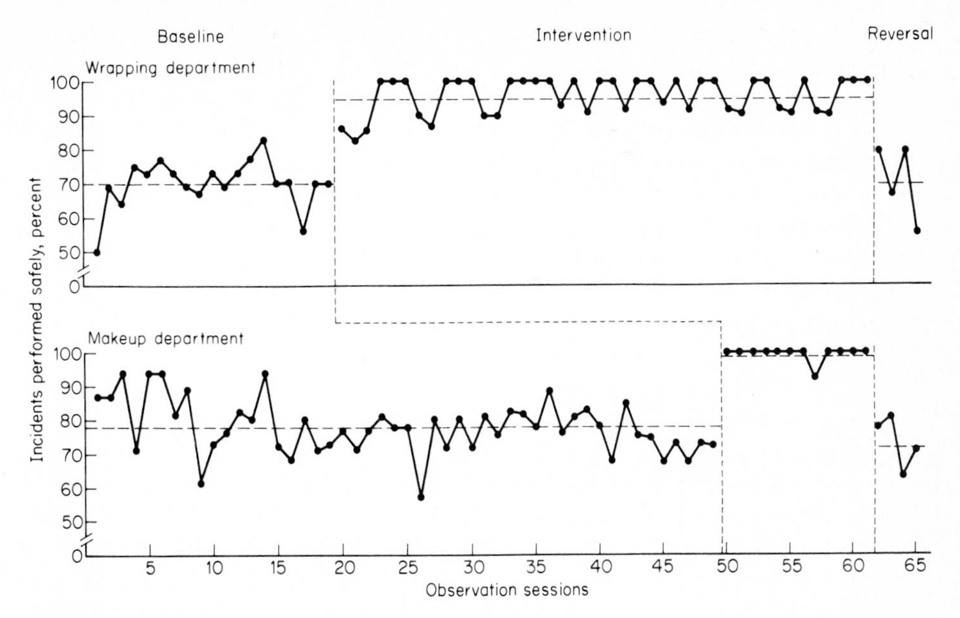

FIGURE 20-9
Percentage of behavioral incidents performed safely by bakery employees during a "baseline" period, an "intervention" period (during which feedback was provided), and a "reversal" period (after removal of feedback). (*Source: Komaki, Barwick, & Scott, 1978, fig. 1, p. 439. Copyright 1978 by the American Psychological Association. Reprinted by permission.*)

As an example of the effectiveness of such a strategy, M. J. Smith, Anger, and Uslan (1978) report the results of their study with a shipyard employing 20,000 workers. They concentrated their attention on the use of reinforcement (what they referred to as *behavioral modification*) in the reduction of eye injuries, which accounted for over 60 percent of all injuries. The program consisted of training five supervisors in the fundamentals of behavior modification: (1) observing worker behaviors (specifically the use of safety equipment); (2) recording worker behaviors; and (3) giving praise for wearing safety equipment. The work groups of these five supervisors had had among the highest eye accident rates. A before and after comparison of the eye accident rates for the subordinates of these five supervisors and for a control group of 39 supervisors is given in Table 20-5.

TABLE 20-5
EYE ACCIDENT RATES OF SHIPYARD WORKERS BEFORE AND AFTER BEHAVIOR MODIFICATION TRAINING

	Eye accident rates		
Work group	**Before**	**After**	**Change**
Experimental (5)	11.8	4.3	−7.5%
Control (39)	5.8	4.7	−1.2%

Source: M. J. Smith, Anger, & Uslan, 1978.

Cohen et al. (1979) report the use of reinforcement strategies in a couple of organizations aimed toward encouraging workers to use ear protection devices more consistently, with encouraging results. In one of those instances the reinforcements consisted of posting of the percent of workers wearing earplugs, social praise, coffee, doughnuts, money, and market goods, and in the other instance it consisted of providing employees with the results of hearing tests. As Cohen et al. (1979) caution us, the potential value of reinforcement strategies depends very much on the careful selection of target behaviors and rewards.

Incentive Programs Incentive safety programs consist of offering some type of incentive to individuals, groups, or supervisors of groups for achieving certain safety records. Among the incentives cited by Cohen et al. (1979) in certain organizations are using group safety records as a part of supervisors' performance evaluations; tokens redeemable for catalogue merchandise; trading stamps given to individual employees with bonus stamps to work groups with no accidents; and direct payment of money to individuals and groups with good safety records. Cohen et al. (1979) cite the experiences of certain organizations that have been very successful in reducing accident rates. However, they caution against excessive preoccupation with or continued use of such programs, on the grounds that a succession of such efforts, each requiring a bigger prize than the one before, would appear unwise. They therefore recommend limited use of this plan.

Propaganda It is probably evident that the distinctions among training, feedback, reinforcement, and incentives as related to safety are intricately intertwined, and do not exist in independent airtight compartments. Propaganda in the form of posters and other forms of safety communications also gets mixed up in this conglomeration. Recognizing these interrelationships, Sell (1977) sets forth the following aims of safety propaganda as such:

- To give more knowledge of safety factors.
- To change the attitudes of work people so that they are more inclined to act safely.
- To ensure that safe behavior takes place. It is obvious that this is the most important aim, and the one at which all propaganda must be directed.

In this regard Sell points out that in the analysis of accidents the typical approach is to trace the cause back to the person on the spot, and to come to the conclusion that that person was careless or lacked attention. But Sell goes on to indicate that in many accidents there are contributory factors that generally are out of control of the individuals involved—factors such as improper design in terms of human factors, fatigue (which individuals cannot avoid), lack of training, and failure by management to ensure that guards and protective clothing are available. In effect he focuses on the fact that safety propaganda directed toward workers can be useful only if the accident-causing factors are under the control of the workers. (The responsibility for factors that workers cannot control should of course fall in the lap of management.)

On the basis of his analysis of various studies dealing with safety propaganda directed toward workers, Sell emphasizes two points: (1) the need to try to modify the attitudes of workers that, in turn, influence their behavior; and (2) the fact that for any communication to have an effect it must be perceived and understood by the individuals in question. Drawing further from his survey he states that there is little doubt that safety posters and other propaganda can be made to produce the desired behaviors. However, he believes that for the propaganda to be really effective it must:

1 Be specific to a particular task and situation.
2 Back up a training program.
3 Give a positive instruction.
4 Be placed close to where the desired action is to take place.
5 Build on existing attitudes and knowledge.
6 Emphasize nonsafety aspects.

The propaganda should not:

1 Involve horror, because in the present state of our knowledge this appears to bring in defense mechanisms in the people at whom the propaganda is most directed.

2 Be negative, because this can show the wrong way of acting when what is required is the correct way.

3 Be general, because almost all people think they act safely. This type of propaganda is thus seen as only relevant to other people.

There are relatively few studies that have provided evidence of the actual modification of behavior following the use of safety posters. In one such study in a steel mill posters were used to encourage crane slingers to hook back the chain slings onto the crane hook when they were not in use, as a safety precaution. Three types of large posters with instructions and illustrations to "Hook that sling" were posted in relevant areas where gantry cranes were used. The percent of slings "hooked back" before and after posting the signs, as shown in Figure 20-10, shows a systematic difference, the

FIGURE 20-10
Percent of crane slings "hooked back" as a safety precaution by crane operators in a British steel mill before and after the posting of safety posters to "Hook that sling." (*Source: Sell, 1977, fig. 3, p. 212. This figure appeared as fig. 3 of Vol. 8, No. 4, p. 212 of* Applied Ergonomics, *published by IPC Science and Technology Press Ltd., Guilford, Surrey, U. K.*)

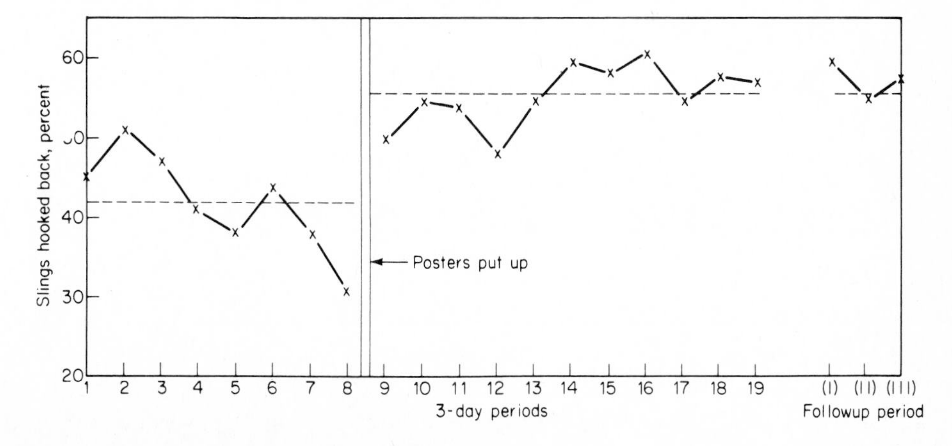

improvement actually carrying over to a follow-up period. Such data suggest that posters can be useful in changing safety behavior if they are used in circumstances in which workers do have the opportunity to control the occurrence of accidents by their own actions.

Discussion of Accidents and Safety

There is no such thing as a completely accident-free circumstance in human life. However, the human race should not assume a completely fatalistic attitude toward the inevitability of accidents, since there are actions that can be taken to reduce the likelihood of such events. These actions include the following: the design of equipment and facilities in terms of human factors considerations that will contribute to safety; the development of procedures that contribute to safety and the training of personnel to follow such procedures; and the consistent use of appropriate protective devices by personnel.

SOME CLOSING REFLECTIONS

Several of the earlier chapters of this text dealt with certain features of human beings as they are relevant to the design of the things people use in their work and everyday lives, including people's sensory and perceptual skills, psychomotor abilities, and anthropometric characteristics. In addition various chapters have dealt with some of the implications of such human characteristics to the design of some of the things people use, as displays, controls, work-space arrangement and layout, and the physical environment. In addition particular attention has been given to the human factors as related to the built environment, vehicles and related facilities, and (in this chapter) to certain specific work-related topics such as monitoring tasks, inspection, and safety.

In reflecting about this coverage it should be kept in mind that the material included represents only a small sample of the relevant material that could have been included. This text has been intended as an overview of the human factors field and not as a complete compendium of relevant information. (Such a compendium would occupy more than the proverbial 5-ft bookshelf.) Realizing that this text then represents a sample of the human factors domain, it has been an objective of the text to help the reader become aware of, or become more sensitive to, the human factors implications of the things people use.

In connection with human factors research and applications to date we could develop quite a laundry list of areas that it has not been feasible to include in this text, or that are touched on very briefly. Such a list could include such areas as consumer products, health services, recreation, the expanding use of computers, law enforcement, exploration of outer space, communication systems, mining, agriculture, and (in the words of the king of Siam in *The King and I*) etcetera, etcetera, etcetera. (It should be added that the human factors inroads into some of these, and other, areas have to date still been fairly limited.)

But now, what of the future? In the first place there are many items of unfinished business relating to the human factors implications of the many facets of present-day work and living that have not yet benefited from the possible application of human factors data and principles. Beyond such concerns, however, there are at least a couple of broad areas of attention that should serve as challenges to the human factors clan.

One of these deals with what (in present-day expressions) is called the *quality of working life*, or more broadly, the *quality of life*. There is some truth to the charges that have been made that human factors efforts have tended toward making work easier for people to do and perhaps toward greater work specialization. In recent years, however, we have heard much about job enlargement and job enrichment programs that, to some degree, appear to be directed toward opposite objectives of broadening work responsibilities, increasing the decision-making aspects of jobs, etc. These efforts have as their aim the enhancement of job satisfaction and the improvement of the quality of working life. Although job enlargement probably is not for everybody, its general goals indeed are on the positive side of the ledger. In recent years several people prominent in human factors affairs have expressed their concern about the quality of working life and have urged the human factors community to work toward such a goal. However, there seems to be a basic need for efforts to blend the human factors and the job enlargement approaches into an integrated effort that would offer reasonable possibilities for people to engage in work activities that are efficient and safe and that also would enhance the quality of working life.

The second area of challenge to the human factors discipline deals with the shape of things to come, and actually is related to the previous discussion of the quality of working life. The future is always somewhat shrouded, but we certainly can expect increased automation in the production of goods and services with increased use of computer-controlled processes, significant changes in technology and in the products and services that such changes may bring, and the development of new products and services that we simply cannot now envision. The "brave new world" of the future should indeed be developed with people–you and us–in mind. Thus, the human factors discipline must be at the cutting edge of future developments to ensure that such developments will, in reality, contribute to the improvement of the quality of working life and of life in general.

REFERENCES

Bergum, B. O., & Klein, I. C. *A survey and analysis of vigilance research* (Research Report No. 8). Alexandria, Va.: Human Resources Research Office, November 1961.

Brown, D. B. Cost/benefit of safety investments using fault tree analysis. *Journal of Safety Research,* 1973, *5*(3), 73–79.

Chaney, F. B. & Teel, K. S. Improving inspector performance through training and visual aids. *Journal of Applied Psychology,* 1967, *51*(4), 311–315.

Christensen, J. M. Human factors in hazard/risk evaluation. In *Proceedings of the symposium–Human factors and industrial design in consumer products.* Medford, Mass.: Tufts University, May 1980, pp. 442–447.

Cohen, A., Smith, M. J., & Anger, W. K. Self-protective measures against workplace hazards. *Journal of Safety Research,* 1979, *11*(3), 121–131.

Colquhoun, W. P. The effect of a short rest pause on inspection efficiency. *Ergonomics,* 1959, *2*, 367–372.

Drury, C. G. Inspection of sheet materials: Model and data. *Human Factors,* 1975, *17* (3), 257-265.

Drury, C. G., & Brill, M. New methods of consumer product accident investigation. In *Proceedings of the symposium–Human factors and industrial design in consumer products.* Medford, Mass.: Tufts University, May 1980, pp. 196-211.

Edwards, D. S., & Hahn, C. P. A chance to happen. *Journal of Safety Research,* 1980, *12*(2), 59-67.

Harris, D. H., & Chaney, F. B. *Human factors in quality assurance.* New York: Wiley, 1969.

Kelly, M. L. A study of industrial inspection by the method of compared comparison. *Psychological Monographs,* 1955, *69*(9), no. 394.

Komaki, J., Barwick, K. D., & Scott, L. R. A behavioral approach to occupational safety: Pinpointing and reinforcing safe performance in a food manufacturing plant. *Journal of Applied Psychology,* 1978, *63*(4), 434-445.

Lion, J. S., Richardson, E., Weightman, D., & Browne, B. C. The influence of the visual arrangement of material, and of working singly or in pairs, upon performance at simulated industrial inspection. *Ergonomics,* 1975, *18*(2), 195-204.

Loeb, M. On the analysis and interpretation of vigilance: Some remarks on two recent articles by Craig. *Human Factors,* 1978, *20*(4), 447-451.

Purswell, J. L., Greenhaw, L. N., & Oats, C. An inspection task experiment. In *Proceedings of the Human Factors Society.* Santa Monica, Calif.: Human Factors Society, 1972, pp. 297-300.

Purswell, J. L., & Hoag, L. L. Strategies for improving visual inspection performance. In *Proceedings of the Human Factors Society.* Santa Monica, Calif.: Human Factors Society, 1974, pp. 397-403.

Ramsey, J. D. Identification of contributory factors in occupational injury. *Journal of Safety Research,* 1973, *5*(4), 260-267.

Ramsey, J. D. Ergonomic support of consumer product safety. Paper presented at the American Industrial Hygiene Association Conference, May 1978.

Sell, R. G. What does safety propaganda do for safety? A review. *Applied Ergonomics,* 1977, *8*(4), 203-214.

Smith, G. L., Jr. Signal detection theory and industrial inspection. In *Proceedings of the Human Factors Society.* Santa Monica, Calif.: Human Factors Society, 1972, pp. 284-290.

Smith, M. J., Anger, W. K., & Uslan, S. S. Behavioral modification applied to occupational safety. *Journal of Safety Research,* 1978, *10*(2), 87-88.

Teichner, W. H. The detection of a simple visual signal as a function of time of watch. *Human Factors,* 1974, *16*(4), 339-353.

Waikar, A. Quality improvement using multiple inspector systems (Unpublished Master's thesis, University of Oklahoma, 1973).

Wiener, E. L. Stimulus presentation rate in vigilance. *Human Factors,* 1977, *19*(3), 301-303.

Williges, R. C., & Streeter, H. Influence of static and dynamic displays on inspection performance. In *Proceedings of the Human Factors Society.* Santa Monica, Calif.: Human Factors Society, 1972, pp. 291-296.

APPENDIXES

LIST OF ABBREVIATIONS

AI	articulation index
AFB	Air Force Base
AFHRL	Air Force Human Resources Laboratory
AFSC	Air Force Systems Command
AMD	Aerospace Medical Division, Air Force Systems Command
AMRL (of USAF)	Aerospace Medical Research Laboratory, Aerospace Medical Division, Air Force Systems Command
ANSI	American National Standards Institute
APA	American Psychological Association
ASD	Aeronautical Systems Division, Air Force Systems Command
ASHA	American Speech and Hearing Association
ASHRAE	American Society of Heating, Refrigerating, and Air-Conditioning Engineers
ASHVE	American Society of Heating and Ventilating Engineers (now ASHRAE)
BCD	borderline between comfort and discomfort
C	Celsius (when used as a measure of temperature)
C	convection (when used as a method of heat exchange)
C	contrast (when used regarding differences in reflectance of visual stimuli)
C/D	control-display
cd	candela
CIE	Commission Internationale de l'Eclairage
clo	clo unit (a measure of insulation of clothing)
cm	centimeter
CNEL	Community Noise Equivalent Level
C/R	control-response

CRT	cathode-ray tube
d	day
dB	decibel
DGR	discomfort glare rating
DVA	dynamic visual acuity
E	evaporation
ECP	evoked cortical potential
Ed(s).	editor(s)
ed.	edition
EEG	electroencephalogram
EMG	electromyogram
EPA	Environmental Protection Agency
EPRI	Electric Power Research Institute
ET	original effective temperature scale
ET*	current (new) effective temperature scale
F	Fahrenheit
fc	footcandle
fL	footlambert
ft	feet
G	gravity (referring to acceleration)
G	gravity (referring to vibration)
g	gram
gal	gallon
GSR	galvanic skin response
H	amount of information in bits
h	hour
HRV	heart rate variability (same as SA)
HST	hand-skin temperature
Hz	Hertz
IEEE	Institute of Electrical and Electronics Engineers, Inc.
IES	Illuminating Engineering Society
in	inch
IRE	Institute of Radio Engineers
ISO	International Organization for Standardization
JND	just-noticeable difference
K	degree Kelvin
kc	kilocycle
kcal	kilocalorie
kg	kilogram
km	kilometer
lb	pound
lm	lumen
lx	lux
M	metabolism
m	meter
mi	mile
min	minute
mL	millilambert
mm	millimeter
ms	millisecond

N	newton
NADC	Naval Air Development Center
NAS	National Academy of Sciences
NASA	National Aeronautics and Space Administration
NC	noise criteria
NIPTS	noise-induced permanent threshold shift
nm	nanometers
NRC	National Research Council
OSHA	Occupational Safety and Health Administration
PLdB	perceived level of noise, decibel
PNdB	perceived noise, decibel
PNL	perceived noise level
PSD	power spectral density
PSIL	preferred-octave speech interference level
PTS	permanent threshold shift (of hearing)
R	radiation (when used as method of heat exchange)
R	reliability (when used with regard to performance)
rh	relative humidity
rms	root-mean-square
s	second
SA	sinus arrhythmia (same as HRV)
SDT	signal detection theory (same as TSD)
SI	International System of Units
SIL	speech interference level
SPL	sound pressure level
SRP	seat reference point
TOT	time on target
TR	Technical Report (term used by various organizations)
TSD	theory of signal detection (same as SDT)
TVSS	Tactile Vision Substitution System
USA	United States Army
USAF	United States Air Force
USASI	United States of America Standards Institute (now ANSI)
USN	United States Navy
USPHS	United States Public Health Service
UV	ultraviolet
VCP	visual comfort probability
VL	visibility reference function or level
VL_{eff}	effective VL
VTE	Visual Task Evaluator
VWF	vibration-induced white finger
W	watt
WADC	Wright Air Development Center, USAF (see AMRL and AFHRL)
WADD	Wright Air Development Division, USAF (see AMRL and AFHRL)
WBGT	wet-bulb globe temperature
μ	micron (10^{-6})
π	pi

APPENDIX **B**

CONTROL DEVICES

Table B-1 presents a brief evaluation of the operational characteristics of certain types of control devices.[1] Table B-2 presents a summary of recommendations regarding certain features of these types of control devices.[2] In the use of these and other recommendations, it should be kept in mind that the unique situation in which a control device is to be used and the purposes for which it is to be used can affect materially the appropriateness of a given type of control and can justify (or virtually require) variations from a set of general recommendations or from general practice based on research or experience. For further information regarding these, refer to the original sources given in the reports from which these are drawn.

COMMENTS REGARDING CONTROLS[3]

- *Hand push button:* Surface concave, or provide friction. Preferably audible click when activated. Elastic resistance plus slight sliding friction, starting low, building up rapidly, sudden drop. Minimize viscous damping and inertial resistance.
- *Foot push button:* Use elastic resistance, aided by static friction, to support foot. Resistance to start low, build up rapidly, drop suddenly. Minimize viscous damping and inertial resistance.
- *Toggle switch:* Use elastic resistance which builds up and then decreases as position is approached. Minimize frictional and inertial resistance.

[1] Adapted largely from A. Chapanis. "Design of Controls," chap. 8 in H. P. Van Cott and R. G. Kinkade, *Human engineering guide to equipment design* (Rev. ed.). Washington, D. C.: U. S. Government Printing Office, 1972.
[2] Adapted from ibid.
[3] Adapted from ibid.

• *Rotary selector switch:* Provide detent for each control position (setting). Use elastic resistance which builds up, then decreases as detent is approached. Minimum friction and inertial resistance. Separation of detents should be at least ¼ in.

• *Knob:* Preferably code by shape if used without vision. Type of desirable resistance depends on performance requirements.

• *Crank:* Use when task involves two rotations or more. Friction (2 to 5 lb) reduces effects of jolting but degrades constant-speed rotation at slow or moderate speeds. Inertial resistance aids performance for small cranks and low rates. Grip handle should rotate.

• *Lever:* Provide elbow support for large adjustment, forearm support for small hand movements, wrist support for finger movements. Limit movement to 90°.

• *Handwheel:* For small movements, minimize inertia. Indentations in grip rim to aid holding. Displacement usually should not exceed ±60° from normal. For displacements less than 120°, only two sections need be provided, each of which is at least 6 in long. Rim should have frictional resistance.

• *Pedal:* Pedal should return to null position when force is removed; hence, elastic resistance should be provided. Pedals operated by entire leg should have 2 to 4 in displacement, except for automobile-brake type, for which 2 to 3 in of travel may be added. Displacement of 3 to 4 in or more should have resistance of 10 lb or more. Pedals operated by ankle action should have maximum travel of 2½ in.

TABLE B-1
COMPARISON OF THE CHARACTERISTICS OF COMMON CONTROLS

Characteristic	Hand push button	Foot push button	Toggle switch	Rotary switch	Knob	Crank	Lever	Hand-wheel	Pedal
Space required	Small	Large	Small	Medium	Small-medium	Medium-large	Medium-large	Large	Large
Effectiveness of coding	Fair-good	Poor	Fair	Good	Good	Fair	Good	Fair	Poor
Ease of visual identification of control position	Poor*	Poor	Fair-good	Fair-good	Fair-good†	Poor‡	Fair-good	Poor-fair	Poor
Ease of nonvisual identification of control position	Fair	Poor	Good	Fair-good	Poor-good	Poor‡	Poor-fair	Poor-fair	Poor-fair
Ease of check reading in array of like controls	Poor*	Poor	Good	Good	Good†	Poor‡	Good	Poor	Poor
Ease of operation in array of like controls	Good	Poor	Good	Poor	Poor	Poor	Good	Poor	Poor
Effectiveness in combined control	Good	Poor	Good	Fair	Good§	Poor	Good	Good	Poor

* Except when control is backlighted and light comes on when control is activated.
† Applicable only when control makes less than one rotation and when round knobs have pointer attached.
‡ Assumes control makes more than one rotation.
§ Effective primarily when mounted concentrically on one axis with other knobs.

TABLE B-2

SUMMARY OF SELECTED DATA REGARDING DESIGN RECOMMENDATIONS FOR CONTROL DEVICES

	Size, in		Displacement		Resistance	
Device	**Minimum**	**Maximum**	**Minimum**	**Maximum**	**Minimum**	**Maximum**
Hand push button						
Fingertip operation	½	None	⅛ in	15 in	10 oz	40 oz
Foot push button	½	None				
Normal operation			½ in			
Wearing boots			1 in			
Ankle flexion only				2½ in		
Leg movement				4 in		
Will *not* rest on control					4 lb	20 lb
May rest on control					10 lb	20 lb
Toggle switch			30°	120°	10 oz	40 oz
Control tip diameter	⅛	1				
Lever arm length	½	2				
Rotary selector switch					10 oz	40 oz
Length	1	3				
Width	½	1				
Depth	½					
Visual positioning			15°	40° *		
Nonvisual positioning			30°	40° *		
Knob, continuous adjustment†						
Finger-thumb						4½–6 in/oz
Depth	½	1				
Diameter	⅜	4				
Hand/palm diameter	1½	3				
Crank†						
For light loads, radius	½	4½				
For heavy loads, radius	½	20				
Rapid, steady turning						
<3–5 in radius					2 lb	5 lb

(Continued)

TABLE B-2 (continued)
SUMMARY OF SELECTED DATA REGARDING DESIGN RECOMMENDATIONS FOR CONTROL DEVICES

	Size, in		Displacement		Resistance	
Device	**Minimum**	**Maximum**	**Minimum**	**Maximum**	**Minimum**	**Maximum**
5–8 in radius					5 lb	10 lb
8-in radius					?	?
For precise settings					2½ lb	8 lb
Lever§						
Fore-aft (one hand)				14 in		
Lateral (one hand)				38 in		
Finger grasp, diam.	½	3			12 oz	32 oz
Hand grasp, diam.	1½	3			2 lb	20–100 lb
Handwheel†				90°–120°	5 lb	30 lb‡
Diameter	7	21				
Rim thickness	¾	2				
Pedal						
Length	3½					
Width	1					
Normal use			½ in			
Heavy boots			1 in			
Ankle flexion				2½ in		10 lb
Leg movement				7 in		180 lb
Will *not* rest on control					4 lb	
May rest on control					10 lb	

* When special requirements demand large separations, maximum should be 90°.
† Displacement of knobs, cranks, and handwheels should be determined by desired control-display ratio.
‡ For two-handed operation, maximum resistance of handwheel can be up to 50 lb.
§ Length depends on situation, including mechanical advantage required. For long movements, longer levers are desirable (so movement is more linear).

APPENDIX **C**

SELECTED REFERENCES

This appendix includes a list of selected books and journals that deal with human factors. A number of the books listed are general references, while others deal with specific aspects of human factors. For additional references that deal with topics that are specific to the content of the individual chapters the reader is referred to the references at the ends of the individual chapters.

BOOKS

Anderson, D., Istance, H., & Spencer, J. (Eds.). *Human factors in the design and operation of ships.* Stockholm, Sweden: Ergonomilaboratoriet AB, 1977.

Anthropometric source book. Vol. I. *Anthropometry for designers* (NASA RP 1024). Houston, Texas: National Aeronautics and Space Administration, 1978.

Bennett, E., Degan, J., & Spiegel, J. (Eds.). *Human factors in technology.* New York: McGraw-Hill, 1963.

Cakir, A., Hart, D., & Stewart, T. *Visual display terminals.* New York: Wiley, 1979.

Chapanis, A. *Research techniques in human engineering.* Baltimore: Johns Hopkins, 1959.

Chapanis, A. *Man-machine engineering.* Belmont, Calif.: Wadsworth, 1965.

Chapanis, A. (Ed.). *Ethnic variables in human factors engineering.* Baltimore: Johns Hopkins, 1975.

Damon, A., Stoudt, H. W., & McFarland, R. A. *The human body in equipment design.* Cambridge, Mass.: Harvard, 1966.

DeGreene, K. B. *Systems psychology.* New York: McGraw-Hill, 1972.

Drury, C. & Fox, J. (Eds.). *Human reliability in quality control.* London: Taylor and Francis, 1975.

Edholm, O. G. *The biology of work.* New York: World University Library, McGraw-Hill, 1967.

Edwards, E. & Lees, F. (Eds.). *The human operator in process control.* London: Taylor and Francis, 1974.

Forbes, T. W. (Ed.). *Human factors in highway traffic safety research.* New York: Wiley, 1972.

Grandjean, E. *Fitting the task to the man: an ergonomic approach.* London: Taylor and Francis, 1980.

Grandjean, E. & Vigliani, E. (Eds.). *Ergonomic aspects of visual display terminals.* London: Taylor and Francis, 1980.

Harris, D. H., & Chaney, F. B. *Human factors in quality assurance.* New York: Wiley, 1969.

Helander, M. (Ed.). *Human factors/ergonomics for building and construction.* New York: Wiley, 1981.

Howell, W. C., & Goldstein, I. L. *Engineering psychology: current perspectives in research.* New York: Meredith Corporation, 1971.

Kelley, C. *Manual and automatic control.* New York: Wiley, 1968.

Konz, S. *Work design.* Columbus, Ohio: Grid Inc., 1979.

Kraiss, K., & Moraal, J. (Eds.). *Introduction to human engineering.* Bonn, Germany: Verlag TUV Rheinland GimgH, 1976.

Mackie, R. (Ed.). *Vigilance: theory, operational performance, and physiological correlates, Vol. 3 Human Factors NATO Conference Series,* New York, Plenum Press, 1977.

Meister, D. *Human factors: Theory and practice.* New York: Wiley, 1971.

Meister, D. *Behavioral foundations of systems development.* New York: Wiley, 1976.

Meister, D., & Rabideau, G. F. *Human factors evaluation in system development.* New York: Wiley, 1965.

Murrell, K. *Ergonomics: man in his working environment.* London: Chapman and Hall, 1969.

Murrell, K. *Men and machines.* London: Methuen and Co., 1976.

National Aeronautics and Space Administration. *Bioastronautics data book* (NASA SP-3006). J. F. Parker, Jr. & V. R. West (Managing Eds.). Washington, D. C.: U. S. Government Printing Office, 1973.

Parsons, H. M. *Man-machine system experiments.* Baltimore: Johns Hopkins, 1972.

Poulton, E. *The environment at work.* Springfield, Ill.: Charles C Thomas, 1979.

Roebuck, J., Kroemer, K., & Thomson, W. *Engineering anthropometry techniques.* New York: Wiley, 1975.

Roscoe, S. (Ed.). *Aviation psychology.* Ames, Iowa: Iowa State University Press, 1980.

Seminara, J., Gonzalez, W., & Parsons, S. *Human factors review of nuclear power plant control room design.* Palo Alto, Calif.: Electric Power Research Institute, 1977.

Shackel, B. (Ed.). *Applied ergonomics handbook.* Surrey, England: IPC Science and Technology Press, 1974.

Sheridan, T., & Ferrell, W. *Man-machine systems: information, control, and decision models of human performance.* Cambridge, Mass.: MIT Press, 1974.

Shurtleff, D. *How to make displays legible.* La Mirada, Calif.: Human Interface Design, 1980.

Singleton, W. *Introduction to ergonomics.* Geneva: World Health Organization, 1972.

Singleton, W., Easterby, R., & Whitfield, D. (Eds.). *The human operator in complex systems.* London: Taylor and Francis, 1971.

Singleton, W., Fox, J., & Whitfield, D. (Eds.). *Measurement of man at work.* London: Taylor and Francis, 1971.

Swain, A., & Guttmann, H. *Handbook of human reliability analysis with emphasis on nuclear power plant applications.* Washington, D. C.: U. S. Nuclear Regulatory Commission, 1980.

Van Cott, H. P., & Kinkade, R. G. *Human engineering guide to equipment design* (Rev. ed.). Washington, D. C.: U. S. Government Printing Office, 1972.

Weiner, J., & Maule, H. (Eds.). *Case studies in ergonomic practice, Vol. I. Human factors in work, design, and production.* London: Taylor and Francis, 1977.

Winter, D. *Biomechanics of human movement.* New York: Wiley, 1979.

Woodson, W. E. *Human factors design handbook.* New York: McGraw-Hill, 1981.

JOURNALS

Applied Ergonomics, IPC House, Guilford, Surrey, England.

Ergonomics, Taylor & Francis, Ltd., London.

Human Factors (Journal of the Human Factors Society, Santa Monica, Calif.).

INDEXES

NAME INDEX

SUBJECT INDEX